D0322225

Engineering Design
A Systematic Approach

Springer

London
Berlin
Heidelberg
New York
Barcelona
Budapest
Hong Kong
Milan
Paris
Santa Clara
Singapore
Tokyo

G. Pahl and W. Beitz

Engineering Design

A Systematic Approach

Translated by Ken Wallace,
Luciënne Blessing and Frank Bauert

Edited by Ken Wallace

With 354 Figures

 Springer

Gerhard Pahl, Dr.-Ing. Dr. h.c. mult
Fachbereich 16 Maschinenbau, Technische Hochschule Darmstadt,
Magdalenenstr.4, 64298 Darmstadt, Germany

Wolfgang Beitz, Dr.-Ing.
Institut für Maschinenkonstruktion, Technische Universität Berlin,
Strasse des 17. Juni 135, 10623 Berlin, Germany

ISBN 3-540-19917-9 Springer-Verlag Berlin Heidelberg New York

British Library Cataloguing in Publication Data
Pahl, Gerhard
    Engineering Design: Systematic Approach. - 2Rev.ed
    I. Title II. Beitz, Wolfgang III. Wallace, Ken
    620.0042
ISBN 3-540-19917-9

Library of Congress Cataloging-in-Publication Data
Pahl, G. (Gerhard), 1925-
    Engineering design: a systematic approach/Gerhard Pahl, Wolfgang Beitz; translated
by Ken Wallace, Luciënne Blessing, and Frank Bauert : edited by Ken Wallace.
        p.   cm.
    Includes bibliographical references and index.
    ISBN 3-540-19917-9 (acid-free paper)
    1. Engineering design. I. Beitz, Wolfgang. II. Title.
TA174.P34   1995
620′.0042--dc20   95-10248

© Springer-Verlag London Limited 1996
Printed in Great Britain

Typeset by Photo-graphics, Honiton, Devon
Printed and bound by Bell & Bain Limited, Glasgow, Scotland
69/3830-543210 Printed on acid-free paper

Contents

Authors' Foreword

The aim of the first two German editions of our book *Kon-struktionslehre (Engineering Design)* was to present a comprehensive, consistent and clear approach to systematic engineering design. The book has been translated into five languages, making it a standard international reference of equal importance for improving the design methods of practising designers in industry and for educating students of mechanical engineering design.

Although the third German edition conveys essentially the same message, it contains additional knowledge based on further findings from design research and from the application of systematic design methods in practice. The latest references have also been included. With these additions the book achieves all our aims and represents the state of the art.

Substantial sections remain identical to the previous editions. The main extensions include:

- a discussion of cognitive psychology, which enhances the creativity of design work;
- enhanced methods for product planning;
- principles of design for recycling;
- examples of well-known machine elements*;
- special methods for quality assurance; and
- an up-to-date treatment of CAD*.

The layout of the book has been changed slightly compared with the previous editions to place more emphasis on general methods for finding and evaluating solutions. These methods can usefully be applied during any phase of the design process, so separating them from the description of the main design phases and bringing them together in their own chapter reinforces this point. Readers familiar with the previous editions, as well as new readers, will quickly appreciate the logic of the new layout. The design methodology presented is based on

---

* These sections are not included in the second English edition (see Editor's foreword)

engineering fundamentals, and provides a comprehensive systematic treatment of the design and development process applicable in many engineering domains.

We gratefully thank and acknowledge colleagues and practising engineers who have encouraged and supported us with suggestions and contributions during the preparation of the third German edition. In particular we extend our sincerest thanks to Dr-Ing K H Beelich and Prof Dr-Ing P Praß for their critical reviews of individual chapters and for their valuable suggestions. Corrections and additional drawing work were undertaken by Herr W Laßhof, Frau C Prokopf, Frau I Sauter and Frau R Walinschus. The text was prepared by Frau S Fraaß and Frau H Möller, supported by Frau D Prentzel. We thank all of them for their enthusiastic cooperation.

We must thank our publisher for excellent advice and assistance, as well as for another very careful production. Our heartfelt thanks go to our wives for their continuous understanding and complete support. Without their support, the third edition would not exist.

Gerhard Pahl, Darmstadt and
Wolfgang Beitz, Berlin, 1993

Editor's Foreword

The first German edition of *Konstruktionslehre* was published in 1977. The first English edition entitled *Engineering Design* was published in 1984 and was a full translation of the German text. Both the German and the English editions of the book rapidly became established as important references on systematic engineering design in industry, research and education. International interest in engineering design grew rapidly during the 1980s and many developments took place. To keep up-to-date with the changes, a second German edition was published in 1986. It was too soon after the publication of the first English edition to consider a second English edition. However, since the translation was being extensively used to support engineering design teaching, both at Cambridge University and at other academic institutions, it became clear that a student edition was required to make the text more widely available. An abbreviated student edition entitled *Engineering Design – A Systematic Approach* was therefore published in 1988.

When preparing the student edition, the opportunity was taken to review the translation and the contents of the first edition. No changes in terminology were thought necessary and the contents were the same as the first English edition except for the removal of two chapters. The reasons for their removal are explained below.

The short chapter on detail design was left out. It must be emphasised that this does not mean that detail design is considered unimportant or lacking in intellectual challenge. Quite the reverse is true. Detail design is far too broad and complex a subject to be covered in a general text. There are many excellent books covering the detail design of specific technical systems and machine elements. For these reasons, the chapter on detail design in the German editions did not discuss technical aspects of detail design, but only dealt with the preparation of production documents and the numbering techniques required to keep track of them.

The second chapter to be removed dealt with CAD. Again, this chapter was clearly not removed because the topic is unimportant. CAD has undergone many changes in recent years and many specialist texts are now available.

In 1993 an updated and extended third German edition of *Konstruktionslehre* was published. It was considered timely to produce a *second* English edition to bring the translation into step with the latest thinking. The overall structure of the second English edition, described later, is not exactly the same as the German third edition: some chapters have again been omitted. The changes are summarised below.

The new layout mentioned in the Authors' Foreword has been incorporated, along with the important discussions of psychology and recycling. The new chapters on design for quality and design for minimum cost have been included. For the same reasons given for the student edition, the chapters on detail design and CAD have been omitted. The third German edition also includes a chapter on the design of well-known machine elements. This chapter has also been omitted. A translation of the German *Dubbel* has recently been published by Springer [*Dubbel*, Springer-Verlag, London, 1994] and this comprehensive reference covers many of the machine elements discussed in the deleted chapter.

It must be stressed that nothing has been deleted that detracts from the main aim of the original German book, that is, to present a comprehensive, consistent and clear approach to systematic engineering design.

The Translation

The aim of the translation has been to render each section of the book comprehensible in its own right and to avoid specialist terminology. Terms are defined as they arise, rather than in a separate glossary, and their meanings should be clear from their usage. On occasions other authors have used slightly different terms, but it is hoped that no misunderstandings arise and that the translation is clear and consistent throughout.

It is hoped that the translation faithfully conveys the ideas of Pahl and Beitz while adopting an English style.

German Standards and Guidelines

There are numerous references to DIN (Deutsches Institut für Normung) standards, VDI (Verein Deutscher Ingenieure) guidelines and German material specifications. In important cases, references to DIN standards and VDI guidelines have been retained in the English text, but elsewhere they have simply been listed along with the other references. In technical examples, DIN standards have been referred to without any attempt to find English equivalents.

In several figures (eg Fig. 7.85) and examples, particular steels have been referred to by their standard German specifications. Often there is no exact English equivalent. However it seemed important to provide a general indication of the type of steel being referred to. The following example explains the code adopted:

German specification   24 Cr Mo V 55

Code adopted   24C / 1.5% Cr – Mo – V

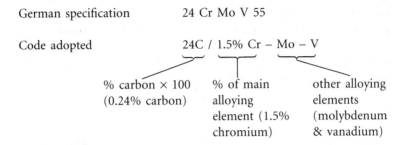

% carbon × 100   % of main   other alloying
(0.24% carbon)   alloying   elements
                      element (1.5%   (molybdenum
                      chromium)   & vanadium)

References

The original text includes many references. Most of these are in German and therefore not of immediate interest to the majority of English readers. However, to have omitted them would have detracted from the authority of the book and its value as an important source of reference. The references have therefore been retained in full but grouped together at the end of the book, rather than at the end of each chapter as in the original text.

A selected English bibliography has been added.

Acknowledgements

First my sincerest thanks must go to Dr Luciënne Blessing who has worked with me so conscientiously and with such careful attention to detail through the new translation. Her extensive knowledge of design methods coupled with a flair for technical translation made her contribution to the second English edition invaluable.

Next I must express my appreciation to Dr-Ing Frank Bauert for applying his engineering expertise and knowledge of both languages to check the technical content of the drafts and to translate many of the new figures.

Those who contributed to the translation of the first edition must be acknowledged. Donald Welbourn strongly encouraged the translation in the first place. Many of the problems with the terminology that arose during the original translation were solved with the help and support of Arnold Pomerans.

Nicholas Pinfield and Imke Mowbray from Springer contributed enormously to the production of the book. Their help and patience is gratefully acknowledged.

Many colleagues, too numerous to mention, assisted in a variety of ways throughout the task and their help has been very much appreciated.

Finally, and most sincerely, I must thank both Professor Pahl and Professor Beitz for their support.

I apologise for any errors, for which I am entirely responsible. I should be grateful to have these pointed out and, in addition, would be pleased to receive any suggestions for improvements.

Ken Wallace
Cambridge, Summer 1995

The Overall Structure of the Second Edition

Because of the complex nature of modern technology, it is now rarely possible for an individual to tackle the design and development of a major new product single-handed. Often a large team is required and this introduces problems of organisation and communication. In order to increase the probability of success of a new venture, the design process must be planned carefully and executed systematically. In particular, an engineering design method must integrate the many different aspects of designing in such a way that the whole process becomes logical and comprehensible. To that end, the design process must be broken down, first into phases and then into distinct steps, each with its own working methods. It is with these aims in mind that Pahl and Beitz have brought together the extensive body of knowledge about systematic design. They present an up-to-date and comprehensive theory of general engineering design.

The Main Phases of the Design Process

Pahl and Beitz split the design process into four main phases which have been translated as:

- Product planning and clarifying the task
- Conceptual design
- Embodiment design
- Detail design.

Of these four phases, only the terminology used for the third, embodiment design, requires some explanation. Other translations, in a similar context, have used *layout design, main design, scheme design* or *draft design*. The input to this third phase is a design concept and the output is a technical description, often in the form of a scale drawing. Depending on the particular company involved, this scale drawing may be called a *general arrangement,* a *layout,* a *scheme,* a *draft,* or a *configuration drawing,* and it defines the arrangement and preliminary shapes of the components in a technical artefact. The most common name is probably general arrangement (drawing); but layout (drawing) is also widely used and, because it has the advantage of brevity, it was chosen for this book. The idea to introduce the term *embodiment design* came from French's book, *Engineering Design: The Conceptual Stage,* published in 1971. In his book French splits the design process into four phases: analysis of problem; conceptual design; embodiment of schemes; and detailing. *Embodiment* is therefore the development of a more or less abstract concept into a more concrete proposal. This is quite general and not tied to any particular branch of engineering. *Embodiment design* incorporates both *layout design* (the arrangement of components and their relative motions) and *form design* (the shapes and materials of individual components). The term is used in BS 7000, *The Management of Product Design,* published by the British Standards Institution in 1989. The term *form design* is widely used in the literature, and its meaning ranges from the overall *form* of a product in an industrial design context, to the more restricted *form* of individual components in an engineering context. This book tends towards the latter usage.

The Structure

The book starts with the historical background to modern design thinking in Germany. The work of influential design researchers and practitioners is reviewed briefly. Although the methods proposed by these authors are all influenced by their own specialist fields, they share the following pattern:

- The requirements are abstracted to identify the crux of the problem and the essential functions.
- Emphasis is placed on selecting appropriate physical processes.
- There is a step-by-step advance from the qualitative to the quantitative.
- There is a deliberate variation and combination of solution elements of different complexity.

In the second chapter the fundamentals of engineering systems and procedures are examined. In all German methods, great emphasis is placed on the establishment of a *function structure* and on the systematic variation and combination of *sub-functions.* To that end, the *overall function* of the *system* must be established and the *system boundary,* with the relevant *inputs* and *outputs,* clearly identified. The flows and conversions that take place inside the system boundary involve *energy*, *material* and *signals*, the last of these being the technical form of *information*. The selected *function structure* is then systematically developed into a *working structure*, a *construction structure* and, if appropriate, a *system structure* (see Fig. 2.15).

Every design must meet both *task-specific* and *general constraints*. To remind designers of these during all stages of the design process, a set of checklists is used throughout the book. There are slight differences in the early headings of the lists for the different phases, but the following general ones are used throughout: *safety, ergonomics, production, quality control, assembly, transport, operation, recycling, maintenance* and *expenditure*. The final part of Chapter 2 discusses a general working methodology, that is, methods and concepts that are generally applicable throughout the design process, including analysis, abstraction and synthesis.

In Chapter 3, the general problem-solving process and the flow of work during the process of planning and designing are described. The authors' overall model of the design process is shown in Fig. 3.3.

Chapter 4 discusses general methods for finding and evaluating solutions. These methods are not linked to any specific design phase or type of product and include a range of intuitive and discursive methods.

The remainder of the book is largely devoted to a detailed elaboration of three of the four main phases. Detail design, as mentioned in the Editor's Foreword, is not covered.

Product Planning and Clarifying the Task (Chapter 5)

This phase involves planning the initial product by collecting information about customer requirements and by generating initial product ideas. It culminates in the drawing up and elaboration of a detailed *requirements list* (design specification).

Conceptual design (Chapter 6)

This phase involves (see Fig. 6.1):

– abstracting to find the essential problems;
– establishing function structures;
– searching for working principles;
– combining working principles into working structures;
– selecting a suitable working structure and firming up into a principle solution (concept).

Embodiment design (Chapter 7)

During this phase, designers start with the selected *concept* and work through the steps shown in Fig. 7.1 to produce a *definitive layout* of the proposed technical product or system in accordance with technical and economic requirements. The selected definitive layout provides a check of function, strength, spatial compatibility and so on. It is at this stage, at the very latest, that the financial viability of the project must be assessed.

About 40% of the book is devoted to this phase and the authors discuss the basic rules, principles and guidelines of embodiment design, all supported by well-chosen examples. The basic rules are *clarity, simplicity* and *safety*. Next they discuss the principles of: *force transmission, division of tasks, self-help, stability and bi-stability,* and *fault free design.* Finally, they provide numerous guidelines, which include design: *to allow for expansion, to allow for creep and relaxation, against corrosion damage, for ergonomics, for aesthetics, for production, for ease of assembly, to standards, for ease of maintenance, for recycling,* and *for minimum risk.* A comprehensive example of the embodiment design of an impulse-loading test rig for shaft-hub connections is included at the end of the chapter.

Three short chapters follow that describe general approaches not specifically aligned with any particular phase of the design process. Chapter 8 discusses *Developing Size Ranges and Modular Products*; Chapter 9 *Design for Quality*; and Chapter 10 *Design for Minimum Cost.* The short final chapter provides a summary of the ideas covered in the book. Figures 11.1, 11.2, and 11.3 provide a quick reference to the main ideas and how they relate to one another.

Suggestions for Using this Book in Design Teaching

Teachers of engineering design in many academic institutions have found the rational structure of the book helpful when planning an engineering design course; and students have found the clearly described approach, supported by well-chosen examples and useful checklists, easy to follow and apply when working on design exercises and projects. Initially, the approach can appear rather overwhelming and time-consuming. It is worth persevering. As familiarity and confidence increase, the approach becomes almost sub-conscious and the techniques can be applied quickly and flexibly to the particular task in hand.

A course in engineering design generally includes three main activities: lectures, exercises and projects. There may also be tests and examinations, though it is quite common for student assessment to be based entirely on coursework exercises and projects.

Lectures

The layout of the book assists with the planning of lecture courses. Chapter 2 (Fundamentals), Chapter 3 (Process Planning and Designing), Chapter 4 (General Methods for Finding and Evaluating Solutions), and Chapter 5 (Product Planning and Clarifying the Task) provide a wealth of material for a general introduction to engineering design that can be used at any level. Chapter 6 (Conceptual Design) provides the basis for an excellent stand-alone course. Chapter 7 (Embodiment Design) describes a set of teachable rules, principles and guidelines that can be used in conjunction with other well-established texts to organise a course covering essential machine elements. Chapter 8 (Developing Size Ranges and Modular Products), Chapter 9 (Design for Quality) and Chapter 10 (Design for Minimum Cost) describe numerous generally applicable methods that can enhance many specialist design courses.

Figures 3.3, 6.1 and 7.1, along with Figs. 11.1, 11.2 and 11.3, link everything together.

Exercises

The book contains many examples. Using these as samples, exercises for class practice, homework and tests can easily be generated using the many technical products that surround us. Particularly valuable as a guide are the conceptual design examples in Chapter 6 and, in particular, the one-handed household water mixing tap in Section 6.6. The examples are split into the steps listed below, and numerous exercises can be set on each step permitting them to be independently practised and assessed.

Step 1 – Clarifying the task and elaborating the requirements list
Step 2 – Abstracting to identify the essential problems
Step 3 – Establishing function structures
Step 4 – Searching for solution principles
Step 5 – Selecting suitable working principles
Step 6 – Firming up into principle solution variants
Step 7 – Evaluating the principle solutions.

Exercises can be based on products from each individual teacher's own area of expertise or on products from a particular area of technology such as transport, power generation, agriculture, health care etc. Before students have selected a particular field for specialisation, the personal environment provides a particularly rich and appropriate source of products on which to base exercises. Examples include: cookers, microwave ovens, toasters, food processors, refrigerators, washing machines, dish washers, kettles, coffee makers, waste disposers, hair dryers, torches, personal stereos, fire extinguishers, table lamps, lawnmowers, garden sprinklers, fruit pickers, glasshouse ventilators, motor cars (horns, hydraulic cylinders, thermostats etc), motor cycles (gearboxes, shock absorbers, speedometers etc), bicycles (gears, dynamos, brakes, bells etc), electric drills, jig saws, power sanders, automatic garage door openers, water taps, shower mixer valves, immersion heaters, gas meters, home security devices, electrical fittings (switches, junction boxes, plugs, sockets etc), staplers, paper punches, scissors, telephones, sailing boats, climbing gear, hang gliders, cameras, exercise equipment, camping gear etc. Environmental issues provide an extensive source of projects connected with environmentally safe design and design for recycling.

Some examples of exercises which have been successfully used in class are given below. Different products can easily be substituted for the ones selected, and each exercise can either be used as it stands or turned into a full project and taken right through to the end of the detail design phase. Each example demonstrates a slightly different formulation. The problem statements are intended to be realistic rather than ideal; that is, the problem statements are not solution-neutral. A point which should be picked up and discussed with students.

Example 1: Domestic Hair Dryer

Your company wishes to enter the highly competitive domestic hair dryer market. The new dryer is to be aimed primarily at teenagers, and an initial market survey suggests that the power rating should be around 1500 W and that particular emphasis should be placed on making it as quiet as possible.

Using the guidelines in Section 5.2.2, and the checklist given in Figure 5.7, prepare a preliminary requirements list (design specification) for the hair dryer. Your requirements list should contain approximately 20 requirements and constraints.

Example 2: Can Crusher

A device to accept and store used aluminium drink cans for subsequent recycling is to be designed. The device is to be located in busy public areas and is to accept cans one at a time from an individual and pay out a small coin as a reward. To reduce storage space, the can is to be crushed to a height of approximately 15 mm. The maximum force required is 2 kN. The original height of a can lies between 115 and 155 mm, a typical diameter is 65 mm and the average mass is 0.02 kg.

(a)   Identify the main problems to be confronted when designing such a device and list 20 important requirements to be included in the preliminary requirements list.

(b)   Identify the demands and wishes, and rank the wishes as being of high, medium or low importance.

(c)   The overall function of the device is to "accept and store cans". Using your list of requirements as the starting point, break down this overall function into approximately 15 sub-functions and arrange these into a suitable function structure. Indicate clearly the system boundary and the relevant flows of material, energy and signals.

Example 3: Woodyard Saw

A power-driven saw is to be designed for use in woodyards which sell timber and boards directly to the public. The saws are to be used to cut and trim boards, such as plywood and chipboard, to customer requirements while they wait. The maximum size of board to be cut is 2400 by 1200 mm, and the maximum thickness is 30 mm.

(a)   Discuss the problems of designing the saw to ensure adequate safety in use, and describe, with appropriate examples, the design guidelines which could be applied. It will be helpful to refer to Section 7.3.3 in the book.

(b)   Prepare a preliminary requirements list for the saw. It should include not more than 25 requirements and constraints.

(c)   Identify the overall function and break this down into a function structure containing between 10 and 20 sub-functions.

(d)   Describe the subsequent steps in the design process to be followed in order to arrive at a suitable concept for further development.

Example 4: Fuel Hose Connector

Large commercial aircraft are refuelled between flights from mobile tankers containing aviation kerosene. The ends of the hoses from the tankers are fitted with special connectors that are attached by the ground crew to sockets under panels on the wings, where the main fuel tanks are located. A hose end connector and socket are to be designed that will allow the attachmentand removal of the hoses to be carried out quickly and safely.

(a)   Considering the requirements and the range of working conditions of the ground crew, prepare a preliminary requirements list, quantifying where possible.

(b)   Identify the overall function and the main inputs and outputs. Prepare a function structure containing at least 10 sub-functions. Systematically vary this function structure to produce three alternative arrangements.

(c)   Starting with one of the function structures, search for solution principles for each sub-function and combine these into three feasible combinations. Represent these combinations by appropriately annotated sketches.

Step 7 in the list above involves the evaluation of principle solutions (concept variants). During the design process, this evaluation is usually carried out before anything has been made and is therefore based on predictions of how the proposed solutions would perform if they were to be made. Clearly, any evaluation is more accurate if prototypes are built and tested. This situation can be simulated easily and cheaply in the classroom by taking advantage of the low cost of many modern mass-produced products. Any product, for example a domestic water tap, is generally available from several different manufacturers. For any type of product, three or four different samples can be bought and given to the students to evaluate using the method described in Section 4.2. It adds considerably to class interest to be able to handle and take to pieces a number of different products, and the exercise also prompts a discussion of modern materials and manufacturing techniques. It is always interesting to compare the differences in embodiment and detail design and discuss how these influence a particular product's appeal. An example of such an exercise is given below.

Example 5: Evaluation

The three samples of water tap/electrical plug/paper punch/etc (many other small products can be substituted) are to be evaluated and the best selected. All these products aim to fulfil the same overall function, but they differ in price and detail design features. After studying the different samples, prepare an evaluation chart, using an objectives tree to determine the weightings of the various evaluation criteria you select. It is suggested that you do not use more than 15 evaluation criteria. Where detailed information is not available, sensible estimates should be made. Select the best product and explain why, in your view, it is superior to the others.

For more complex exercises, involving products which are too expensive to buy, valuable sources of comparative data are consumer magazines (freezers, motor cars, video recorders etc), trade journals (pumps, gearboxes, electric motors etc) and specialist magazines (cameras, outboard motors, home computers etc).

Projects

The exercises just described permit the individual steps of the design process to be practised and assessed, but *design projects* bring everything together and are therefore an essential part of any design course. The purpose of design projects is to develop the students' ability and confidence to work through the complete design process, ending up with a feasible design solution. It is clearly an advantage if the resulting designs can be built and tested, but time and resources seldom permit this for any but the simplest projects. Students frequently get stuck when working on their projects. When this happens, it is an advantage to be able to recommend a specific course of action to help remove the block. Here the book comes to the aid of the teacher. The comprehensive way in which the overall approach is described, alongwith specific methods to help with each step, means that one can always suggest to the students that they try a specific method and then come back and discuss the results. By doing this the students continue to confront the problem and this is often sufficient to remove the block. However, if this fails the resulting discussion usually succeeds. The approach described in the book has been successfully used for all sizes of project, ranging from six-hour first-year introductory projects to a two-year postgraduate project, which took a complex design through from the initial market survey to the commissioning of the resulting product in the customer's premises.

It is generally easier to find projects for the later years of a design programme than for the earlier ones. In the later years the students are committed to design and their ability and confidence are reasonably well developed. At this later stage, projects can easily be generated from an individual teacher's field of expertise and from links with industry. Finding suitable projects for the first year of a design programme can prove difficult. Here the aim is to encourage an enthusiasm for design in the students. If inappropriate projects are selected, students can easily be put off. The products listed in the section on exercises above provide a source of useful ideas. To supplement these, some suggested guidelines and examples of class-tested projects are provided below.

The following guidelines for selecting six-hour introductory projects are suggested:

- easily understood context;
- several different solutions possible;
- scope for novel ideas;
- simple order-of-magnitude calculations can be carried out;
- problem statement not to exceed 10 lines; and
- interesting and fun.

Example 6: Water Sampler

Design an independent mechanical device to be used from a rowing boat by a research worker who wishes to collect samples of water from fresh-water lakes at known depths down to a maximum of 500 m.

After release, the device must not be attached to the boat and must descend to within 10 m of an easily adjustable pre-determined depth. It must return to the surface with a 0.5 litre sample of water from that depth and then float on the surface until picked up.

Example 7: Fence Post Extractor

A portable human-powered device is required which will extract fence posts in remote areas. The fence posts are made of wood of square cross-section with sides of between 50 and 100 mm, are between 2 and 2.5 m long, and may have been sunk up to one metre into the ground. An initial vertical force of up to 2 kN may be required to extract the posts, which must be in are usable condition afterwards.

Example 8: Fire Escape

A mechanical system is required which, in the event of a fire, will enable people to escape from a six-storey building by lowering themselves to the ground from windows. The system, which might make use of a 5 mm diameter steel cable, must be capable of lowering either a small child or a heavy adult at approximately the same constant speed.

Example 9: Rope Tensioner

Heavy loads of various sizes and shapes have to be securely retained on the backs of trucks during transport. Ropes are frequently used but are difficult to tie and tension, and can come loose during a journey. Design a mechanical means of rapidly connecting and tensioning commercially available ropes of between 5 and 20 mm diameter. It would be an advantage if the device could be used in as many situations as possible, for example connecting and tensioning steel cables and chains as well as ropes.

Example 10: Rescue Device

Design a mechanical means of accurately, reliably and safely getting a rope to a drowning person. The aim is to assist someone in situations where it would be unwise to swim out to help and where the person is too far away to throw a rope accurately by hand. It can be assumed that the person in trouble is conscious and can participate in the rescue. The portable device is to be stored in locations where people might get into difficulty in the sea, a river or a lake. It might also be carried by life savers patrolling on foot.

The device must reach a person in difficulty at least 50 m away from the rescuer, but an even greater distance would be an advantage. The device must be entirely human powered, and operated by one person.

It is hoped that the above suggestions, along with the many examples in the book, will help design teachers create suitable lectures, exercises and projects for their classes. Any comments you might have on your experiences of using the book as a teaching text are of considerable interest to me. In addition, any suggestions for improvements which might be incorporated in future editions of the book would be welcomed. Please write to me directly: Ken Wallace, Department of Engineering, University of Cambridge, Trumpington Street, Cambridge, CB2 1PZ, UK.

1 Introduction

1.1 The Scope of Design

1.1.1 Tasks and Activities

The main task of engineers is to apply their scientific and engineering knowledge to the solution of technical problems, and then to optimise those solutions within the requirements and constraints set by material, technological, economic, legal, environmental and human-related considerations (see 2.1.7). Problems become concrete tasks after the clarification and definition of the problems which engineers have to solve to create new technical products (artefacts). The mental creation of a new product is the task of design or development engineers, whereas its physical realisation is the responsibility of manufacturing engineers. In this book, *designer* is used synonymously for design and development engineer. Designers contribute to finding solutions and developing products in a very specific way. They carry a heavy responsibility since their ideas, knowledge and skills determine in a decisive way the technical, economic and ecological properties of the product.

Design is an engineering activity that:

- affects almost all areas of human life;
- uses the laws and insights of science;
- builds upon special experience; and
- provides the prerequisites for the physical realisation of solution ideas [1.84].

Dixon [1.25] and later Penny [1.104] placed the work of engineering designers at the centre of two intersecting cultural and technical streams (see Fig. 1.1). Other models are also possible:

In *psychological* respects, designing is a creative activity that calls for a sound grounding in mathematics, physics, chemistry, mechanics, thermodynamics, hydrodynamics, electrical engineering, production engineering, materials technology and design theory, as well as knowledge and experience of the domain of interest. Initiative, resolution, economic insight, tenacity, optimism and teamwork are qualities that stand all designers in good stead and are indispensable to those in responsible positions [1.97] (see 2.2.1).

In *systematic* respects, designing is the optimisation of given objectives within

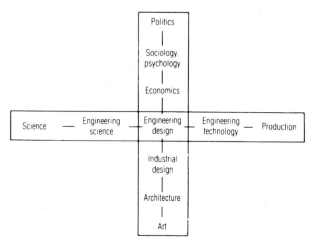

Figure 1.1.   The central activity of engineering design, from [1.25, 1.104].

partly conflicting constraints. Requirements change with time, so that a particular solution can only be optimised for a particular set of circumstances.

In *organisational* respects design is an essential part of the product life cycle. This cycle is triggered by a market need or a new idea. It starts with product planning and ends, when the product's useful life is over, with recycling or environmentally safe disposal (see Fig. 1.2). This cycle represents a process of converting raw materials into economic products of high added value. Designers must undertake their tasks in close co-operation with specialists in a wide range of disciplines and with different skills (see 1.1.2).

The tasks and activities of designers are influenced by several characteristics (see Fig. 1.3).

Origin of the task: Projects related to mass production and batch production are usually started by a product planning group after carrying out a thorough analysis of the market (see 5.1). The requirements established by the product planning group usually leave a large solution space for designers.

In the case of a customer order for a specific one-off or small batch product, however, there are usually tighter quantitative requirements to fulfil. In these cases it is wise for designers to base their solutions on the existing company know-how that has been built up from previous developments and orders. Such developments usually take place in small incremental steps in order to limit the risks involved.

If the development involves only part of a product (assembly or module), the requirements and the design space are even tighter and the need to interact with other design groups is very high.

When it comes to the production of a product, there are design tasks related to production machines, jigs and fixtures, and inspection equipment. For these tasks, fulfilling the functional requirements and technological constraints is especially important.

Organisation: The organisation of the design and development process depends in the first instance on the overall organisation of the company. In product-ori-

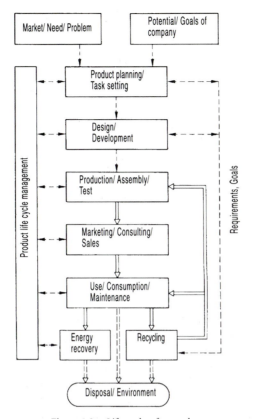

Figure 1.2.   Life cycle of a product.

ented companies, responsibility for product development and subsequent production is split between separate divisions of the company based on specific product types (eg rotary compressor division, piston compressor division, accessory equipment division).

Problem-oriented companies split the responsibility according to the way the task is broken down into partial tasks (eg mechanical engineering, control system, materials selection, stress analysis). In this arrangement the project manager must pay particular attention to the co-ordination of the project as the work passes from group to group.

Other organisational structures are possible, for example based on: the particular phase of the design process (conceptual design, embodiment design, detail design); the domain (mechanical engineering, electrical engineering, software development); the stage of the product development process (research, design, development, pre-production) (see 3.3). In large projects with clearly distinct domains, it is often necessary to develop individual modules of the product in parallel.

Novelty: New tasks and problems that are realised by *original designs* incorporating new solution principles. These can be realised either by selecting and combining known principles and technology, or by inventing completely new technology. The

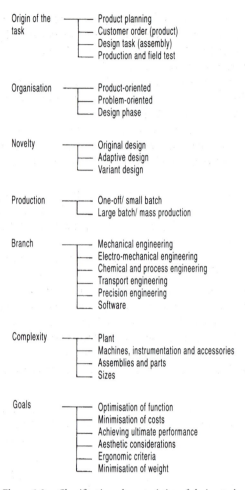

Figure 1.3.   Classification characteristics of design tasks.

term original design is also used when existing or slightly changed tasks are solved using new solution principles.

Original designs usually proceed through all design phases; depend on physical and process fundamentals; and require a careful technical and economic analysis of the task.

Original designs can involve the whole product or just assemblies or components.

In *adaptive design* one keeps to known and established solution principles and adapts the embodiment to changed requirements. It may be necessary to undertake original designs of individual assemblies or components. In this type of design the emphasis is on geometrical, analytical (strength, stiffness etc), production and material issues.

In *variant design* the sizes and arrangements of parts and assemblies are varied within the limits set by previously designed product structures (eg size ranges and modular products, see 8). Variant design requires original design effort only once. It includes designs in which only the dimensions of individual parts are changed to meet a specific order (in [1.95, 1.125] this type of design is referred to as "principle design" or "design with fixed principle").

In practice it is often not possible to define precisely the boundaries between the three types of design and they must be considered only as a broad classification.

Production: The design of one-off and small batch products requires particularly careful design of all physical processes and embodiment of details to minimise risk. In these cases it is usually not economic to produce development prototypes. Often functionality and reliability have a higher priority than economic optimisation.

Products to be made in large quantities (large batch or mass production) must have their technical and economic characteristics fully checked prior to full-scale production. This is achieved using models and prototypes and often requires several development steps (see Fig. 1.4).

Branch: Mechanical engineering covers a wide range of products. For example, food processing machines have to fulfil specific requirements regarding hygiene; machine tools have to fulfil specific requirements regarding precision and operating speed; prime movers have to fulfil specific requirements regarding power-to-weight ratio and efficiency; agricultural machines have to fulfil specific requirements regarding functionality and robustness; and office machines have to fulfil specific requirements regarding ergonomics and noise levels.

Electro-mechanical engineering products for heavy current applications, for

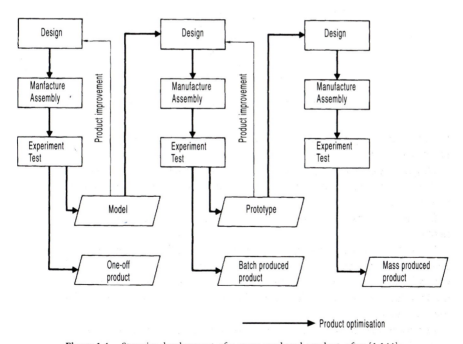

Figure 1.4. Stepwise development of a mass-produced product, after [1.144].

example, have to cover an ever widening range of applications, work at higher voltages and provide greater power outputs.

Chemical and process engineering products (plant and machinery) require the interdisciplinary application of fundamental chemical science and process technology as well as knowledge of large-scale devices. Safety and economics play dominant roles.

Transport engineering products (road vehicles, ships, aircraft) not only include specially developed machine elements and production processes, but also require specialist knowledge (eg composite material technology) to cope with demanding strength and safety requirements.

Precision engineering products are undergoing more rapid change than any other branch of engineering because of the introduction of micro- and opto-electronics. The impact of these new technologies is large and many new design principles are involved.

Software engineering products require systematic design steps similar to those for physical products, but involve only relationships between data elements.

Complexity: When planning the layout of process plant, the emphasis is on the combination of commercially available or specially developed machines, modules and components, and the design of the control system. Because of the variety of domains and system elements involved, generic methods are required for designing and optimising from receipt of order to final commissioning.

When designing machines and devices, tasks of similar complexity to those in process engineering have to be solved. However, detail design usually plays a greater role because of its crucial importance in optimising the performance of this type of product. One has to distinguish between a special one-off order and a machine selected from a product range offered for general sale.

When designing repeat parts or bought-in parts, the constraints and spatial requirements (interfaces with other assemblies) have to be specified comprehensively and in great detail to ensure their compatibility over a wide area of application.

The size of components influences the approach adopted and the embodiment possibilities, because size directly affects production tools, raw and semi-finished materials, transport, assembly and testing.

Goals: Design tasks must be directed towards meeting the optimisation goals taking into account the given restrictions.

Optimisation of function requires the application of computational and experimental optimisation procedures. Typical applications are aerospace products, machine tools and test equipment.

Minimisation of costs within a given framework requires: generation of alternative solutions; determination of production costs; close co-operation with purchasing and production planning; and careful attention to detail (see 10).

To achieve ultimate performance (eg very high precision, critical safety, extreme performance etc) special design principles and advanced computer-based analysis methods have to be used.

Aesthetic considerations are important for products that can be seen, and essential in the case of consumer products.

Ergonomic criteria must be taken into account for human-operated production equipment, office machinery and consumer products.

Minimisation of weight, for example in automobile engineering, requires special design strategies and machine elements.

To cope with this wide variety of tasks, designers have to adopt different approaches; use a wide range of skills and tools; have broad design knowledge; and consult specialists on specific problems. This becomes easier if designers master a general working procedure (see 2.2.2); understand generation and evaluation methods (see 4); and are familiar with well known technical solutions to existing problems.

The activities of designers can be roughly classified into:

- Conceptualising, ie searching for solution principles. Generally applicable methods can be used along with the special methods described in [1.6].
- Embodying, ie engineering a solution principle by determining the general arrangement and preliminary shapes and materials of all components. The methods described in [1.7] and [1.10] are useful.
- Detailing, ie finalising production and operating details. The methods described in [1.8] are useful.
- Computing, drawing and information collecting. These occur during all phases of the design process.

Another common classification is the distinction between *direct* design activities (eg conceptualising, embodying, detailing, computing), and *indirect* design activities (eg collecting and processing information, attending meetings, co-ordinating staff). One should aim to keep the proportion of the indirect activities as low as possible.

In the design process, the required design activities have to be structured in a purposeful way, that is in a clear sequence of main phases and individual working steps, so that the flow of work can be planned and controlled (see 3).

1.1.2 The Position of the Design Process Within a Company

The design department is of a central importance in any company. Designers determine the properties of every product in terms of function, safety, ergonomics, production, transport, operation, maintenance, recycling and disposal. In addition, designers have a large influence on production and operating costs, on quality and on production lead times. Because of this heavy responsibility, designers have to continuously reappraise the general goals of the task in hand (see 2.1.7).

A further reason for the central role of designers in the company is the position of design and development in the overall product development process. The links and information flows between departments are shown in Fig. 1.5, from which it can be seen that production and assembly depend fundamentally on information from product planning, design and development. However, design and development are strongly influenced by knowledge and experience from production and assembly. Because of current market pressures to increase product performance,

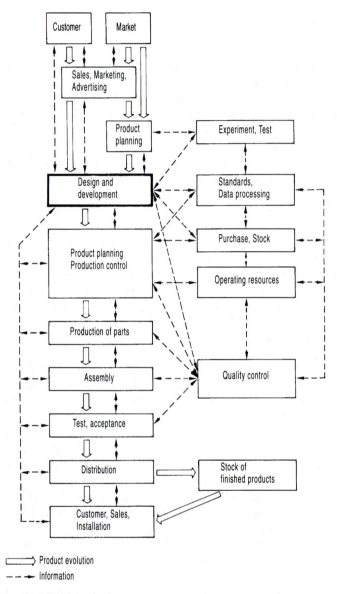

Figure 1.5. Information flows between departments.

lower prices and reduce time-to-market, product planning, sales and marketing must draw increasingly upon specialised engineering knowledge. Because of their key position in the product development process, it is therefore particularly important to make full use of the theoretical knowledge and product experience of designers (see 5).

Current product liability legislation [1.7] demands not only professional and responsible product development using the best technology but also the highest possible production quality.

1.1.3   Trends

The most important impact in recent years on the design process, and on the activities of designers, has come from computer-based data processing. Computer Aided Design (CAD) is influencing design methods, organisational structures, the division of work, eg between conceptual designers and detail designers, as well as the creativity and thought processes of individual designers. New staff, eg system managers, CAD assistants etc, are being introduced into the design process. In the future, routine tasks, such as variant designs, will be largely undertaken by the computer, leaving designers free to concentrate on original designs and customer-specific one-off products. These tasks will be supported by computer tools that enhance the creativity, engineering knowledge and experience of designers.

The development of knowledge-based systems (so called expert systems) [1.53, 1.54, 1.82, 1.132] will increase the ease with which information can be retrieved, including specific design data, details of standard components and information about existing products. These systems will also aid the analysis, optimisation and combination of solutions, but they will not replace designers. On the contrary, the decision ability of designers will be even more crucial because of the very large number of solutions it will be possible to generate, and also because of the need to co-ordinate the inputs from the many specialists involved in modern multi-disciplinary projects.

Computer Integrated Manufacturing (CIM) has consequences for designers in terms of company organisation and information exchange. The design manage-ment system within a CIM structure makes better planning and control of the design process necessary and possible. The same holds true for "simultaneous engineering" where development times are reduced by focusing on the flexible and partially parallel activities of product optimisation, production optimisation and quality optimisation [1.139]. The trend is to bring production planning forward into the design process through the application of computers.

Apart from these developments which influence the working methods of design-ers, designers increasingly have to take into account rapid technological develop-ments (eg new production and assembly procedures, micro-electronics and software) and new materials (eg composites, ceramics and recyclable materials). The integration of mechanical, electronic and software engineering (mechatronics) has led to many exciting product developments.

In summary it can be concluded that the pressures on designers are high and will increase further. This requires continuous further education for existing designers. However, the initial education of designers must take account of the many changes taking place. It is essential that future designers not only understand traditional science and engineering fundamentals (physics, chemistry, mathematics, mech-anics, thermodynamics, fluid mechanics, electronics, electrical engineering, materials science); but also specific domain knowledge (instrumentation, control, machine elements, transmission technology, production technology, electrical drives, electronic controls, ergonomics); and general organisational and business methods (systematic design, business administration, project management, CAD and CIM).

1.2   The Necessity for Systematic Design

1.2.1   Requirements and Need for Systematic Design

In line with the variety of problems and tasks in the development of technical products (see 1.1.1), design activities are many sided. First of all they rely on basic scientific and engineering knowledge, but also on special experience in the specific product area. The activities cannot be forced into rigid organisational or procedural templates.

In view of the central responsibility of designers for the technical and economic properties of a product, and the commercial importance of timely and efficient product development, it is important to have a defined design procedure to find good solutions. This procedure must flexible, and at the same time be capable of being planned, optimised and verified.

Such a procedure, however, can only be realised if the designers have the necessary domain knowledge and are required to work in a systematic way.

Nowadays one distinguishes between design science and design methodology [1.67]. Design science uses scientific methods to analyse the structure of technical systems and their relationships with the environment. The aim is to derive rules for the development of these systems from the system elements and their relationships.

Design methodology, however, is a concrete course of action for the design of technical systems that derives its knowledge from design science and cognitive psychology, and from practical experience in different domains. It includes: plans of action to link working steps and design phases according to content and organisation (see 3.3); strategies, rules and principles to achieve general and specific goals (see 7–10); and methods to solve individual design problems or partial tasks (see 4, 6).

A *design methodology*, therefore, must:

- encourage a problem-directed approach; ie it must be applicable to every type of design activity, no matter in what specialist field;
- foster inventiveness and understanding; ie facilitate the search for optimum solutions;
- be compatible with the concepts, methods and findings of other disciplines;
- not rely on finding solutions by chance;
- facilitate the application of known solutions to related tasks;
- be compatible with electronic data processing;
- be easily taught and learned; and
- reflect the findings of cognitive psychology and modern ergonomics; ie reduce workload, save time, prevent human error, and help to maintain active interest.

Such an approach will lead designers to possible solutions more quickly and directly than any other. As other disciplines become more scientific, and as the use of computers calls increasingly for logical data preparation, so designing, too, must become more logical, more sequential, more transparent, and more open to correction [1.32]. An enhanced appreciation of the contribution and status of designers and the attraction of more highly gifted and scientifically interested engineers into the field is only possible when their methods and style of work are in line with current developments in scientific and industrial practice.

This is not meant to detract from the importance of *intuition* or experience, quite the contrary—the additional use of systematic procedures can only serve to increase the output and inventiveness of talented designers. Any logical and systematic approach, however exacting, involves a measure of intuition; that is, an inkling of the overall solution. No real success is likely without intuition.

Design methodology should therefore foster and guide the abilities of designers, encourage creativity, and at the same time drive home the need for objective evaluation of the results. Only in this way is it possible to raise the general standing of designers and the regard in which their work is held. Systematic procedures help to render designing comprehensible and also enable the subject to be taught. However, what is learned and recognised about design methodology should not be taken as so many dogmas. Systematic procedures merely try to steer the efforts of designers from unconscious into conscious and more purposeful paths. As a result, when they collaborate with other engineers, designers will not merely be holding their own, but will be able to take the lead [1.97].

Systematic design provides an effective way to *rationalise* the design and production processes. In original design an ordered and stepwise approach, even if this is on a partially abstract level, will provide solutions that can be used again.

Structuring the problem and task makes it easier to recognise application possibilities for established solutions from previous projects and to use design catalogues. The approach of developing size ranges and modular products is an important start to rationalisation in the design area but especially important for the production process (see 8). The stepwise concretisation of established solution principles makes it possible to select and optimise in an early stage with a small amount of effort.

A design methodology is also a prerequisite for flexible and continuous *computer support* of the design process using product models stored in the computer. Without this methodology it is not possible to: develop knowledge-based systems; model using features; use stored data and methods; link separate programs, especially geometric modellers with analysis programs; ensure the continuity of data flow; and link data from different company divisions (CIM). Systematic procedures also make it easier to divide the work between designers and computers in a meaningful way.

A rational approach must also cover cost and quality considerations. More accurate and speedy preliminary calculations with the help of better data are a necessity in the design field, as is the early recognition of weak points in a solution. All this calls for systematic processing of the design documentation.

1.2.2 Historical Background

All developments have antecedents. They mature when there is a need for them, when the right technology is available, and when they are economically feasible. This also applies to the development we have called "systematic design".

It is difficult to determine its real origins. Can we trace it back to Leonardo da Vinci? Anyone looking at the sketches of this early master must be surprised to see—and the modern systematist delights in discovering—the great extent to which Leonardo used systematic variation of possible solutions [1.92]. Right up to the industrial era, designing was closely associated with arts and crafts.

With the rise of mechanisation, as Redtenbacher [1.110] pointed out early on in his "Prinzipien der Mechanik und des Maschinenbaus" (*Principles of Mechanics and of Machine Construction*), attention became increasingly focused on a number of characteristics and principles that continue to be of great importance, namely: sufficient strength, sufficient stiffness, low wear, low friction, minimum use of materials, low weight, easy assembly and maximum rationalisation.

Redtenbacher's pupil Reuleaux [1.111] developed these ideas but, in view of their often conflicting requirements, suggested that the assessment of their relative importance must be left to the intelligence and discretion of individual designers.

Important contributions to the development of engineering design were also made by Bach [1.6] and Riedler [1.114], who realised that the selection of materials, the choice of production methods and the provision of adequate strength are of equal importance and that they influence one another. Rötscher [1.119] mentions the following essential characteristics of design: specified purpose, effective load paths, and efficient production and assembly. Loads should be conducted along the shortest paths, and if possible by axial forces rather than by bending moments. Longer load paths not only waste materials and increase costs but also require considerable changes in shape. Calculation and laying out must go hand in hand. Designers start with what they are given and with ready-made assemblies. As soon as possible, they should make scale drawings to ensure the correct spatial layout. Calculation can be used to obtain either rough estimates for the preliminary layout, or precise values for checking the detail design.

Laudien [1.81], examining the load paths in machine parts, gives the following advice: for a rigid connection, join the parts in the direction of the load; if flexibility is required, join the parts along indirect load paths; do not make unnecessary provisions; do not over-specify; do not fulfil more demands than are required; save by simplification and economical construction.

Modern systematic ideas were pioneered by Erkens [1.30] in the 1920s. He insists on a *step-by-step approach* based on *constant testing and evaluation,* and also on the *balancing of conflicting demands,* a process that must be continued until a network of ideas—the design—emerges.

A more comprehensive account of the "technique of design" has been presented by Wögerbauer [1.151] who divides the *overall task* into *subsidiary tasks,* and these into operational and implementational tasks. He also examines, but fails to present in systematic form, the numerous interrelationships between the identifiable constraints designers must take into account. Wögerbauer himself does not proceed to a systematic elaboration of solutions. His systematic search starts with a solution discovered more or less intuitively and varied as comprehensively as possible in respect of the basic form, materials and method of production. The resulting *profusion of possible solutions* is then reduced by tests and *evaluations,* cost being a crucial criterion. Wögerbauer's very comprehensive list of *characteristics* helps in the search for an optimum solution and also in testing and evaluating the results.

Franke [1.39] discovered a comprehensive structure for transmission systems using a logical–functional analogy based on elements with different physical effects (electrical, mechanical, hydraulic effects for the same logical functions guiding, coupling and separating). For this reason he is regarded as a representative of those working on the functional comparison of physically different solution elements. Rodenacker in particular further developed this analogical approach [1.116].

Though some need for improving and rationalising the design process was felt even before World War II, progress was impeded by the following factors:

- the absence of a reliable means of representing abstract ideas; and
- the widespread view that designing is a form of art, not a technical activity like any other.

The rise of systematic design had therefore to wait until these obstacles had been cleared and for the wider adoption of systematic techniques, not least in non-technical areas, and the emergence of modern data-processing methods. A period of staff shortages [1.143] provided a further impetus.

Modern ideas of systematic design were given a great boost by Kesselring, Tschochner, Niemann, Matousek and Leyer. These men were not merely important pioneers; their work continues to provide most useful suggestions for handling the individual phases and steps of systematic design.

Kesselring [1.71] first explained the basis of his method of successive approximations in 1942 (for a summary see [1.73] and [1.147]). Its salient feature is the evaluation of form variants according to *technical* and *economic criteria*. In his theory of technical composition [1.72], Kesselring presents—in addition to a series of basic ideas on the technical contribution of designers and on their conduct, attitude and responsibility—an account of the underlying *scientific principles* (the mathematical and physical relationships) and the *economic constraints* (the production costs). In the theory of form design which he derives from the above, he mentions five overlying principles:

- the principle of minimum production costs;
- the principle of minimum space requirement;
- the principle of minimum weight;
- the principle of minimum losses; and
- the principle of optimum handling.

The design and optimisation of individual parts and simple technical artefacts is the object of the theory of form design. It is characterised by the simultaneous application of physical and economic laws, and leads to a determination of the shape and dimensions of components and an appropriate choice of materials, production methods etc. If selected optimisation characteristics are taken into account, the best solution can be found with the help of mathematical methods.

Tschochner [1.133] mentions four fundamental design factors, namely *working principle*, *material*, *form* and *size*. They are interconnected and dependent on the requirements, the number of units, costs etc. Designers start from the working principle, determine the other fundamental factors—material and form—and match them with the help of the chosen dimensions.

Niemann [1.91] starts out with a scale layout of the overall design showing the main dimensions and the general arrangement. Next he divides the overall design into parts that can be developed in parallel. He proceeds from a *definition of the task* to a systematic *variation of possible solutions* and finally to a *formal selection of the optimum solution*. These steps are in general agreement with those used in more recent methods. Niemann also draws attention to the then lack of methods for arriving at new solutions. He must be considered a pioneer of systematic design inasmuch as he consistently demanded and encouraged its development.

Matousek [1.85] lists four essential factors: *working principle, material, manufacture* and *form* design, and then, following Wögerbauer [1.151], elaborates an overall working plan based on these four factors. He adds that, if the cost aspect is unsatisfactory, all four factors have to be re-examined in an iterative manner.

Leyer [1.83] is mainly concerned with form design. He distinguishes three main design phases. In the first, the working principle is laid down with the help of an idea, an invention, or established facts; the second phase is that of actual design; the third phase is that of implementation. His second phase is essentially that of embodiment, that is, layout and form design supported by calculations. During this phase, principles or rules have to be taken into account—for instance the principle of constant wall thickness, the principle of lightweight construction, the principle of shortest load paths and the principle of homogeneity. Leyer's rules of form design are so valuable because, in practice, failure is still far less frequently the result of bad working principles than of poor detail design.

These preliminary attempts made way for intensive developments of methods, mainly by university professors who had learned the fundamentals of design in practical contact with technical products of increasing complexity. They realised that greater reliance on physics, mathematics and information theory, and the use of systematic methods, were not only possible but, with the growing division of labour, quite indispensable. Needless to say, these developments were strongly affected by the requirements of the particular industries in which they originated. Most came from precision, power transmission and electromechanical engineering, in which systematic relationships are more obvious than in heavy engineering.

Hansen and other members of the *Ilmenau School* (Bischoff, Bock) first put forward their systematic design proposals in the early 1950s [1.12, 1.15, 1.57]. Hansen presented a more comprehensive design system in the second edition of his standard work published in 1965 [1.58].

Hansen's approach is defined in a so-called *basic system*. The four working steps in this approach are applied in the same way in conceptual, embodiment and detail design. Hansen begins with the analysis, critique, and specification of the task, which leads him to the *basic principle* of the development (the crux of the task). The basic principle encompasses the overall function that has been derived from the task, the prevailing conditions, as well as the required measures. The overall function (goal and restrictions) and the context (elements and properties) constitute the crux of the task together with the given constraints.

The second working step is a systematic search for solution elements and their combination into *working means,* ie *working principles.*

Hansen attaches great importance to the third step, in which any shortcomings of the developed working means are analysed with respect to their properties and quality characteristics, and then, if necessary, improved.

In the fourth and last step, these improved working means are evaluated to determine the optimum working means for the task.

In 1974 Hansen published another work, entitled "Konstruktionswissenschaft" (*Science of Design*) [1.59]. In it, he uses systems analysis and information theory to define the design process and the nature of technical artefacts. He dwells on the various types of structure and function and their interrelationships, and also discusses the problem of data storage. The book is more concerned with theoretical fundamentals than with rules of practical design.

Similarly, Müller [1.86] in his "Grundlagen der systematischen Heuristik" (*Fundamentals of Systematic Heuristics*) presents a theoretical and abstract picture of the design process. This book offers essential foundations of design science. Further important publications are [1.87, 1.88].

After Hansen, it is Rodenacker [1.116, 1.117] who became pre-eminent by developing an original design method. His approach is characterised by solving the required *working structure* stepwise through *logical, physical* and *embodiment* relationships. He emphasises the recognition and suppression of disturbing influences and failures as early as possible during formulation of the physical process; the adoption of a general selection strategy from simple to complex; and the evaluation of all parameters of the technical system against the criteria *quantity, quality* and *cost.* Other characteristics of his method are the emphasis on logical function structures based on *binary logic* (connecting and separating), and on a conceptual design stage based on the recognition that product optimisation can only take place once a suitable solution principle has been found. The most important aspect of Rodenaker's systematic design approach is undoubtedly his emphasis on establishing the physical process. Based on this, he not only deals with the systematic processing of concrete design tasks, but also with a methodology for inventing new technical systems. For the latter he starts with the question: For what new application can a known physical effect be used? He then searches systematically to discover completely new solutions.

The fundamental works of Kuhlenkamp [1.80], Richter [1.112, 1.113] and Brader [1.16] stem from the field of precision engineering. They propose combination algorithms, mathematical simulations and optimisation procedures with which the selection of solutions and the optimisation of a product can be supported using the computer. This is possible because they deal mainly with electro-mechanical systems for signal processing. In such systems, function structures, physical processes and links between the building blocks can often be described unambiguously by physical equations. In particular Richter proposes a function-oriented design synthesis based on system dynamics.

In addition to the methods we have been describing, there is a view that a one-sided emphasis on discursive methods does not present the complete picture. For that reason various attempts have been made to develop design methods with the help of automatic control techniques involving feedback. Such methods not only help to elucidate the interrelationship between designers and the environment, but also throw light on the general scope of human thought processes.

Thus Wächtler [1.148, 1.149] argues, by analogy with cybernetic concepts such as control and learning, that creative design is the most complex form of the "learning process". Learning represents a higher form of control, one that involves not only quantitative changes at constant quality (control), but also changes in the quality itself. Similarly, designing changes technical quantities as well as working principles. In structural terms, learning and control can, despite qualitative differences, be considered as comparable circular processes.

What matters for the purpose of optimisation is that the design process should not be treated statically, but dynamically as a control process in which the information feedback must be repeated until the information content has reached the level at which the optimum solution can be found. The learning process thus keeps increasing the level of information and hence facilitates the search for a solution.

1.2.3   Current Methods

Many of the original systematic design methods of Leyer, Hansen, Rodenacker, Kuhlenkamp and Wächtler are still being applied today, having been integrated into the more recent developments in design methodology. The methods described in the following sections are either still under development or are directed towards special goals.

1   Design Methods

Since 1965 Roth has been developing a design methodology. In 1988 he was succeeded by his former pupil Franke. This methodology is characterised by a set of integrated and logically structured *design catalogues* [1.122]. A particular goal is the partitioning of the design process into small steps that can be described by algorithms. It should then be possible to process these steps completely and continuously using the computer. To that end "product representing models" are defined that can be described unambiguously by "product defining data" [1.123].

Roth divides the design process into a *problem formulation* phase, a *functional* phase and an *embodiment* phase. Corresponding to this he defines three product-representing models: "problem-representing models", "function-representing models", and "embodiment-representing models". The three phases are divided into smaller steps. The result of the problem formulation phase is a requirements list identifying the main tasks and requirements. The functional phase results in: logical function structures; cybernetic function structures (general function structures) with material, energy and information functions; physical function structures with physical, chemical and other effects; and principle solutions. The embodiment phase results in: geometrical working structures; embodied principle solutions; modular structures based on pre-designed modules and functional elements; and overall designs with all the details necessary for the final product documentation. The reasons for this division are to provide as many starting points as possible for systematic variation, particularly of geometry during the embodiment phase, and to permit greater use of computers in the design process.

Franke continues to develop Roth's ideas, but he emphasises more strongly the iterative nature of the design process and the importance of the individual creativity of designers [1.38]. He is therefore tending to move away from the very strong emphasis on an algorithmic approach to design.

Gierse, who comes from the field of transmission technology, underlines in his systematic procedure the basic steps of Value Analysis. He not only applies these steps to reduce the cost of products and components, but also as a general problem solving method. As a consequence, the identification of functions and function structures plays an important role in his procedure [1.47, 1.48].

Since 1970 Koller has been working at Aachen on a design methodology that aims to make the design process more algorithmic and thus permit more extensive computer support [1.76, 1.78]. He divides the design process into three main phases: function synthesis, qualitative synthesis and quantitative synthesis. To permit computer support, these phases are further divided into several steps and algorithms defined for their execution. In deriving his functions, Koller starts with the fact that in technical systems only the states and flows of energy, material, and signals can be changed in magnitude and direction. Together with physical inputs and outputs, this provides 12 functions and their inverses, which Koller calls *basic operations*. In his function synthesis, logical and physical basic operations are linked together in a function structure. The realisation of these basic operations is undertaken in the qualitative synthesis phase using physical effects, effect carriers, principle solutions, and system and embodiment variants; and in the quantitative synthesis phase using dimensioned layouts and production documents. In his latest work, Koller emphasises the development of general rules for embodiment synthesis; the development of product ranges and product groupings; and the restrictions placed on the selection of solutions [1.77].

Jung, working at Stuttgart, emphasises "functional embodiment" which he applies to the design of devices and instruments [1.69].

Hubka belongs to the group of German-speaking design scientists whose goal is to define a complete and unambiguous structure and ordering for technical systems and their environment. He started his work on design science in former Czechoslovakia in the sixties, but for the past two decades he has been working in Zürich. He has made significant contributions to establishing the fundamentals of a comprehensive design science and deriving guidelines for the activities of designers [1.65, 1.66]. This includes establishing unambiguous symbols and terminology for the different levels of abstraction of the design process, and defining the relationships between design steps, solution elements and external influences.

Schregenberger, also at Zürich, includes elements of psychology, theory of science, decision theory and management in his *Programme of Methodical Conscious Problem Solving in Design* [1.127].

Ehrlenspiel, the successor of Rodenacker at Munich, places great emphasis on design for cost [1.27]. His aim is to develop a method to determine costs at an early stage in the design process and he made important contributions to VDI Guideline 2235: *Economic Decision Making in Design*. His systematic approach is based strongly on the fundamentals of systems engineering. His work on computer-based support of design concentrates increasingly on the development of knowledge-based systems for individual design tasks [1.28] and on the integration of production-related criteria in the decision making process.

As Pahl's successor at Darmstadt, Birkhofer continues developing the methodologies of Pahl and Beitz, and of Roth, under whom he studied. He applied these approaches successfully in his industrial practice thereby acquiring considerable experience of their application [1.10]. Of particular relevance in practice is the selection and evaluation of designs [1.11]. Birkhofer is currently developing knowledge-based systems for designers, improving the systematic selection of standard and bought-in components, and undertaking research into embodiment design methods.

The emphasis of the work of Seeger is on industrial design (see 7.5.6) which he is integrating into a systematic design methodology [1.128].

In former East Germany, similar developments in design methodology have taken place since the early 1960s, starting with the work of Hansen, Bischoff and Bock (Illmenau School [1.58, 1.59]). Compared to the design research in former West Germany, cognitive psychology and heuristics were introduced from the start to gain a better understanding of the behaviour of designers [1.89]. Because of the earlier shortage of computers, it is only in the last few years that these researchers have turned their attention to CAD. Based on their publications on design methodology and CAD developments, the following have made significant contributions: Müller for many contributions to creativity and heuristics as well as to engineering science [1.89]; Höhne from the Illmenau School [1.63] and Heinrich from Dresden [1.60, 1.61] for their work on systematic methods and the use of computers in precision engineering design; Hennig from Dresden for his methods of designing machine tools [1.62]; Frick from Halle for a systematic approach focusing, in particular, on aesthetics [1.44, 1.45]; Franz for his work on CAD/CAM systems [1.41]; Klose from Dresden for his systematic approach to CAD systems [1.74, 1.75]; Rugenstein from Magdeburg for his definition of design principles [1.124]; and Schlottmann from Rostock for his design methodology [1.126]. The contributions mentioned above represent only a selection of those covered in more detail in [1.89].

A number of interesting contributions to design methodology have been made outside Germany.

In the United Kingdom the following design methodologists have become well known through their books and papers, and through their contributions at international conferences. One of the earliest contributors was Archer. He investigated the structure of the design process and systematic design methods, and compared these with those found in other disciplines. He discussed the limitations of systematic design methods in [1.5]. In his book [1.21], Cross explains the basics of systematic design using clear examples. Flursheim discusses the complex relationships between engineering design and related domains, paying particular attention to industrial design [1.36, 1.37]. French places great emphasis on fundamental solution strategies for conceptual design; on developing basic design principles; and on using ideas and analogies from nature [1.42, 1.43]. In his books, Glegg made a number of original contributions to design theory [1.49–51]. Gregory focused, in particular, on methods to encourage creativity [1.52]. Pugh is recognised for his contributions to a methodology encompassing the total design and development process (Total Design), which emphasises, in particular, the product design specification, multi-disciplinary teamwork and the importance of links with other company departments and customers (marketing, sales) [1.109]. Wallace contributed significantly to defining the terminology of design methodology in the English language through his translation of the book *Engineering Design* by Pahl and Beitz [1.101] and the VDI Guideline 2221 *Systematic Approach to the Design of Technical Systems and Products* [1.142]. In addition he undertook an empirical investigation into the application of a systematic design approach to an industrial project, which led to a number of interesting conclusions about the application of design methods in practice [1.55, 1.150].

Important contributions to design methodology have also come from researchers in other European countries. In Denmark Andreasen and Hein have proposed a systematic approach for the whole product development process [1.3], and Andreasen has reported several problems with the application of systematic design methods in practice [1.4]. Andreasen has also made significant contributions to design for assembly [1.2]. In Finland Konttinen spread the message of systematic design through his translation of the book by Pahl and Beitz [1.102]. In Italy Pighini works on the systematic design of machine elements [1.105]. In Croatia Kostelic focuses on design for quality [1.79]. In the Netherlands a novel design methodology has been developed by Eekels and Roozenburg [1.121], focusing, in particular, on evaluation and decision processes [1.120]. Also in the Netherlands van den Kroonenberg has worked on the integration of design methodology and CAD [1.138]. In Norway Jakobsen focuses on the functional requirements in the design process [1.68]. In addition the following must be mentioned: in Poland, Dietrych [1.24] and Gasparski [1.46]; in Sweden, Bjärnemo [1.13]; in Switzerland, Flemming [1.35] and Breiing [1.18] (the contributions of Hubka and Schregenberger have already been mentioned); in Poland Walczak translated the book by Pahl and Beitz [1.100]; and in Hungary Kozma and Straub, under the editorial control of Bercsey and Varga, translated the book by Pahl and Beitz [1.99].

Design methodology in the USA is characterised by a stronger orientation on basic theory and the application of computers. This probably stems from their theoretically based educational system. In the last few years the number of publications on design methodology has increased due to the funding initiatives of government and industry. From the large number of publications, only those by authors with international reputations will be mentioned. In 1966, Dixon published a fundamental book about design methodology [1.25]. In [1.26] he gives a comprehensive overview of research into systematic design. For over 20 years Nadler has contributed to design methodology [1.90]. Kannapan and Marshek focus on a theoretical description of the design process based on transformation rules for functions and information [1.70]. In 1977 Ostrofsky published a comprehensive book on design, planning and development methodology [1.96]. Rinderle has contributed to a number of different areas of design methodology [1.115]. Suh has become known for his work on design for manufacture [1.130]. Ullman, Stauffer and Dietterich investigated the design process in practice [1.134]. Ullman has also proposed his own design approach [1.135] and evaluated other methods [1.136]. The work of Ulrich and Seering focuses on the systematic understanding of conceptual design and an abstract description of design [1.29, 1.137]. O'Grady discusses a theme of increasing importance—concurrent engineering [1.94].

Only a few works on design methodology have emerged from Japan. Hongo has worked on a computer-supported design methodology, but his main focus is on engineering design education [1.64]. Taguchi emphasises methods to optimise quality [1.131]. Yoshikawa works on knowledge processing methods in the design process [1.152].

In the former USSR, Polovinkin has undertaken research into the complete design process including the application of computers [1.106, 1.107]. Odrin worked on the morphological synthesis of systems [1.93]. Altshuler developed a systematic approach for invention based on the analysis of a large number of patents [1.1].

The following publications provide overviews of international contributions to design science and methodology: Müller [1.89] focuses on the German language literature and on publications from Eastern Europe; Hales [1.56] and de Boer [1.23] focus on the literature from the Western industrial nations; and Finger and Dixon [1.34] cover a wide range of international contributions, except those from the former USSR and other East European countries.

A truly comprehensive overview of international engineering design research and education activities from 1981 until 1993 can be found in the proceedings of the ICED series of conferences (International Conference on Engineering Design) [1.108].

2 Systems Theory

In socio–economic–technical processes, procedures and methods of *systems theory* are increasingly important. Systems theory as an inter-disciplinary science uses special methods, procedures and aids for the analysis, planning, selection and optimum design of complex systems [1.8, 1.9, 1.14, 1.19, 1.20, 1.22, 1.103, 1.153].

Technical artefacts, including the products of light and heavy engineering industry, are artificial, concrete and mostly dynamic systems consisting of sets of ordered elements, interrelated by virtue of their properties. A system is also characterised by the fact that it has a boundary which cuts across its links with the environment (see Fig. 1.6). These links determine the external behaviour of the system, so that it is possible to define a function expressing the relationship between inputs and outputs, and hence changes in the magnitudes of the system variables (see 2.1.3).

From the idea that technical artefacts can be represented as systems, it was a short step to the application of systems theory to the design process, the more so as the objectives of systems theory correspond very largely to the expectations we have of a good design method as specified at the beginning of this chapter [1.9]. The systems approach reflects the general appreciation that complex problems are best tackled in fixed steps, each involving analysis and synthesis (see 2.2.2).

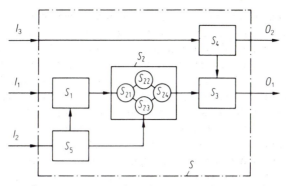

Figure 1.6. Structure of a system. S: system boundary; S_1-S_5: sub-systems of S; $S_{21}-S_{24}$: sub-systems or elements of S_2; I_1-I_3: inputs; O_1-O_2: outputs.

Figure 1.7 shows the steps of the systems approach. The first of these is the gathering of information about the system under consideration by means of market analyses, trend studies or specific tasks. In general this step can be called problem analysis. The aim here is the clear formulation of the problem (or sub-problem) to be solved, which is the actual starting point for the development of the system. In the second step, or perhaps even during the first step, a programme is drawn up to give formal expression to the goals of the system (problem formulation). Such goals provide important criteria for the subsequent evaluation of solution variants and hence for the discovery of the optimum solution. Several solution variants are then synthesised on the basis of the information acquired during the first two steps. Before these variants can be evaluated, the performance of each must be analysed for its properties and behaviour. In the evaluation that follows, the performance of each variant is compared with the original goals, and on the

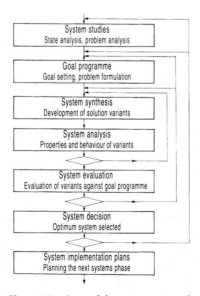

Figure 1.7. Steps of the systems approach.

basis of this a decision is made and the optimum system selected. Finally, information is given out in the form of system implementation plans. As Fig. 1.7 shows,
the steps do not always lead straight to the final goal, so iterative procedures may
be needed. Built-in decision steps facilitate this optimisation process, which constitutes a transformation of information.

In a systems theory process model [1.14, 1.39] the steps repeat themselves in
so-called life cycle phases of the system in which the chronological progression of
a system goes from abstract to concrete (see Fig. 1.8).

An important sphere of application of the systems approach is function-oriented
synthesis. On the basis of a known or a developed solution concept it is possible
to produce a function model (function structure, function chain) whose inputs
and outputs, together with their links, can be subjected to mathematical variation
and optimised to satisfy the demands of the problem. A necessary requirement for
the use of such mathematical statements about the quantitative behaviour of solution concepts is the availability of mathematical relations expressing the static
and dynamic behaviour of the elements, for instance in the form of transfer functions, and the formulation of a goal function. Such mathematical laws for the
description of systems models have been established, above all, for signal-pro-
cessing equipment. Richter [1.112, 1.113] and Findeisen [1.33] are developing this
method, paying particular attention to the optimisation of dynamic systems.

3   Value Analysis

The main aim of *Value Analysis* described in DIN 69910 is to reduce cost (see
10.4). To that end a systematic overall approach is proposed which is applicable,

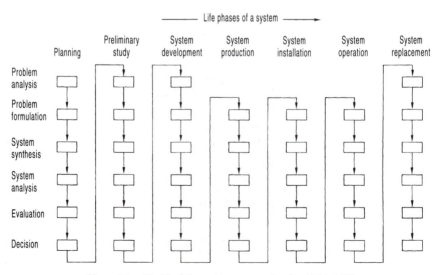

Figure 1.8.   Model of the systems approach, after [1.14, 1.39].

in particular, to the further development of existing products. Figure 1.9 shows the basic working steps of Value Analysis. In general a start is made with an existing design, which is analysed with respect to the required functions and costs. Solution ideas are then proposed to meet the new targets. Because of its emphasis on functions and the stepwise search for better solutions, Value Analysis has much in common with systematic design.

4   VDI Guidelines

VDI Guideline 2222 [1.145, 1.146] defines an overall approach and individual methods for the conceptual design of technical products and is therefore particularly suitable for the development of new products. The VDI Guideline 2221 [1.144] (English translation [1.142]) proposes a systematic approach to the design of technical systems and products and emphasises the general applicability of the approach in the fields of mechanical, precision, control, software and process engineering. The approach (see Fig. 1.10) includes seven basic working steps that accord with the fundamentals of technical systems (see 2.1) and company strategy (see 3). Both guidelines have been developed by a VDI Committee comprising senior designers from industry and many of the previously mentioned design methodologists from former West Germany. Because the aim is for general applicability, the design process has been only roughly structured thus permitting product-

Prepare project
- Assemble team
- Define scope of Value Analysis
- Define organisation and procedure

Analyse actual state
- Recognise functions
- Determine function costs

Determine target state
- Define target functions
- Identify additional requirements
- Match target costs with target functions

Develop solution ideas
- Collect existing ideas
- Search for new ideas

Determine solutions
- Evaluate ideas
- Develop ideas into solutions
- Evaluate and decide upon solutions

Realise solutions
- Detail chosen solutions
- Plan their implementation

Figure 1.9.   Basic working steps of Value Analysis (after DIN 69910).

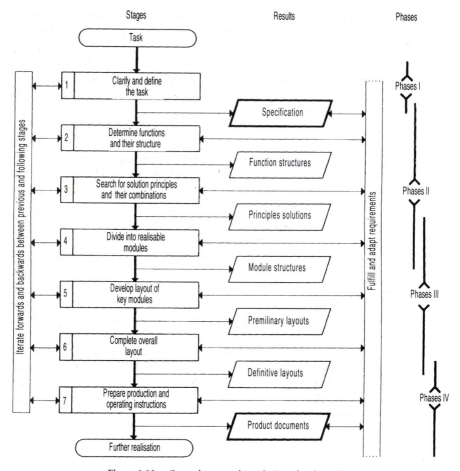

Figure 1.10. General approach to design after [1.144].

specific and company-specific variations. Figure 1.10 should therefore be regarded as a guideline to which detailed working procedures can be assigned. Special emphasis is on the iterative nature of the approach and the sequence of the steps must not be considered rigid. Some steps might be omitted, and others repeated frequently. Such flexibility is in accordance with practical design experience and is very important for the application of all design methods.

1.2.4 Authors' Own Aims

On closer examination the methods we have been describing have been strongly influenced by their authors' specialist fields. They nevertheless resemble one another far more closely than the various concepts and terms would suggest. VDI Guidelines 2222 and 2221 confirm these resemblances as they were developed in collaboration with a wide range of experienced contributors.

The proposals of individual methodologists and schools of systematic design, however, have different emphases according to the different contexts and requirements of the product groups they represent. For example, precision engineering products are characterised by electronic and software issues, and a focus on function; heavy engineering products by strength, stiffness and production issues; automobile engineering by tuning individual components within the overall system, by mass production issues, by light-weight engineering and by external shape; and process plant engineering by the combination of devices developed elsewhere (see 1.1.1).

Based on our experience in the heavy machinery industry and more than 20 years in engineering design education at undergraduate and graduate levels, this book sets out a comprehensive design methodology for all phases of the product planning, design and development processes for technical systems. Most of the arguments are elaborations of a series of papers we published [1.98] and which we have since discussed at length with practising designers and research engineers, tested repeatedly in practice, and changed and amplified accordingly. It should be emphasised that between the publication of the first edition of *Engineering Design* (1976) and of this edition none of the statements had to be dropped because they were overtaken. However, extensions were necessary because of new research findings in the domains of cognitive psychology and design methodology, as well as new technological developments, in particular the use of computers.

As before our own theory of design still does not claim to be the final word on the subject—it simply tries to combine various methods in a coherent and practicable way. We hope that it may serve as an introduction and springboard for the learner; as a help and illustration for the teacher; and as a source of information, and of further learning, for the practitioner.

2 Fundamentals

Designing is a many-sided and wide-ranging activity. It is based not only on mathematics, physics and their branches—mechanics, thermodynamics etc—but also on production technology, materials science, machine elements, industrial management and computer science, which are not discussed in this book.

To develop a theory of design that can serve as a strategy for the development of solutions, we must first examine the fundamentals of technical systems and procedures. Only when that has been done is it possible to make detailed recommendations for design work.

2.1 Fundamentals of Technical Systems

2.1.1 System, Plant, Equipment, Machine, Assembly and Component

Technical tasks are performed with the help of *technical artefacts* and, after the introduction of electronic data processing, to an increasing degree with the help of *software*. Mechanical engineering designers design and develop technical artefacts that may include software developed by specialists. The basic structure of software and its design and development processes are similar to those of classical technical artefacts, though the terminology used is rather different. In 2.1.7, and in the VDI Guideline 2221 [2.39] (English edition [1.142]), these similarities are identified and it is demonstrated how systematic methods can also be applied to the development of technical software.

Technical artefacts include plant, equipment, machines, assemblies and components, listed here in the approximate order of their complexity. These terms may not have identical uses in different fields. Thus, a piece of equipment (reactor, evaporator) is sometimes considered to be more complex than plant, and artefacts described as "plant" in certain fields may be described as "machines" in others.

A machine consists of assemblies and components. Control equipment is used in plant and machines alike and may be made up of assemblies and components, and perhaps even of small machines. The various uses of these terms reflect historical developments and application areas. There are attempts to define standards in which energy transforming technical artefacts are referred to as machines, material transforming artefacts as apparatus and signal transforming artefacts as devices. It is evident that a clear division on the basis of these characteristics is not always possible and that the current terminology is not ideal.

Hubka [2.17, 2.18] has drawn up a comprehensive list of possible classifications of technical artefacts based on such criteria as function, working means, complexity, production, product etc. It is, however, impossible to agree on a generally acceptable system of classification — the tasks, applications and forms are much too varied and complex. Hence there is much to be said for Hubka's suggestion that technical artefacts should be treated as *systems* connected to the environment by means of *inputs* and *outputs*. A system can be divided into sub-systems. What belongs to a particular system is determined by the *system boundary*. The inputs and outputs cross the system boundary (see 1.2.3). With this approach it is possible to define appropriate systems at every stage of abstraction, analysis or classification. As a rule such systems are parts of larger, superior systems.

A concrete example is the combined coupling shown in Fig. 2.1. It can be considered as a system "coupling" which within a machine, or joining two machines, can be considered an assembly. This coupling assembly can be treated as two *sub-systems* — a "flexible coupling" and a "clutch". Each sub-system can, in turn, be subdivided into system elements, in this case components.

The system depicted in Fig. 2.1 is based on its mechanical construction. It is, however, equally possible to consider it in terms of its functions (see 2.1.3). In that case, the total system "coupling" can be split up into the sub-systems "damp-

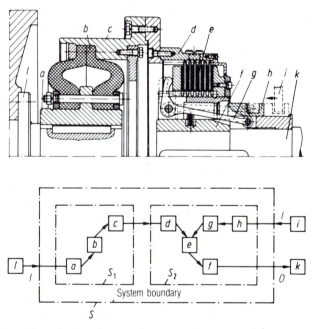

Figure 2.1. System: "Coupling" $a...h$ system elements; $i...l$ connecting elements; S overall system; S_1 subsystem "flexible coupling"; S_2 subsystem "clutch"; I inputs; O outputs.

ing" and "clutching"; the second sub-system into the further sub-systems "changing clutch operating force into normal force" and "transferring torque". Thus the system element g could equally well be treated as a sub-system whose function is to convert the actuating force into a larger normal force acting on the friction surfaces, and through its flexibility provide some equalisation of the wear.

Depending on their use, any number of such subdivisions may be made. Designers have to establish particular systems for particular purposes, and must specify their various inputs and outputs and fix their boundaries. In doing this they may use what terminology they like or is customary in their particular field.

2.1.2   Conversion of Energy, Material and Signals

One encounters matter in many shapes and forms. Its natural form, or the form impressed upon it, provides information about its possible uses. Matter without form is inconceivable—form is a primary source of information about the state of matter. With the development of physics, the concept of force became essential. Force was conceived as being the means by which the motion of matter was changed. Ultimately this process was explained in terms of energy. The theory of relativity postulated the equivalence of energy and matter. Weizsacker [2.44] lists energy, matter and information as basic concepts. If change or flow is involved, time must be introduced as a fundamental quantity. Only by reference to time does the physical event in question become comprehensible, and can the interplay of energy, matter and information be adequately described.

In the technical sphere the previous terminology is usually linked to concrete physical or technical representations. *Energy* is often specified by its manifest form. We speak of mechanical, electrical, optical energy etc. For matter, it is usual to substitute *material* with such properties as weight, colour, condition etc. The general concept of information is generally given more concrete expression by means of the term *signal*—that is, the physical form in which the information is conveyed. Information exchanged between people is often called a message [2.20].

The analysis of technical systems—plant, equipment, machine, device, assembly or component—makes it clear that all of them involve technical processes in which energy, material and signals are channelled and/or converted. Such conversions of energy, material and signals have been analysed by Rodenacker [2.33].

Energy can be converted in a variety of ways. An electric motor converts electrical into mechanical and thermal energy, a combustion engine converts chemical into mechanical and thermal energy, a nuclear power station converts nuclear into thermal energy, and so on.

Materials too can be converted in a variety of ways. They can be mixed, separated, dyed, coated, packed, transported or reshaped. Raw materials are turned into part-finished and finished products. Mechanical parts are given particular shapes and surface finishes and some are destroyed for testing purposes.

Every plant must process information in the form of *signals*. Signals are received, prepared, compared or combined with others, transmitted, displayed, recorded, and so on.

In technical processes, one type of conversion (of energy, material or signals) may prevail over the others, depending on the problem or the type of solution. It is useful to consider these conversions as flows, and the prevailing one the *main flow*. It is usually accompanied by a second type of flow, and quite frequently all three come into play. Thus there can be no flow of material or signals without an accompanying flow of energy, however small.

The flow of energy is often associated with the flow of material, even though no such flow may be visible (as in a nuclear, compared with a coal-fired, power station). The associated flow of signals constitutes an important subsidiary flow for the control and regulation of the entire process.

However, numerous measuring instruments receive, transform or display signals without any flow of material. In many cases energy has to be specially provided for this purpose; in other cases latent energy can be drawn upon directly. Every flow of signals is associated with a flow of energy, though not necessarily with a flow of material.

In what follows, we shall be dealing with:

- Energy: mechanical, thermal, electrical, chemical, optical, nuclear ..., also force, current, heat ...
- Material: gas, liquid, solid, dust ..., also raw material, test sample, workpiece ..., end product, component ...
- Signals: magnitude, display, control impulse, data, information ...

In every type of proposed conversion, *quantity* and *quality* must be taken into consideration if rigorous criteria for the definition of the task, for the choice of solutions and for an evaluation, are to be established. No statement is fully defined unless its quantitative as well as its qualitative aspects have been taken into account. Thus, the statement: "100 kg/s of steam at 80 bar and 500°C" is not a sufficient definition of the input of a steam turbine unless there is the further specification that these figures refer to a nominal quantity of steam and not, for instance, to the maximum flow capacity of the turbine; and unless the admissible fluctuations of the state of the steam are fixed at, say, 80 bar ± 5 bar and 500°C ± 10°C, that is, extended by a qualitative aspect.

In many applications, it is also essential to stipulate the cost or value of the inputs and the maximum permissible costs of the outputs (see [2.33] Categories: Quantity—Quality—Costs).

All technical systems, therefore, involve the conversion of energy, material and/or signals which must be defined in quantitative, qualitative and economic terms (see Fig. 2.2).

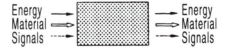

Figure 2.2. The conversion of energy, material and signals. Solution not yet known; task or function described on the basis of inputs and outputs.

2.1.3 Functional Interrelationship

1 Fundamentals

In order to solve a technical problem we need a system with a clear and easily reproduced relationship between inputs and outputs. In the case of material conversions, for instance, we require identical outputs for identical inputs. Also, between the beginning and the end of a process, for instance filling a tank, there must be a clear and reproducible relationship. Such relationships must always be planned—that is, designed to meet a specification. For the purpose of describing and solving design problems, it is useful to apply the term *function* to the general input/output relationship of a system whose purpose is to perform a task.

For static processes it is enough to determine the inputs and outputs; for processes that change with time (dynamic processes), the task must be defined further by a description of the initial and final magnitudes. At this stage there is no need to stipulate what solution will satisfy this kind of function. The function thus becomes an abstract formulation of the task, independent of any particular solution.

If the *overall task* has been adequately defined—that is, if the inputs and outputs of all the quantities involved and their actual or required properties are known—then it is possible to specify the *overall function*.

An overall function can often be divided directly into identifiable *sub-functions* corresponding to sub-tasks. The relationship between sub-functions and overall function is very often governed by certain constraints, inasmuch as some sub-functions have to be satisfied before others.

On the other hand it is usually possible to link sub-functions in various ways and hence to create variants. In all such cases, the links must be compatible.

The meaningful and compatible combination of sub-functions into an overall function produces a so-called *function structure*, which may be varied to satisfy the overall function.

To that end it is useful to make a block diagram in which the processes and sub-systems inside a given block (black box) are at first ignored (see Fig. 2.3).

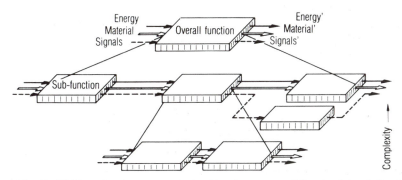

Figure 2.3. Establishing a function structure by breaking down an overall function into sub-functions.

The symbols for representing functions in a function structure are summarised in Fig. 2.4.

Functions are usually defined by statements consisting of a verb and a noun, for example "increase pressure", "transfer torque" and "reduce speed". They are derived from the conversions of energy, material and signals discussed in 2.1.2. So far as is possible, all these data should be accompanied with specifications of the physical quantities. In most mechanical engineering applications, a combination of all three types of conversion is usually involved, with the conversion either of material or of energy influencing the function structure decisively.

It is useful to distinguish between main and auxiliary functions. While *main functions* are those sub-functions that serve the overall function directly, *auxiliary functions* are those that contribute to it indirectly. They have a supportive or complementary character and are often determined by the nature of the solution. These definitions are derived from Value Analysis [2.7, 2.42, 2.43]. Although it may not always be possible to make a clear distinction between main and auxiliary functions, the terms are nevertheless useful.

It is also necessary to examine the relationship between the various sub-functions, and to pay particular attention to their logical sequence or required arrangement.

As an example, consider the packing of carpet tiles, stamped out of a length of carpet. The first task is to introduce a method of control so that the perfect tiles can be selected, counted and packed in specified lots. The main flow here is that of material shown in the form of a block diagram in Fig. 2.5, which, in this case, is the only possible sequence. On closer examination we discover that this chain of sub-functions requires the introduction of auxiliary functions because:

- the stamping-out process creates offcuts that have to be removed;
- rejects must be removed separately and reprocessed; and
- packing material must be brought in.

The result is the function structure shown in Fig. 2.6. It will be seen that the sub-function "count tiles" can also give the signal to pack the tiles into lots of a specified size.

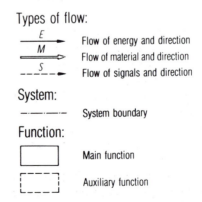

Figure 2.4. Symbols for representing sub-functions in a function structure.

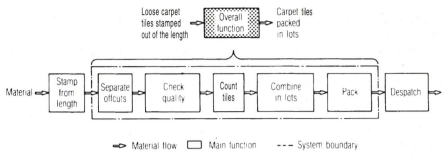

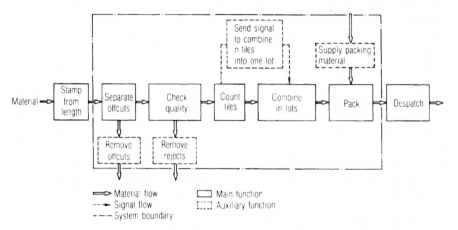

Figure 2.5. Function structure for the packing of carpet tiles.

Figure 2.6. Function structure for the packing of carpet tiles as in Fig. 2.5 with auxiliary functions added.

Outside the design domain the term function is sometimes used in a broader sense, sometimes in a narrower sense, depending on the context.

Brockhaus [2.31] has defined functions in general as activities, effects, goals and constraints. In mathematics, a function is the association of a magnitude y with a magnitude x so that a unique value (single-valued function) or more than one value (multi-valued function) of y is assigned for every value of x. According to the Value Analysis definition given in [2.7], all functions are determined by objectives (tasks, activities, characteristics).

Various writers on design methods (see 1.2.3) have put forward wider or stricter definitions of generally applicable functions.

In theory, it is possible to classify functions so that the lowest level of the function structure consists exclusively of functions that cannot be sub-divided further while remaining generally applicable.

Rodenacker [2.33] has defined *generally valid functions* in terms of binary logic, Roth [2.35] in terms of their general applicability, and Koller [2.21, 2.22] in terms of the required physical effects. Krumhauer [2.24] has examined general functions in the light of possible computer applications during the conceptual design phase, paying special attention to the relationship between inputs and outputs after changes in type, magnitude, number, place and time. By and large, he arrives at the same functions as Roth, except that by "change" he refers exclusively to changes in the type of input and output, while by "increase or decrease" he refers exclusively to changes in magnitude.

As Fig. 2.7 shows, all these definitions are compatible if it is remembered that Rodenacker uses "connect" and "separate" to refer to the logical connections only.

In the context of the design methodology presented here, the generally valid functions of Krumhauer will be used (see Fig. 2.8).

The function chain shown in Fig. 2.5 can be represented using generally valid functions as shown in Fig. 2.9.

A comparison between the functional representations in Fig. 2.5 and Fig. 2.9 shows that the description using generally valid functions is at a higher level of abstraction. For this reason, it leaves open all possible solutions and makes a systematic approach easier. However, using generally valid functions can represent a problem at such an abstract level that it sometimes hinders the direct search for solutions.

2   Logical Considerations

The logical analysis of functional relationships starts with the search for the essential ones that must necessarily appear in a system if the overall problem is to be solved. It may equally well be the relationships between sub-functions as those between inputs and outputs of particular sub-functions.

Let us first of all look at the relationships between sub-functions.

As we have pointed out, certain sub-functions must be satisfied before another sub-function can be meaningfully introduced. The so-called "if-then" relationship helps to clarify this point: if sub-function A is present, then sub-function B can come into effect, and so on. Often several sub-functions must all be satisfied simultaneously before another sub-function can be put into effect. The arrangement of sub-functions thus determines the structure of the energy, material and signal conversions under consideration. For example, during a test of tensile strength, the first sub-function—"load specimen"—must be satisfied before the other sub-functions—"measure force" and "measure deformation"—can be deployed. The last two sub-functions, moreover, must be satisfied simultaneously. Attention must be paid to consistency and order within the flow under consideration, and this is done by the unambiguous combination of the sub-functions.

Logical relationships, moreover, must also be established between the inputs and outputs of a particular sub-function. In most cases there are several inputs and outputs whose relationships can be treated like propositions in binary logic. For that purpose *elementary logical links* of the input and output magnitudes exist. In binary logic these are statements such as true/untrue, yes/no, in/out, fulfilled/not fulfilled, present/not present, which can be computed using Boolean algebra.

We distinguish between AND-functions, OR-functions and NOT-functions, and

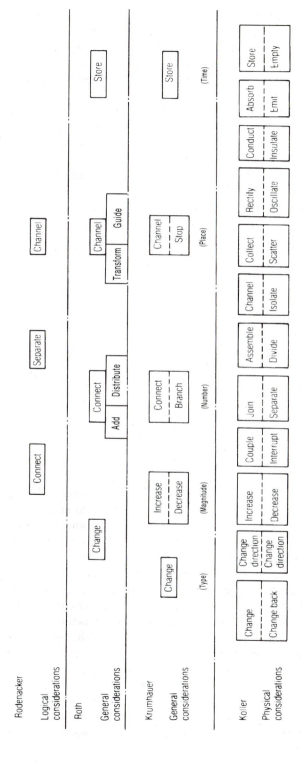

Figure 2.7. Comparison of generally valid functions according to various authors [2.21, 2.23, 2.24, 2.33, 2.35].

Characteristic Input (I)/Output (O)	Generally valid functions	Symbols	Explanations
Type	Change	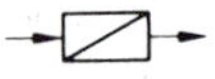 ⊘	Type and outward form of I and O differ
Magnitude	Vary	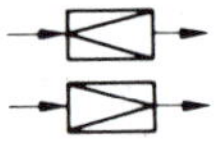	I < 0 I > 0
Number	Connect	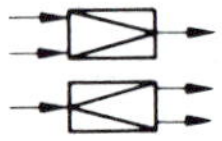	Number of I > 0 Number of I < 0
Place	Channel	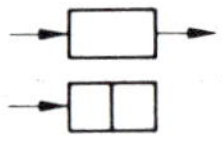	Place of I ≠ 0 Place of I = 0
Time	Store	 ⊙	Time of I ≠ 0

Figure 2.8. Generally valid functions derived from the characteristics type, magnitude, number, place and time in respect of the conversion of energy, material and signals.

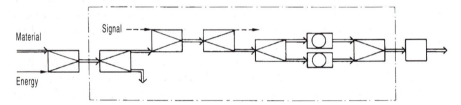

Figure 2.9. Function structure as in Fig. 2.5 but represented using generally valid functions as defined in Fig. 2.8.

also between their combination into more complex NOR-functions (OR with NOT), NAND-functions (AND with NOT) and storage functions with the help of flip-flops [2.4, 2.13, 2.32, 2.33]. All these are called *logical functions*.

In the case of AND-functions, all signals on the input side must have the same validity if a valid signal is to appear on the output side.

In the case of OR-functions, one signal only on the input side must be valid if a valid signal is to appear on the output side.

In the case of NOT-functions, the signal on the input side is negated so that the negated signal appears on the output side.

All these logical functions can be expressed by standard symbols [2.4]. The logical validity of any signal can be read off from the truth table shown in Fig. 2.10, in which all the inputs are combined systematically to yield the relevant outputs. The Boolean equations have been added for the sake of completeness. Using logical functions it is possible to construct complex switchings and thus to increase the safety and reliability of control and communication systems.

Designation	AND-function (Conjunction)	OR-function (Disjunction)	NOT-function (Negation)
Symbol	X_1 —[&]— Y X_2 —	X_1 —[≧1]— Y X_2 —	X —[1]o— Y
Truth table	X_1 | 0 | 1 | 0 | 1 X_2 | 0 | 0 | 1 | 1 Y | 0 | 0 | 0 | 1	X_1 | 0 | 1 | 0 | 1 X_2 | 0 | 0 | 1 | 1 Y | 0 | 1 | 1 | 1	X | 0 | 1 Y | 1 | 0
Boolean algebra (Function)	$Y = X_2 \land X_1$	$Y = X_1 \lor X_2$	$Y = \overline{X}$

Figure 2.10. Logical functions. X independent statement (signal); Y dependent statement; "0", "1" value of statement, eg "off", "on".

Figure 2.11 shows two mechanical clutches with their characteristic logical functions. The workings of the clutch on the left can be represented by a simple AND-function (signal must be sent and clutch engaged before the torque can be transmitted). The clutch on the right has been so constructed that, when the operating signal is given, the clutch is disengaged so that X_1 must be negative if the torque is to be transmitted. In other words, only X_2 may be present or positive if the desired effect is to be produced.

As our next example, we shall consider the catch mechanism of a car door from Gerber [2.13] (see Fig. 2.12). Here, too, the logical relationship is a simple AND-function, because the catch operating lever C can only be activated by the input force F acting on the lever A if the locking lever B is at "1". Should the specification contain further demands affecting the logical connection, then the function structure will grow correspondingly more complex [2.13].

Figure 2.13 shows a logical system for monitoring the bearing lubrication system of a multi-bearing machine shaft involving AND- and OR-functions. Every bearing position is monitored for oil pressure and oil flow by a comparison of a specified or target value with the actual value. However, only one positive value for each bearing position is needed to allow the system to operate.

2.1.4 Working Interrelationship

Establishing a function structure facilitates the discovery of solutions because it simplifies the general search for them and also because solutions to sub-functions can be elaborated separately.

Individual sub-functions, originally represented by "black boxes", must now be replaced with more concrete statements. Sub-functions are usually fulfilled by physical, chemical or biological processes—mechanical engineering solutions are based mainly on physical processes whereas process engineering solutions are based

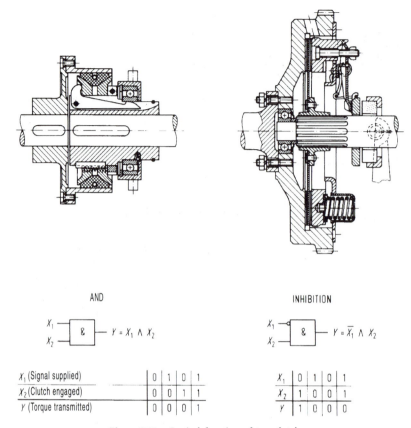

Figure 2.11. Logical function of two clutches.

mainly on chemical and biological processes. If, in what follows, we refer to *physical processes*, we tacitly include the effects of possible chemical or biological processes.

A physical process realised by the selected *physical effects* and the determined *geometric* and *material characteristics* results in a *working interrelationship* that fulfils the function in accordance with the task.

1 Physical Effects

Physical effects can be described quantitatively by means of the physical laws governing the physical quantities involved. Thus the friction effect is described by Coulomb's law, $F_F = \mu F_N$; the lever effect by the lever law $F_A \cdot a = F_B \cdot b$; and the expansion effect by the expansion law $\Delta l = \alpha \cdot l \cdot \Delta\theta$ (see Fig. 2.14). Rodenacker [2.33] and Koller [2.21], in particular, have collated such effects.

Several physical effects may have to be combined in order to fulfil a sub-function. Thus the operation of a bi-metallic strip is the result of a combination of two effects, namely thermal expansion and elasticity.

A sub-function can often be fulfilled by one of a number of physical effects.

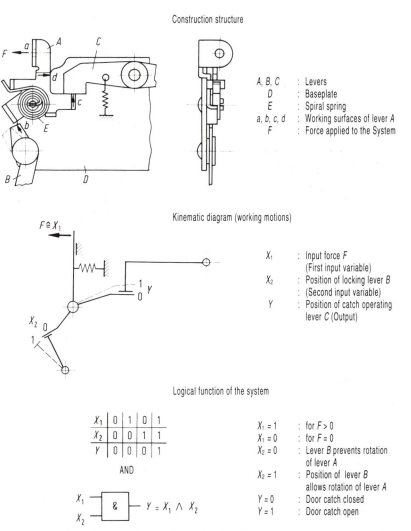

Figure 2.12. Catch mechanism of a car door (after [2.13]) with construction structure, kinematic diagram and logical function of the system.

Thus a force can be amplified by the lever effect, the wedge effect, the electromagnetic effect, the hydraulic effect etc. The physical effect chosen for a particular sub-function must, however, be compatible with the physical effects of other, related sub-functions. A hydraulic amplifier, for instance, cannot be powered directly by an electric battery. Moreover, a particular physical effect can only fulfil a sub-function optimally under certain conditions. Thus a pneumatic control system will be superior to a mechanical or electrical control system only in particular circumstances.

As a rule compatibility and optimal fulfilment can only be realistically assessed in relation to the overall function once the geometric and material characteristics have been established more concretely.

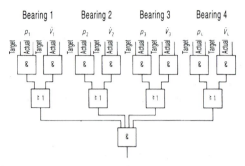

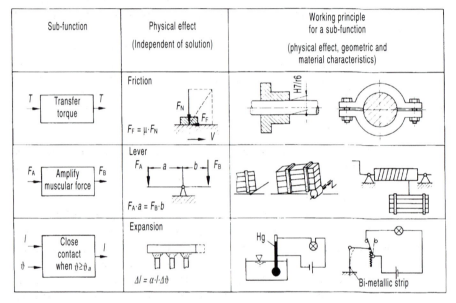

Figure 2.13. Logical functions for monitoring a bearing lubrication system. A positive signal for every bearing (oil present) permits operation. Monitor pressure p; monitor oil flow $\dot{V}$.

Sub-function	Physical effect (Independent of solution)	Working principle for a sub-function (physical effect, geometric and material characteristics)
$T \rightarrow$ Transfer torque $\rightarrow T$	Friction F_N F_F $F_F = \mu \cdot F_N$ $\longrightarrow V$	
$F_A \rightarrow$ Amplify muscular force $\rightarrow F_B$	Lever $F_A \vdash a \longrightarrow b \dashv F_B$ $F_A \cdot a = F_B \cdot b$	
$l \rightarrow$ Close contact when $\vartheta \geq \vartheta_a$ $\rightarrow l$ $\vartheta \rightarrow$	Expansion $\Delta l = \alpha \cdot l \cdot \Delta \vartheta$	Hg Bi-metallic strip

Figure 2.14. Fulfilling sub-functions by working principles built up of physical effects and geometric and material characteristics.

2 Geometric and Material Characteristics

The place where the physical process actually takes effect is the *working location* (active location). A function is fulfilled by the physical effect, which is realised by the *working geometry*, ie the arrangement of *working surfaces* (or working spaces) and by the choice of *working motions* [2.33].

The working surfaces are varied in respect of and determined by:

• Type
• Shape
• Position
• Size

- Number [2.34].

Similarly the required working motions (kinematics) are determined by:

- Type: translation — rotation
- Nature: regular — irregular
- Direction: in x, y, z-directions and/or about x, y, z-axes
- Magnitude: velocity etc
- Number: one, several etc.

In addition, we need a general idea of the *type of material* with which the working surfaces are to be produced, for example, whether it is solid, liquid or gaseous; rigid or flexible; elastic or plastic; stiff, hard or tough; or corrosion-resistant. A general idea of the final embodiment is often insufficient; the *main material properties* must be specified before a working interrelationship can be formulated adequately (see Fig. 4.11).

Only the combination of the physical effect with the geometric and material characteristics (working surfaces, working motions and materials) allows the principle of the solution to emerge. This interrelationship is called the *working principle* (Hansen [2.14] refers to this as the working means), and it is the first concrete step in the implementation of the solution.

Figure 2.14 shows some examples:

- Transferring the torque by friction against a *cylindrical working surface* in accordance with Coulomb's law will, depending on the way in which the normal force is applied, lead to the selection of a shrink fit or a clamp connection as the working principle.
- Amplifying muscular force with the help of a lever in accordance with the lever law after determining the *pivot and force application points (working geometry)* and considering the necessary *working motion* will lead to a description of the working principle (lever solution, eccentric solution, winch solution etc).
- Making electric contact by bridging a gap using the expansion effect, applied in accordance with the linear expansion law, only leads to an overall working principle after determination of the *sizes (eg diameter and length)* and *positions* of the *working surfaces* needed for the *working motion* of the expanding medium, a *material* (mercury) expanding by a fixed amount or a bi-metallic strip serving as a switch.

To satisfy the overall function, the working principles of the various sub-functions, have to be combined. There are obviously several ways in which this can be done. Guideline VDI 2222 [2.40] calls each combination a *combination of principles* (see 4.1.4).

The combination of several working principles results in the *working structure* of a solution. It is through this combination of working principles that the solution principle for fulfilling the overall task can be recognised. The working structure derived from the function structure thus represents how the solution will work at the fundamental principle level. Hubka refers to the working structure as the organ structure [2.17, 2.18].

For known elements, a circuit diagram or a flow chart is sufficient as a means of representing a working structure. Mechanical artefacts can be effectively represented as line drawings, though new or uncommon elements may require additional explanatory sketches (see Fig. 2.14).

Often the working structure alone will not be concrete enough to evaluate the solution principle. It may need to be quantified, for example by calculations and rough scale drawings, before the solution principle can be fixed. The result is called a *principle solution*.

2.1.5   Constructional Interrelationship

The working interrelationship established in the working structure is the starting point for further concretisation leading to the *construction structure*. The construction structure takes into account the needs of production, assembly, transport etc. From this interrelationship the modules, assemblies and machines, along with their related links, emerge to form a concrete technical artefact or system. Figure 2.15 shows the fundamental interrelationships for the clutch shown in Fig. 2.1. The increasing levels of concretisation can be seen clearly.

The concrete elements of a construction structure must satisfy the requirements of the selected working structure plus any other requirements necessary for the technical system to operate as intended. To fulfil these requirements fully, it is usually necessary to consider the system interrelationship.

2.1.6   System Interrelationship

Technical artefacts or systems do not operate in isolation and are, in general, part of a superior system. To fulfil its overall function, such a system often involves human beings who influence it through *input effects* (operating, controlling etc). The system returns *feedback effects* or signals that lead to further actions (see Fig. 2.16). Thereby human beings support or enable the *intended effects* of the technical system.

Apart from desired inputs, undesired ones from the environment and from neighbouring systems can affect a technical system. Such *disturbing effects* (eg excess temperatures) can cause undesired *side effects* (eg deviations from shape or shifts in position). Further it is possible that in addition to the desired working interrelationship (intended effects) unwanted phenomena occur (eg vibrations) as side effects from individual components within the system or from the overall system itself. These side effects can have an adverse effect on humans or the environment.

In accordance with Fig. 2.16 it is useful to make the following distinctions (after [2.41]):

Intended effect: Functionally desired effect in the sense of system operation.
Input effect: Functional relationship due to human action on a technical system.

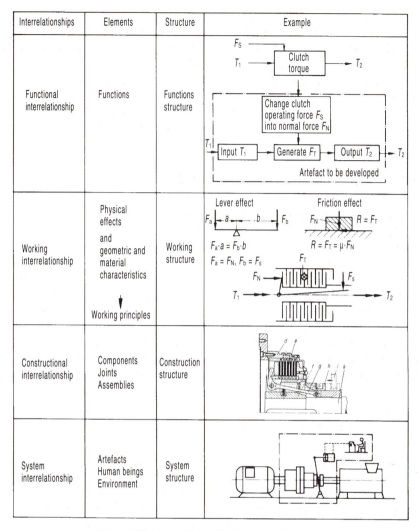

Figure 2.15.   Interrelationships in technical systems.

Feedback effect: Functional relationship due to the action of a technical system on a human or another technical system.

Disturbing effect: Functionally undesired influence from outside on a technical system or human that makes it difficult for a system to fulfil its function.

Side effect: Functionally undesired and unintended effect of a technical system on a human or on the environment.

The overall interrelationship of all these effects has to be carefully considered during the development of technical systems. To help recognise them in time, so that desired effects can be used and undesired ones avoided, it is useful to take note of the general objectives and constraints in 2.1.8.

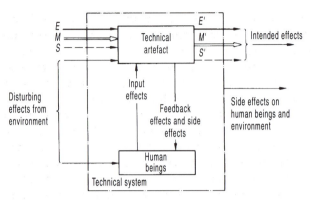

Figure 2.16. Interrelationships in technical systems under the influence of human beings.

2.1.7 Data Processing Systems

At the beginning of this chapter it was emphasised that technical tasks are often solved with the help of computers. Software is used not only to monitor and control production systems, process plant and machine tools, but also to support designers during the design process in the form of CAD and knowledge-based systems.

These systems are divided into *sub-systems* that must have compatible interfaces. These sub-systems are referred to as programs, sub-programs or modules. The way programs are linked is represented by program structures showing sub-systems and system elements. The *data flow* indicates how data is changed, transferred and outputted, ie processed, by means of algorithms (operations) using inputted or retrieved data.

The desired functions (operations) are connected by a *functional interrelationship*. These functions, and the data flows linking them, can be fulfilled by a variety of solution principles (working principles). This requires algorithms. According to [2.3] the *"working principle"* of a software solution comprises:

- an organising principle in the form of an algorithm for data management, data transfer and data storage;
- an operating principle in the form of an algorithm for transforming input data through mathematical or logical operations; and
- a communicating principle in the form of an algorithm for preparing internal data to present it to the user, and for preparing external data for internal processing.

An integrated set of function modules can therefore be represented by organisation, operation, and communication data modules. According to VDI 2221 [2.93], working principles consist of algorithms and data structures. Data structures are data elements that are combined into a structure through their relationships. The complete set of working principles to fulfil the overall function forms a *working interrelationship*. This interrelationship is represented by data flow charts, program flow charts and structure diagrams.

The next, more concrete, phase is the programming of the functional–logical interrelationships of the working principle or the working structure using the appropriate software language. Individual program modules are combined into a program structure (construction structure) that results in an executable program representing the constructional interrelationship. Modules from such a program structure can often be used in other software solutions.

Finally the *system interrelationship* between the environment and human operators is important in the development and use of software. The influences shown in Fig. 2.16 are particularly important for the design of the user interface and are determined by the communication data module. When networking computers, great care must be given to the way the system is structured with regard to the hardware and software interfaces and the standardisation of data transfer.

2.1.8 Objectives, Constraints and Guidelines

The solution of technical tasks is determined by the general objectives and constraints.

The *fulfilment of the technical function*, the *attainment of economic feasibility* and the *observance of safety requirements* for humans and the environment can be considered as general objectives. The fulfilment of the technical function alone does not complete the task of designers; it would simply be an end in itself. Economic feasibility is another essential requirement, and concern with human and environmental safety must impose itself for ethical reasons. Every one of these objectives has direct repercussions on the rest.

In addition, the solution of technical tasks imposes certain constraints or requirements resulting from ergonomics, production methods, transport facilities, intended operation etc, no matter whether these constraints are the result of the particular task or the general state of technology. In the first case we speak of task-specific constraints, in the second of general constraints that, though often not specified explicitly, must nevertheless be taken into account.

Hubka [2.17, 2.18] separates the properties affected by the constraints into categories based variously on operational, ergonomic, aesthetic, distribution, delivery, planning, design, production and economic factors.

Besides satisfying the functional and working interrelationships, a solution must also satisfy certain general or task-specific *constraints*. These can be classified under the following headings:

- Safety: also in the wider sense of reliability and availability
- Ergonomics: human–machine context, also aesthetics
- Production: production facilities and type of production
- Quality control: throughout the production process
- Assembly: during and after the production of parts
- Transport: inside and outside the factory
- Operation: intended use, handling
- Maintenance: upkeep, inspection and repair

- Recycling: reuse, reprocess, disposal, final storage
- Expenditure: costs, schedules and deadlines

The characteristics that can be derived from these constraints, that in general are formulated as requirements (see 5.2.2), affect the function, working and construction structures, and also influence one another. Hence they should be treated as *guidelines* throughout the design process, and adapted to each level of embodiment (see 11.1).

It is advisable to consider them even during the *conceptual phase*, at least in essence. During the *embodiment phase*, when the layout and form design of the more or less qualitatively elaborated concept is first quantified, both the objectives of the task and also the general and task-specific constraints must be considered in concrete detail. This involves several steps—the collection of further information, layout and form design, and the elimination of weak links, together with a fresh, if limited, search for solutions to a variety of sub-tasks until, finally, in the *detail phase* the elaboration of detail drawings and production documents brings the design process to a conclusion.

2.2   Fundamentals of the Systematic Approach

Before we deal with the specific steps and rules of systematic design, we must first discuss cognitive psychological relationships and general methodical principles. These help to structure the proposed procedures and individual methods so that they can be applied to the solution of design tasks in a purposeful way. The ideas come from a host of different disciplines, mainly non-technical ones, and are usually built on inter-disciplinary fundamentals. Psychology, philosophy and the science of human factors are among the main inspirations, which is not surprising when we consider that methods designed to improve working procedures impinge on the qualities, capacities and limitations of human thought.

2.2.1   Psychology of Problem Solving

Designers are often confronted with tasks containing problems they cannot solve immediately. Problem solving in novel areas of application and at different levels of concretisation is a characteristic of their work. Designing, which involves a continuous search for solutions, places high demands on the thinking ability of designers. Researching the essence of human thinking is the focus of cognitive psychology. The results of this research have to be taken into account in engineering design so that the procedures and methods recommended match the way humans think; and so that the thinking processes of designers are supported in ways that help them find solutions more easily. The following sections are based largely on the work of Dörner [2.8].

1 Problem Characteristics

A *problem* is characterised by three components:

- an undesirable initial state, ie an unsatisfactory situation exists;
- a desirable goal state, ie realising a satisfactory situation; and
- obstacles that prevent a transformation from the undesirable initial state to desirable goal state at a particular point in time.

Obstacles that prevent a transformation can arise from the following:

- the means to overcome the obstacles are unknown and have to be found (synthesis problem);
- the means are known, but they are so numerous or involve so many combinations that a systematic investigation is impossible (interpolation problem, combination and selection problem);
- the goals are only known vaguely or are not formulated clearly. Finding a solution involves continuous deliberation and the removal of conflicts until a satisfactory situation is reached (dialectic problem, search and application problem).

The difficulties become more pronounced if the characteristics of the problem area change with time, especially if the changes are unpredictable.

Problems also have the following important characteristics:

- *complexity:* many components are involved and these components, through links of different strength, influence each other;
- *uncertainty:* not all requirements are known; not all criteria are established; the effect of a partial solution on the overall solution or on other partial solutions is not fully understood, or only emerges slowly.

These characteristics distinguish a problem from a *task*. A task imposes mental requirements for which various means and methods are available to assist. An example is the design of a shaft with given loads, connecting dimensions and production methods.

Tasks and problems occur in design in a number of ways, often combined and not clearly separable initially. A specific design task can, for example, turn out to be a problem when looked at more closely. Many large tasks can be divided into sub-tasks, some of which can reveal difficult sub-problems. On the other hand it is sometimes possible for a problem to be solved by fulfilling several sub-tasks in a previously unknown sequence.

2 Means of Problem Solving – Thought Structures

Thinking processes take place in the memory and involve changes in memory content. When thinking the contents of the memory, and the way in which they are linked, play an important role.

In simple terms, one can say that to start solving a problem humans need a certain amount of *factual knowledge* about the domain of the problem. In cognitive psychology, knowledge that has been transferred into memory is called the *epistemic structure*.

Humans also need certain *procedures* (methods) to find solutions and to find these effectively. This aspect involves the *heuristic structure* of human thought.

It is possible to distinguish between short-term and long-term memory. Short-term memory is a kind of working storage. It has limited capacity and can only retain about seven arguments or facts at the same time. Long-term memory probably has unlimited capacity and contains factual and heuristic knowledge stored in a structured way.

Thereby humans are able to recognise specific relationships in many possible ways, to use these relationships and to create new ones. Such relationships are very important in the technical domain, for example:

- *concrete–abstract relationship*
  eg angular contact bearing—ball bearing—rolling element bearing—bearing—guide—force transfer and component location.
- *whole–part relationship* (hierarchy)
  eg plant—machine—assembly—part.
- *space–time relationship*
  eg arrangement: front—back, below—above,
  sequence: this first—that next.

The memory can be thought of as a semantic network with nodes (knowledge) and connections (relationships) which can be modified and extended. Figure 2.17 shows a possible, though not necessarily complete, semantic network related to the term "bearing". In this network it is possible to recognise the relationships mentioned above as well as others, such as property relationships and ones indicating opposites (polar relationships).

Thinking involves building and restructuring such semantic networks, and the thinking process itself can proceed intuitively or discursively.

Intuitive thinking is strongly associated with flashes of inspiration. The actual thinking process takes place to a large extent unconsciously. Insights appear in the conscious mind suddenly, caused by some trigger or association. This is referred to as primary creativity [2.1, 2.23] and involves processing quite complex relations. In this context Müller [2.28] refers to "silent knowledge", that is common and background knowledge. This is also the knowledge that is available when one deals with episodic memories, vague concepts and imprecise definitions. It is activated by both conscious and unconscious thinking activities.

Generally time is needed for undisturbed and unconscious "thinking" before sudden insights appear. The length of this incubation period cannot be predetermined. Insights can be triggered, for example, by producing freehand sketches or technical drawings of solution ideas. According to [2.11] these manual activities focus concentration on the subject, but still leave space in the mind that can be used by unconscious thinking processes, which can also be stimulated by such activities.

Discursive thinking is a conscious process that can be communicated and influ-

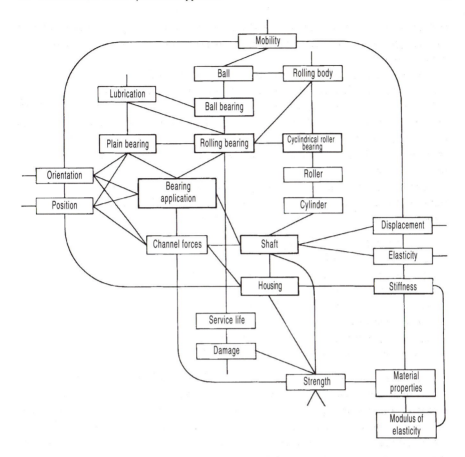

Figure 2.17. Extract of a semantic network related to bearings.

enced. Facts and relationships are consciously analysed, varied, combined in new ways, checked, rejected, and considered further. In [2.1, 2.23] this is referred to as secondary creativity. This type of thinking involves checking exact and scientific knowledge and building this into a knowledge structure. In contrast to intuitive thinking, this process is slow and involves many small conscious steps.

In the memory structure explicit and consciously acquired knowledge cannot be separated precisely from the vaguer common or background knowledge. Besides, the two types of knowledge influence each other. For knowledge to be easily retrieved and combined, it is thought that an ordered and logical structure of factual knowledge in the mind of the problem solver is decisive, and that this is true whether the thinking process is intuitive or discursive.

The *heuristic structure* includes explicit knowledge (ie knowledge that can be explained) as well as non-explicit knowledge. This is necessary in order to organise the sequence of thinking operations, including modifying operations (searching and finding) and testing operations (checking and assessing). It appears that problem solvers often start without a fixed plan in the hope of immediately finding a solution from their knowledge bases without much effort. Only when this approach fails, or when contradictions emerge, do they adopt a more clearly planned or systematic sequence of thinking operations.

The so-called TOTE model [2.26] represents an important fundamental sequence for thinking processes (see Fig. 2.18). It consists of two processes: a modification process and a testing process. The TOTE model shows that before an operation of change an operation of testing (*Test*) takes place to analyse the initial state. Only then is the chosen operation of change (*Operation*) executed. This is followed by another operation of testing (*Test*), during which the resulting state is checked. If the result is satisfactory, the process is exited (*Exit*); if not, the modification process is adapted and repeated.

In more complex thinking processes, the TOTE sequences are linked in a chain or several modification processes are executed before a testing process takes place. Thus when linking mental processes, many combinations and sequences are possible, but all of them can be mapped onto the basic TOTE model.

Thinking processes involve linking structured factual knowledge, purposeful modifications and testing operations. This implies that these processes operate within a semantic network, generating new insights and knowledge that are again stored in the epistemic structure.

3 Characteristics of Good Problem Solvers

The following statements are the result of the work of Dörner [2.9] and of the research which has been undertaken with him by Ehrlenspiel and Pahl. The results of the research led by Ehrlenspiel and Pahl can be found in the publications of Rutz [2.37], Dylla [2.10] and Fricke [2.12].

Intelligence and Creativity

In general *intelligence* is thought to involve a certain cleverness, combined with the ability to understand and judge. Analytical approaches are often emphasised.

Creativity is an inspirational force that generates new ideas or produces novel combinations of existing ideas, leading to further solutions or deeper understanding. Creativity is often associated with an intuitive, synthesising approach.

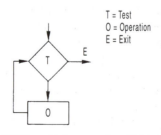

T = Test
O = Operation
E = Exit

Figure 2.18.   Basic TOTE model for organising thinking processes [2.8, 2.26].

Intelligence and creativity are personal characteristics. Up until now it has not been possible to come up with precise scientific definitions of, or a clear distinction between, intelligence and creativity. Attempts have been made to measure the level of intelligence of individuals using intelligence tests. The resulting Intelligence Quotients provide measures compared to the average of a large sample. Because of the different forms in which intelligence appears, various tests are needed to capture a complete picture and draw tentative conclusions.

For problem solving, a minimum level of intelligence is required and it appears that people with high Intelligence Quotients are more likely to be good problem solvers. However, according to [2.8, 2.9] intelligence tests on their own do not give much insight into which combination of factors makes a particular individual a good problem solver. The reason, according to Dörner [2.8], is that intelligence tests use tasks or problems that only require a few thinking steps to find a solution, so the sequence of steps seldom becomes conscious. Few intelligence tests require a large number of steps to be organised into a specific problem solving procedure. Such organisation requires switching between the different levels and possibilities of a general problem solving procedure, and is essential for the execution of long-term thinking activities.

The same is true for creativity tests. Often these are of such a low level that they do not address complex problem solving which involves planning and guiding one's own approach. Furthermore, in engineering design, creativity is always focused on a specific goal. A purely unfocused generation of ideas and variants might support a specific phase of the problem solving process [2.1], however it can sometimes hinder the overall process.

Decision Making Behaviour

Apart from having well structured factual knowledge, applying a systematic approach, and using focused creativity, designers have to master decision making processes. For decision making the following mental activities and skills are essential:

- *Recognising dependencies*
  In complex systems the dependencies between the individual elements can vary in strength. Recognising the types and strengths of such dependencies is an essential prerequisite for dividing the problem into more manageable, less complex sub-problems or sub-goals so that these can be addressed separately. However those working on each separate sub-problem must check to see how their own decisions influence the overall design.
- *Estimating importance and urgency*
  Good problem solvers know how to recognise *importance* (factual significance) and *urgency* (temporal significance) and how to use this information to modify their approach to problems.
  They try to resolve the most important things first and then tackle the dependent sub-problems. They have the courage to be satisfied with sub-optimal

solutions for less significant problems if they have good or acceptable solutions for the most significant ones. By doing this they avoid immersing themselves in less relevant issues and thereby losing valuable time.

The same is true for estimating the urgency. Good problem solvers estimate the time they need accurately. They prepare a demanding, but not impossible, time plan. Janis and Mann [2.19] have concluded that mild, ie bearable, stress is important for creativity. Therefore realistic time planning has a positive effect on thinking processes and new developments should take place under reasonable time pressure. But, of course, individuals react differently to time pressure.

- *Persistence and flexibility*
  Persistence means an continuous focus on achieving the goals, but there is a danger that excessive focus leads to a rigid approach. Flexibility means a ready ability to adapt to changing requirements. However, this should not lead to purposeless jumping from one approach to another.

  Good problem solvers find a suitable balance between persistence and flexibility. They demonstrate a continuous and consistent, but at the same time, flexible behaviour. They stick to the given goals despite any hold ups and difficulties they encounter. On the other hand they adapt their approach immediately when the situation changes and when new problems occur.

  They consider heuristics, procedures and instructions first of all as guidelines and not as rigid prescriptions. Dörner states [2.8]: "Heuristic plans should not degenerate into automatic procedures. Individuals should learn to develop what they have learnt. Heuristics should not be misinterpreted as prescriptions, but should be treated as guidelines that can, and often should, be developed."

- *Failures cannot be avoided*
  In complex systems with strong internal dependencies, at least partial failures are difficult to avoid because it is not possible to recognise all the possible effects simultaneously. In recognising such failures the most important thing is the way one reacts. Being flexible is crucial, supported by the ability to analyse one's own approach and the ability to make decisions that lead to corrective actions.

The results of cognitive psychology research are summarised below.
Good problem solvers:

- have a sound and structured technical knowledge, ie they have a well structured model in their minds;
- find an appropriate balance between concreteness and abstraction, depending on the situation;
- can deal with uncertainty and fuzzy data; and
- continuously focus on the goals while adopting a flexible decision making behaviour.

Apart from knowledge about methods and facts, good problem solvers also possess a further ability which can be called *heuristic competence* [2.12]. This involves:

- activating goal directed creativity;
- recognising importance and urgency; and

- planning, guiding and controlling their work.

This heuristic competence depends largely on personal characteristics, but can be developed through training.

A design methodology has to adopt these findings and offer designers guidance, without restricting individual abilities and approaches.

4 Problem Solving as Information Conversion

When we discussed the basic ideas of the systems approach (see 1.2.3) we found that problem solving demands a large and constant flow of information. Dörner [2.8] also views problem solving as information processing. The most important terms used in the theory of information processing are described in [2.5, 2.6].

Information is *received, processed* and *transmitted* (see Fig. 2.19).

Information is *received* from market analyses, trend studies, patents, technical journals, research, licenses, inquiries from customers, concrete assignments, design catalogues, analyses of natural and artificial systems, calculations, experiments, analogies, general and in-house standards and regulations, stock sheets, delivery instructions, computer data, test reports, accident reports, and also through "asking questions". Data collection is an essential element of problem solving [2.2].

Information is *processed* by analysis and synthesis, the development of solution concepts, calculation, experiment, the elaboration of layout drawings and also the evaluation of solutions.

Information is *transmitted* by means of sketches, drawings, reports, tables, production documents, assembly manuals, user manuals etc. Quite often provision must also be made for information to be *stored*.

In [2.25] some *criteria for characterising information* are given and these can be used for formulating user information requirements. They include:

- Reliability: the probability of the information being trustworthy and correct.
- Sharpness: the precision and clarity of the information content.
- Volume and density: an indication of the number of words and pictures needed for the description of a system or process.
- Value: the importance of the information to the recipient.
- Actuality: an indication of the point in time when the information can be used.
- Form: the distinction between graphic and alpha-numeric data.
- Originality: an indication of whether or not the original character of the information must be preserved.

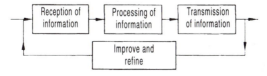

Figure 2.19. The conversion of information with iteration.

- Complexity: the structure of, or similarity between, information symbols and information elements, units or complexes.
- Degree of refinement: the quantity of detail in the information.

Information conversion is usually a very complicated process. Solving problems requires information of different types, content and range. In addition, to raise the level of information and improve it, it may be necessary to reiterate certain steps.

Iteration is the process by which a solution is approached step-by-step. In this process one or more steps are repeated, each time at a higher level of information based on the results of the previous loop. Only in this way is it possible to obtain the information to refine a solution and ensure continuous improvement (see Fig. 2.19). Such iterations occur frequently at all stages of the problem solving process.

2.2.2   General Working Methodology

A general working methodology should be widely applicable, independent of discipline and not require specific technical knowledge from the user. It should support a structured and effective thinking process. The following general ideas appear time and time again in specific approaches, either directly or slightly amended to adapt them to the special requirements of developing technical systems. The purpose of this section is to provide a general introduction to systematic procedures.

The following procedures are based not only on our own professional experience and on the findings of cognitive psychology mentioned in 2.2.1, but also on the work of Holliger [2.15, 2.16], Nadler [2.29, 2.30], Müller [2.27, 2.28] and Schmidt [2.38]. They are also known as "heuristic principles" (a *heuristic* is a method for generating ideas and finding solutions) or "creativity techniques".

The following conditions must be satisfied by anyone using a systematic approach:

- *Define the goals* by formulating the overall goal, the individual sub-goals and their importance. This ensures the motivation to solve the task and supports insight into the problem.
- *Clarify the boundary conditions* by defining the initial and marginal constraints.
- *Dispel prejudice* to ensure the most wide-ranging possible search for solutions and to avoid logical errors.
- *Search for variants,* that is find a number of possible solutions or combinations of solutions from which the best can be selected.
- *Evaluate* based on the goals and the requirements
- *Make decisions.* This is facilitated by objective evaluations. Without decisions there can be no progress.

To make these general methods work, the following *thinking* and *acting operations* have to be considered.

1 Purposeful Thinking

As described in 2.2.1, intuitive and discursive thinking are possible. The former tends to be more unconscious, the latter more conscious.

Intuition has led to a large number of good and even excellent solutions. The prerequisite is, however, always a very conscious and intensive involvement with the given problem. Nevertheless a purely intuitive approach has the following disadvantages:

- the right idea rarely comes at the right moment since it cannot be elicited and elaborated at will;
- the result depends strongly on individual talent and experience; and
- there is a danger that solutions will be circumscribed by preconceived ideas based on one's special training and experience.

It is therefore advisable to use more deliberate procedures that tackle problems step by step. Such procedures are called *discursive*. Here the steps are chosen intentionally; they can be influenced and communicated. Usually individual ideas or solution attempts are consciously analysed, varied and combined. It is an important aspect of this procedure that a problem is rarely tackled as a whole, but is first divided into manageable parts and then analysed.

It must, however, be stressed that intuitive and discursive methods are not opposites. Experience has shown that intuition is stimulated by discursive thought. Thus while complex assignments must always be tackled one step at a time, the subsidiary problems involved may, and often should, be solved in intuitive ways.

Holliger [2.16] distinguishes between unconscious, preconscious and conscious thought and prescribes the transformation of stereotypes, of aimless and unconscious procedures, and of disorderly and fantasy-charged preconscious procedures into a *conscious or deliberate approach*. This can be done with the help of *systematic rules, clear task formulation* and *structured procedures*.

A further aid to conscious thought is the *association of ideas*, ie linking several ideas. Such associations can be addressed, activated and modified by so-called *central ideas*. Central ideas demonstrate characteristics that are typical of a large group of ideas. Examples are the "ordering viewpoints" of Hansen [2.14]. One should, however, avoid set complexes of ideas because these may turn out to be too inflexible, and such complexes should be deliberately dissolved. It is obvious that systematic thought is needed more for original design than for routine tasks, which can generally be performed successfully even if the underlying thought processes remain unconscious.

Another important characteristic of human thought is the *inevitability of errors*, for which allowances should, if possible, be made from the start. This can, for example, be done by preferring those solutions that deal with the recognised failure possibilities. In this connection, Holliger speaks of "catastrophe analysis" and Müller [2.28] insists on a "failure analysis" in the early stages of a product development. One should, therefore, try to minimise errors or the weak points in a solution resulting from these errors. This can be done by:

- clearly defining the requirements and constraints of a particular task;

- not forcing intuitive solutions but using a discursive approach;
- avoiding fixed ideas;
- adapting methods, procedures and tools to the task in hand; and
- deliberately using methods for failure identification.

In addition, it has to be realised that creativity can be inhibited or encouraged by different influences [2.1]. It is, for example, often necessary to encourage intuitive thinking by interrupting the activity to provide some periods for incubation (see 2.2.1). On the other hand, too many interruptions can be disturbing and thereby inhibit creativity. A systematic approach including discursive elements encourages creativity when one has to cope with many different viewpoints. Examples include using different solution methods; moving between abstract and concrete ideas; collecting information using solution catalogues; and dividing work between team members. Furthermore according to [2.19] realistic planning encourages rather than inhibits motivation and creativity.

2 Analysis

Analysis is the resolution of anything complex into its elements and the study of these elements and of their interrelationships. It calls for identification, definition, structuring and arrangement. The acquired information is transformed into knowledge.

If errors are to be minimised, then problems must be formulated clearly and unambiguously. To that end, they have to be analysed. *Problem analysis* means separating the essential from the inessential and, in the case of complex problems, preparing a discursive solution by resolution into individual, more transparent, subsidiary problems. If the search for the solution proves difficult, a new formulation of the problem may provide a better starting point. The reformulation of statements is often an effective means of finding new ideas and insights. Experience has shown that careful analysis and formulation of problems are among the most important steps of the systematic approach.

The solution of a problem can also be brought nearer by *structure analysis*, that is, the search for hierarchical structures or logical connections. In general, this type of analysis can be said to aim at the demonstration of similarities or repetitive features in different systems (see 4.1.1).

Another helpful approach is *weak spot analysis*. It is based on the fact that every system has weaknesses caused by ignorance, mistaken ideas, external disturbances, physical limitations and manufacturing errors. During the development of a system it is therefore important to analyse the design concept or design embodiment for the express purpose of discovering possible weak spots and prescribing remedies. To that end special selection and evaluation procedures (see 4.2) and weak spot identification methods (see 9.1) have been developed. Experience has shown that this type of analysis may not only lead to specific improvements of the chosen solution principle, but may also trigger off new solution principles.

3 Abstraction

Through *abstraction* it is possible to find a higher level interrelationship, that is, one which is more generic and comprehensive. Such a procedure reduces complexity and emphasises the essential characteristics of the problem and thereby provides an opportunity to search for and find other solutions containing the identified characteristics. At the same time new structures emerge in the minds of designers and these assist with the organisation and retrieval of the many ideas and representations. So abstraction supports both creativity and systematic thinking. It makes possible the definition of a problem in such a way that a coincidental solution path is avoided and a more generic solution is found (see 6.2).

4 Synthesis

Synthesis is the putting together of parts or elements to produce new effects and to demonstrate that these effects create an overall order. It involves search and discovery, and also composition and combination. An essential feature of all design work is the combination of individual findings or sub-solutions into an overall working system—that is the association of components to form a whole. During the process of synthesis the information discovered by analyses is processed as well. In general, it is advisable to base synthesis on a *global* or *systems approach;* in other words to bear in mind the general task or course of events while working on sub-tasks or individual steps. Unless this is done, there is the grave risk that, despite the optimisation of individual assemblies or steps, no suitable overall solution will be reached. Appreciation of this fact is the basis of the inter-disciplinary method known as Value Analysis which proceeds from the analysis of the problem and function structure to a global approach involving the early collaboration of all departments concerned with product development. A global approach is also needed in large-scale projects, and especially in preparing schedules by such techniques as Critical Path Analysis. The entire systems method is strongly based on the global approach, which is particularly important in the evaluation of solution proposals because the value of a particular solution can only be gauged after overall assessment of all the requirements and constraints (see 4.2).

5 Generally Applicable Methods

The following general methods provide further support for systematic work, and are widely used [2.16].

The Method of Persistent Questions

When using systematic procedures it is often a good idea to keep *asking questions* as a stimulus to fresh thought and intuition. A standard list of questions also fosters the discursive method. In short, asking questions is one of the most important methodological tools. This explains why many authors have drawn up special questionnaires (checklists).

The Method of Negation

The method of *deliberate negation* starts from a known solution, splits it into individual parts or describes it by individual statements, and negates these statements one by one or in groups. This deliberate inversion often creates new solution possibilities. Thus, when considering a "rotating" machine element one might also examine the "static" case. Moreover, the mere omission of an element can be tantamount to a negation. This procedure is also known as "systematic doubting" [2.16].

The Method of Forward Steps

Starting from a *first solution attempt,* one follows *as many paths as possible* yielding further solutions. This method is also called the method of divergent thought. It is not necessarily systematic, but frequently starts with an unsystematic divergence of ideas. The method is illustrated in Fig. 2.20 for the development of a shaft–hub connection. The arrows indicate the direction of the thinking process.

Such a thinking process can be improved (see Fig. 4.11) if the flow of thoughts follows closely a systematic variation of the characteristics (see Fig. 4.14). Where variation is done without conscious thought, even with well structured representations, the identified characteristics are not used to their full potential.

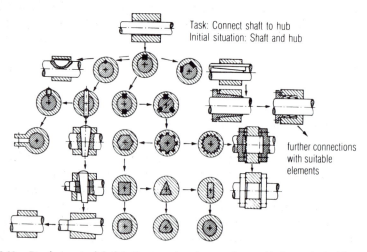

Task: Connect shaft to hub
Initial situation: Shaft and hub

further connections
with suitable
elements

Figure 2.20. Development of shaft-hub connections in accordance with the method of forward steps.

The Method of Backward Steps

The starting point for this method is the goal rather than the initial problem. Beginning with the final objectives of the development, one *retraces all the possible paths* that may have led up to it. This method is also called the method of convergent thought, because only such ideas are developed as converge on the ultimate goal.

The method is particularly useful for drawing up production plans and developing systems for the production of components.

It is similar to the method of Nadler [2.29], who has proposed the construction of an *ideal system* that will satisfy all demands. This system is not developed in practice but formulated in the mind. It demands optimum conditions such as an ideal environment which causes no external disturbances. Having formulated such a system, there follows a step-by-step investigation of what concessions must be made to turn this purely theoretical and ideal system into a technologically feasible one, and finally into one that meets all the concrete requirements. Unfortunately it is rarely possible to specify in advance the ideal system, because the ideal state of all functions, system elements and modules is difficult to specify in advance especially if they are linked together in a complex system.

The Method of Factorisation

Factorisation involves breaking down a complex interrelationship or system into manageable, less complex and more easily definable individual elements (factors). The overall problem or task is divided into separate sub-problems or sub-tasks, that are to a certain degree independent (see Fig. 2.21). Each of these sub-problems or sub-tasks can initially be solved on its own, though the links between them in the overall structure must be kept in mind. Factorisation not only creates more manageable sub-tasks, but it also clarifies their importance and influence in the overall structure, allowing priorities to be set. This approach is used in systematic design to divide an overall function into sub-functions and to develop function structures (see 6.3); to search for working principles for sub-functions (see 6.4); and to plan the working steps during embodiment design (see 3.1 to 3.3).

The Method of Systematic Variation

Once the required characteristics of the solution are known, it is possible, by *systematic variation*, to develop a more or less complete solution field. This involves the construction of a generalised classification, that is, a schematic representation of the various characteristics and possible solutions (see 4.1.3). From the viewpoint of human factors, too, it is obvious that the discovery of solutions is assisted by the construction and use of classification schemes. Nearly all authors consider systematic variation one of the most important methodical procedures.

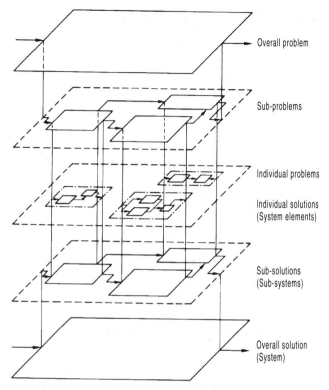

Figure 2.21. Dividing the overall problem into sub-problems; finding individual solutions; combining solutions into an overall solution [2.39]. Sub-problems can become sub-tasks.

6 Division of Labour and Collaboration

An essential finding of the science of human factors is that the implementation of large and complex tasks calls for the *division of labour*, the more so as specialisation increases. This is also demanded by the increasingly tight schedules of modern industry. Nowadays, division of labour implies inter-disciplinary *collaboration* which, in its turn, involves special organisational and staff arrangements along with appropriate staff attitudes, including receptiveness to the ideas of others. It must, however, be stressed that inter-disciplinary collaboration and teamwork also demand a rigorous allocation of responsibility. Thus the product manager should be in sole charge of the development of a particular product, regardless of departmental boundaries.

Systematic design, in combination with methods that make use of group dynamics, such as brainstorming, gallery methods (see 4.1.2) and group evaluation (see 4.2), can overcome lack of information exchange caused by the division of work and can also help the search for solutions by stimulating ideas between team members.

3 Process of Planning and Designing

In the previous two chapters we examined the fundamentals on which design work should be built to best advantage. They form the basis of a systematic approach which practising designers can follow, regardless of their speciality. The approach is not based on one method but applies known and less well known methods where they are most suitable and useful for specific tasks and working steps.

3.1 General Problem-Solving Process

An essential part of our own problem-solving method involves step-by-step *analysis* and *synthesis*. In it we proceed from the *qualitative* to the *quantitative*, each new step being more concrete than the last.

In this context we refer to the thinking and acting operations identified in cognitive psychology: *modifying* a situation (operating) and *testing* the results (evaluating) [3.2] (see 2.18). The work of Dörner [3.2, 3.3] shows that in order to solve problems successfully, it is necessary to develop an approach adapted to the specific problem and modified appropriately as the problem solving proceeds (see 2.2.1). When searching for solutions, it is very effective to view the problem from different perspectives, such as different levels of concretisation. One can, for example, by analysing a concrete idea be stimulated to develop a more abstract perspective that leads to new ideas and vice versa.

Independently of these suggestions we propose in the next sections *plans* and *procedures* that should be regarded as mandatory for the general problem-solving process of planning and designing technical products, and as guidance for the more concrete phases of the design process. These plans and procedures assist in identifying what in principle has to be done, but they must, of course, be adapted to specific problem situations.

All procedural plans proposed in this book have to be considered as *operational guidelines for action* based on the pattern of technical product development and the logic of stepwise problem solving. According to Müller [3.7] they are process models that are suitable for describing in a rational way the approach necessary to make complex processes comprehensible and transparent. Thus these procedures are not descriptions of individual thinking processes as described in Section 2.2.1 and are not determined by personal characteristics.

In a practical application of these procedural plans, the operational guidelines for action blend with *individual thinking processes*. This results in a set of individual planning, acting and controlling activities based on general procedures, specific problem situations and individual experiences.

As discussed in 2.2.1, the suggested procedural plans are meant to be guidelines and not rigid prescriptions. However they have to be regarded as essentially sequential because, for example, a solution cannot be evaluated before it has been found or elaborated. On the other hand, the procedural plans have to be adapted to specific situations in a flexible manner. It is, for example, possible to leave out certain steps or order them in another sequence. It may be necessary or useful to repeat certain steps at a higher information level; and special procedures, adapted from the more general plans proposed here, may be appropriate in specific product domains.

Given the complexity of the product development process and the many methods that have to be applied, not adopting a procedural plan would leave designers with an unmanageable number of possible approaches. It is necessary, therefore, for designers to learn about the design process and the application of individual methods, as well as the working steps proposed in the procedural plans.

The activity of planning and designing was described in 2.2.1 as information processing. After each information output, it might become necessary to improve or increase the value of the result of the last working step. That is to repeat the working step at a higher information level, or to execute other working steps until the necessary improvements have been achieved.

Repeating working steps is the *process of iteration* by which one approaches a solution step-by-step until the result seems satisfactory. The so-called iteration loop can also be observed in the basic thinking process, for example in the TOTE model (see 2.2.1). Such iteration loops are almost always required and occur continuously within and between steps. The reasons for this are that the interrelationships are often so complex that the desired solution cannot be achieved in one step and that information is frequently needed from a subsequent step. The iteration arrows in procedural plans refer clearly to this fact. It is therefore important that the procedural plans proposed are not considered rigid and purely sequential.

A systematic approach aims to keep the iteration loops as small as possible in order to make design work effective and efficient. It would be a disaster, for example, if the design team had to start again at the beginning having reached the end of a product development. This would correspond to an iteration loop covering the whole of the product development process.

The division into working and decision making steps ensures the necessary and permanent links between *objectives, planning, execution* (organisation) and *control* [3.1, 3.14]. With these links we can, following Krick [3.4] and Penny [3.9], construct a general process for finding solutions to tasks and problems (see Fig. 3.1).

Every task involves first of all a *confrontation* of the problem with what is known or not (yet) known. The intensity of this confrontation depends on the knowledge, ability and experience of the designers, and on the particular field in which they

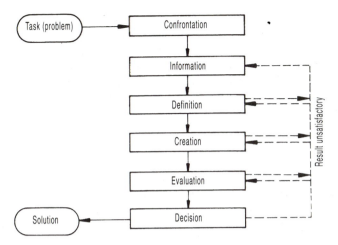

Figure 3.1. General process for finding solutions.

are engaged. In all cases, however, more detailed *information* about the task itself, about the constraints, about possible solution principles, and about known solutions for similar problems is extremely useful as it clarifies the precise nature of the requirements. This information can also reduce confrontation and increase confidence for finding solutions.

Next comes the *definition* of the essential problems (the crux of the task) on a more abstract plane, in order to fix the objectives and main constraints. Such solution-neutral definitions open the way to an unconstrained search for solutions because the abstract definition encourages a search for more unconventional solutions.

The next step is *creation*, when solutions are developed by various means and then varied and combined using methodical guidelines. If the number of variants is large there must also be *evaluation* followed by a *decision* on the basis of which the best variant is selected. Because each step of the design process must be evaluated, evaluation serves as a check on progress towards the overall objective.

Decision making involves the following considerations (see Fig. 3.2):

- If the results of the previous step meet the objective, the next step can be started if the results are incompatible with the objective, the next step may not be taken.
- If resources permit repetition of the previous step (or if necessary several preceding steps) and good results can be expected, the step must be repeated on a higher information level.
- If the answer to the previous question is no, the development must be stopped.

Even if the results of a particular step do not meet the objectives, they might nevertheless prove useful if the objectives were wholly or partly changed. In that case it should be investigated if the objectives can be changed or if the results can be used for other applications.

This whole process, leading from confrontation through creation to decision,

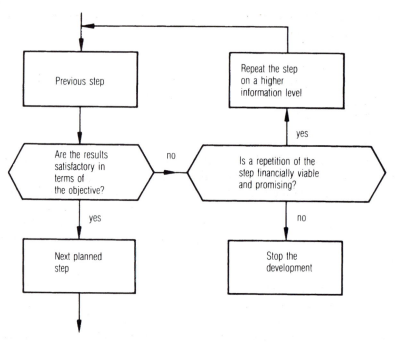

Figure 3.2.   General decision-making process.

must be repeated in each successive, increasingly concrete, phase of the design process.

3.2   Flow of Work During the Process of Planning and Designing

The flow of work during the process of planning and designing has been described in both general terms as well as domain and product-specific terms in VDI Guidelines 2221 and 2222 [3.10, 3.11] (see Fig. 1.10). In line with these Guidelines, to which the authors contributed substantially, the next sections provide an extensive description of this flow of work, focused on mechanical engineering. The description is essentially based on the fundamentals of technical systems (see 2.1), the fundamentals of the systematic approach (see 2.2), and the general problem-solving process (see 3.1). The aim is to adapt the general statements to the requirements of the mechanical engineering design process and to incorporate the specific working and decision making steps for this domain. In principle the planning and design process proceeds from the planning and clarification of the task, through the identification of the required functions, the elaboration of principle solutions, the construction of modular structures, to the final documentation of the complete product.

It is useful and common to divide the planning and design process into the following *main phases*:

- Planning and clarifying the task: specification of information
- Conceptual design: specification of principle
- Embodiment design: specification of layout (construction)
- Detail design: specification of production.

As we will see later on, it is not always possible to draw a clear borderline between these main phases. For example, aspects of the layout might have to be addressed during conceptual design, or it might be necessary to determine some production processes in detail during the embodiment phase. Neither is it possible to avoid backtracking, for example during embodiment design when new auxiliary functions may be discovered for which principle solutions have to be found. Nevertheless the division of the planning and control of a development process into main phases is always helpful.

The working steps proposed for each of the main phases must be considered the main working steps (see Fig. 3.3). The results of these main working steps provide the basis for the subsequent working steps. To realise these results many lower level working steps are required such as collecting information, searching for solutions, calculating, drawing and evaluating. Each of these working steps is accompanied by indirect activities such as discussing, classifying and preparing. The *operational main working steps* listed in the procedural plans proposed in this chapter are considered to be the most useful strategic guidelines for a technical domain. Guidelines that are not listed include, for example, those related to basic problem solving, collecting information and verifying results. This is because they can usually only be recommended in relation to a specific problem and a particular designer. Recommendations for such elementary working steps will, where possible, be given in the sections describing individual methods and those dealing with practical applications.

After the main phases, and some of the more important main working steps, *decision making steps* are required. The decision steps listed are the main ones, that is those that end a main phase or working step, which after an appropriate assessment of the results, allow the main flow of work to proceed. It is also possible, because the result of a decision making step was unsatisfactory, that certain steps have to be repeated. The smallest possible iteration loop is desirable. Again, the individual test and decision making steps (see for example the TOTE model in 2.2.1) that are necessary for every single action have not been listed separately. This would have been impossible because such decisions are determined by the approach of individual designers and by particular problem situations.

The decision to stop a development that ceases to be viable, as discussed in 3.1, is not mentioned explicitly in the individual decision making steps of the procedural plans. One should, however, always explicitly consider this possibility because an early and clear decision to halt a hopeless situation will, in the end, minimise disappointment and cost.

In all cases, procedural plans should be applied in a flexible manner and adapted

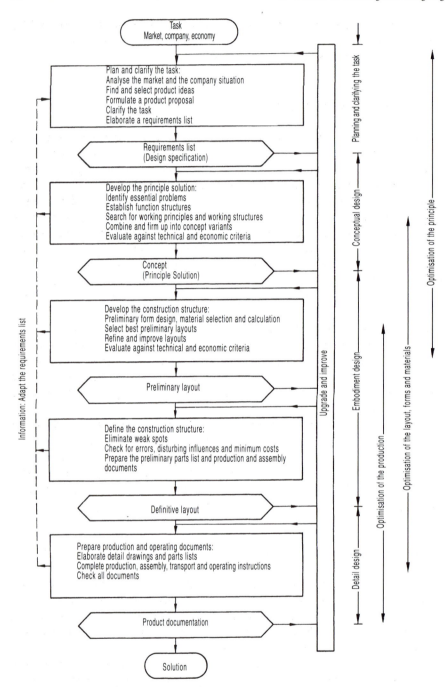

Figure 3.3.   Steps of the planning and design process.

to the particular problem situation. At the end of each main working and decision step, the overall approach should be assessed and adjusted if necessary.

The four main phases are outlined below.

1 Planning and Clarifying the Task

To start a product development, a product idea is needed that looks promising given the current market situation, company needs and economic outlook.

The planning of products can take place as described in 5.1 as part of a systematically executed *product planning process*. However, product ideas can also arise through less formal discussions and idea generation sessions within development teams. Notwithstanding the method, a successful product planning process always takes into account the market, the company and the economy. Several product ideas will be found and will need to be discussed in order to select suitable ones. The end result is a more detailed product proposal. In this sense, product planning is similar to the search for solutions that takes place later, and therefore the same methods are used. The difference being that it takes place on a more abstract and preliminary level.

Irrespective of whether the task is based on a product proposal stemming from a product planning process or a specific customer order, it is necessary to clarify the given task in more detail before starting product development. The purpose of this *clarification of the task* is to collect information about the requirements that have to be fulfilled by the product, and also about the existing constraints and their importance.

This activity leads to the formulation of a *requirements list* that focuses on, and is tuned to, the interests of the design process and subsequent working steps (see 5.2).

The conceptual design phase and subsequent phases should be based on this document that has to be updated continuously (this is indicated by the information feedback loop in Fig. 3.3). The result of this phase is the *specification of information* in a requirements list.

2 Conceptual Design

After completing the task clarification phase, the conceptual design phase determines the principle solution. This is achieved by abstracting the essential problems, establishing function structures, searching for suitable working principles and then combining those principles into a working structure. Conceptual design results in the *specification of principle*.

Often, however, a working structure cannot be assessed until it is transformed into a more concrete representation. This concretisation involves selecting preliminary materials, producing a rough dimensional layout, and considering technological possibilities. Only then, in general, is it possible to assess the essential aspects of a solution principle and review the objectives and constraints (see 2.1.8). It is possible that there will be several principle solution variants.

The representation of a principle solution (solution principle) can take many forms. For existing building blocks a schematic representation in the form of a function structure, a circuit diagram or a flow chart may be sufficient. In other cases a line sketch might be more suitable, and sometimes a rough scale drawing is necessary.

The conceptual design phase consists of several steps (see 6), none of which may be skipped if the most promising principle solution is to be reached. In the subsequent embodiment and detail design phases it is extremely difficult or impossible to correct fundamental shortcomings of the solution principle. A lasting and successful solution is more likely to spring from the choice of the most appropriate principles than from exaggerated concentration on technical details. This claim does not conflict with the fact that even in the most promising solution principles or combinations of principles problems may emerge during the detail design phase.

The solution variants that have been elaborated must now be evaluated. Variants that do not satisfy the demands of the requirements list have to be eliminated; the rest must be judged by the methodical application of specific criteria. During this phase, the chief criteria are of a technical nature, though rough economic criteria also begin to play a part (see 6.5.2). On the basis of the evaluation the best solution concept can now be selected.

It may be that several variants look equally promising, and that a final decision can only be reached on a more concrete level. Moreover, various form designs may satisfy one and the same solution principle. The design process now continues on a more concrete level referred to as embodiment design.

3   Embodiment Design

During this phase, designers, starting from a concept (working structure, principle solution), determine the construction structure (overall layout) of a technical system in line with technical and economic criteria. Embodiment design results in the *specification of layout.*

It is often necessary to produce several *preliminary layouts* to scale simultaneously or successively in order to obtain more information about the advantages and disadvantages of the different variants.

After sufficient elaboration of the layouts, this design phase also ends with an evaluation against technical and economic criteria. This results in new knowledge on a higher information level. Frequently, the evaluation of individual variants may lead to the selection of one that looks particularly promising but which may nevertheless benefit from, and be further improved by, incorporating ideas and solutions from the others. By appropriate combination and the elimination of weak links, the best layout can then be obtained.

That *definitive layout* provides a check of function, strength, spatial compatibility etc, and it is also at this stage, at the very latest, that the financial viability of the project must be assessed. Only then should work start on the detail design phase.

4   Detail Design

This is the phase of the design process in which the arrangement, forms, dimensions and surface properties of all the individual parts are finally laid down, the materials specified, production possibilities assessed, costs estimated and all the drawings and other production documents produced [3.13] (see also [3.12]). The result of the detail design phase is the *specification of production.*

It is important that designers should not relax their vigilance at this stage, otherwise their ideas and plans might change out of recognition. It is a mistake to think that the detail design poses subordinate problems lacking in importance or interest. As we said earlier, difficulties frequently arise from lack of attention to detail. Quite often corrections must be made during this phase and the preceding steps repeated, not so much with regard to the overall solution, as for the improvement of assemblies and components.

In the flow diagram (see Fig. 3.3), the crucial activities are:

- optimisation of the principle;
- optimisation of the layout, forms and materials; and
- optimisation of the production.

They influence each other and, as the figure shows, overlap to a considerable extent. It is obvious that important production criteria (such as maximum size, available production methods etc) play a crucial role during the conceptual design phase, as do the range of possible materials and the spatial requirements, ie the embodiment characteristics. In general, however, the optimisation of the form designs and hence of the production processes assumes growing importance as embodiment proceeds.

The main phases of the design process cannot always be clearly delimited. Thus even a conceptual decision may require a scale drawing for the purpose of deciding on possible layouts. Conversely, the preliminary layout selected during the embodiment design phase may involve nothing more than rough sketches [3.5]. Moreover, certain optimisations may be postponed until the detail design phase. Such variations of the design process, caused by the specific task and type of product, in no way detract from the value of the general scheme.

Figure 3.3 does not include producing models and prototypes because the information they supply may be needed at any point in the design process and cannot therefore be fitted into any particular slot. In many cases, models and prototypes have to be developed even during the conceptual phase, particularly when they are intended to clarify fundamental questions in, say, the precision engineering, electronics and mass production industries. In heavy engineering, on the other hand, if prototypes are needed at all, they must often be preceded by an almost complete run through of the detail design phase.

Figure 3.3 also does not indicate when work has to be sub-contracted because this depends upon the type of product. It must also be stressed that the execution of orders need not be part of the design process, especially in the case of size ranges and modular products where it can take place quite late in the product development process.

The activity of executing orders using computer support is often regarded as an activity outside the actual design process. On receiving an order, existing documents are used, and production instructions, sub-contractor orders, parts-lists etc, only have to be compiled. Apart from tender drawings, layout drawings and assembly plans no further design work is needed, and, in many cases, this can be done automatically using variant design software.

On looking at Fig. 3.3, and after reading about the methods described in the following chapters, practising designers may well object that they lack the time to go through every one of the many steps. They should bear in mind that:

- most of the steps are taken in any case, albeit unconsciously; though they are often carried out too quickly leading to unforeseen consequences;
- the deliberate step-by-step procedure, on the other hand, ensures that nothing essential has been overlooked or ignored, and is therefore indispensable in the case of original designs;
- in the case of adaptive designs, it is possible to resort to time-tested approaches and to reserve the step-by-step procedure for where it offers special benefits, for example, when improving a specific detail;
- if designers are expected to produce better results, then they must be given the extra time the systematic approach demands, though experience has shown that only a little extra time is needed for a stepwise procedure (see 11.2); and
- scheduling becomes more accurate if the step-by-step method is followed rigorously.

The next chapters provide the methods and tools that can be used in the various working steps.

4 General Methods for Finding and Evaluating Solutions

In the planning and product development process (see 3.2), many methods are used. They can be divided into general methods and problem-specific methods. The former are not linked to a specific design phase or type of product, but support the search for solutions and the evaluations that take place throughout the design process. The latter, on the other hand, can only be used for specific tasks, for example to calculate stiffnesses or to estimate costs.

In order to emphasise their breadth and frequency of application, the general methods will be discussed before a more detailed discussion of the main phases of the design process. The problem-specific methods, on the other hand, will be discussed in those sections in subsequent chapters that describe the main phases and working steps.

For educational and methodological reasons, the general methods are divided into conventional, intuitive and discursive methods. They are not mutually exclusive, but complement each other in many ways. The results from cognitive studies (see 2.2.1) indicate that changing between searching methods and searching levels has many advantages. Examples are the switching between abstract and concrete, and between intuitive and discursive. Which method should be applied in a specific case depends on the particular problem, the available information, the state of progress, and the skills and experience of the designers. Solutions found using conventional methods can be extended and completed effectively using intuitive and discursive methods. These methods greatly increase the chances of finding new solutions.

4.1 Solution Finding Methods

4.1.1 Conventional Methods

1 Literature Search

For designers, up-to-date technological data provide a wealth of important information. Such data can be found in textbooks and technical journals, in patent files and in brochures published by competitors. They provide a most useful survey of known solution possibilities. Increasingly, this type of information is fed into computer databases and stored for future use.

2   Analysis of Natural Systems

The study of natural forms, structures, organisms and processes can lead to very useful and novel technical solutions. The connections between biology and technology are investigated by bionics and biomechanics. Nature can stimulate the creative imagination of designers in a host of different ways [4.18].

Technical applications of the design principles of natural forms include light-weight structures employing honeycombs, tubes and rods, and the profiles of aeroplanes, ships etc. Of great importance are lightweight structures in the form of thin stems (see Fig. 4.1). Another technical application is sandwich construction. Figure 4.2 shows a few derivations of the natural principles that have proved useful in aircraft.

The hooks of a burr provided the solution incorporated in the Velcro fastener (see Figs. 4.3 and 4.4).

3   Analysis of Existing Technical Systems

The analysis of existing technical systems is one of the most important means of generating new or improved solution variants step-by-step.

This analysis involves the mental or even physical dissection of finished products. It may be considered a form of structure analysis (see 2.2.2) aimed at the discovery of related logical, physical and embodiment design features. Figure 6.10 shows an example of this type of analysis. Here, sub-functions were derived from the existing configuration. From them, further analysis led to the identification of the physical

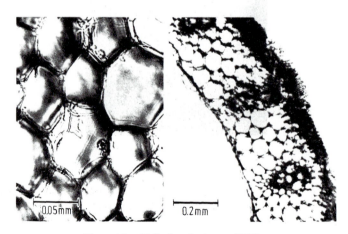

0.05mm

0.2mm

Figure 4.1.   Wall of a wheat stem [4.18].

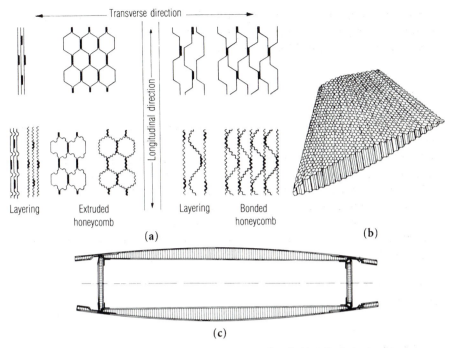

Figure 4.2.   Sandwich construction for lightweight structures [4.19]. (a) A few honeycomb structures. (b) Completed honeycomb structure. (c) Sandwich box girder.

Figure 4.3.   Hooks of a burr, after [4.18].

effects involved which, in turn, might have suggested new solution principles for corresponding sub-functions. It is also possible to adopt solution principles discovered during the analysis.

Existing systems used for analysis might include:

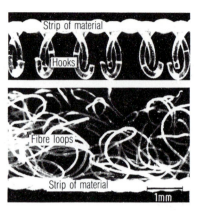

Figure 4.4.   Velcro fastener, after [4.18].

- products or production methods of competing companies;
- older products and production methods of one's own company; and
- similar products or assemblies in which some sub-functions or parts of the
  function structure correspond with those for which a solution is being sought.

Because the only systems to be analysed are those having some bearing on the new
problem as a whole or on parts of it, we may call this way of collecting information
the systematic exploitation of proven ideas, or of experience. It will prove particularly helpful in finding a first solution concept as a starting point for further variations. It must, however, be said that this approach carries the danger of causing
designers to stick with known solutions instead of pursuing new paths.

4   Analogies

In the search for solutions and in the analysis of system properties, it is often
useful to substitute an analogous problem (or system) for the one under consideration, and to treat it as a model. In technical systems, analogies may be obtained,
for instance, by changing the type of energy used [4.2, 4.43]. Analogies chosen
from the non-technical sphere may prove very useful as well.

Besides helping in the search for solutions, analogies are also helpful in the study
of the behaviour of a system during the early stages of its development by means
of simulation and model techniques, and in the subsequent identification of essential new sub-solutions and the introduction of early optimisations.

If the model is to be applied to systems of markedly different dimensions and
conditions, a supportive similarity (dimensional) analysis should be undertaken
(see 8.1.1).

5   Measurements and Model Tests

Measurements on existing systems, model tests supported by similarity analyses
and other experimental studies are among the most important sources of infor-

mation. Rodenacker [4.37] in particular lays great stress on the importance of experimental studies, arguing that design can be interpreted as the reversal of physical experiment.

In precision engineering and mass production industries, including the development of micro-mechanisms and electronic products, experimental investigations are an important and established means of arriving at solutions. This approach has organisational repercussions since, in the creation of such products, experimental development is often incorporated within the design activity (see Fig. 1.4).

In a similar way, the testing and subsequent modification of software solutions belong to this empirically based group of methods.

4.1.2   Intuitive Methods

Designers often seek and discover solutions for difficult problems by intuition — that is, solutions come to them in a flash after a period of search and reflection. These solutions suddenly appear as conscious thoughts and often their origins cannot be traced. As Galtung of the International Peace Research Institute in Oslo has put it: "The good idea is not discovered or undiscovered; it comes, it happens". It is then developed, modified and amended, until such time as it leads to the solution of the problem.

Good ideas are always scrutinised by the subconscious or preconscious in the light of expert knowledge, experience and the task in hand, and often the simple impetus resulting from the association of ideas suffices to force them into consciousness. That impetus can also come from apparently unconnected external events or discussions. Frequently, a sudden idea will hit the bull's eye, so that all that needs doing is to make changes or adaptations that lead straight to a final solution. If that is indeed the case and a successful product is created, then this represents the optimum procedure. Very many good solutions are born in that way and developed successfully. A good design method, far from trying to eliminate this process, should serve to back it up.

An industrial company should nevertheless beware of exclusive reliance on the intuition of its designers and its designers themselves should not leave everything to chance or rare inspiration. Purely intuitive methods have the following disadvantages:

- The right idea does not always come at the right time, since it cannot be forced.
- Current conventions and personal prejudices may inhibit original developments.
- Because of inadequate information, new technologies or procedures may fail to reach the consciousness of designers.

These dangers increase with specialisation, the division of tasks and with time pressure.

There are several methods of encouraging intuition and opening new paths by the association of ideas. The simplest and most common of these involves critical discussions with colleagues. Provided that such discussions are not allowed to stray too far and are based on the general methods of persistent questions, negation, forward steps etc (see 2.2.2), they can be very helpful and effective.

Methods with an intuitive bias such as Brainstorming, Synectics, Gallery Method, Delphi Method, Method 635 and many others involve group dynamics to generate the widest possible range of ideas. One of the effects of group dynamics is the uninhibited exchange of associated ideas between the members. Most of these techniques were originally devised for the solution of non-technical problems. They are, however, applicable to any field that demands new, unconventional ideas.

1   Brainstorming

Brainstorming can be described as a method of generating a flood of new ideas. It was originally suggested by Osborn [4.31] and provides conditions in which a group of open-minded people from as many different spheres of life as possible bring up any thoughts that occur to them and thus trigger off new ideas in the minds of the other participants [4.51]. Brainstorming relies strongly on stimulation of the memory and on the association of ideas that have never been considered in the current context or have never been allowed to reach consciousness.

For maximum effect, brainstorming sessions should be run on the following lines:

Composition of the Group

- The group should have a leader and consist of a minimum of five and a maximum of 15 people. Fewer than five constitute too small a spectrum of opinion and experience, and hence produce too few stimuli. With more than 15, close collaboration may decline because of individual passivity and withdrawal.
- The group must not be confined to experts. It is important that as many fields and activities as possible are represented, the involvement of non-technical members adding a rich new dimension.
- The group should not be hierarchically structured but, if possible, made up of equals to prevent the censoring of such thoughts as might give offence to superiors or subordinates.

Leadership of the Group

- The leader of the group should only take the initiative in dealing with organisational problems (invitation, composition, duration and evaluation). Before the actual brainstorming session the leader must outline the problem and, during the session, must see to it that the rules are observed and, in particular, that the atmosphere remains free and easy. To that end the leader should start the session by expressing a few absurd ideas, or mentioning an example from another brainstorming session; but should never lead in the expression of ideas.

On the other hand the flow of new ideas should be encouraged whenever the productivity of the group slackens. The leader must ensure that no one criticises the ideas of other participants, and appoint one or two members to take minutes.

Procedure

- All participants must try to shed their intellectual inhibitions—that is, they should avoid rejecting as absurd, false, embarrassing, stupid or redundant any ideas expressed spontaneously by themselves or by other members of the group.
- No participant may criticise ideas that are brought up, and everyone must refrain from using such killer phrases as "We've heard it all before", "It can't be done", "It will never work" and "It has nothing to do with the problem". New ideas are taken up by the other participants, who may change and develop them at will. It is also useful to combine several ideas into new proposals.
- All ideas should be written down, sketched out, or recorded on tape.
- All suggestions should be concrete enough to allow the emergence of specific solution ideas.
- The practicability of the suggestions should be ignored at first.
- A session should not generally last for more than 30 to 45 minutes. Experience has shown that longer sessions produce nothing new and lead to unnecessary repetitions. It is better to make a fresh start with new ideas or with other participants later.

Evaluation

- The results should be reviewed by experts to find potential solution elements. If possible, these should be classified and graded in order of feasibility and then developed further.
- The final result should be reviewed with the entire group to avoid possible misunderstandings or one-sided interpretations on the part of the experts. New and more advanced ideas may well be expressed or developed during such a review session.

Brainstorming [4.33] is indicated whenever:

- No practical solution principle has been discovered.
- The physical process underlying a possible solution has not yet been identified.
- There is a general feeling that deadlock has been reached.
- A radical departure from the conventional approach is required.

Brainstorming is useful even in the solution of sub-problems arising in known or existing systems. Moreover, it has a beneficial side effect: all the participants are supplied with new data, or at least with fresh ideas on possible procedures, applications, materials, combinations etc, because the group represents a broad spectrum of opinion and expertise (for instance, designers, production engineers, salespeople, materials experts and buyers). It is astonishing what a profusion and range of ideas such a group can generate. The designers will remember the ideas brought up during brainstorming sessions on many future occasions. Brainstorming triggers off new lines of thought, stimulates interest and represents a break in the normal routine.

It should, however, be stressed that no miracles must be expected from brainstorming sessions. Most of the ideas expressed will not be technically or economically feasible, and those that are will often be familiar to the experts. Brainstorming is meant first of all to trigger off new ideas, but it cannot be expected to produce ready-made solutions because problems are generally too complex and too difficult to be solved by spontaneous ideas alone. However, if a session should produce one or two useful new ideas, or even some hints in what direction the solution might be sought, it will have achieved a great deal.

An example of a solution obtained by brainstorming will be found in 6.6, which also shows how the resulting ideas were evaluated and how classifying criteria for the subsequent search for solutions were derived from them.

2 Method 635

Brainstorming has been developed into Method 635 by Rohrbach [4.38].

After familiarising themselves with the task and after careful analysis, each of six participants is asked to write down three rough solutions in the form of keywords. After some time, the solutions are handed to the participant's neighbour who, after reading the previous suggestions, enters three further solutions or developments. This process is continued until each original set of three solutions has been completed or developed through association by the five other participants. Hence the name of the method.

Method 635 has the following advantages over brainstorming:

- A good idea can be developed more systematically.
- It is possible to follow the development of an idea and to determine more or less reliably who originated the successful solution principle, which might prove advisable for legal reasons.
- The problem of group leadership hardly arises.

The method has the following disadvantage:

- Reduced creativity by the individual participants owing to isolation and lack of stimulation in the absence of overt group activity.

3 Gallery Method

The gallery method developed by Hellfritz [4.16] combines individual work with group work, and is particularly suitable for embodiment problems because solution proposals in the form of sketches are easily included. The organisation and team building are similar to brainstorming. The method consists of the following steps.

Introduction Step: The group leader presents the problem and explains the context.

Idea Generation Step 1:   For 15 minutes the individual group members create solutions intuitively and without prejudice using sketches supported, where necessary, by text.

Association Step: The results of idea generation step 1 are hung on a wall as in an art gallery so that all group members can see and discuss them. The purpose of this 15 minute association step is to find new ideas or to identify complementary or improved proposals through negation and reappraisal.

Idea Generation Step 2:   The ideas and insights from the association step are further developed individually by each of the group members.

Selection Step: All generated ideas are reviewed, classified and, if necessary, finalised. Promising solutions are then selected (see 4.2.1). It is also possible to identify potential solution characteristics that can be developed later using a discursive method (see 4.1.3).

The gallery method has the following advantages:

- Intuitive group working takes place without unduly lengthy discussions.
- An effective exchange of ideas using sketches is possible.
- Individual contributions can be identified.
- Documentary records are easily assessed and stored.

4   Delphi Method

In this method, experts in a particular field are asked for written opinions [4.3].
The requests take the following form:

First Round: What starting points for solving the given problem do you suggest? Please make spontaneous suggestions.

Second Round: Here is a list of various starting points for solving the given problem. Please go through this list and make what further suggestions occur to you.

Third Round: Here is the final evaluation of the first two rounds. Please go through the list and write down what suggestions you consider most practicable.

This elaborate procedure must be planned very carefully and is usually confined to general problems bearing on fundamental questions or on company policy. In the field of engineering design, the Delphi method should be reserved for fundamental studies of long-term developments.

5   Synectics

Synectics is a word derived from Greek and refers to the activity of combining various and apparently independent concepts. Synectics is comparable to brainstorming, with the difference that its aim is to trigger off fruitful ideas with the help of analogies from non-technical or semi-technical fields.

The method was first proposed by Gordon [4.14]. It is more systematic than brainstorming with its arbitrary flow of ideas. For the rest, both methods call for complete frankness and lack of inhibition or criticism.

A synectics group should consist of no more than seven members, lest the ideas expressed run away with themselves. The leader of the group has the additional task of helping the group to develop the proposed analogies by guiding them through the following steps:

- Presentation of the problem.
- Familiarisation with the problem (analysis).
- Grasp of the problem.
- Rejection of familiar assumptions with the help of analogies drawn from other spheres.
- Analysis of one of the analogies.
- Comparison of the analogy with the existing problem.
- Development of a new idea from that comparison.
- Development of a possible solution.

If the result is unsatisfactory, the process may have to be repeated with a different analogy.

An example may help to illustrate this method. In a seminar set up for the purpose of discovering the best method of removing urinary calculi from the human body, several mechanical devices for gripping, holding and extracting these stones were mentioned. The device would have to stretch and open up inside the urethra. The keywords "stretch" and "open up" suggested the idea of an umbrella to one of the participants (see Fig. 4.5).

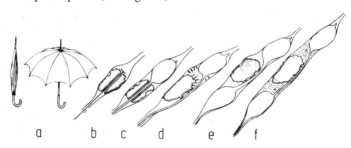

Figure 4.5.   Step-by-step development of a solution principle for the removal of urinary calculi based on an analogy.

Question: How can the umbrella analogy (a) in Fig. 4.5 be applied?

Possible Answer 1: By (b) drilling through the stone, pushing the umbrella through the hole and opening it up. Not very feasible.

Possible Answer 2: By (c) pushing a tube through the hole and blowing it up (balloon) behind the stone. Drilling of hole not feasible.

Possible Answer 3: By (d) pushing the tube past the stone. When the tube is withdrawn the resistance may seriously damage the urethra.

Possible Answer 4: By (e) adding a second balloon as a guide and by (f) embedding the stone in a gel between the two balloons and then pulling it out. This was found to be the best solution.

This example shows the association with a semi-technical analogy (umbrella) from which a solution was developed that took into account the existing special constraints. The solution shown here is not the final solution resulting from the seminar but represents an example of how the method was used.

Characteristic of this approach is the unrestricted use of analogies which, in the case of technical problems, are selected from non-technical or semi-technical spheres. Such analogies will generally suggest themselves quite spontaneously at the first attempt but, during subsequent development and analysis, they will generally be derived more systematically.

6 Combination of Methods

Any one of these methods taken by itself may not lead to the required goal. Experience has shown that:

- The group leader of, or another participant in, a brainstorming session may, when the flow of ideas dries up, introduce synectic procedures — deriving analogies, systematic negation etc — to release a new flood of ideas.
- A new idea or an analogy may radically change the approach and ideas of the group.
- A summary of what has been agreed so far may lead to new ideas.
- The explicit use of the methods of negation and reappraisal and of forward steps (see 2.2.2) can enrich and extend the variety of ideas.

In the seminar we mentioned, the presentation of the idea "destroy stone" produced a host of new suggestions such as drilling, smashing, hammering, ultrasonic disintegration and so on. When the flow of ideas eventually dried up, the group leader asked, "How does nature destroy?" which immediately evoked a number of new suggestions including weathering, heating and cooling, decay, putrefaction, bacterial action, ice expansion and chemical decomposition. A combination of the two principles "clasp stone" and "destroy stone" provoked the question, "What else?". This produced the answer "contact", which in turn threw up such new ideas as sucking, gluing, and applying various contact forces.

The different methods should be combined so as best to meet particular cases. A pragmatic approach ensures the best results.

4.1.3 Discursive Methods

Methods with a discursive bias provide solutions in a deliberate step-by-step approach that can be influenced and communicated. Discursive methods do not exclude intuition, which can make its influence felt during individual steps and in the solution of individual problems, but not in the direct implementation of the overall task.

1 Systematic Study of Physical Processes

If the solution of a problem involves a known physical (chemical, biological) effect represented by an equation, and especially when several physical variables are involved, various solutions can be derived from the analysis of their interrelationships—that is, of the *relationship* between a dependent and an independent variable, all other quantities being kept constant. Thus, if we have an equation in the form $y = f(u,v,w)$ then, according to this method, we investigate solution variants for the relationships $y_1 = f(u,\underline{v},\underline{w})$, $y_2 = f(\underline{u},v,\underline{w})$ and $y_3 = f(\underline{u},\underline{v},w)$, the underlined quantities being kept constant.

Rodenacker has given several examples of this procedure, one of which concerns the development of a capillary viscometer [4.37]. Four solution variants can be derived from the well known law of capillary action $\eta \sim \Delta p \cdot r^4/(\dot{V} \cdot l)$. They are shown schematically in Fig. 4.6.

1. A solution in which the differential pressure Δp serves as a measure of the viscosity: $\Delta p \sim \eta$ ($\dot{V}$, r and l = constant).
2. A solution based on the diameter of the capillary tube: $\Delta r \sim \eta$ ($\dot{V}$, Δp and l = constant).
3. A solution based on changes in the length of the capillary tube: $\Delta l \sim \eta$ (Δp, $\dot{V}$ and r = constant).
4. A solution based on changes in the volume flow rate: $\Delta \dot{V} \sim \eta$ (Δp, r and l = constant).

Another way of obtaining new or improved solutions by the analysis of physical equations is the resolution of known *physical effects* into their *individual components*. Rodenacker, in particular, has used this approach in the design of new devices or the development of new applications for existing devices.

By way of example, let us look at the development of a frictional thread locking device, based on the analysis of the equation governing the torque needed to release a threaded fastener:

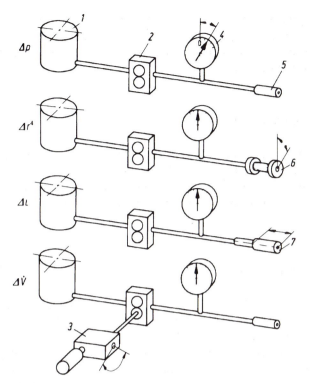

Figure 4.6. Schematic representation of four viscometers, after [4.37]. 1, Container; 2, gear pump; 3, variable drive; 4, pressure guage; 5, fixed capillary tube; 6, capillary tube with variable diameter; 7, capillary tube with variable length.

$$T = P[(d/2) \tan(\phi_v - \beta) + (D/2) \mu_f] \tag{1}$$

The torque given by equation (1) is made up of the following components:
  Frictional torque in the thread:

$$T_t \sim P(d/2) \tan\phi_v = P(d/2) \mu_v \tag{2}$$

where

$$\tan\phi_v = \mu_t/\cos(\alpha/2) = \mu_v$$

Frictional torque on the bolt head or nut face:

$$T_f = P(D/2) \tan\phi_f = P(D/2) \mu_f \tag{3}$$

Release torque of the thread due to pre-load and thread pitch:

$$T_r \sim P(d/2) \tan(-\beta) = -P \cdot \frac{p}{2\pi} \tag{4}$$

(p = thread pitch, β = helix angle, d = mean thread (t) diameter, P = pre-load, D = mean face (f) diameter, μ_v = virtual (v) coefficient of friction in the thread,

μ_t = actual coefficient of friction in the thread, μ_f = coefficient of friction on the head or nut face, α = flank angle, ϕ = angle of friction).

To discover solution principles for the improvement of the locking properties of a threaded fastener, we must analyse the physical relationships further so as to identify the physical effects involved. The individual effects involved in equations (2) and (3) are:

- the friction effect (Coulomb friction)

$$F_t = \mu_v P \text{ and } F_f = \mu_f P$$

- the lever effect

$$T_t = F_t \, d/2 \text{ and } T_f = F_f \, D/2$$

- the wedge effect

$$\mu_v = \mu_t/\cos (\alpha/2)$$

The individual effects in equation (4) are:

- the wedge effect

$$F_r \sim P \tan(-\beta)$$

- the lever effect

$$T_r = F_r \, d/2$$

An examination of the individual physical effects will yield the following solution principles for the improvement of the locking properties of the fastener:

- Use of the wedge effect to reduce the tendency to loosen by decreasing the helix angle β.
- Use of the lever effect to increase the frictional moment on the head or nut face by increasing the mean face diameter D.
- Use of the friction effect to increase the frictional forces by increasing the coefficient of friction μ.
- Use of the wedge effect to increase the frictional force on the face by means of conical surfaces ($P\mu_f/\sin \gamma$ with included angle = 2γ). This method is used with automobile wheel attachment nuts.
- Increase of the flank angle α to increase the virtual coefficient of friction in the thread.

2   Systematic Search with the Help of Classification Schemes

In 2.2.2 we showed that the systematic presentation of data is helpful in two respects. On the one hand it stimulates the search for further solutions in various directions; on the other hand it facilitates the identification and combination of essential solution characteristics. Because of these advantages a number of classification schemes have been drawn up, all with a similar basic structure. Dreibholz

[4.6] has published a comprehensive survey of the possible applications of such classification schemes.

The usual two-dimensional scheme consists of rows and columns of parameters used as classifying criteria. Figure 4.7 illustrates the general structure of classification schemes: (a) when parameters are provided for both the rows and the columns; and (b) when parameters are provided for the rows only, because the columns cannot be arranged in any apparent order. If necessary, the classifying criteria can be extended by a further breakdown of the parameters or characteristics (see Fig. 4.8), which process, however, often tends to confuse the general picture. By allocating the column parameters to the rows it is possible to transform every classification scheme based on row and column into a scheme in which only the row parameters are retained, and the columns are merely numbered (see Fig. 4.9).

Such classification schemes help the design process in a great many ways. In particular, they can serve as design catalogues during all phases of the search for a solution, and they can also help in the combination of sub-solutions into overall solutions (see 4.1.4). Zwicky [4.54] has referred to them as "morphological matrices".

The choice of *classifying criteria* or of their parameters is of crucial importance. In establishing a classification scheme it is best to use the following step-by-step procedure:

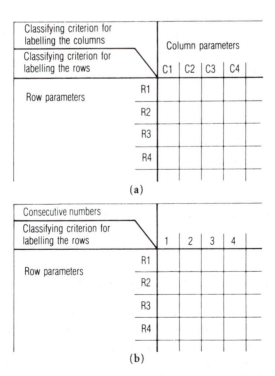

Figure 4.7. General structure of classification schemes, after [4.6].

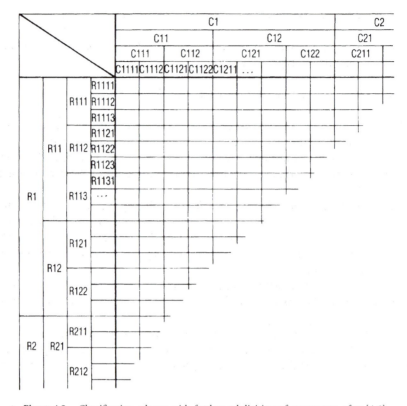

Figure 4.8. Classification scheme with further subdivision of parameters, after [4.6].

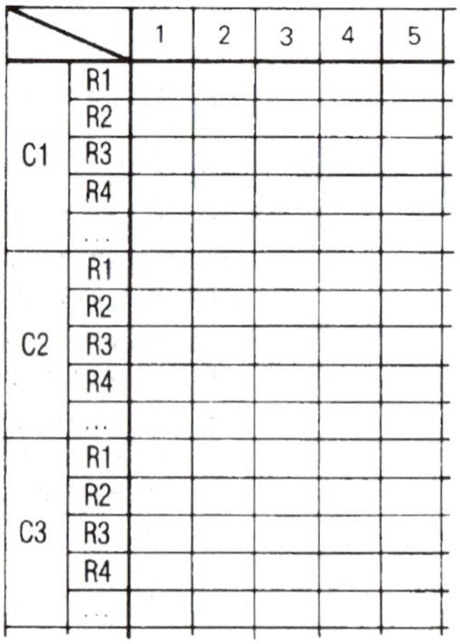

Figure 4.9. Modified classification scheme, after [4.6].

Step 1: Solution proposals are entered in the rows in random order.

Step 2: These proposals are analysed in the light of the headings (characteristics) such as type of energy, type of motion etc.

Step 3: They are classified in accordance with these headings.

These characteristics or classifying criteria can also be obtained from an earlier use of intuitive methods to analyse known solutions or solution ideas.

This procedure not only helps with the identification of compatible combinations but, more importantly, encourages the opening up of the widest possible solution fields.

The classifying criteria and characteristics listed in Figures 4.10 and 4.11 can be useful when searching systematically for solutions and varying solution ideas for technical systems. They refer to types of energy, physical effects, shape, as well as the characteristics of the working geometry, working motion, and the basic material properties (see 2.1.4).

Figure 4.12 provides a simple example of searching for a solution to satisfy a *sub-function*. Here the answer was obtained by varying the type of energy against a number of working principles.

Figure 4.13 is an example of variation based on working motions.

Figure 4.14 shows the variation of the working surfaces in the design of shaft-hub connections. Thanks to such arrangements, the multiplicity of solutions

Classifying criteria:	
Types of energy, physical effects and outward appearances	
Headings:	*Examples:*
Mechanical	Gravitation, intertia, centrifugal force
Hydraulic	Hydrostatic, hydrodynamic
Pneumatic	Aerostatic, aerodynamic
Electrical	Electrostatic, electrodynamic, inductive, capacitative, piezo-electric, transformation, rectification
Magnetic	Ferromagnetic, electromagnetic
Optical	Reflection, refraction, diffraction, interference, polarisation, infra-red, visible, ultra violet
Thermal	Expansion, bimetal effect, heat storage, heat transfer, heat conduction, heat insulation
Chemical	Combustion, oxidation, reduction, dissolution, combination, transformation, electrolysis, exothermic and endothermic reaction
Nuclear	Radiation, isotopes, source of energy
Biological	Fermentation, putrefaction, decomposition

Figure 4.10. Classifying criteria and headings (characteristics) for variation in the physical search area.

Classifying criteria

Working surfaces, working motions and basic material properties

Working geometry	
Headings	*Example:*
Type	Point, line, surface, body
Shape	Curve, circle, ellipse, hyperbola, parabola Triangle, square, rectangle, pentagon, hexagon, octagon Cylinder, cone, rhomb, cube, sphere Symmetrical, asymmetrical
Position	Axial, radial, tangential, vertical, horizontal Parallel, sequential
Size	Small, large, narrow, broad, tall, low
Number	Undivided, divided Simple, double, multiple

Working motions	
Headings	*Examples*
Type	Stationary, translational, rotational
Nature	Uniform, non-uniform, oscillating Plane or three-dimensional
Direction	In x,y,z direction and/or about x,y,z axis
Magnitude	Velocity
Number	One, several, composite movements

Basic material properties	
Headings	*Examples*
State	Solid, liquid, gaseous
Behaviour	Rigid, elastic, plastic, viscous
Form	Solid body, grains, powder, dust

Figure 4.11. Classifying criteria and headings (characteristics) for variation in the form design search area.

obtained, for instance by the method of forward steps (see 2.2.2 and Fig. 2.20), can be put into order and completed.

To sum up the following recommendations are given:

- Classification schemes should be built up step-by-step and as comprehensively as possible. Incompatibilities should be discarded, and only the most promising solution proposals pursued. In so doing, designers should try to determine which classifying criteria contribute to the discovery of a solution, and to examine further variations of these.
- The most promising solutions should be chosen and labelled by a special selection procedure (see 4.2.1).
- If possible, the most comprehensive classification schemes should be drawn up, that is, schemes for repeated use, but systems should never be built for the sake of systematics alone.

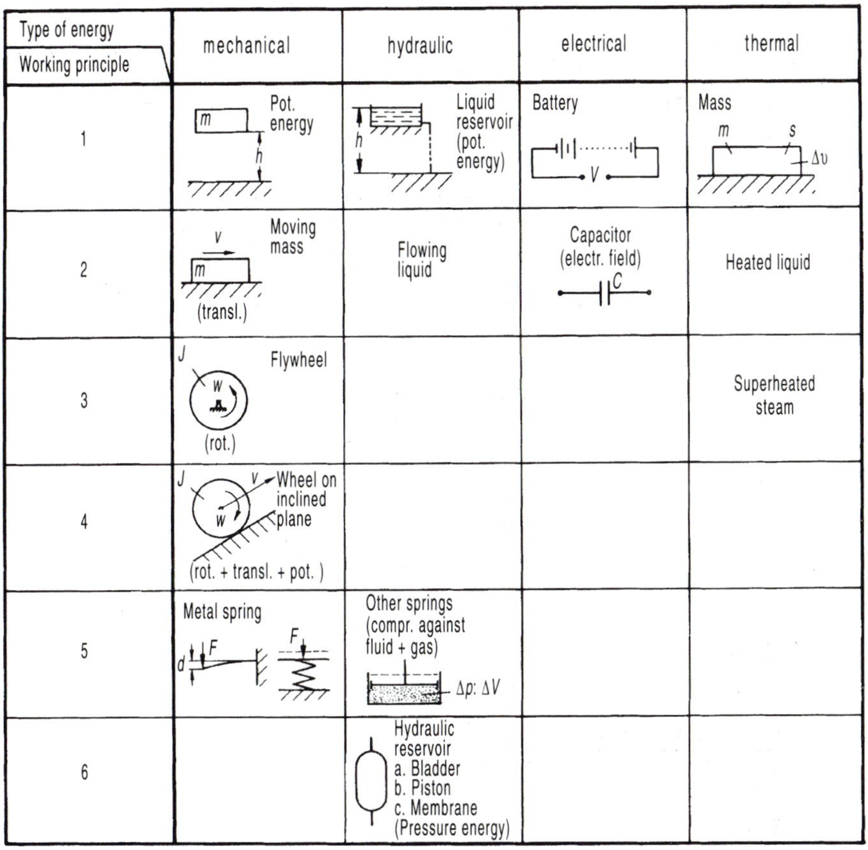

Type of energy / Working principle	mechanical	hydraulic	electrical	thermal
1	m — Pot. energy — h	Liquid reservoir (pot. energy) — h	Battery — V	Mass m — s — Δv
2	v — Moving mass — m (transl.)	Flowing liquid	Capacitor (electr. field) — C	Heated liquid
3	J — Flywheel — w (rot.)			Superheated steam
4	J — v Wheel on inclined plane — w (rot. + transl. + pot.)			
5	Metal spring — d F — F	Other springs (compr. against fluid + gas) — Δp: ΔV		
6		Hydraulic reservoir a. Bladder b. Piston c. Membrane (Pressure energy)		

Figure 4.12. Different working principles to satisfy the function "store energy" by varying the type of energy.

3 Use of Design Catalogues

Design catalogues are collections of known and proven solutions to design problems. They contain data of various types and solutions of distinct levels of embodiment. Thus they may cover physical effects, working principles, principle solutions, machine elements, standard parts, materials, bought-out components etc. In the past, such data were usually found in textbooks and handbooks, company catalogues, brochures and standards. Some of these contained, apart from purely objective data and suggested solutions, examples of calculations, solution methods and other design procedures. It is also possible to imagine catalogue-like collections for these methods and procedures. Design catalogues should provide:

- quicker, more problem-oriented access to the accumulated solutions or data;
- the most comprehensive range of solutions possible, or, at the very least, the most essential ones, which can be extended later;
- applications independent of specific company or discipline; and

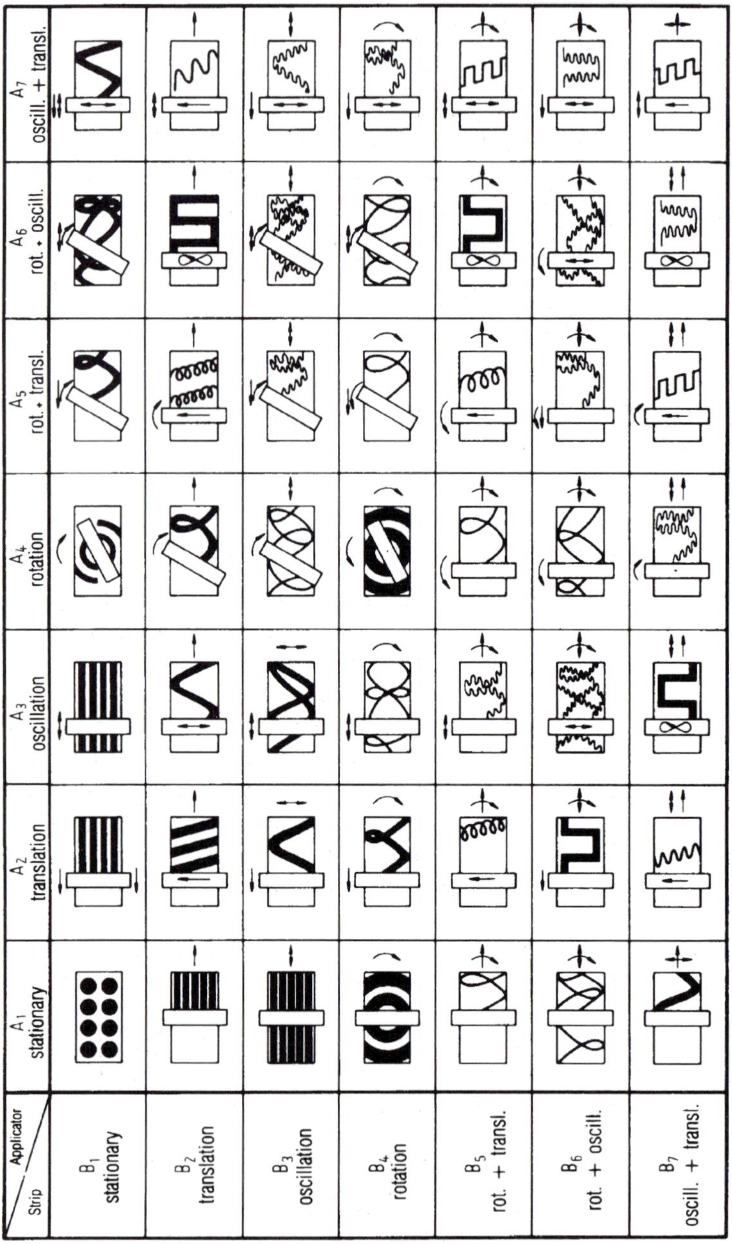

Figure 4.13. Means of coating the backs of carpets by combining motions of the carpet (strip) and of the applicator.

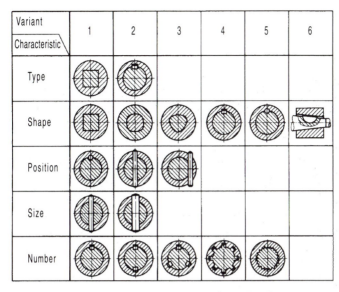

Figure 4.14. Variations of working geometry for shaft-hub connections.

- data for conventional design procedures as well as for computer-aided methods.

The construction of design catalogues has been studied, above all, by Roth and collaborators [4.40]. Roth suggests that a design catalogue of the type shown in Fig. 4.15 is most likely to satisfy all the demands listed above.

The *classifying criteria* determine the structure of the catalogue. They influence the ease with which catalogues can be handled and reflect the level of complexity of particular solutions, and also their degree of embodiment. In the conceptual design phase, for instance, it is advisable to select as classifying criteria the functions to be fulfilled by the solutions. This is because the conceptual design is based on the underlying sub-functions. For classifying characteristics it is best to choose generally valid functions (see 2.1.3), which help to elicit the most product-independent solutions.

Further classifying criteria might include the types and characteristics of energy (mechanical, electrical, optical etc), of materials or signals, of working principles,

Classifying criteria			Solutions			Solution characteristics					Remarks		
1	2	3	1	2	Nr.	1	2	3	4	5	1	2	3
					1			▨					
					2			▨	▨				
					3	▨	▨						
					4								
					5			▨					
					6								
					7	▨							

Figure 4.15. Basic construction of a design catalogue, after [4.40].

Table 4.1. Available design catalogues

Application	Object	Author and reference
General	Construction of catalogues	Roth [4.40]
	List of available catalogues and solutions	Roth [4.40]
Principle solutions	Physical effects	Roth [4.40]
	Solutions to functions	Koller [4.22]
Connections	Types of connections	Roth [4.40]
	Connections	Ewald [4.9]
	Fixed connections	Roth [4.40]
	Welded joints for steel profiles	Wölse and Kastner [4.52]
	Riveted joints	Roth [4.40]
		Kopowski [4.24]
		Grandt [4.15]
	Adhesive joints	Fuhrmann and Hinterwalder [4.12]
	Clamping elements	Ersoy [4.7]
	Principles of threaded joints	Kopowski [4.24]
	Threaded fasteners	Kopowski [4.24]
	Elimination of backlash in threaded joints	Ewald [4.9]
	Elastic joints	Gießner [4.13]
	Shaft-hub-connections	Roth [4.40], Diekhöner and Lohkamp [4.5], Kollmann [4.23]
Guides and bearings	Linear guides	Roth [4.40]
	Rotational guides	Roth [4.40]
	Plain and roller bearings	Diekhöner [4.4]
	Bearings and guides	Ewald [4.9]
Power generation, power transmission	Electric motors (small)	Jung and Schneider [4.20]
	Drives (general)	Schneider [4.44]
	Power generators (mechanical)	Ewald [4.9]
	Effects to generate power	Roth [4.40]
	Single stage power multiplication	Roth [4.40], VDI 2222 [4.47]
	Lifting mechanisms	Raab and Schneider [4.34]
	Screw drives	Kopowski [4.24]
	Friction systems	Roth [4.40]

of working surfaces and of working motions. In the case of design catalogues intended for the embodiment design phase, useful classifying criteria include the properties of materials and the characteristics of particular machine elements, such as couplings and clutches.

The *solutions D column* is the main part of the catalogue and contains the solutions. Depending on the level of abstraction, the solutions are represented as sketches, with or without physical equations, or as more or less complete drawings or illustrations. The type and completeness of the information given once again

Table 4.1. Continued

Application	Object	Author and reference
Kinematics, mechanisms	Solving motion problems using mechanisms	VDI 2727, part 2 [4.49]
	Chain drives and mechanisms	Roth [4.40]
	4-bar mechanisms	VDI 2222, part 2 [4.47]
	Logical inverse mechanisms	Roth [4.40]
	Logical conjunctive and disjunctive mechanisms	Roth [4.40]
	Mechanical flip-flops	Roth [4.40]
	Mechanical non-return safety devices	Roth [4.40], VDI 2222, part 2 [4.47]
	Lifting mechanisms	Raab and Schneider [4.34]
	Uniform-motion transmissions	Roth [4.40]
	Handling devices	VDI 2740 [4.50]
Gearboxes	Spur gear	VDI 2222, part 2 [4.47], Ewald [4.9]
	Mechanical single-stage gearboxes with constant gear ratio	Diekhöner and Lohkamp [4.5]
	Elimination of backlash in spur gears	Ewald [4.9]
Safety technology	Danger situations	Neudörfer [4.28]
	Protective barriers	Neudörfer [4.29]
Ergonomics	Indicators, controls	Neudörfer [4.27]
Production processes	Casting	Ersoy [4.7]
	Drop forging	Roth [4.40]
	Press forging	Roth [4.40]

depends on the intended application. It is important that all data are of the same level of abstraction and omit side issues.

Important for the choice of solutions is the column covering the *solution characteristics*.

The *remarks column* can be used for information about the origin of the data and for additional comments.

The characteristics used for selection may involve a great variety of properties — for instance, typical dimensions, reliability, response, number of elements etc. They help designers in the preliminary selection and evaluation of solutions and, in the case of computer-based catalogues, they can also be used in the final selection and evaluation.

Another important requirement of design catalogues is that they should have uniform and clear definitions and symbols.

The more concrete and detailed the stored information, the more direct but also the more limited is the application of a catalogue. With increasing degree of embodiment, data for a given solution become more comprehensive. However, the chances of arriving at a complete solution field decreases because the number of details, for example embodiment variants, increases rapidly. Thus it may be possible to provide a full list of physical effects fulfilling the function "channel", but it would hardly be possible to list all the potential embodiments of bearings (channelling a force from a rotating to a stationary system).

Table 4.1 lists the currently available design catalogues that satisfy the requirements and structure described above. Therefore in what follows we include just a few examples of, or extracts from, available design catalogues.

Figure 4.16 shows a catalogue of physical effects associated with the functions "change energy" and "vary energy component". It is based on Koller [4.22] and Krumhauer [4.25]. The catalogue makes it possible to derive these effects from the classifying criteria, "inputs and outputs". The characteristics on which the selection is based must be derived from the technical literature.

Figure 4.17 shows an extract of a catalogue for shaft-hub connections based on [4.40]. In this, unlike the previous catalogue, the solutions are concrete enough, thanks to the specification of the form design features, for the embodiment design phase to start with a scale layout drawing.

4.1.4 Methods for Combining Solutions

As described in 2.2.2 it is often useful to divide problems, tasks and functions into sub-problems, sub-tasks and sub-functions and to solve these individually (factorisation method) (see also 6.3). Once the solutions for sub-problems, sub-tasks or sub-functions are available, they have to be *combined* in order to arrive at an overall solution.

The methods we have been describing, particularly those with an intuitive bias, are intended to lead to the discovery of suitable combinations. However, there are special methods for arriving at such syntheses more directly. In principle, they must permit a clear combination of solution principles with the help of the associated physical and other quantities and the appropriate geometrical and material characteristics. When analysing combinations it is important to identify and use appropriate solution characteristics.

The main problem with such combinations is ensuring the physical and geometrical compatibility of the solution principles to be combined, which in turn ensures the smooth flow of energy, material and signals, and avoids geometrical interference in mechanical systems. For information systems, the main problem is the compatibility requirements of the information flow.

A further problem is the selection of technically and economically favourable combinations of principles from the large field of theoretically possible combinations. This aspect will be discussed at greater length in 4.2.2.

1 Systematic Combination

For the purpose of systematic combination, the classification scheme to which Zwicky [4.54] refers as the "morphological matrix" (see Fig. 4.18) is particularly

Function	Input	Output	Physical effects							
$E_{mech} \rightarrow E_{mech}$	Force, pressure, torque	Length, angle	Hooke (Tension/compression/bending)	Shear, torsion	Upthrust Poisson's effect	Boyle-Mariotte	Coulomb I and II		...	...
		Speed	Energy Law	Conservation of momentum	Conservation of angular momentum				...	...
		Acceleration	Newton's Law	...	...	...	...	...	...	...
	Length, angle	Force, pressure, torque	Hooke	Shear, torsion	Gravity	Upthrust	Boyle-Mariotte	Capillary		
			Coulomb I and II	...	...	...	...	...	...	
		Speed	Coriolis force	Conservation of momentum	Magnus-effect	Energy law	Centrifugal force	Eddy current		
		Acceleration	Newton's Law	...	...	...	...	...	...	
$E_{mech} \rightarrow E_{hyd}$	Force, length, speed, pressure	Speed, pressure	Bernoulli	Viscosity (Newton)	Toricelli	Gravitational pressure	Boyle-Mariotte	Conservation of momentum	...	
$E_{hyd} \rightarrow E_{mech}$	Speed	Force, length	Profile lift	Turbulence	Magnus-effect	Flow resistance	Back pressure	Reaction principle	...	
$E_{mech} \rightarrow E_{therm}$	Force, speed	Temperature, quantity of heat	Friction (Coulomb)	1st law	Thomson-Joule	Hysteresis (damping)	Plastic deformation		...	
$E_{therm} \rightarrow E_{mech}$	Temperature, heat	Force, pressure, length	Thermal expansion	Steam pressure	Gas Law	Osmotic pressure	...	...	...	
$E_{electr} \rightarrow E_{mech}$	Voltage, current, magn field	Force, speed, pressure	Biot-Savart-effect	Electro-kinetic effect	Coulomb I	Capacitance effect	Johnsen-Rhabeck-effect	Piezoeffect	...	
$E_{mech} \rightarrow E_{electr}$	Force, length, speed, pressure	Voltage, current	Induction	Electro-kinetics	Electro-dynamic effect	Piezoeffect	Frictional electricity	Capacitance effect	...	
$E_{electr} \rightarrow E_{therm}$	Voltage, current	Temperature, heat	Joule heating	Peltier-effect	Electric arc	Eddy current	...	...	...	
$E_{therm} \rightarrow E_{electr}$	Temperature, heat	Voltage, current	Electr. conduction	Thermo-effect	Thermionic emission	Pyroelectricity	Noise-effect	Semiconductor. Super-conductor	...	
$E_{mech} \rightarrow E_{mech}$	Force, length, pressure, speed	Force, length, pressure, speed	Lever	Wedge	Poisson's effect	Friction	Crank	Hydraulic effect	...	
$E_{hyd} \rightarrow E_{hyd}$	Pressure, speed	Pressure, speed	Continuity	Bernoulli	...	...	...	...	...	
$E_{therm} \rightarrow E_{therm}$	Temperature, heat	Temperature, heat	Heat conduction	Convection	Radiation	Condensation	Evaporation	Freezing	...	
$E_{electr} \rightarrow E_{electr}$	Voltage, current	Voltage, current	Transformer	Valve	Transistor	Transducer	Thermogalvanometer	Ohm's law	...	
...	...	...	...	...	...	...	...	...	...	

Figure 4.16. Design catalogue of physical effects based on [4.22, 4.25] for the generally applicable functions "change energy" and "vary energy component". Also applicable to flow of signals.

Classifying criteria		Solutions				Solution characteristics														Remarks		
Type of interface	Type of force transmission	Equation	Name	Configuration	No.	Transferable torque	Torque transmission depending on	Stress concentration / Axial forces	Applicable for	Behaviour at overload	Centering possible	Unbalanced force	Axial displacement of hub	Hub moveable	Joint adjustable	Shaft diameter (mm)	Material	Manufacturing effort	Assembly effort	Standard (DIN)	Application examples	Remarks
1	2	1	2	3	No.	1	2	3	4	5	6	7	8	9	10	11	12	13	14	15	16	17
Normal (form fit)	Direct	$T \le K \cdot d \cdot \dfrac{A_u \cdot \tau_{max}}{2}$ or $T \le K \cdot d \cdot \dfrac{A_p \cdot \sigma_{max}}{2}$	spline shaft		1	large	h	large / —	pulsating or alternating load	fracture	yes	no	clearance fit	yes	no	10-150		high	small	5461/63 5471/72	toothed wheels	exterior, flank, interior centering possible
			involute spline shaft		2		e	—	pulsating or alternating load											5480 5482		short hub possible
			serrated shaft		3			—								150-500	shaft: 37 Cr 4, 41 Cr 4, 42 CrMo4	small, special machinery necessary		5481		used for short and thin hubs, conical shaft end possible, broaching or grinding necessary
			3-polygon-shaft		4		$\dfrac{d_p}{D}$	med-ium	—		self-centering	no	clearance fit without load		yes for taper	10-100		medium		—		
			4-polygon-shaft		5		d_p	med-ium	—				—			10-100				—		
	Indirect		transverse pin	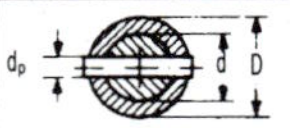	6	small	$\dfrac{d_p}{D}$	large / —	—		yes	yes	—	no	yes for taper pin	0.5-50	pin: 4D, 5S, 6S, 8G, 9S, 20K	medium	medium	1.7, 1470-77, 1481, 6324, 7346	power stretcher, machine tools, vehicles	taper and grooved pin possible
			tangential pin		7		d_p	—	—			yes	—				St 50K, St 70, St 60					
			in line pin		8		$h - b$	large	—				clearance fit		no	5-500						
			key joint		9			—	—								spring St 60		small	6885		
			Woodruff key	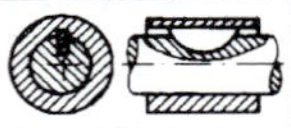	10			—	—								shaft,hub: GG, GS St			6888		

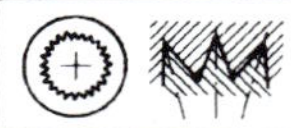

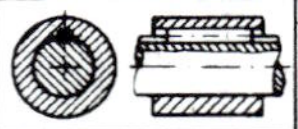

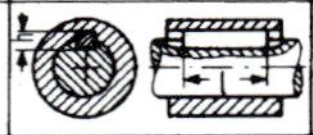

Figure 4.17.　Extract of a catalogue for shaft-hub connections, after [4.40].

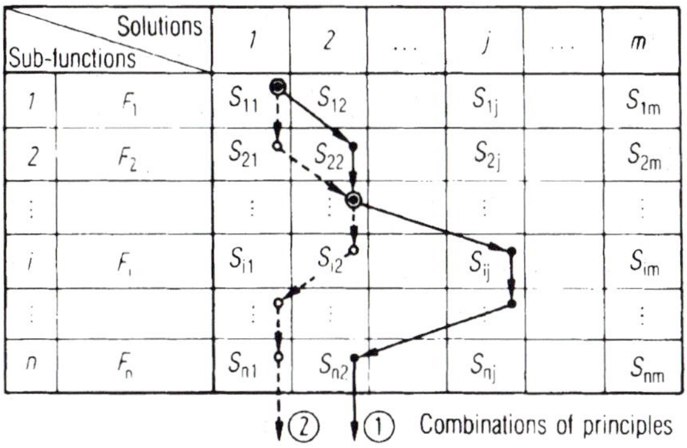

Solutions Sub-functions		1	2	...	j	...	m
1	F_1	S_{11}	S_{12}		S_{1j}		S_{1m}
2	F_2	S_{21}	S_{22}		S_{2j}		S_{2m}
⋮		⋮	⋮		⋮		⋮
i	F_i	S_{i1}	S_{i2}		S_{ij}		S_{im}
⋮		⋮	⋮				⋮
n	F_n	S_{n1}	S_{n2}		S_{nj}		S_{nm}

② ① Combinations of principles

Figure 4.18. Combining solution principles into combinations of principles: Combination 1: $S_{11} + S_{22} + \cdots S_{n2}$; Combination 2: $S_{11} + S_{21} + \cdots S_{n1}$.

useful. Here, the sub-functions, usually limited to the main functions, and the appropriate solutions (solution principles) are entered in the rows of the scheme.

If this scheme is to be used for the elaboration of overall solutions, then at least one solution principle must be chosen for every sub-function (that is, for every row). To provide the overall solution, these principles (sub-solutions) must then be combined systematically into an overall solution. If there are m_1 solution principles for the sub-function F_1, m_2 for the sub-function F_2 and so on, then after complete combination we have $N = m_1 \cdot m_2 \cdot m_3 \cdots m_n$ theoretically possible overall solution variants.

The main problem with this method of combination is to decide which solution principles are compatible, that is, to narrow down the theoretically possible search field to the practically possible search field.

The identification of compatible sub-solutions is facilitated if:

- the sub-functions are listed in the order in which they occur in the function structure, if necessary separated according to flow of energy, material and signals;
- the solution principles are suitably arranged with the help of additional column parameters, for example the type of energy;
- the solution principles are not merely expressed in words but also in rough sketches; and
- the most important characteristics and properties of the solution principles are recorded as well.

The verification of compatibilities, too, is facilitated by classification schemes. If two sub-functions to be combined—for instance, "change energy" and "vary mechanical energy component"—are entered respectively in the column and row heading of a matrix with their characteristics in the appropriate cells, then the compatibility of the sub-solutions can be verified more easily than it could be were such examinations confined to the designer's head. Figure 4.19 illustrates this type of compatibility matrix.

Further examples of this method of combination will be found in 6.4.2 (Figs. 6.18 and 6.21).

Change energy / Vary mechan. energy component		Electric motor	Oscillating solenoid	Bimetal spiral in hot water	Oscillating hydraulic piston	
		1	2	3	4	
Four-bar linkage	A	if A capable of rotating	slow motion	yes	additional lever linkage but only for low piston speeds	
Chain drive Spur gear drive	B	yes	slow rotation only through additional elements (free wheeling etc.) difficult to reverse direction	Gear segments suffice, depending on angle of rotation	with a rack and swivel, but only for low piston speeds	
Maltese drive	C	yes look out for shock loads	see B2	yes (when angle of rotation is small, lever with sliding block)	Lever with sliding block, but only for low piston speeds	
Friction wheel drive	D	yes	see B2	Large forces because of torque during slow movement, imprecise positioning	see D3	

⊠ very difficult to apply (do not pursue further)

☒ can only be applied under certain circumstances (defer)

Figure 4.19. Compatability matrix for combination possibilities of the sub-functions "change energy" and "vary mechanical energy component", from [4.6].

To sum up:

- Only combine compatible sub-functions.
- Only pursue such solutions as meet the demands of the requirements list and look like falling within the proposed budget (see selection procedures in 4.2.1).
- Concentrate on promising combinations and establish why these should be preferred above the rest.

In conclusion, it must be emphasised that what we have been discussing is a generally valid method of combining sub-solutions into overall solutions. The method can be used for the combination of working principles during the conceptual phase, and of sub-solutions or even of components and assemblies during the embodiment phase. Because it is essentially a method of information processing, it is not confined to technical problems but can also be used in the development of management systems and in other fields.

2 Combining with the Help of Mathematical Methods

Mathematical methods and computers should only be used for the combination of solution principles if real advantages can be expected from them. Thus, at the relatively abstract conceptual phase, when the nature of the solution is not yet fully understood, a quantitative elaboration—that is, a mathematical combination along with an optimisation—is quite out of place and can be misleading. The exceptions are combinations of known elements and assemblies, for instance in variant or circuit design. In the case of purely logical functions, combinations can be performed with the help of Boolean algebra [4.11, 4.37] in, say, the layout of safety systems or the optimisation of electronic or hydraulic circuits.

In principle, the combination of sub-solutions into overall solutions with the help of mathematical methods calls for knowledge of the characteristics or properties of the sub-solutions that are expected to correspond with the relevant properties of the neighbouring sub-solutions. These properties must be unambiguous and quantifiable. In the formation of principle solutions (for example working structures), data about the physical relationships may be insufficient, since the geometrical relationships may have a limiting effect and hence may, in certain circumstances, lead to incompatibilities. In that case, physical equation and geometrical structure must first be matched mathematically, and this is not generally possible except for systems of low complexity. For systems of higher complexity, by contrast, such correlations often become ambiguous, so that designers must once again choose between variants. We may, accordingly, speak of dialogue systems in which the process of combination consists of mathematical and creative steps.

This makes it clear that though, with increasing physical realisation or embodiment of a solution, it becomes simpler to establish quantitative combination rules, the number of properties increases and with them the number of constraints and optimisation criteria, so that the mathematical effort required becomes very great.

4.2 Selection and Evaluation Methods

4.2.1 Selecting Solution Variants

For the systematic approach, the solution field should be as wide as possible. By paying regard to all possible classifying criteria and characteristics, designers are often led to a larger number of possible solutions. This profusion constitutes the strength and also the weakness of the systematic approach. The very great, theoretically admissible, but practically unattainable, number of solutions must be reduced at the earliest possible moment. On the other hand, care must be taken not to eliminate valuable working principles, because often it is only through their combination with others that an advantageous working structure will emerge. While there is no absolutely safe procedure, the use of a systematic and verifiable selection procedure greatly facilitates the choice of promising solutions from a wealth of proposals [4.32].

This *selection procedure* involves two steps, namely *elimination* and *preference*.

First, all totally unsuitable proposals are eliminated. If too many possible solutions still remain, those that are patently better than the rest must be given preference. Only these solutions are further elaborated and evaluated.

If faced with a large number of solution proposals, the designer should compile a *selection chart* (see Fig. 4.20). In principle, after every step—that is, even after establishing function structures—only such solution proposals should pursued as:

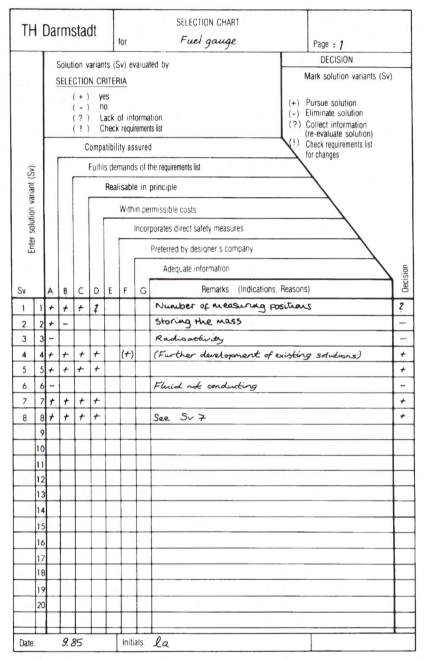

Sv		A	B	C	D	E	F	G	Remarks (Indications, Reasons)	Decision
1	1	+	+	+	?				Number of measuring positions	?
2	2	+	–						Storing the mass	–
3	3	–							Radioactivity	–
4	4	+	+	+	+		(+)		(Further development of existing solutions)	+
5	5	+	+	+	+					+
6	6	–							Fluid not conducting	–
7	7	+	+	+	+					+
8	8	+	+	+	+				See Sv 7	+
	9									
	10									
	11									
	12									
	13									
	14									
	15									
	16									
	17									
	18									
	19									
	20									

Date: 9.85   Initials: la

Figure 4.20. Systematic selection chart: 1, 2, 3, etc are solution variants of the proposals made in Table 4.2. The column reserved for remarks lists reasons for lack of information or elimination.

- are compatible with the overall task and with one another (Criterion A);
- fulfil the demands of the requirements list (Criterion B);
- are realisable in respect of performance, layout etc (Criterion C); and
- are expected to be within permissible costs (Criterion D).

Unsuitable solutions are eliminated in accordance with these four criteria applied in the correct sequence. Criteria A and B are suitable for yes/no decisions and their application poses relatively few problems. Criteria C and D often need a more quantitative approach, which should only be used once criteria A and B have been satisfied.

Since criteria C and D involve quantitative considerations, they may lead not only to the elimination of proposed solutions with too small an effect or too high a cost, but also to preferences based on large effects, small space requirements and low cost.

A preference is justified if, among the very large number of possible solutions, there are some that:

- incorporate direct safety measures or introduce favourable ergonomic conditions (Criterion E); or
- are preferred by the designer's company, that is, can be readily developed with the available know-how, materials, and procedures, and under favourable patent conditions (Criterion F). Additional selection criteria can be used if they are helpful in coming to a decision.

It must be stressed that selection based on preferential criteria is only advisable when there are so many variants that a full evaluation would involve too much time and effort.

If, in the suggested sequence, one criterion leads to the elimination of a proposal, then the other criteria need not be applied to it there and then. At first, only such solution variants should be pursued as satisfy all the criteria. Sometimes, however, it is impossible to settle the issue because of lack of information. In the case of promising variants that satisfy criteria A and B, the gap will have to be filled by a re-evaluation of the proposal, which will ensure that no good solutions are passed over.

The criteria are listed in the order shown above as a labour-saving device, and not in order of importance.

The selection procedure has been systematised for easier implementation and verification (see Fig. 4.20). Here, the criteria are applied in sequence and the reasons for eliminating any solution proposal are recorded. Experience has shown that the selection procedure we have described can be applied very quickly, that it gives a good picture of the reasons for selection, and that it provides suitable documentation in the form of the selection chart.

If the number of solution proposals is small, elimination may be based on the same criteria, but less formally recorded.

The example we have chosen concerns solution proposals for a fuel gauge in accordance with the requirements in Fig. 6.2. An extract from the list of proposals is given in Table 4.2.

A further example of a selection chart can be found in 6.4.3 (see Fig. 6.22).

Table 4.2. Extract from a list of solutions for a fuel gauge

No.	Solution principle	Signal
	1. *Measuring the quantity of fluid*	
	1.1. *Mechanical, static*	
1.	Fix container at three points. Measure vertical forces (weight). (Measuring at one support may be sufficient)	Force
2.	Mutual attraction. The force is proportional to the masses and therefore to the fluid mass	Force
	1.2. *Atomic*	
3.	Distribution of radioactive material in the fluid	Concentration of radiation intensity
	2. *Measuring the fluid level*	
	2.1. *Mechanical, static*	
4.	Float with or without lever effect. Lever output: linear or angular displacement. Potentiometer resistance represents fluid level within the container	Displacement
	2.2. *Electrical*	
5.	Resistance wire: hot in air, cold in fluid. Level of fluid determines: overall resistance, volume (dependent on temperature and length of wire)	Ohmic resistance
6.	Fluid as ohmic resistance (level dependent). Changing the level of the (conducting) fluid changes the resistance	Ohmic resistance
	2.3. *Optical*	
7.	Photocells in the container. Fluid covers more or less photocells. The number of light signals is a measure of the fluid level	Light signal (discrete)
8.	Light transmission or light reflection. Transmission in presence of fluid. Total reflection in presence of air	Light signal (discrete)

4.2.2 Evaluating Solution Variants

The promising solutions that result from the selection procedure have usually to be firmed up before a final evaluation using criteria that are more detailed and possibly quantified. This evaluation involves an assessment of technical, safety, environmental and economic values. For this purpose evaluation procedures have been developed that can be used to evaluate technical and non-technical systems, and can be applied in all phases of product development.

Evaluation procedures are by their very nature more elaborate than selection procedures and are therefore only applied at the end of the main working steps to determine the current value of a solution. This occurs, in general, when preparing for a fundamental decision concerning the direction of the solution path, or at the end of the conceptual and embodiment phases [4.39].

1 Basic Principles

An *evaluation* is meant to determine the "value", "usefulness" or "strength" of a solution with respect to a given objective. An objective is indispensable since the value of a solution is not absolute, but must be gauged in terms of certain requirements. An evaluation involves a comparison of concept variants or, in the case of a comparison with an imaginary ideal solution, a "rating" or degree of approximation to that ideal.

The evaluation should not be based on individual aspects such as production cost, safety, ergonomics or environment, but should in accordance with the overall aim (see 2.1.8) consider all aspects in the appropriate balance.

Hence there is a need for methods that allow a more comprehensive evaluation, or in other words cover a broad spectrum of objectives (task-specific requirements and general constraints). These methods are intended to elaborate not only the quantitative but also the qualitative properties of the variants, thus making it possible to apply them during the conceptual phase, with its low level of embodiment and correspondingly low state of information. The results must be reliable, cost-effective, easily understood and reproducible. The most important methods to date are Use-Value Analysis* (UVA) based on the systems approach [4.53], and the combined technical and economic evaluation technique specified in Guideline VDI 2225 [4.48], which essentially originates from Kesselring [4.21].

In what follows, we shall outline a basic evaluation procedure incorporating the concepts of use-value analysis and of Guideline VDI 2225. At the end the similarities and differences between both methods will be discussed.

Identifying Evaluation Criteria

The first step in any evaluation is the drawing up of a set of objectives from which evaluation criteria can be derived. In the technical field, such objectives are mainly derived from the requirements list and from general constraints (see guidelines in 2.1.8), which are identified while working on a particular solution.

A set of objectives usually comprises several elements that not only introduce a variety of technical, economic and safety factors, but that also differ greatly in importance.

A range of objectives should satisfy as far as possible the following conditions:

- The objectives must cover the decision-relevant requirements and general constraints as *completely* as possible, so that no essential criteria are ignored.
- The individual objectives on which the evaluation must be based should be as *independent* of one another as possible, that is, provisions to increase the value of one variant with respect to one objective must not influence its values with respect to the other objectives.

---

* The term "use-value analysis" is a direct translation from the original German text. A similar, but more general, term for this type of technique is " cost-benefit analysis".

- The properties of the system to be evaluated must, if possible, be expressed in concrete *quantitative* or at least *qualitative* (verbal) terms.

The tabulation of such objectives depends very much on the purpose of the particular evaluation—that is, on the design phase and the relative novelty of the product.

Evaluation criteria can be derived directly from the objectives. Because of the subsequent assignment of values, all criteria must be given a positive formulation, for example:

"low noise" not "loudness level",
"high efficiency" not "magnitude of losses",
"low maintenance" not "maintenance requirements".

Use-value analysis systematises this step by means of an objectives tree, in which the individual objectives are arranged in hierarchical order. The sub-objectives are arranged vertically into levels of decreasing complexity, and horizontally into objective areas—for instance, technical, economic—or even into major and minor objectives (see Fig. 4.21). Because of their required independence, sub-objectives of a higher level may only be connected with an objective of the next lower level. This hierarchical order helps designers to determine whether or not all decision-relevant sub-objectives have been covered. Moreover, it simplifies the assessment of the relative importance of the sub-objectives. The evaluation criteria (called objective criteria in use-value analysis) can then be derived from the sub-objectives of the stage with the lowest complexity.

Guideline VDI 2225, on the contrary, introduces no hierarchical order for the evaluation criteria, but derives a list of them from minimum demands and wishes and also from general technical properties.

Weighting Evaluation Criteria

To establish evaluation criteria, we must first assess their relative contribution (weighting) to the overall value of the solution, so that relatively unimportant

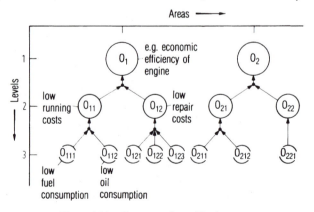

Figure 4.21.   Structure of an objectives tree.

criteria can be eliminated before the evaluation proper begins. The evaluation criteria retained are given "weighting factors" which must be taken into consideration during the subsequent evaluation step. A weighting factor is a real, positive number. It indicates the relative importance of a particular evaluation criterion (objective).

It has been suggested that such weightings should also be assigned to the wishes recorded in the requirements list [4.40, 4.42], but that is only possible if such wishes can be ranked in order of importance when the requirements list is first drawn up. That, however, rarely happens at this early stage—experience has shown that many evaluation criteria emerge during the development of a solution, and that their relative importance changes. It is nevertheless most helpful to include rough estimates of the importance of wishes when drawing up the requirements list, because, as a rule, all the persons concerned are available at that time (see 5.2.2).

In use-value analysis, weightings are based on factors ranging from 0 to 1 (or from 0 to 100). The sum of the factors of all evaluation criteria (sub-objectives at the lowest stage) must be equal to 1 (or 100) so that a percentage weighting can be attached to all the sub-objectives. The drawing up of an objectives tree greatly facilitates this process. Figure 4.22 illustrates the procedure. Here the objectives have been set out on four levels of decreasing complexity and provided with weighting factors. The evaluation proceeds step-by-step from a level of higher complexity to the next lower level. Thus the three sub-objectives O_{11}, O_{12} and O_{13} of the second level are first weighted with respect to the objective O_1 (in this particular case the weightings are 0.5, 0.25, and 0.25). The sum of the weighting factors for any one level must always be equal to $\Sigma w_i = 1.0$. Next comes the weighting of the objectives of the third level with respect to the sub-objectives of the second level. The relative weightings of O_{111} and O_{112} with respect to the higher objective O_{11} were fixed at 0.67 and 0.33. The remaining objectives are treated in similar fashion. The relative weighting of an objective at a particular level with respect to the

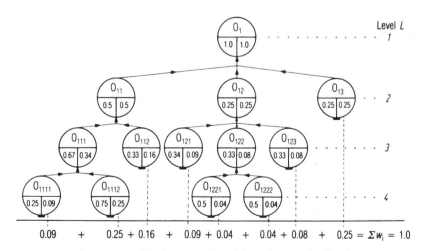

Figure 4.22.   Objectives tree with weighting factors, after [4.53].

objective O_1 is found by multiplication of the weighting factor of the given objective level by the weighting factors of the higher objective levels. Thus the sub-objective O_{1111}, which has a weighting of 0.25 with respect to the sub-objective O_{111} of the next higher level, has a weighting of $0.25 \times 0.67 \times 0.5 \times 1 = 0.09$ with respect to O_1.

Such step-by-step weighting generally produces a realistic ranking because it is much easier to weight two or three sub-objectives with respect to an objective on a higher level than to confine the weighting to one particular level only, especially the lowest. Figure 6.33 gives a concrete example of the recommended procedure.

Guideline VDI 2225 tries to dispense with weightings and relies instead on evaluation criteria of approximately equal importance. Weighting factors (2×, 3×) are, however, used for pronounced differences. Kesselring [4.21], Lowka [4.26] and Stahl [4.46] have examined the influences of such weighting factors on the overall value of a solution. Their conclusion was that they exert a significant influence whenever the variants to be evaluated have very distinct properties, and whenever the corresponding evaluation criteria have a great importance.

Compiling Parameters

The setting up of evaluation criteria and the determination of their importance is followed, as a next step, by the assignment to them of known (or analytically determined) parameters. These parameters should either be quantifiable or, if that is impossible, be expressed by statements framed as concretely as possible. It has proved very useful to assign such parameters to the evaluation criteria in an evaluation chart before proceeding to the actual evaluation. Figure 4.23 shows an example of such a chart for an internal combustion engine, with appropriate magnitudes entered in the relevant variant columns. The reader will see that the verbal formulation of the evaluation criteria strongly resembles that of the parameters.

In use-value analysis these parameters are referred to as objective parameters (objective criteria) that are compiled with evaluation criteria in a parameter matrix. A concrete example is given in Fig. 6.34.

In Guideline VDI 2225, by contrast, evaluation follows immediately upon the setting up of evaluation criteria (see Fig. 6.51).

Assessing Values

The next step is the assessment of values and hence the actual evaluation. These "values" derive from a consideration of the relative scale of the previously determined parameters, and are thus more or less subjective in character.

The values are expressed by points. Use-value analysis employs a range from 0 to 10; Guideline VDI 2225 a range from 0 to 4 (see Fig. 4.24). The advantage of the wider range is that, as experience has shown, classification and evaluation are greatly facilitated by the use of a decimal system and percentages. The advantage of the smaller range is that, in dealing with what are so often no more than inad-

No.	Evaluation criteria	Wt.	Objective Parameters	Unit	Variant V_1 (e.g. $Eng._1$)			Variant V_2 (e.g. $Eng._2$)			...	Variant V_j			...	Variant V_m		
					Magn. m_{i1}	Value v_{i1}	Weighted value wv_{i1}	Magn. m_{i2}	Value v_{i2}	Weighted value wv_{i2}		Magn. m_{ij}	Value v_{ij}	Weighted value wv_{ij}		Magn. m_{im}	Value v_{im}	Weighted value wv_{im}
1	Low fuel consumption	0.3	Fuel consumption	$\frac{g}{kWh}$	240			300			...	m_{1j}			...	m_{1m}		
2	Lightweight construction	0.15	Mass per unit power	$\frac{kg}{kW}$	1.7			2.7			...	m_{2j}			...	m_{2m}		
3	Simple production	0.1	Simplicity of components	-	low			average			...	m_{3j}			...	m_{3m}		
4	Long service life	0.2	Service life	km	80 000			150 000			...	m_{4j}			...	m_{4m}		
...	...	...	...	...	...			...				...				...		
i	...	w_i	...	...	m_{i1}			m_{i2}			...	m_{ij}			...	m_{im}		
...	...	...	...	...	...			...				...				...		
n	...	w_n	...	...	m_{n1}			m_{n2}			...	m_{nj}			...	m_{nm}		
		$\sum\limits_{i=1}^{n} w_i = 1$																

Figure 4.23.   Correlation of evaluation criteria and parameters in an evaluation chart.

Value scale			
Use-value analysis		Guideline VDI 2225	
Pts.	Meaning	Pts.	Meaning
0	absolutely useless solution	0	unsatisfactory
1	very inadequate solution		
2	weak solution	1	just tolerable
3	tolerable solution		
4	adequate solution	2	adequate
5	satisfactory solution		
6	good solution with few drawbacks	3	good
7	good solution		
8	v. good solution	4	very good (ideal)
9	solution exceeding the requirement		
10	ideal solution		

Figure 4.24.   Points awarded in use-value analysis and guideline VDI 2225.

equately known characteristics of the variants, rough evaluations are sufficient and, indeed, may be the only meaningful approach. They involve the following assessments:

- far below average
- below average
- average
- above average
- far above average.

It is useful to begin with a search of variants with extremely good and bad qualities and to assign appropriate points to them. Points 0 and 4 (or 10) should only be awarded if the characteristics are really extreme—that is, unsatisfactory or very good (ideal). Once these extreme points have been assigned, the remaining variants are relatively easy to fit in.

Before points can be assigned to the parameters of the variants, the evaluator must at least be clear about the assessment range and the shape of the so-called "value function" (see Fig. 4.25). A value function connects values and parameter magnitudes, and its characteristic shape is determined either with the help of the known mathematical relationship between the value and the parameter or, more frequently, by means of estimates [4.17].

It is useful to draw up a chart in which the parameter magnitudes are correlated step-by-step with the value scale. Figure 4.26 shows such a scheme, incorporating the point system of use-value analysis and also of VDI 2225.

All in all, therefore, the assignment of a value, the selection of a value function

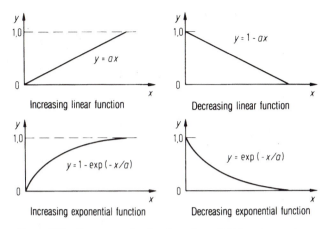

Figure 4.25. Common value functions, from [4.53]; $x \triangleq m_{ij}$, $y \triangleq v_{ij}$.

Value scale		Parameter magnitudes			
Use-value analysis Pts	VDI 2225 Pts	Fuel consumption g/kWh	Mass per unit power kg/kW	Simplicity of components -	Service life km
0	0	400	3,5	extremely complicated	$20 \cdot 10^3$
1	0	380	3,3	extremely complicated	30
2	1	360	3,1	complicated	40
3	1	340	2,9	complicated	60
4	2	320	2,7	average	80
5	2	300	2,5	average	100
6	3	280	2,3	simple	120
7	3	260	2,1	simple	140
8	4	240	1,9	extremely simple	200
9	4	220	1,7	extremely simple	300
10	4	200	1,5	extremely simple	$500 \cdot 10^3$

Figure 4.26. Chart correlating parameter magnitudes with value scales.

and the setting up of an assessment scheme may involve strong subjective influences. Cases with a clear, or even experimentally verified, correlation between the values and the parameters are few and far between. One such exception is the evaluation of machine noise, where the correlation between the value (that is, the protection of the human ear) and the parameter (noise level in dB) is clearly defined by ergonomics.

The values v_{ij} of every solution variant established in respect of every evaluation criterion are added to the list shown in Fig. 4.23 (see also Fig. 4.27).

Whenever the evaluation criteria are of different importance to the overall value of a solution, the weighting factors determined during the second step must also be taken into consideration. To that end, sub-value v_{ij} is multiplied by the weight-

Evaluation criteria			Parameters		Variant V_1 (e.g. $Eng._1$)			Variant V_2 (e.g. $Eng._2$)			...	Variant V_j			...	Variant V_m		
No.		Wt.		Unit	Magn. m_{i1}	Value v_{i1}	Weighted value wv_{i1}	Magn. m_{i2}	Value v_{i2}	Weighted value wv_{i2}		Magn. m_{ij}	Value v_{ij}	Weighted value wv_{ij}		Magn. m_{im}	Value v_{im}	Weighted value wv_{im}
1	Low fuel consumption	0.3	Fuel consumption	$\frac{g}{kWh}$	240	8	2.4	300	5	1.5	...	m_{1j}	v_{1j}	wv_{1j}	...	m_{1m}	v_{1m}	wv_{1m}
2	Lightweight construction	0.15	Mass per unit power	$\frac{kg}{kW}$	1.7	9	1.35	2.7	4	0.6	...	m_{2j}	v_{2j}	wv_{2j}	...	m_{2m}	v_{2m}	wv_{2m}
3	Simple production	0.1	Simplicity of components	-	compli-cated	2	0.2	average	5	0.5	...	m_{3j}	v_{3j}	wv_{3j}	...	m_{3m}	v_{3m}	wv_{3m}
4	Long service life	0.2	Service life	km	80 000	4	0.8	150 000	7	1.4	...	m_{4j}	v_{4j}	wv_{4j}	...	m_{4m}	v_{4m}	wv_{4m}
⋮	⋮	⋮	⋮	⋮	⋮	⋮	⋮	⋮	⋮	⋮		⋮	⋮	⋮		⋮	⋮	⋮
i		w_i			m_{i1}	v_{i1}	wv_{i1}	m_{i2}	v_{i2}	wv_{i2}	...	m_{ij}	v_{ij}	wv_{ij}	...	m_{im}	v_{im}	wv_{im}
⋮	⋮	⋮	⋮	⋮	⋮	⋮	⋮	⋮	⋮	⋮		⋮	⋮	⋮		⋮	⋮	⋮
n		w_n			m_{n1}	v_{n1}	wv_{n1}	m_{n2}	v_{n2}	wv_{n2}	...	m_{nj}	v_{nj}	wv_{nj}	...	m_{nm}	v_{nm}	wv_{nm}
		$\sum_{i=1}^{n} w_i = 1$				OV_1 R_1	OWV_1 WR_1		OV_2 R_2	OWV_2 WR_2			OV_j R_j	OWV_j WR_j			OV_m R_m	OWV_m WR_m

Figure 4.27. Evaluation chart completed with values (see Figure 4.23).

ing factor w_i ($wv_{ij} = w_i \cdot v_{ij}$). Figure 6.34 gives a practical example. The use-value analysis refers to the unweighted values as *objective values* and the weighted ones as *use values*.

Determining Overall Value

The sub-values for every variant having been determined, the overall value must now be calculated.

For the evaluation of technical products, the summation of sub-values has become the usual method of calculation but can only be considered accurate if the evaluation criteria are independent. However, even when this condition is only satisfied approximately, the assumption that the overall value has an additive structure seems to be justified.

The overall value of a variant j can then be determined.

Unweighted: $OV_j = \displaystyle\sum_{i=1}^{n} v_{ij}$

Weighted: $OWV_j = \displaystyle\sum_{i=1}^{n} w_i \cdot v_{ij} = \sum_{i=1}^{n} wv_{ij}$

Comparing Concept Variants

On the basis of the summation rule it is possible to assess variants in several ways.

Determining the maximum overall value: In this procedure that variant is judged best which has the maximum *overall value:*

$$OV_j \rightarrow \max \qquad \text{or} \qquad OWV_j \rightarrow \max$$

What we have here is a relative comparison of the variants. This fact is made use of in use-value analysis.

Determining the Rating: If a relative comparison of the variants is considered to be insufficient and the absolute *rating* of a variant has to be established, then the overall value must be referred to an imaginary ideal value which results from the maximum possible value.

Unweighted: $R_j = \dfrac{OV_j}{v_{max} \cdot n} = \dfrac{\displaystyle\sum_{i=1}^{n} v_{ij}}{v_{max} \cdot n}$

Weighted: $WR_j = \dfrac{OWV_j}{v_{max} \cdot \displaystyle\sum_{i=1}^{n} w_i} = \dfrac{\displaystyle\sum_{i=1}^{n} w_i \cdot v_{ij}}{v_{max} \cdot \displaystyle\sum_{i=1}^{n} w_i}$

If the available information about the properties of all the concept variants allows cost estimates, then it is advisable to proceed to a separate determination of the *technical rating* R_t and the *economic rating* R_e. The technical rating is calculated in accordance with the rule we have given—that is, by division of the technical overall value of the given variant by the ideal value—and the economic rating is calculated similarly, but by reference to comparative costs. The latter procedure is suggested in VDI 2225, which relates the manufacturing costs determined for a variant to the comparative manufacturing costs C_o. In that case, the economic rating becomes $R_e = (C_o/C_{variant})$. It is possible to put, say, $C_o = 0.7 \times C_{admissible}$ or $C_o = 0.7 \times C_{minimum}$ of the cheapest variant. If the technical and economic ratings have been determined separately, then the determination of the "overall rating" of a particular variant may prove interesting. For that purpose, Guideline VDI 2225 suggests a so-called s-diagram (strength diagram) with the technical rating R_t as the abscissa and the economic rating R_e as the ordinate (see Fig. 4.28). Such diagrams are particularly useful in the appraisal of variants during further developments, because they show up the effects of design decisions very clearly.

In some cases it is useful to derive the overall rating from these partial ratings and to express it in numerical form, for instance for computer processing. To that end, Baatz [4.1] has proposed two procedures, namely:

- the straight line method, based on the arithmetic mean:

$$R = \frac{R_t + R_e}{2}$$

and

- the hyperbolic method which involves multiplying both ratings and then reducing to values between 0 and 1:

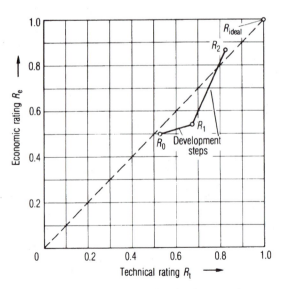

Figure 4.28.   Rating diagram, after [4.21, 4.48].

$$R = \sqrt{R_t \times R_e}$$

The two methods have been combined in Fig. 4.29.

Where there are great differences between the technical and economic ratings, the straight-line method can be used to compute a higher overall rating than is the case with low but balanced partial ratings. Because balanced solutions should be preferred, however, the hyperbolic method is the better of the two; it helps to balance great differences in rating by its progressive reduction effect. The greater the imbalance, the greater the reduction effect on the lower overall values.

Rough Comparison of Solution Variants: The method we have described relies on differentiated value scales. It is useful whenever the "objective" parameters can be stated with some accuracy and whenever clear values can be assigned to them. If these conditions cannot be satisfied, relatively fine evaluations based on a differentiated value scale constitute a questionable and expensive method. The alternative here is a rough evaluation involving the application of a particular evaluation criterion to two variants at a time and the selection of the better in every case. The results are entered in a so-called dominance matrix [4.10] (see Fig. 4.30). From the sum of the columns it is possible to establish a ranking order. If such matrices of individual criteria are combined into an overall matrix, an overall ranking order can be established, either by addition of the preference frequencies or by addition of all the column sums.

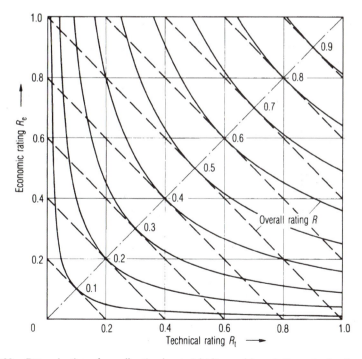

Figure 4.29.   Determination of overall rating by straight-line and hyperbolic methods, after [4.1].

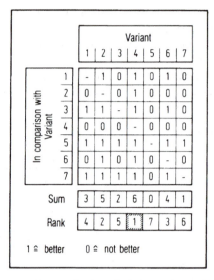

Figure 4.30.   Binary evaluation of solution variants, after [4.10].

While this method is comparatively easy and quick, it is not nearly as informative as the other procedures we have discussed.

Estimating Evaluation Uncertainties

The possible errors or uncertainties of the proposed evaluation methods fall into two main groups, namely subjective errors and inherent shortcomings of the procedure.

Subjective errors can arise through:

- Abandonment of the neutral position, that is, through bias and partiality. The bias may be hidden from designers, for instance when they compare their own designs with those of rivals. Hence an evaluation by *several persons*, if possible from various departments, is always advisable. It is equally important to refer to the different variants in *neutral terms*, for instance as A, B, C, rather than as "Smith's Proposal" etc, since otherwise unnecessary identifications and emotional overtones may be introduced. Extensive systematisation of the procedure also helps to reduce subjective influences.
- Comparison of variants by the application of the same evaluation criteria not equally suited to all the variants. Such mistakes arise even during the determination of the parameters and their association with the evaluation criteria. If it is impossible to determine the parameter magnitudes of individual variants for certain evaluation criteria, then these criteria must be *reformulated* or *dropped* lest they lead to mistaken evaluations of the individual variants.
- The evaluation of variants in isolation instead of successively by application of the established evaluation criteria. Each criterion must be applied to *all the*

variants in turn (row by row in the evaluation chart) to eliminate any bias in favour of a particular variant.

- Pronounced interdependence of the evaluation criteria.
- Choice of unsuitable value functions.
- Incompleteness of evaluation criteria. This defect can be minimised if one of the *checklists for design evaluation* appropriate to the relevant design phase is followed (see Fig. 6.29 and Fig. 7.147).

Procedure-inherent shortcomings of the recommended evaluation methods are the result of the almost inevitable "prognostic uncertainty" arising from the fact that the predicted parameter magnitudes and also the values are not precise, but subject to uncertainty and to random variation. These mistakes can be greatly reduced by estimates of the mean error.

With regard to prognostic uncertainty it is therefore advisable not to express the parameters in figures unless this can be done with some accuracy. Otherwise, it is preferable to use verbal estimates (for instance high, average, low) which do not claim to be precise. Numerical values, by contrast, are dangerous because they introduce a false sense of certainty.

A more detailed analysis of evaluation procedures for the purpose of judging their reliability and also for comparative purposes has been carried out by Feldmann [4.10] and Stabe [4.45]. The latter also provides an extensive bibliography. If there is an adequate number of evaluation criteria, and if the sub-values of a particular variant are fairly balanced, then the overall value will be subject to a balancing statistical effect, and partly too optimistic and partly too pessimistic individual values will more or less balance out.

Searching for Weak Spots

Weak spots can be identified from below-average values for individual evaluation criteria. Careful attention must be paid to them, particularly in the case of promising variants with good overall values, and they ought if possible to be eliminated during further development. The identification of weak spots may be facilitated by graphs of the sub-values—for instance, by the so-called value profiles illustrated in Fig. 4.31. The lengths of the bars correspond to the values and the thicknesses to the weightings. The areas of the bars then indicate the weighted sub-values, and the cross-hatched area the overall weighted value of a solution variant. It is clear that, in order to improve a solution, it is essential to improve those sub-values that provide a greater contribution to the overall value than the rest. This is the case with the evaluation criteria that have an above average bar thickness (great importance) but a below average bar length. Apart from a high overall value, it is important to obtain a *balanced value profile*, with no serious weak spots. Thus, in Fig. 4.31, variant 2 is better than variant 1, although both have the same overall weighted value.

There are also cases in which a minimum permissible value is stipulated for all sub-values—that is, any variant that does not fulfil this condition has to be rejected,

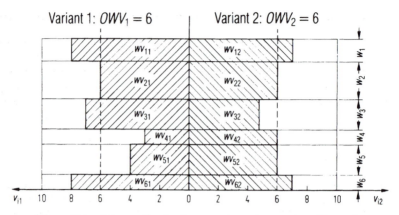

Figure 4.31.   Value profiles for the comparison of two variants ($\Sigma w_i = 1$).

and all variants that do are developed further. In the literature this procedure is described as the "determination of satisfactory solutions" [4.53].

2   Comparison of Evaluation Procedures

Table 4.3 lists the individual steps in the evaluation procedures we have described and also the similarities and differences between use-value analysis and Guideline VDI 2225, which are based on similar principles.

The individual steps of use-value analysis are more highly differentiated and more clear cut, but involve more work than those of Guideline VDI 2225. The latter is more suitable when there are relatively few and roughly equivalent evaluation criteria, which is frequently the case during the conceptual phase, and also for the evaluation of certain form design areas during the embodiment phase.

The essence of evaluation procedures has been described on the basis of existing evaluation methods. However, these methods have been extended and the terms clarified. Specific suggestions for the use of these methods during the conceptual phase are given in 6.5.2, and during the embodiment phase in 7.6.

Table 4.3. Individual steps in evaluation, and comparison between use-value analysis and Guideline VDI 2225

Step		Use–value analysis	VDI Guideline 2225
1	Identification of objectives or evaluation criteria for the evaluation of concept variants with the aid of the requirements list and a checklist	Construction of a hierarchically related system of design objectives (objectives tree) based on the requirements list and other general requirements	Compilation of important technical characteristics and also of the minimum demands and wishes of the requirements list.
2	Analysis of the evaluation criteria for the purpose of determining their weighting to the overall value of the solution. If necessary, determination of weighting factors	Step by step weighting of the objective criteria (evaluation criteria) and if necessary elimination of unimportant criteria	Determination of weighting factors only if evaluation criteria differ markedly in importance
3	Compilation of parameters applicable to the concept variants	Construction of an objective parameter matrix	Not generally included
4	Assessment of the parameter magnitudes and assignment of values (0–10 or 0–4 points)	Construction of objective value matrix with the help of a points system or value functions; 0–10 points	Assessment of characteristics by points (0–4 points)
5	Determination of the overall value of the individual concept variants, generally by reference to an ideal solution (rating)	Construction of a use–value matrix with due regard to the weightings; determination of overall values by summation	Determination of a technical rating by summation, with or without weightings based on an ideal solution. If necessary determination of an economic rating based on manufacturing costs
6	Comparison of concept variants	Comparison of overall use–values	Comparison of the technical and economic ratings. Construction of an s-(strength) diagram
7	Estimation of evaluation uncertainties	Estimation of objective parameter scatter and use–value distribution	Not explicitly included
8	Search for weak spots for the purpose of improving selected variants	Construction of use–value profiles	Identification of characteristics with a few points only

5 Product Planning and Clarifying the Task

Assignments are set not only by clients but increasingly, and especially in the case of original designs, they originate in the special planning departments of companies. In that case, designers are bound by the planning ideas of others. Even then, however, the special skills of designers will prove most useful in the medium- and long-term planning of products. The senior staff of the design department should therefore maintain close contacts not only with the production department, but also with the product planning department.

Planning can also be done by outside bodies, for instance, by clients, by authorities, by planning committees etc.

As discussed in 3.2 (see Fig. 3.3), the design process for original designs starts with conceptualisation based on a requirements list (design specification). This preliminary list is usually based on requirements identified by product planning. It is therefore important for designers to know the essential points and steps of the product planning process. This will help them to understand the origin of the requirements and if necessary to add to the list. If there has not been a formal product planning phase, designers can organise the relevant steps using their own knowledge about product planning, or can undertake this phase themselves using simpler procedures.

In this chapter, and therefore also in Fig. 3.3, product planning and clarifying the task are consciously combined into one main phase. This is to emphasise the importance of integrating both activities. This remains important even when product planning and clarifying the task are undertaken separately within an organisation.

5.1 Product Planning

5.1.1 Task and General Approach

Before a commercial product can be designed there has to be a product idea; that is, an idea that promises to lead to a technically and economically viable product. This is not necessary if the task comes directly from a client.

According to [5.12, 5.22], *product planning*, based on the company's goals, is the systematic search for, and the selection and development of, promising product ideas. In many companies, accordingly, the product planning department is expected to follow the development of a product idea through the design and manufacturing departments, and then to watch over its market behaviour. This includes monitoring the financial position and market success of the product and, if necessary, taking appropriate corrective measures (see Fig. 1.2). In this book we shall only be dealing with product planning in the narrower sense, that is as a preparation for product development.

Companies deal with product planning in different ways. In many cases, it is left to the good sense of a director, or an individual member of staff, to develop and introduce the right product idea at the right time. Nowadays, however, systematic procedures are frequently used to find new product ideas. An important aspect of this systematic approach is its potential to monitor more accurately the time and cost of product planning and development.

The basis and starting point for product planning is *marketing*. Marketing provides the interface with customers, that is to the market, and ensures that "the voice of the customer influences product planning and vice versa" [5.19]. Any company operating in a *consumer market* in which customers have choice has to survive competition, so the most important goal is to achieve, maintain, defend and renew competitive advantage. All disciplines within the company have to contribute to this. Marketing, alongside product planning, design, production and sales, monitors the fulfilment of customer requirements and acts as the agent of the customer within the company [5.19]. This customer focus means:

- to know and understand the customer;
- to listen to the customer; and
- to respond to the customer in an appropriate manner.

Marketing and product planning cannot really be separated, they are methodically and organisationally linked. In this chapter, we focus on product planning because it leads immediately to design and the steps in both influence one another.

Several systematic product planning approaches exist [5.2, 5.6, 5.9, 5.10, 5.12, 5.15, 5.22] and all of them have much in common (see Fig. 5.1).

The stimuli for product plans come from outside, that is, from the *market* or *other sources*, or from inside, that is, from the *company* itself. These stimuli are usually identified by marketing.

Stimuli from the *market* include:

- the technical and economic position of the company's products in the market, in particular when changes occur, such as a reduction in turnover or a drop in market share;
- changes in market requirements, for example new functions or fashions;
- suggestions and complaints of customers; and
- technical and economic superiority of competing products.

Stimuli from *other sources* include:

- economic and political changes, for example oil price increases, resource shortages, transport restrictions;

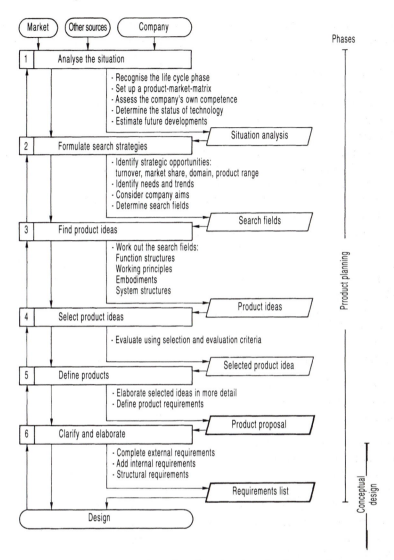

Figure 5.1.   Procedure of product planning, after [5.15, 5.22].

- new technologies and research results, for example micro-electronic replacing mechanical solutions or laser cutting replacing flame cutting; and
- environmental and recycling issues.

Stimuli from within the *company* include:

- new ideas and results of company research applied in development and production;
- new functions to extend or satisfy the market;
- introduction of new production methods;
- rationalisation of product range and production; and

- increasing degree of product diversification, that is a range of products with life cycles that are planned to overlap.

These external and internal stimuli initiate five main working steps that will be discussed in more detail in the following sections.

- *Analysing the situation* of the company and its products, using knowledge from the market and other sources. The result is a *situation analysis.*
- *Formulating search strategies* taking into account the goals, strengths and weaknesses of the company as well as market niches and needs. The result is a set of promising *search fields.*
- *Finding product ideas* by searching within each field for new functions, working principles or geometries based on existing or extended energy, material and signal flows. The result is a set of *product ideas.*
- *Selecting product ideas* using a selection procedure that takes into account the company's goals, strengths and market. The result is a *product idea.*
- *Defining products* by elaborating and evaluating product ideas, for example through the formulation of a first requirements list (see 5.2). The result is a *product proposal.*

These main working steps relate strongly to the general working methods described in 2.2.2 and more or less conform to systematic conceptual design (see 6 and Fig. 3.3).

5.1.2   Situation Analysis

The situation at the beginning of the product planning stage involves several aspects and these must be clarified through a number of investigations each with a different aim.

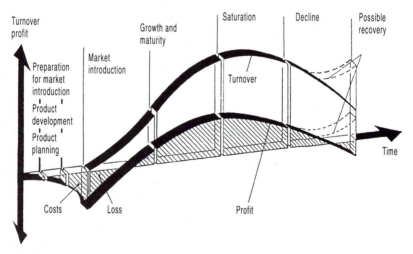

Figure 5.2.   Life cycle of a product, after [5.15].

Recognising the life cycle phase. Every product has a life cycle (see Fig. 1.2), which is illustrated in Fig. 5.2. This is based on an economic viewpoint showing turnover, as well as profit and loss. The *cycle time* depends strongly on the type of product and the branch of engineering, but in general cycles times are becoming shorter. When the saturation phase has been reached, at the latest, measures to reactivate the market or generate new products have to be introduced. The introduction of these measures is an important task of product monitoring. A related activity in this context is the development of *market share*.

A life cycle analysis can also be used to recognise the need for diversification, that is the phased development and sale of several different products. This will help to realise a balance of overlapping life cycles.

Setting up a product-market-matrix. Recognising and clarifying the status of existing products from the company and from competitors in the various markets (field I in Fig. 5.3) with respect to turnover, profit and market share, should reveal the strengths and weaknesses of each of the products. A comparison with strong competitors is of particular interest.

Assessing the company's own competence. This part of the analysis extends the

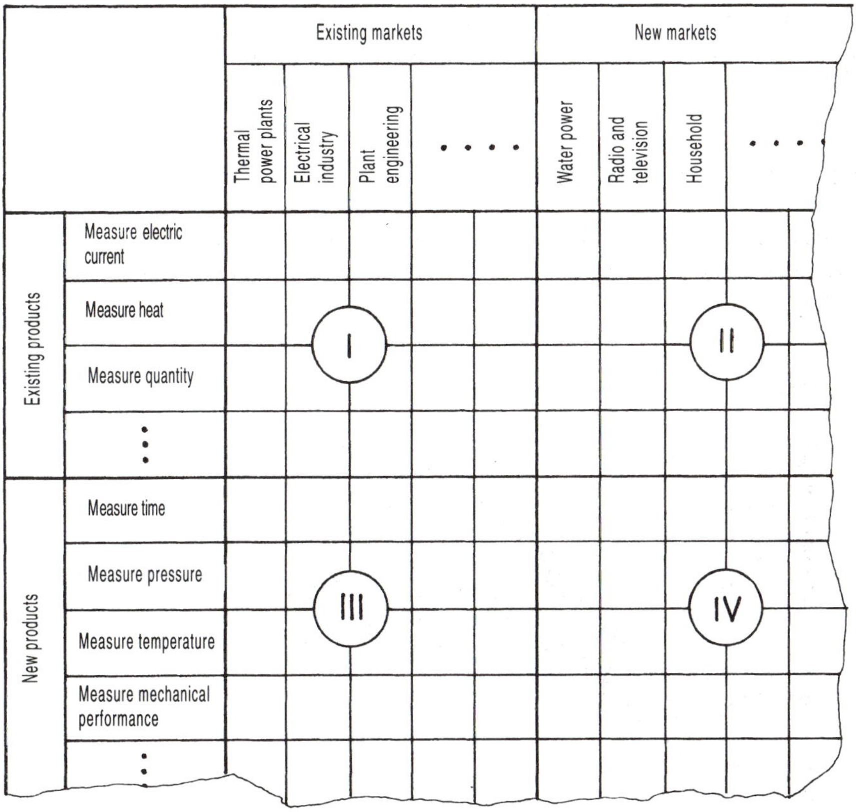

Figure 5.3. Product-market-matrix, after [5.4] and [5.12], for a company producing measuring devices for industry.

previous one and provides the reasons for the current market position through an assessment of the company's technical weaknesses and through a comparison with competing companies (Fig. 5.4). This analysis should not only be based on orders, because these already represent a selection which is profitable for the company, but also on customer enquiries and complaints, as well as installation and test reports.

Determining the status of technology. This includes reviewing the products of the company, related technologies, concepts and products in the literature and patents, as well as competitors' products. In addition the latest standards, guidelines and regulations are important.

Estimating future developments. Guidance can be obtained from knowledge of future projects, expected customer behaviour, technological trends, environmental requirements and the results of fundamental research.

Situation analysis determines the search strategies and the search fields that have to be addressed.

5.1.3   Formulating Search Strategies

Identifying strategic opportunities. It is possible that during the situation analysis some gaps in the current product range or in the market are identified. The task now is to determine which strategy to adopt: to introduce new products into the current market (field III in Fig. 5.3); to open new markets with existing products (field II); or even to enter into new markets with new products (field IV). The latter involves the highest risk.

Criteria	Competitors				
	A	B	C	D	.
Turnover					
Market share%					
Market situation					
– Conditions					
– Service					
– Delivery times					
–					
Product					
Management					
Product programme					

= same as us; + better; – worse

Figure 5.4.   Analysis of competing companies, after [5.14].

A promising gap, which will determine this search field [5.2, 5.9], has to be found taking into account the company's goals, strengths and market (see Table 5.1). Kramer [5.13] calls these strategic opportunities. They can relate to profit, market share, type of industry and product range. The weightings listed in Table 5.1 indicate that company goals are the most important criteria.

Identifying needs and trends. Most important for determining search fields is the identification of customer needs and market trends. Clues for this come from changing customer behaviour caused, for example, by social developments such as environmental awareness, disposal problems, reduction in working week, and transport problems. A further starting point can be changes in the manufacturing supply chain which can lead to new markets for suppliers. An often used tool is the *need-strength-matrix* [5.12] (see Fig. 5.5). In this matrix, one axis lists customer needs in decreasing order of importance, the other lists the strengths and competences of the company. The crossed fields in the top left corner of the matrix are the preferred search fields which are used in the preparation of the search field proposal. Another tool is a *client-problem analysis* [5.16]. The application of these tools does not automatically determine the most suitable search fields, but supports or extends the number of ideas, dreams and possibilities.

Considering company aims. Table 5.1 lists the aims and strengths of the company which have to be used to select a search field. The matrix in Fig. 5.5 also emphasises the importance of the strengths and competences of the company in the selection of a worthwhile search field.

Table 5.1.   Decision criteria for product planning

Criteria	Weighting
Company goals	≥50%
Adequate financial cover	
High turnover	
High market growth	
Large market share (market leader)	
Short-term market opportunity	
Large functional advantages for users and excellent quality	
Differentiation from competitors	
	≥30%
Company strengths	
Extensive know-how	
Favourable extension to range and/or product programme (diversification)	
Strong market position	
Limited need for investment	
Few sourcing problems	
Favourable rationalisation potential	
	≥20%
Market and other sources	
Low danger of substitution	
Weak competition	
Favourable patent status	
Few general restrictions	

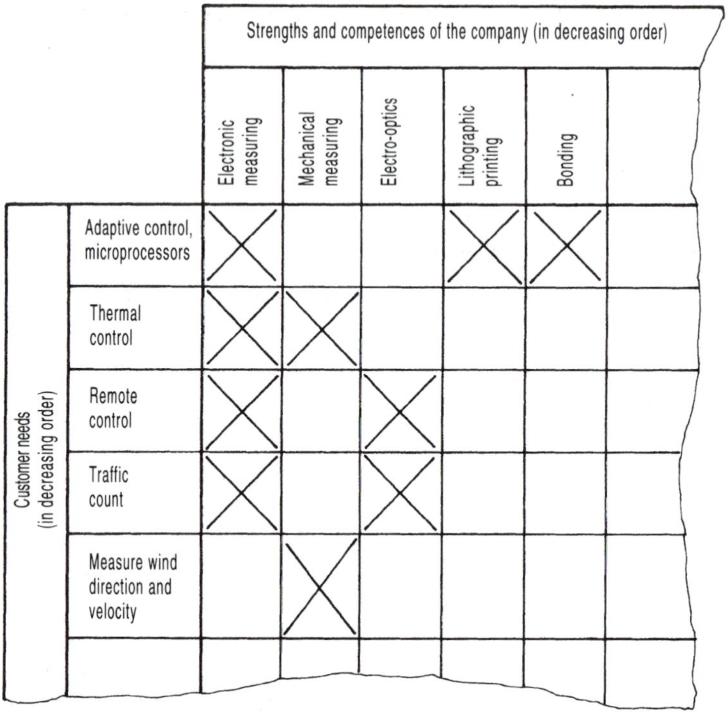

	Strengths and competences of the company (in decreasing order)						
Customer needs (in decreasing order)	Electronic measuring	Mechanical measuring	Electro-optics	Lithographic printing	Bonding		
Adaptive control, microprocessors	X			X	X		
Thermal control	X	X					
Remote control	X		X				
Traffic count	X		X				
Measure wind direction and velocity		X					

Figure 5.5.   Need-strength-matrix developed by a company, based on Figure 5.3 [5.12].

Determining search fields. The previously described steps of this product planning stage should lead, after a selection process, to a limited number of search fields on which to concentrate the further search for products (see [5.5] 3 through 5).

5.1.4   Finding Product Ideas

The preferred search fields are now investigated in more detail using known search methods such as those that are used in product development (see 4.1 and 6.4). These include: considering functions; intuitive methods such as Brainstorming (see the so-called idea finding workshops in [5.5]); discursive methods such as ordering schemes, morphological charts and systematic combination.

When working out the search fields, a directed search for product ideas can be encouraged by the general relationships in technical products and their particular level of concretisation (see 2.1). Depending on the degree of novelty, the starting point for new products can be new product functions; other working principles; new embodiments; and rearrangements of an existing or new system structure. For a company producing measuring instruments, for example (see Figs. 5.3 and 5.5), worthwhile product ideas can emerge from: new measuring functions; new physical effects (eg Laser effect) to fulfil known functions; or new embodiment goals (eg miniaturisation or improved ergonomics).

The considerations follow the known interrelationship between function, working principle and embodiment:

Function:

- Which function does the client require?
- Which functions do we already fulfil?
- What complements existing functions?
- Which functions represent a generalisation of the existing ones?

For example, until now we only transport unit loads overland.

- What can we do in future?
- Shall we also use boats?
- Shall we start transporting very large, heavy items?
- Shall we also transport bulk goods?
- Shall we try to solve transport problems in general?

Working principle:

Existing products are based on a specific working principle. Would a change of working principle lead to better products?
Characteristics to look for are the types of energy and physical effects.
For example, should a temperature-dependent flowrate controller be based on the principle of fluid expansion, the bi-metallic effect or use microprocessor-controlled temperature probes?

Embodiment:

- Is the space used still appropriate?
- Should we focus on miniaturisation?
- Is the shape still appealing?
- Could the ergonomics be improved?

For example, is it still appropriate to use laces in shoes? Would Velcro or hooks be more appealing and more comfortable?
The answers to these questions determine the novelty of the product ideas and thereby the development risks.

5.1.5   Selection of Product Ideas

The product ideas generated are first subjected to a selection procedure (see 4.2.1). For this initial selection, the criteria linked to the company's goals are sufficient (see Table 5.1). At the very least, high turnover, large market share and functional advantages for the customer should be taken into account. A more detailed selection involves the other criteria. To identify promising product ideas, it is often sufficient, in the sense of an efficient application of selection procedures, to work only with binary values (yes/no).

5.1.6   Product Definition

In this step product ideas that seem promising are elaborated more concretely and in more detail. It is useful to consider the headings of the requirements list used in product development (see 5.2). During this step at the latest, sales, marketing, research and development, and design should work actively together. This can be encouraged by involving these groups in the evaluation and selection of product ideas.

Product ideas, after elaboration, are then subjected to an evaluation in which all the criteria listed in Table 5.1, as far as they are known, are used.

Often some criteria, such as investment needs or sourcing problems, cannot be assessed because they are solution dependent. In these cases they will not be considered during this step. The best product definitions are given to the product development department as a *product proposal* together with a preliminary requirements list. The product development department then develops the actual product, using, for example, the systematic approach we propose.

The *product proposal* should:

- Describe the intended *functions.*
- Contain a *preliminary requirements list* that should have been compiled as far as possible using the headings that are used later on during product development to clarify the task and finalise the requirements list.
- Formulate all *requirements* in a *solution-neutral* way. The working principle should only be determined in so far as it is really necessary from the point of view of the overall functionality. For example the same working principle will be specified when an existing product range is being extended. Suggestions for working principles, however, should always be indicated; in particular, when suitable solution principles have emerged during the idea finding step. These should not prejudice product development (see also the solution-neutral formulation of requirements).
- Indicate a *cost target* or a *budget* linked to the company's goals which clarifies future intentions such as production volume, extensions to product range, new suppliers, etc.

This concludes the product planning phase. By using the listed decision criteria, only those proposals that are likely to fit the company's goals and strengths, and match the macro- and micro-economic situation, will enter the development stage. The development of the requirements list using the method applied in product development ensures an easy and seamless transition from product planning to product development.

For successful product planning and development, it is important that both groups work together using the same methods and similar evaluation criteria. At the latest, product development should be actively involved when product ideas are selected and the product is defined. Together they should also develop the requirements list in a format suitable for product development (see 5.2).

5.1.7   Product Planning in Practice

Because of strong competition, new products have to meet market needs closely; be produced at the lowest cost; and be economical to use. In addition, requirements have emerged that relate to disposal and recycling, and to low environmental impact during manufacture and use. Products with such complex requirements need to be *planned systematically* to meet these demands. Just relying on spontaneous ideas or incremental developments to existing products will not, in general, fulfil these demands. Systematic product planning often uses the same methods as concept development, and staff can usefully be exchanged between the two departments.

The following guidelines are also important:

- The size of the company determines whether or not it is possible to set up inter-disciplinary project groups or departments. In smaller companies it might be necessary to involve *external consultants* to supply expertise that is missing in the company.
- To use *company expertise*, however, can involve less risk and often increases client confidence.
- If product planning focuses on *existing product lines*, that is, further development or systematic variation, the development department responsible for the product line can monitor the new product, or this can be done by a special planning group that includes members from that department.
- When product planning takes place *outside an existing product line*, that is, the focus is on completely new products or diversification of the product programme, it is better to set up a new planning group. This group works on "innovative planning" and can either be set up as a permanent department or as a temporary working group.
- *More elaborate analysis* and conscious thought is required when planning for *new markets* than when dealing with known sales channels and existing client circles.
- When the *starting situation is complex*, it can be useful to undertake product planning and development using a stepwise and iterative approach. Acquisition of information and the decision making steps should be scheduled so that the anticipated effort and success can be reviewed and planned.
- Even when product ideas have been generated *intuitively*, a situation analysis and a feasibility study using the search strategies should still be done.
- To *identify customer problems*, it is useful to have an intensive collaboration with a few leading clients, referred to as "lead users" [5.5].
- *When new products are introduced*, technical failures and weaknesses can have a far reaching impact on the reputation of such products. Part of a careful product planning process, therefore, should include sufficient time for testing and the calculation of risks (see 7.5.12).
- *Entry into the market* later than announced can also have a negative effect on reputation because it suggests technical problems.
- During the planning and introduction of new products, it is useful to have a

powerful product champion, eg a Board member, who identifies personally with the new product. This helps *overcome a potential lack of interest and conventional resistance* [5.5].

- Finally, it has to be said that the *procedure shown in Fig. 5.1* does not represent a straight path with sequential steps, but a guideline for an essentially purposeful approach. The practical application of this approach will require an iterative procedure in which forward and backward steps on higher levels of abstraction are necessary. This is quite normal in successful product finding.

5.2   Clarifying the Task

5.2.1   The Importance of Clarifying the Task

The work of designers starts with a particular problem. Every task involves certain constraints that may change with time but must be fully understood if the optimum solution is to be found. Whether or not this phase has been preceded by product planning resulting in a preliminary requirements list, designers must still define the task as fully and clearly as possible so that amplifications and corrections during its subsequent elaboration can be confined to the most essential. To that end, and also as a basis for subsequent decisions, a requirements list (design specification) [5.17, 5.18] should always be drawn up and consulted. It is indispensable in the case of original designs, even when these address sub-tasks.

The task is generally presented to the design or development department in one of the following forms (see 1.1.1):

- as a development order (from outside or from the product planning department in the form of a product proposal);
- as a definite order; or
- as a request based on, for instance, suggestions and criticisms by sales, research, test or assembly staff, or originating in the design department itself.

Without close contact between the client or proposer on the one hand and those in charge of the design department on the other, no optimum solution can be expected because the problem, as presented to the design department, often does not contain all the necessary information. A phase of further data collection must then be initiated. This phase must answer the following questions:

- What is the problem really about?
- What implicit wishes and expectations are involved?
- Do the specified constraints actually exist? and
- What paths are open for development?

Fixed solution ideas or concrete indications implicit in the task formulation often have an adverse effect on the final outcome. Only the required function with the appropriate inputs and outputs and the task-specific constraints should be specified right at the start. For that purpose the following questions must be asked:

- What objectives is the intended solution expected to satisfy?
- What properties must it have? and
- What properties must it not have?

As far as this has not already been done in product planning (see 5.1) the development department should undertake the situation analysis described in 5.1.2 to specify the product situation and to identify future developments.

Once all the necessary data have been collected, it is advisable to combine them into a system based on the established steps of the design process. For that purpose a requirements list should be drawn up which is more detailed than the one supplied by the client or the product planning group.

5.2.2   The Requirements List (Design Specification)

1 Contents

When preparing a detailed requirements list it is essential to state whether the individual items are demands or wishes.

Demands are requirements that must be met under all circumstances, in other words, requirements without whose fulfilment the solution is not acceptable (for instance such qualitative demands as "suitable for tropical conditions", "splashproof" etc). Minimum demands must be formulated as such (for example $P > 20$ kW; $L < 400$ mm).

Wishes are requirements that should be taken into consideration whenever possible, perhaps with the stipulation that they only warrant limited increases in cost, for example central locking, less maintenance etc. It is advisable to classify wishes as being of major, medium or minor importance [5.20].

The distinction between demands and wishes is also important at the evaluation stage, since selection (see 4.2.1) depends on the fulfilment of demands, while evaluation (see 4.2.2) bears on only such variants as already meet the demands.

Even before a certain solution is adopted, a list of demands and wishes should be drawn up and the quantitative and qualitative aspects tabulated. Only then will the resulting information be adequate:

Quantity: All data involving numbers and magnitudes, such as number of items required, maximum weight, power output, throughput, volume flow rate etc.

Quality: All data involving permissible variations or special requirements such as waterproof, corrosion proof, shock proof etc.

Requirements should, if possible, be quantified and, in any case, defined in the clearest possible terms. Special indications of important influences, intentions or procedures may also be included in the requirements list, which is thus an internal digest of all the demands and wishes expressed in the language of the various departments involved in the design process. As a result, the requirements list not only reflects the initial position but, since it is continually reviewed, also serves as an up-to-date working document. In addition it is a record that can, if necessary, be presented to the Board and the sales department so that they may make their objections known before the actual work is started.

2 Format

For a recommended layout of a requirements list, see Fig. 5.6.

The top left field identifies the company or department responsible, the middle field the name of the project or the product, the top right field is used for identification or classification and for the page number. Also in the top right corner is shown the version number and the date.

The first column is used to indicate a modification and the date. The next column indicates whether the requirement is a demand or a wish. The middle column contains a structured list of requirements. In the last column the group or individual responsible for the requirement is listed.

The format of the requirements list should be agreed with the company's standards office so that it can be used, elaborated and adopted in as many departments as possible. Figure 5.6 is thus no more than a suggestion that can, of course, be modified at will.

It may prove useful to draw up the requirements list based on sub-systems (functions or assemblies) where these can be identified, or else based on checklist headings (see Fig. 5.7). With established solutions, in which the assemblies to be developed or improved are already determined, the requirements list must be arranged in accordance with these. Special design groups are usually put in charge of the development of each assembly. With motor cars, for instance, the requirements list can be subdivided into engine, transmission and bodywork development.

In the case of essential and also of less obvious requirements it is extremely useful to *record the source* of specific demands or wishes. It is then possible to go back to the proposers of requirements and to enquire into their reasons. This is

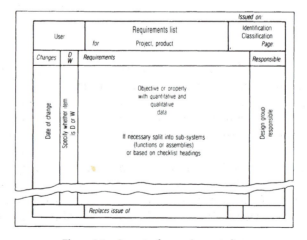

Figure 5.6.   Layout of a requirements list.

Main headings	Examples
Geometry	Size, height, breadth, length, diameter, space requirement, number, arrangement, connection, extension.
Kinematics	Type of motion, direction of motion, velocity, acceleration.
Forces	Direction of force, magnitude of force, frequency, weight, load, deformation, stiffness, elasticity, stability, resonance.
Energy	Output, efficiency, loss, friction, ventilation, state, pressure, temperature, heating, cooling, supply, storage, capacity, conversion.
Material	Physical and chemical properties of the initial and final product, auxiliary materials, prescribed materials (food regulations etc).
Signals	Inputs and outputs, form, display, control equipment.
Safety	Direct safety principles, protective systems, operational, operator and environmental safety.
Ergonomics	Man-machine relationship, type of operation, clearness of layout, lighting, aesthetics.
Production	Factory limitations, maximum possible dimensions, preferred production methods, means of production, achievable quality and tolerances.
Quality control	Possibilities of testing and measuring, application of special regulations and standards.
Assembly	Special regulations, installation, siting, foundations.
Transport	Limitations due to lifting gear, clearance, means of transport (height and weight), nature and conditions of despatch.
Operation	Quietness, wear, special uses, marketing area, destination (for example, sulphurous atmosphere, tropical conditions).
Maintenance	Servicing intervals (if any), inspection, exchange and repair, painting, cleaning.
Recycling	Reuse, reprocessing, waste disposal, storage.
Costs	Maximum permissible manufacturing costs, cost of tooling, investment and depreciation.
Schedules	End date of development, project planning and control, delivery date.

Figure 5.7.   Checklist for drawing up a requirements list.

particularly important when the question arises of whether or not the demands can be changed in the light of subsequent developments.

Such *changes in*, and *additions to*, the original task as might result from a better understanding of solution possibilities or from possible changes in emphasis must always be entered in the requirements list, which will then reflect the progress of the project at any one time.

Responsibility for this work is vested in the chief designer. The updated requirements list should be circulated among all departments concerned with the development of the product (management, sales, accounts, research etc). The requirements list should only be changed or extended by a decision of those in charge of the overall project.

3 Listing the Requirements

As a rule designers have some difficulty in drawing up their first requirements list. Experience, however, will greatly facilitate the compilation of subsequent ones. It is useful to head all requirements lists with a description of the overall task and some characteristic data, for example "Induction motor, rating 63 kW, 4-pole". This helps to convey some idea of the nature and scope of the problem.

Further data are collected with the help of a checklist reflecting the general and specific objectives and constraints (see Fig. 5.7). By applying this checklist to the task in hand and then asking what questions need to be answered, designers may elicit a most beneficial association of ideas.

Franke [5.3] has drawn up a very detailed checklist, based on a search matrix. Checklists and questionnaires are particularly useful if they cover no more than a limited field, if they do not date too quickly, and if they can be taken in at a glance. In this book we shall deliberately refrain from presenting detailed questionnaires. It is our considered opinion that easily memorised checklists with regular headings will help designers to hit upon the essential questions automatically, and without laborious tools.

The first step in the clarification of the task is the elucidation of the necessary functions and task-specific constraints. This is done by reference to the following headings: geometry–kinematics–forces–energy–material–signals. The combination of the relevant concepts produces a welcome redundancy and hence an important check that nothing essential has been forgotten.

The remaining general or task-specific constraints follow the guidelines listed in 2.1.8 to which we refer frequently. Examples in Fig. 5.7 clarify the kind of issues that could be raised.

Once the data have been gathered, they must be combined in a sensible way. To that end, numbering of individual items may prove useful.

In the light of the arguments advanced in this chapter, the following general method of compiling a requirements list can now be recommended:

1 Identify the requirements

- Pay attention to the main headings of the checklist (see Fig. 5.7) and determine the quantitative and the qualitative data.
- Ask:
  What objectives must the solution satisfy?
  What properties must it have?
  What properties must it not have?
- Collect further information.
- Specify demands and wishes clearly.
- If possible, rank wishes as being of major, medium or minor importance.

2 Arrange the requirements in clear order

- Define the main objective and the main characteristics.

- Split into identifiable sub-systems, functions, assemblies etc, or in accordance with the main headings of the checklist.

3 Enter the requirements list on standard forms   and circulate among interested departments, licensees, directors etc.

4 Examine objections and amendments   and, if necessary, incorporate them in the requirements list.

Once the task has been adequately clarified and the relevant departments are satisfied that the listed requirements are technically and economically attainable, the way is clear for the conceptual design phase.

Computer programs have been developed to help draw up requirements lists [5.1, 5.7]. Such programs are based on checklists similar to the one suggested in Fig. 5.7.

4 Examples

Figure 5.8 shows a requirements list for a printed circuit board positioning machine illustrating the main characteristics of the content and format of all requirements lists. It has been structured according to the main headings given in Fig. 5.7. The requirements have been split into demands and wishes, and, where possible and necessary, quantified. Modifications and amendments with their dates are also shown. The latter were the result of an intensive discussion of a first draft of the requirements list (first version 21 April 1988).

In Figs. 6.2, 6.37 and 6.53, requirements lists based on the above recommendations are provided as further examples.

5 Further Applications

Even when the design is not original and the solution principle as well as the layout are fixed so that nothing more than *adaptations* or *dimensional changes* have to be made in a familiar area, orders should nevertheless be executed on the basis of requirements lists which can then take the form of printed forms or questionnaires. They should be constructed in such a way that information for electronic data processing and quality control can be read off directly. As a result, requirements lists become sources of information for direct action.

Beyond that, requirements lists, once compiled, are an invaluable *store of information* about the required or desired properties of the product, and hence extremely helpful for further developments, negotiations with suppliers etc.

Drawing up requirements lists for existing products can also provide a very valuable source of information for the subsequent development and rationalisation of those products.

The examination of a requirements list during project conferences or before assessing various designs is an extremely useful procedure. All those involved are

SIEMENS		Requirements list for a printed circuit board positioning machine	Issued on 27/ 4/ 88 Page: 1
Changes	D W	Requirements	Resp.
		1. Geometry: dimensions of the test sample Circuit board:	Langner's group
	D	Length = 80 – 650 mm	
	D	Breadth = 50 – 570 mm	
	W	Height = 0.1 – 10 mm	
	D	Required height = 1.6 – 2 mm	
	W	Clearance between basic grid boards ≤ 120 mm	
	D	"Clamping area" ≤ 2 mm (3 edges of the board)	
		2. Kinematics:	
27. 4. 88	D	Precise positioning of the test sample	
27. 4. 88	D	Minimum of 2 mm displacement of the test sample normal to the board	
27. 4. 88	D	Feedback to transfer position	
	W	Separate stations for input and output	
	D	Design of clearance zone	
	W	Minimum handling time (as fast as possible)	
		3. Forces:	
	D	Weight of the test sample ≤ 1.7 kg	
27. 4. 88	W	Maximum weight of the test sample ≤ 2.5 kg	
		4. Energy:	
	D	Electrical and/or pneumatic (6–8 bar)	
		5. Material:	
	D	Free from rust	
	D	Isolation between test sample and testing device	
27. 4. 88	W	Thermal expansion of testing device adjusted to expansion of printed circuit	
27. 4. 88	D	Consideration of influence of temperature	
27. 4. 88	D	Temperature range: 15–40°C	
27. 4. 88	D	Humidity: 65%	
27. 4. 88	W	Circuit boards: epoxy–fiberglass sheet	
27. 4. 88	D	No condensation	
		6. Safety:	
27. 4. 88	D	Operator safety	
		7. Production:	
	D	Consideration of tolerance build up	
		8. Operation:	
	D	No contamination inside the testing device	
	D	Destination: production line	
		9. Maintenance:	
	W	Maintenance interval > 10^6 test operations	
		10. Schedule:	
	D	Embodiment finished by July 1988	
		Replaces issue of 21/ 4/ 88	

Figure 5.8. Requirements list for a printed circuit board positioning machine (Siemens AG).

put in possession of all the available information and all salient evaluation criteria are brought home to them.

Requirements lists stored in databases or design support systems can assist the retrieval of old orders and designs, and thus encourage the use of existing documentation as the basis for new projects.

5.2.3 Practical Application of Requirements Lists

In the last few years it has been shown that, at least for original designs, the formulation of a requirements list is a very efficient method for solution development and has been broadly adopted by industry. When used in practice, however, several difficulties often arise. These difficulties are listed below:

- *Obvious requirements*, such as low-cost manufacture and ease of assembly, are often not included in the requirements list. One should take care that not only are these issues addressed but that they are expressed precisely.
- In an early stage of a project it is not always possible to make *precise statements* in the requirements list. The statements have to be amended or corrected during the design and development process.
- A *stepwise development* of the requirements list is very useful when tasks are poorly defined. In these cases the requirements should be formulated more precisely as soon as possible.
- During the formulation of requirements lists or related discussions, *functions or solution ideas* are often mentioned. This is not wrong. They can encourage a clearer formulation of the requirements and even lead to the identification of new requirements. The solution ideas or proposals generated should be recorded so that they can be used later in the systematic search for solutions. They should not be included in the requirements list.
- The identification of *deficiencies and failures* can initiate requirements which then have to be formulated in a solution-neutral way. Failure analysis is often the starting point for a requirements list.
- For *adaptive or variant design*, designers should still make requirements lists for themselves, even when the task is small.
- Drawing up requirements lists should not be formalised too strictly. *Guidelines and forms* prevent forgetting important issues and provide a supporting structure. If one diverts from the recommendations in this book, one should a least consider the main characteristics and distinguish between demands and wishes.

6 Conceptual Design

Conceptual design is that part of the design process in which, by the identification of the essential problems through abstraction, by the establishment of function structures and by the search for appropriate working principles and their combination, the basic solution path is laid down through the elaboration of a solution principle. Conceptual design *determines the principle* of a solution.

From Fig. 3.3 we can see that the conceptual phase is preceded by a decision. The purpose of this decision is to answer the following questions based on the requirements list agreed upon during task clarification:

- Has the task been clarified sufficiently to allow development of a solution in the form of a design?
- Must further information about the task be acquired?
- Is it possible to reach the chosen objective within the given financial restrictions?
- Is a conceptual elaboration really needed, or do known solutions permit direct progress to the embodiment and detail design phases?
- If the conceptual stage is indispensable, how and to what extent should it be developed systematically?

6.1 Steps of Conceptual Design

According to the procedural plan outlined in 3.2, the conceptual design phase follows the clarification of the task. Figure 6.1 shows the steps involved, correlated in such a way as to satisfy the principles of the general problem solving process set out in 3.1.

The reasons for the individual steps have been examined in Chapter 3 and need not be further discussed here. It should, however, be mentioned that refinements of any one of the steps by reiteration on a higher information level should be made whenever necessary. The loops involved have been omitted from Fig. 6.1 for the sake of greater clarity.

The individual steps and the appropriate working methods will now be examined in detail.

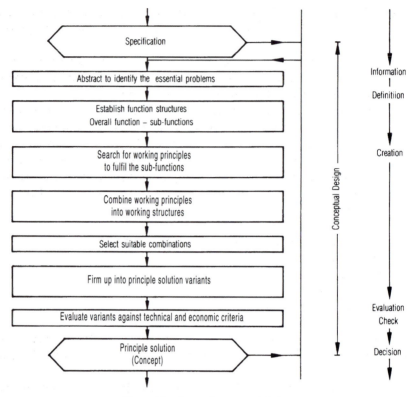

Figure 6.1. Steps of conceptual design

6.2   Abstracting to Identify the Essential Problems

6.2.1   Aim of Abstraction

Solution principles or designs based on traditional methods are unlikely to provide optimum answers when new technologies, procedures, materials, and also new scientific discoveries, possibly in new combinations, hold the key to better solutions.

Every industry and every design office is a store of experiences as well as of prejudices and conventions which, coupled to the wish to minimise risks, stand in the way of better and more economic but unconventional solutions. The client, customer or product planning group might have included specific proposals for a solution in the requirements list. It is also possible that during the discussion of individual requirements, ideas and suggestions for realising a solution emerge. In the unconscious, at least, certain solutions might exist. Perhaps even very concrete ideas exist (fixation).

In their search for optimum solutions, designers, far from allowing themselves to be influenced by fixed or conventional ideas, must therefore examine very carefully whether novel and more suitable paths may not be open to them. To solve the problem of fixation and sticking with conventional ideas, *abstraction* is used. This means ignoring what is particular or incidental and emphasising what is general and essential.

Such generalisation leads straight to the crux of the task. If it is properly formulated, then the overall function and the essential constraints become clear without in any way prejudicing the choice of a particular solution.

As an example, consider the improvement of a labyrinth seal in accordance with a set of requirements. The task is described in detail by means of a requirements list and the formulation of the goal to be achieved. In the abstracting approach, the crux of the task would not so much be the design of a labyrinth seal as that of a shaft seal without physical contact, due regard being paid to certain operating and spatial constraints, and also to cost limits and delivery times. Specifically, the designer should ask whether the crux is:

- to improve the technical functions, eg sealing quality or safety;
- to reduce weight or space;
- to lower costs;
- to shorten delivery time; or
- to improve production methods.

New developments involving a proven solution principle, coupled to modifications in production methods, are often imposed by the need to lower costs and shorten delivery times.

All the questions listed above might have to be satisfied by the overall solution, but their importance may differ from case to case. Nevertheless, due regard must be paid to each of them, since any one is likely to provide the impetus for the discovery of a new and better solution principle.

Once the crux of the task has been clarified to some extent, it becomes much easier to formulate the overall task in terms of the essential sub-problems as they emerge.

Thus if, in the example we have mentioned, an improvement in the sealing properties were the crucial requirement, new sealing systems would have to be found. This would mean studying the flow of fluids in narrow passages and, from the knowledge acquired, providing for better sealing properties, while also satisfying the other sub-problems we have mentioned.

If, on the other hand, cost reduction were the crucial point, then, after an analysis of the cost structure, one would have to see whether the same physical effects could be produced by the use of cheaper materials, by reducing the number of components or by using a different production process. It is also possible to search for new concepts to achieve a better or at least similar sealing performance for lower cost.

It is the identification of the crux of the task with the functional connections and the task-specific constraints that throws up the essential problems for which solutions have to be found [6.5, 6.10, 6.15].

6.2.2   Abstraction and Problem Formulation

The clarification of the task with the help of a requirements list will have helped to focus attention on the problems involved and will have greatly increased the particular level of information. Elaborating the requirements list may thus be said to have prepared the way for the next step. When setting out to tackle an assignment, designers have no ready solution, or only an inadequate one. Depending on their knowledge, experience and familiarity with previous designs, the task will be more or less new to them.

The first step is to *analyse the requirements list* in respect of the required function and essential constraints to establish the crux of the problem. Roth [6.14] advises that the functional relationships contained in the requirements list be formulated explicitly and arranged in order of their importance.

That analysis, coupled to a step-by-step abstraction, will reveal the general aspects and essential features of the task, as follows:

Step 1. Eliminate personal preferences.

Step 2. Omit requirements that have no direct bearing on the function and the essential constraints.

Step 3. Transform quantitative into qualitative data and reduce them to essential statements.

Step 4. Generalise the results of the previous step.

Step 5. Formulate the problem in solution-neutral terms.

Depending on either the nature of the task or the size of the requirements list, or both, certain steps may be omitted.

Table 6.1 illustrates abstraction based on these steps using the requirements list shown in Fig. 6.2. The general formulation makes it clear that, with respect to the functional relationships, the problem is the measurement of quantities of liquid, and that this is subject to the essential conditions that the quantity of liquid is changing continuously and that the liquid is in containers of unspecified size and shape.

This analysis thus leads to a definition of the objective on an abstract plane without laying down any particular solution.

6.2.3   Systematic Broadening of Problem Formulation

Once the crux of the task has been identified by correct problem formulation, a step-by-step enquiry must be initiated to discover if an extension of, or even a change in, the original task might lead to promising solutions.

An excellent illustration of this procedure has been given by Krick [6.9]. The task he used as an example was an improved method of filling, storing and loading bags of animal feed. An analysis gave the situation shown in Fig. 6.3. It would have been a grave mistake to begin immediately by thinking of possible improvements to the existing situation. By proceeding in this way one is likely to ignore other, more useful and more economic solutions.

Table 6.1. Procedure during abstraction: motor vehicle fuel gauge based on requirements list given in Fig. 6.2

Result of Steps 1 and 2

— Volumes: 20 to 160 litres
— Shape of container: fixed or unspecified (rigid)
— Top or side connection
— Height of container: 150 mm to 600 mm
— Distance between container and indicator: $\neq 0$ m, 3 m to 4 m
— Petrol and diesel, temperature range: $-25°C$ to $65°C$
— Ouput of transmitter: unspecified signal
— External energy: DC at 6 V, 12 V, 24 V. Variation -15% to $+25\%$
— Output signal accuracy at maximum $\pm 3\%$ (together with indicator error $\pm 5\%$)
— Response sensitivity: 1% of maximum signal output
— Possibility of signal calibration
— Minimum measurable content: 3% of maximum value

Result of Step 3

— Various volumes
— Various container shapes
— Various connections
— Various contents (liquid levels)
— Distance between container and indicator: $\neq 0$ m
— Quantity of liquid varies with time
— Unspecified signal
— (with outside energy)

Result of Step 4

— Various volumes
— Various container shapes
— Transmission over various distances
— Measure continuous changes in quantity of liquid

Result of Step 5 (Problem formulation)

— Measure continuously changing quantities of liquid in containers of unspecified size and shape and indicate the measurements at various distances from the containers.

In principle, the following problem formulations are possible, each representing a higher level of abstraction than the last:

1. Filling, weighing, closing and stacking bags of feed.
2. Transferring feed from the mixing silo to stacked bags in the warehouse.
3. Transferring feed from the mixing silo to bags on the delivery truck.
4. Transferring feed from the mixing silo to the delivery truck.
5. Transferring feed from the mixing silo to a delivery system.
6. Transferring feed from the mixing silo to the consumer's storage bins.
7. Transferring feed from ingredient containers to consumer's storage bins.
8. Transferring feed ingredients from their source to the consumer.

Krick has incorporated some of these formulations in a diagram (see Fig. 6.4).

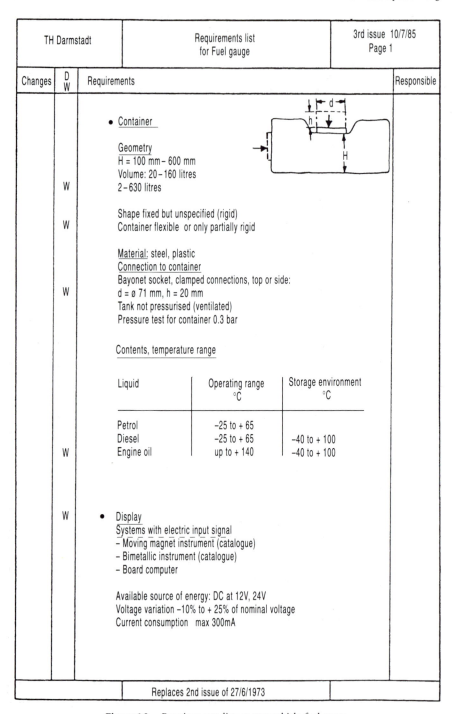

TH Darmstadt	Requirements list for Fuel gauge	3rd issue 10/7/85 Page 1

Changes	D W	Requirements		Responsible

- Container

Geometry
H = 100 mm – 600 mm
Volume: 20 – 160 litres
2 – 630 litres

W

Shape fixed but unspecified (rigid)
W   Container flexible or only partially rigid

Material: steel, plastic
Connection to container
Bayonet socket, clamped connections, top or side:
W   d = ø 71 mm, h = 20 mm
Tank not pressurised (ventilated)
Pressure test for container 0.3 bar

Contents, temperature range

Liquid	Operating range °C	Storage environment °C
Petrol	–25 to + 65	
Diesel	–25 to + 65	–40 to + 100
Engine oil	up to + 140	–40 to + 100

W

W   • Display
Systems with electric input signal
– Moving magnet instrument (catalogue)
– Bimetallic instrument (catalogue)
– Board computer

Available source of energy: DC at 12V, 24V
Voltage variation –10% to + 25% of nominal voltage
Current consumption   max 300mA

	Replaces 2nd issue of 27/6/1973	

Figure 6.2.   Requirements list: motor vehicle fuel gauge.

TH Darmstadt	Requirements list for Fuel gauge	3rd issue 10/7/85 Page 2

Changes	D W	Requirements	Responsible
		• System to be developed	
		Geometry Consider connection constraints to container	
	W	Kinematics No moving parts	
		Energy (see display)	
		Material (see container)	

		Signal
		   o Input
		      Minimum measurable content: 3% of maximum value
		      Reserve tank contents by special signal	

	W	Signal unaffected by angle of liquid surface Possibility of signal calibration	
	W	Possibility of signal calibration with full container	
	W	o Output Output of transmitter: electric signal Output signal accuracy at max. value ±3%	
	W	±2%	
		(together with indicator error ±5%) Under normal conditions, horizontal level, v = const. Able to withstand shocks of normal driving Response sensitivity: 1% of maximum output signal	
	W	0.5% of maximum output signal	
		o Connection between input and output Distance container – display: ≠zero m, 3 m – 4 m	
	W	1 m – 20 m	
		Separate power possible	
		Production Large–scale production	

| | | Replaces 2nd issue of 27/6/1973 | |

Figure 6.2.   Continued.

TH Darmstadt		Requirements list for Fuel gauge	3rd issue 10/7/85 Page 3	
Changes	D W	Requirements		Responsible

	W	Test requirements
		Operating conditions of vehicle
		    Forward acceleration ±10 m/s^2
		    Sideways acceleration ±10 m/s^2
		    Upward acceleration (vibration), up to 30 m/s^2
		    Shocks in forward direction without damage, up to 30 m/s^2
		    Forward tilt up to ±30°
		    Sideways tilt max 45°
		Salt spray tests for inside and outside components according to client's
		requirements (DIN 90905 to be considered)
		Must conform with heavy vehicle regulations
		
		Operation, Maintenance
		Installation by non-specialist
		Life expectancy 10^4 level changes (full/empty)
		Minimum of 5 years service life
		Fuel gauge replaceable
		Fuel gauge maintenance-free
		Fuel gauge simply modified to suit different container sizes
		
		Regulations
		No regulations relating to explosion safety
		
		Quantity
		10000/day of the adjustable type
		5000/day of the most popular type
		
		Costs
		Manufacturing costs ≤ DM 6.00 each (without display)		

	W			
	W			

| | | Replaces 2nd issue of 27/6/1973 | | |

Figure 6.2.   Continued.

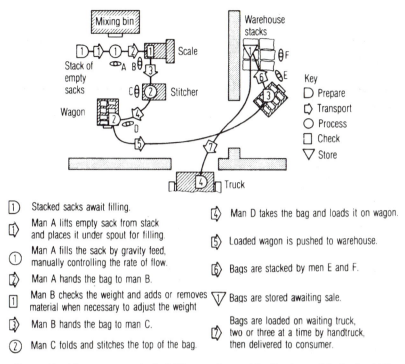

Figure 6.3. The present method of filling, storing, and loading bags of feed, after [6.9]

It is characteristic of this approach that the problem formulation is made *as broad as possible* in successive steps. In other words, the current or obvious formulation is not accepted at face value but *broadened systematically*, which may conflict with decisions already taken, but opens up new perspectives. Thus Formulation 8 above is the broadest, the most general and the least circumscribed.

The crux of the task, in fact, is the transport of the correct quantity and quality of feed from the producer to the consumer and not, for instance, the best method of closing or stacking bags, or moving them into the warehouse. With a broader formulation, solutions may appear that render the filling of bags and storing them in the warehouse unnecessary.

How far this process of abstraction is continued depends on the constraints. In the case under consideration, Formulation 8 must be rejected on technical, seasonal and meteorological grounds: the consumption of feed is not confined to harvest time; for various reasons consumers will not want to store feed for a whole year; moreover, they may be reluctant to mix the required ingredients themselves. However, the transfer of feed on demand, for instance, with delivery trucks taking it directly from the mixing silo to the consumers' storage bins (Formulation 6), is more economical than intermediate storage in a warehouse and the transport of smaller quantities in bags. In this connection, the reader might recall the development in a different field which culminated in the delivery of ready-mixed concrete direct to building sites in special vehicles.

We have tried to show how comprehensive problem formulation on an abstract

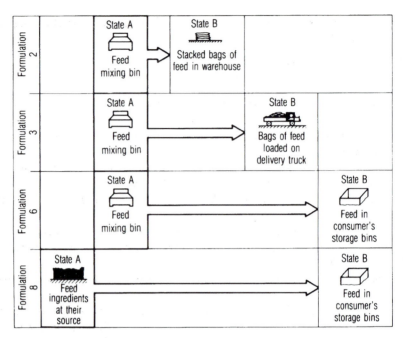

Figure 6.4. Alternative formulations of the feed distribution problem, illustrating progressively broader formulations of a problem, after [6.9]. A = initial state; B = final state.

plane opens the way for better solutions. This approach, furthermore, helps to raise the influence and responsibility of engineers by giving them an overview of the problem and thus involving them in, for instance, environmental protection and recycling.

At this stage, at the very latest, all but *genuine constraints* ought to have been eliminated. Unfortunately, *fictitious constraints* often continue to impede designers or proposers, albeit unconsciously. Thus a solution of the feed-transfer problem in accordance with Formulation 6 would have been impossible had the engineer concerned set the fictitious constraint that the transfer might only be made in bags.

Another example: technical staff of a company that has been exclusively making or using hydraulic control systems will all too easily accept the fictitious constraint that, in the future as in the past, all technical control problems must be solved using the hydraulic principle. This restriction only becomes a genuine constraint if it is decided, after due consideration and a deliberate decision, that the hydraulic control system must be retained, for instance in order to increase the firm's turn-over or in order to unify storekeeping and maintenance.

In principle, all paths must be left open until such time as it becomes clear which solution principle is the best. Thus designers must question all the constraints they are given and work out with the client or proposer whether or not they should be retained as genuine restrictions. In addition, designers must learn to discard fictitious constraints that they themselves have come to accept, and to that end ask critical questions and test all their presuppositions. Here it may be useful to ask the questions we mentioned in connection with the compilation of the requirements list, namely:

- What properties must the solution have?
- What properties must the solution not have?

Abstraction helps to identify fictitious constraints and to eliminate all but genuine restrictions.

We shall conclude this section with a few examples of abstraction and problem formulation:

Do not design a garage door, but look for means of securing a garage in such a way that the car is protected from thieves and the weather.

Do not design a keyed shaft, but look for the best way of connecting gear wheel and shaft.

Do not design a packing machine, but look for the best way of despatching a product safely or, if the constraints are genuine, of packing a product compactly and automatically.

Do not design a clamping device, but look for a means of keeping the workpiece firmly fixed.

From the above formulations, and this is very helpful for the next step, the final formulation can be derived in a way that does not prejudice the solution, ie is *solution-neutral,* and at the same time turns it into a *function*:

"Seal shaft without contact" – and not "Design a labyrinth seal".

"Measure quantity of fluid continuously" – and not "Gauge height of liquid with a float".

"Measure out feed" – and not "Weigh feed in sacks".

6.3 Establishing Function Structures

6.3.1 Overall Function

According to 2.1.3, the requirements determine the function, that is, the relationship between the inputs and outputs of a plant, machine or assembly. In 6.2 we explained that problem formulation obtained by abstraction does much the same. Hence, once the crux of the overall problem has been formulated, it is possible to indicate an *overall function* that, based on the *flow of energy, material and signals* can, with the use of a *block diagram,* express the relationship between *inputs and outputs* independently of the solution. That relationship must be specified as precisely as possible (see Fig. 2.3).

In our example of a fuel gauge (see Fig. 6.2), quantities of liquid are introduced into and removed from a container, and the problem is to measure and indicate the quantity of liquid found in the container at any one time. The result, in the liquid system, is a flow of material with the function "store liquid" and, in the measuring system, a flow of signals with the function "measure and indicate quantity of liquid". The second is the overall function of the specific task under consideration, that is, the development of a fuel gauge (see Fig. 6.5). That overall function can be divided into sub-functions in a further step.

6.3.2 Breaking Down into Sub-functions

Depending on the complexity of the problem, the resulting overall function will in turn be more or less complex. By complexity we mean the relative lack of transparency of the relationships between inputs and outputs, the relative intricacy of the necessary physical processes, and the relatively large number of assemblies and components involved.

Just as a technical system can be divided into sub-systems and elements (see 2.1.3), so a complex *overall function* can be broken down into *sub-functions* of lower complexity. The combination of individual sub-functions results in a *function structure* representing the overall function.

The object of breaking down complex functions is:

• the determination of sub-functions facilitating the subsequent search for solutions; and
• the combination of these sub-functions into a simple and unambiguous function structure.

The optimum method of breaking down an overall function – that is, the number of sub-function levels and also the number of sub-functions per level – is determined by the relative novelty of the problem and also by the method used to search for a solution.

In the case of *original designs*, neither the individual sub-functions nor their relationships are generally known. In that case, the search for and establishment of an optimum function structure constitute some of the most important steps of the conceptual design phase. In the case of *adaptive designs*, on the other hand, the general structure with its assemblies and components is much better known, so that a function structure can be obtained by the analysis of the product to be developed. Depending on the special demands of the requirements list, that function structure can be modified by the variation, addition or omission of individual sub-functions or by changes in their combination. Function structures are of great

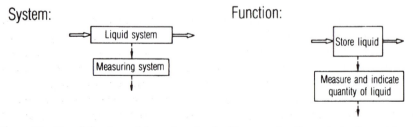

Figure 6.5.   Overall functions of the systems involved in measuring the contents of a container.

importance in the development of modular systems. For this type of *variant design*, the physical structure – that is, the assemblies and individual components used as building blocks and also their relationships – must be reflected in the function structure.

A further advantage of setting up a function structure is that it allows a *clear definition* of existing sub-systems or of those to be newly developed, so that they can be *dealt with separately*. If existing assemblies can be assigned directly as complex sub-functions, the subdivision of the function structure can be discontinued at a fairly high level of complexity. In the case of new assemblies or those requiring further development, however, the division into sub-functions of decreasing complexity must be continued until the search for a solution seems promising. As a result, function structures may save a great deal of time and money.

Apart from helping in the search for a solution, function structures or their sub-functions can also be used for purposes of classification. Examples are the "classifying criteria" of classification schemes (see 4.1.3) and the subdivision of design catalogues.

It may prove expedient not only to set up task-specific functions, but also to elaborate the function structure from generally valid sub-functions (see Fig. 2.8). The latter recur in technical systems and may be helpful in the search for a solution because they may lead to the discovery of task-specific sub-functions and because design catalogues may list solutions for them. Defining generally valid functions can also be of use in varying function structures, for example to optimise the energy, material and signal flows.

In many cases in industry it may not be expedient to build up a function structure from generally valid sub-functions because they are, in fact, too general and thus do not provide a sufficiently concrete picture of the relationships for the subsequent search for solutions. In general, a clear picture only emerges after adding more task-specific details (see 6.3.3).

It is useful to start by determining the *main flow* in a technical system, if this is clear. The *auxiliary flows* should only be considered later. When a basic function structure, including the most important links, has been found, it is easier to undertake the next step, that is, to consider the auxiliary flows with their sub-functions and to achieve a further subdivision of complex sub-functions. For this step it is helpful to create a temporary working structure or a solution for the *basic function structure*, without, however, prejudicing the final solution.

Examples

Figures 6.6 and 6.7 show the function structure of a tensile testing machine with a relatively complex flow of energy, material and signals. In this type of overall function, the function structure is built up step by step from sub-functions, attention at first being focused on essential main functions. Thus, on a first functional level, only such sub-functions are specified as directly satisfy the required overall function. In addition, such complex sub-functions as "change energy into force and movement" and "load specimen" are also formulated at the start, because they help to establish a basic function structure.

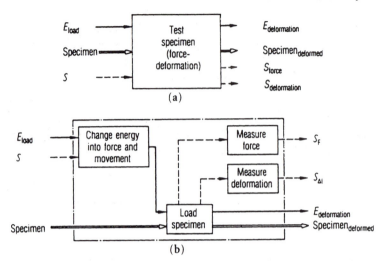

Figure 6.6. Overall function (a) and sub-functions (main functions) (b) of a testing machine.

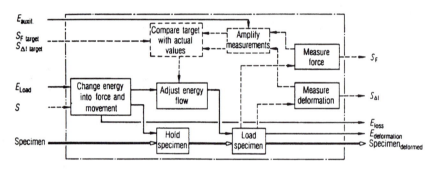

Figure 6.7. Completed function structure for the overall function set out in Fig. 6.6.

In the problem under consideration, the energy and signal flows are of roughly equivalent importance in the search for a solution, while the flow of material – that is the exchange of specimens – is only essential for the holding function which was added to Fig. 6.7. Hence it is impossible to specify just one main flow. In Fig. 6.7 an adjusting function for the load magnitudes and, at the output of the system, the energy lost during the energy flow were added because both clearly affect the design. The energy required to deform the specimen is lost. Moreover, the auxiliary functions "amplify measurements" and "compare target with actual values" proved indispensable for the adjustment of the energy level.

As a further example let us consider the function structure of a dough-shaping machine used, for instance, in the manufacture of biscuits. To satisfy the overall function of this conversion of material in accordance with Fig. 6.8, appropriate or necessary sub-functions of the main flow have to be found. The most important can be deduced, often without great difficulty, from the technology or manufacturing process involved – in this case the manufacture of confectionery. For the rest, the function structure can be completed by asking appropriate questions. Thus a

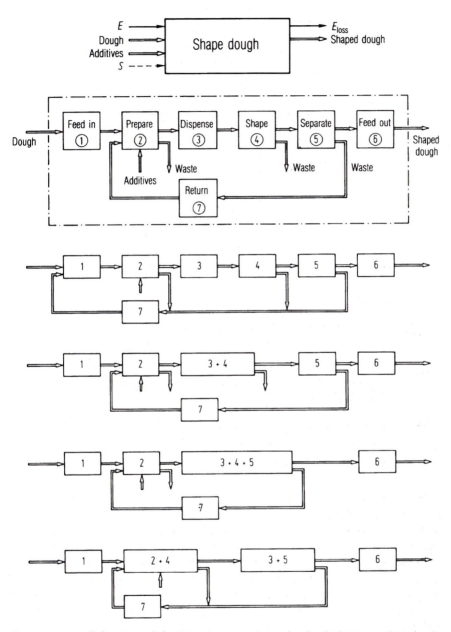

Figure 6.8. Overall function and function structure variants of a dough-shaping machine for the manufacture of biscuits (in respect of the main flow only).

negative answer to the question "May the waste be mixed directly with new dough?" introduces the necessary sub-function "prepare". The operation under consideration was found to involve seven sub-functions which could be combined into a number of function-structure *variants*. Figure 6.8 shows further how several functions can be fused together to provide what are often simple and economic solutions.

There are, however, some problems in which variation of the main flow alone cannot lead to a solution because *auxiliary flows* have a *crucial* bearing on the design. As an example, let us consider the function structure of a potato harvesting machine. Figure 6.9a shows the overall function and the function structure based on the flow of material (the main flow) and the auxiliary flows of energy and signals. In Fig. 6.9b, by comparison, the function structure is represented by means of generally valid functions, to emphasise the clear interrelationship of the different flows. When generally valid functions are used, the separation into sub-functions is as a rule more pronounced than it is in the case of task-specific sub-functions. Thus, in the present example, the sub-function "separate" is replaced with the generally valid functions "connect energy with material mix" and "separate material mix" (the reverse of "connect").

Our next example is meant to illustrate the derivation of function structures by the *analysis of existing systems*. This method is particularly suitable for developments in which at least one solution with the appropriate function structure is

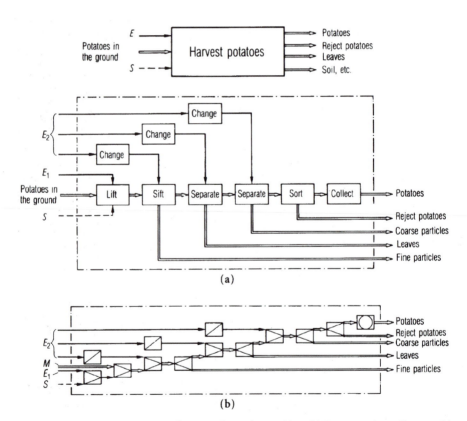

Figure 6.9. (a) Function structure of a potato harvesting machine. (b) For comparison: diagram with generally valid functions based on Fig. 2.8.

known, and the main problem is the discovery of better solutions. Figure 6.10 shows the steps used in the analysis of a flow control valve (a typical on-off switch), showing the individual tasks of the various elements and the sub-functions satisfied by the system. From the sub-functions, the function structure can be derived and then varied for purposes of improving the product.

The function structure examined in Section 6.6 shows clearly that the study of function structures may prove extremely useful, even after the physical effect has been selected, in determining the behaviour of the system at a very early stage of its development, and hence in identifying the structure that best suits the problem under consideration.

6.3.3 Practical Uses of Function Structures

When establishing function structures, we must distinguish between original and adaptive designs. In the case of *original designs*, the basis of a function structure is the *requirements list* and the *abstract formulation of the problem*. Among the

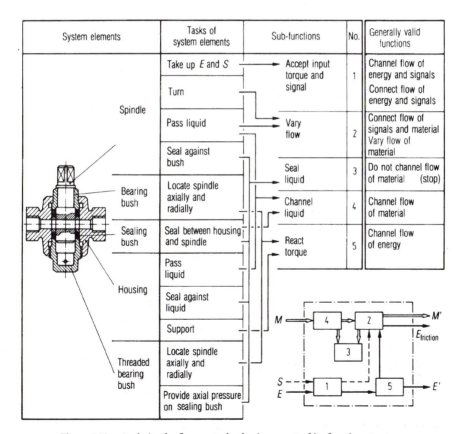

Figure 6.10.   Analysis of a flow control valve in respect of its function structure.

demands and wishes, we are able to identify functional relationships, or at least the sub-functions at the inputs and outputs of a function structure. It is helpful to write out the functional relationships arising from the requirements list in the form of sentences and to arrange these in the order of their anticipated importance or in some other logical order [6.14].

In the case of *adaptive designs*, the starting point is the *function structure of the existing solution* obtained by the analysis of the elements. It helps to develop variants so as to open the path for other solutions, for subsequent optimisation and to develop modular products. The identification of functional relationships can be facilitated by asking the right questions.

In modular systems, the function structure has a decisive influence on the modules and their arrangement (see 8.2). Here, the function structure and that of the assembly is affected not only by functional considerations, but also, and increasingly so, by production needs.

Anyone setting up a function structure ought to bear the following points in mind:

1. First derive a rough function structure with a few sub-functions from what functional relationships you can identify in the requirements list, and then break this rough structure down, step by step, by the resolution of complex sub-functions. This is much simpler than starting out with more complicated structures. In certain circumstances, it may be helpful to substitute a first solution idea for the rough structure and then, by analysis of that first idea, to derive other important sub-functions. It is also possible to begin with sub-functions whose inputs and outputs cross the assumed system boundary. From these we can then determine the inputs and outputs for the neighbouring functions, in other words, work from the system boundary inwards.

2. If no clear relationship between the sub-functions can be identified, the search for a first solution principle may, under certain circumstances, be based on the mere *enumeration of important sub-functions* without logical or physical relationships, but if possible, arranged according to the extent to which they have been realised.

3. *Logical relationships* may lead to function structures through which the logical elements of various working principles (mechanical, electrical etc) can be anticipated.

4. Function structures are not complete unless the existing or expected flow of energy, material and signals can be specified. Nevertheless, it is useful to begin by focusing attention on the *main flow* because, as a rule, it determines the design and is more easily derived from the requirements. The auxiliary flows then help in the further elaboration of the design, in coping with faults, and in dealing with problems of power transmission, control etc. The complete function structure, comprising all flows and their relationships, can be obtained by iteration, that is, by looking first for the structure of the main flow, completing that structure by taking the auxiliary flows into account, and then establishing the overall structure.

5. In setting up function structures it is helpful to know that, in the conversion

of energy, material and signals, several *sub-functions recur* in most structures and should therefore be introduced first. Essentially, these are the generally valid functions of Fig. 2.8, and they can prove extremely helpful in the search for task-specific functions.

Conversion of energy:

- Changing energy – for instance, electrical into mechanical energy.
- Varying energy components – for instance, amplifying torque.
- Connecting energy with a signal – for instance, switching on electrical energy.
- Channelling energy – for instance, transferring power.
- Storing energy – for instance, storing kinetic energy.

Conversion of material:

- Changing matter – for instance, liquefying a gas.
- Varying material dimensions – for instance, rolling sheet metal.
- Connecting matter with energy – for instance, moving parts.
- Connecting matter with signal – for instance, cutting off steam.
- Connecting materials of different type – for instance, mixing or separating materials.
- Channelling material – for instance, mining coal.
- Storing material – for instance, keeping grain in a silo.

Conversion of signals:

- Changing signals – for instance, changing a mechanical into an electrical signal, or a continuous into an intermittent signal.
- Varying signal magnitudes – for instance, increasing a signal's amplitude.
- Connecting signals with energy – for instance, amplifying measurements.
- Connecting signals with matter – for instance, marking materials.
- Connecting signals with signals – for instance, comparing target values with actual values.
- Channelling signals – for instance, transferring data.
- Storing signals – for instance, in data banks.

6. For the application of micro-electronics, it is useful to consider signal flows as shown in Fig. 6.11 [6.10]. This results in a function structure that suggests clearly the modular use of elements to detect (sensors), to activate (actuators),

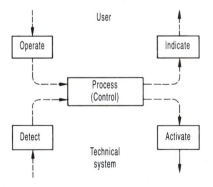

Figure 6.11. Basic signal flow functions for modular use in micro-electronics, after [6.10].

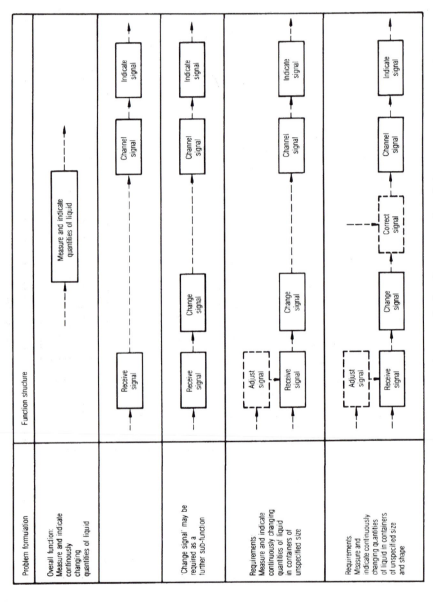

Figure 6.12. Development of a function structure for a fuel gauge.

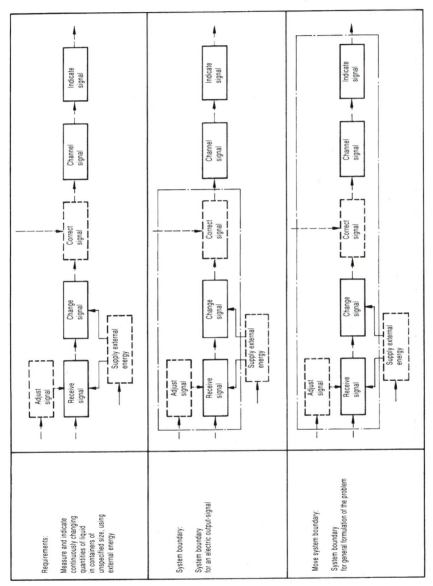

Figure 6.12.   Continued.

to operate (controllers), to indicate (displays) and, in particular, to process signals using microprocessors.

7. From a rough structure, or from a function structure obtained by the analysis of known systems, it is possible to derive further *variants* and hence to optimise the solution, by:

    *   breaking down or combining individual sub-functions;
    *   changing the arrangement of individual sub-functions;
    *   changing the type of switching used (series switching, parallel switching or bridge switching); and
    *   shifting in the system boundary.

    Because varying the function structure introduces distinct solutions, the setting up of function structures constitutes a first step in the search for solutions.

8. Function structures should be kept as *simple* as possible, so as to lead to simple and economical solutions. To this end, it is also advisable to aim at the combination of functions for the purpose of obtaining integrated function carriers. There are, however, some problems in which discrete functions must be assigned to discrete function carriers, for instance, when the requirements demand separation, or when there is a need for extreme loading and quality. In this connection, the reader is referred to our discussion of the division of tasks (see 7.4.2).

9. In the search for a solution, none but *promising function structures* should be introduced, for which purpose a *selection procedure* (see 4.2.1) should be employed, even at this early stage.

10. For the *representation* of function structures it is best to use the *simple and informative symbols* shown in Fig. 2.4, supplemented with task-specific verbal clarifications.

11. An *analysis* of the function structure leads to the identification of those subfunctions for which new working principles have to be found, and of those for which known solutions can be used. This encourages an efficient approach. The search for solutions (see 4.1) then focuses on the sub-functions that are essential for the solution and on which the solutions of other subfunctions depend (see the example in Fig. 6.12).

Function structures are intended to facilitate the discovery of solutions: they are not ends in themselves. It depends very much on the novelty of the task and the experience of designers to what degree they are detailed.

Moreover, it should be remembered that function structures are seldom completely free of physical or formal presuppositions, which means that the number of possible solutions is inevitably restricted to some extent. Hence it is perfectly legitimate to conceive a preliminary solution and then abstract this by developing and completing the function structure by a process of iteration.

Let us return to the example of the fuel gauge (see 6.2.2). Figure 6.12 shows the development and variation of a function structure in accordance with the suggestions presented in this section.

The flow of signals has been treated as the main flow. Associated sub-functions are developed in two steps. Since the specification also provides for measurements in containers of different sizes, holding varying initial quantities of liquid, an adjustment of the signal to the respective size of the container is expedient, and is accordingly introduced as an auxiliary function. Measurements in containers of various unspecified shapes will, in certain circumstances, demand the correction of the signal as another auxiliary function. The measuring operation may require a supply of external energy, which must then be introduced as a further flow. Finally, consider the system boundary. If existing indicating instruments are to be used, the device will have to emit an electric output signal. If they are not, then the sub-functions "channel signal" and "indicate signal" must be included in the search for a solution. An important sub-function that must be satisfied first, and on the working principle of which the others clearly depend, is "receive signal" (see Fig. 6.12). The initial search for a solution will focus on this sub-function and the solution to this will largely decide to what extent individual sub-functions can be changed round or omitted.

6.4 Developing Working Structures

6.4.1 Searching for Working Principles

Working principles have to be found for the various sub-functions and these principles must eventually be combined into a working structure. The concretisation of the working structure will lead to the solution principle. A working principle must reflect the physical effect needed for the fulfilment of a given function and also its geometric and material characteristics (see 2.1.4). In many cases, however, it is not necessary to look for new physical effects, the form design (geometry and materials) being the sole problem. Moreover, in the search for a solution it is often difficult to make a clear mental distinction between the physical effect and the form design features. Designers therefore usually search for working principles that include the physical process along with the necessary geometric and material characteristics, and combine these into a working structure. Theoretical ideas about the nature and form of the function carriers are usually presented by way of diagrams or freehand sketches.

It should be emphasised that the step we are now discussing is intended to lead to several solution variants, that is a solution field. A solution field can be constructed by variation of the physical effects and of the form design features. Moreover, to satisfy a particular sub-function, several physical effects may be involved in one or several function carriers.

In Section 4.1 we discussed methods and tools for finding solutions. In the search for working principles the same methods can be used. Of particular importance, however, are literature search, methods for analysing natural and known technical systems, and intuition-based methods (see 4.1.2). If preliminary solution ideas are available from product planning or through intuition, the systematic analysis of physical processes and the use of classification schemes are also helpful (see 4.1.3). The last two methods usually provide several solutions.

Other important tools are design catalogues, in particular those proposed by Roth and Koller for physical effects and working principles (see 4.1.3) [6.7, 6.14, 6.17]. When solutions have to be found for *several sub-functions*, it is expedient to select the functions as classifying criteria, that is, the sub-functions become the row headings and the possible solution principles are entered in the columns. Figure 6.13 illustrates the structure of such a classification scheme where the sub-functions are represented by F_i and the solution elements by S_{ij}. Depending on the level of concretisation these solution elements can be physical effects or even working principles with geometric and material details.

An example is given in Fig. 6.14 which shows a solution field with working principles to fulfil the generally valid functions "transform energy", "store energy", "control energy", and "change energy component". These sub-functions occur often in test equipment and other mechanical systems (see 6.3.3). The working principles shown represent the appropriate physical effect and preliminary embodiment. The example does not show task-specific formulations for the specific sub-functions.

A further example will clarify the frequently occurring situation when solutions for sub-function have to be generated by the systematic variation of geometry and material properties without changing the physical process. In Fig. 6.15 solutions are to be developed for the function "form support wire" for concrete reinforcing rods. The working motions are varied in three dimensions.

The possible motions of the forming tools are represented in Fig. 6.16a-c. The wires are formed by the coordinated movements of two tools (punch and die). By variation of the basic motions shown in (a) it was possible to correlate 20 punch and die motions (b), and by employing the motions in and round all the co-ordinates to finish up with 239 practicable combinations, some of which are shown in (c). Figure 6.17 shows the selected working principles.

To sum up: the search for working principles for sub-functions should be based on the following:

- Preference should be given to such main sub-functions as determine the prin-

Sub-Functions \ Solutions		1	2	...	j	...	m
1	F_1	S_{11}	S_{12}		S_{1j}		S_{1m}
2	F_2	S_{21}	S_{22}		S_{2j}		S_{2m}
⋮		⋮	⋮		⋮		⋮
i	F_i	S_{i1}	S_{i2}		S_{ij}		S_{im}
⋮		⋮	⋮		⋮		⋮
n	F_n	S_{n1}	S_{n2}		S_{nj}		S_{nm}

Figure 6.13. Basic structure of a classification scheme with the sub-functions of an overall function and associated solutions.

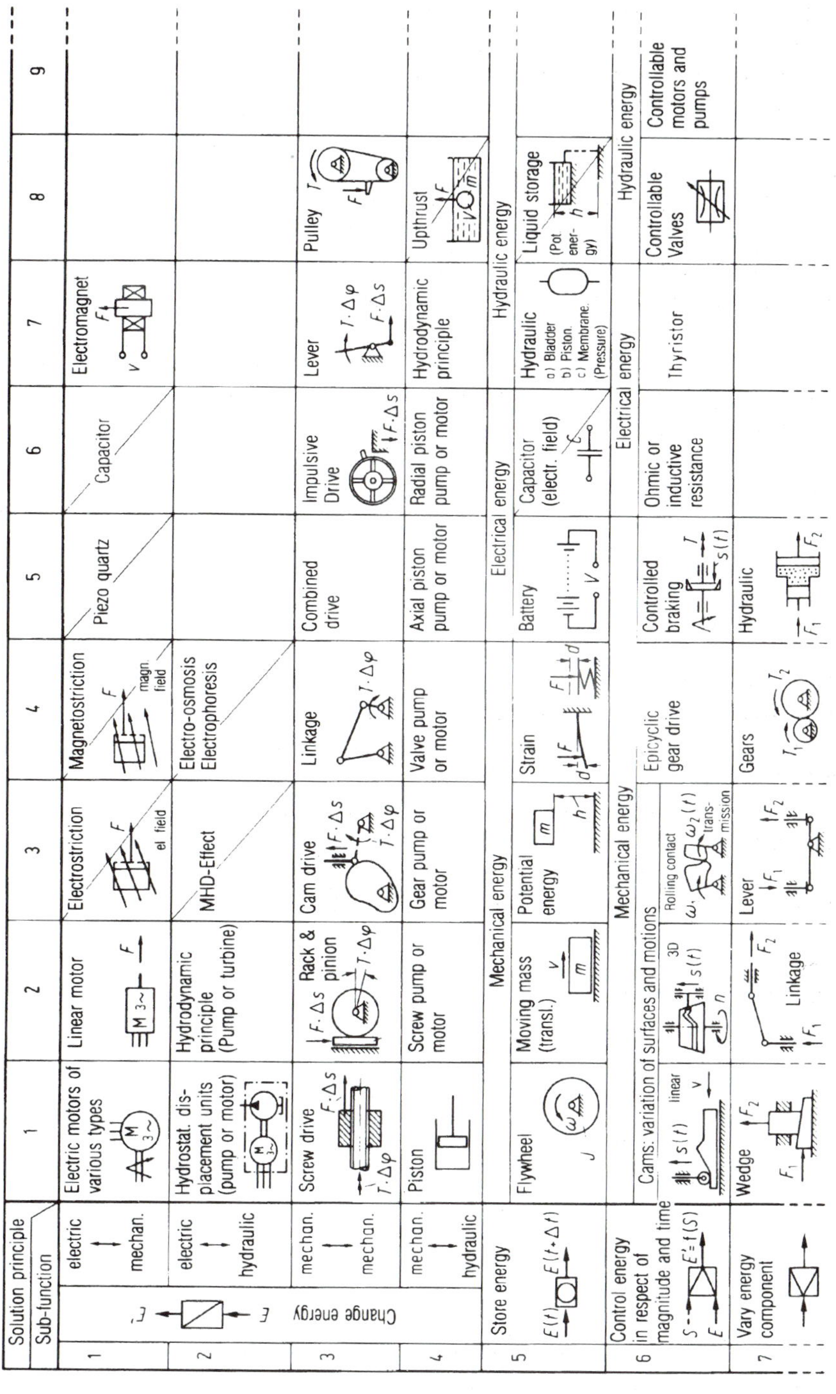

Solution principle / Sub-function		1	2	3	4	5	6	7	8	9
1. Change energy $E \rightarrow \boxed{} \rightarrow E'$	electric → mechan.	Electric motors of various types	Linear motor	Electrostriction	Magnetostriction	Piezo quartz	Capacitor	Electromagnet		
	electric → hydraulic	Hydrostat. displacement units (pump or motor)	Hydrodynamic principle (Pump or turbine)	MHD-Effect	Electro-osmosis Electrophoresis					
	mechan. → mechan.	Screw drive	Rack & pinion	Cam drive	Linkage	Combined drive	Impulsive Drive	Lever	Pulley	
	mechan. → hydraulic	Piston	Screw pump or motor	Gear pump or motor	Valve pump or motor	Axial piston pump or motor	Radial piston pump or motor	Hydrodynamic principle	Upthrust	
5. Store energy $E(t) \rightarrow \boxed{} \rightarrow E(t+\Delta t)$		Flywheel	Moving mass (transl.)	Potential energy	Strain	Battery	Capacitor (electr. field)	Hydraulic a) Bladder b) Piston c) Membrane (Pressure)	Liquid storage (Pot. energy)	
6. Control energy in respect of magnitude and time $S \rightarrow \boxed{} \rightarrow E' = f(S)$	Cams: variation of surfaces and motions	linear $s(t)$	3D $s(t)$	Rolling contact $\omega_1 \cdot \omega_2(t)$ transmission	Epicyclic gear drive	Controlled braking	Ohmic or inductive resistance	Thyristor	Controllable Valves	Controllable motors and pumps
7. Vary energy component		Wedge	Linkage	Lever	Gears	Hydraulic				

(Diagonal band labels within the scheme: Mechanical energy, Electrical energy, Hydraulic energy)

Figure 6.14. Extract from a classification scheme for impulse-loading test rig (see example in 7.7).

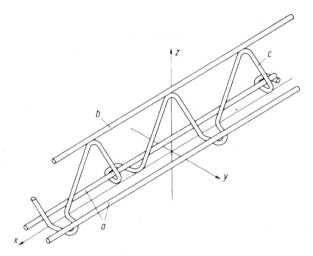

Figure 6.15. Support wire for concrete reinforcing rods [6.6]. *a* lower rods; *b* upper rod; *c* support wire.

ciple of the overall solution and for which no solution principle has yet been discovered.

- Classifying criteria and associated parameters (characteristics) should be derived from identifiable relationships between the energy, material and signal flows or from associated systems.
- If the working principle is unknown, it should be derived from the physical effects and, for instance, from the type of energy. If the physical effect has been determined, appropriate form design features (surfaces, motions and materials) should be chosen and varied. Checklists should be used to stimulate new ideas (see Figs. 4.10 and 4.11).
- Designers should analyse which key classifying criteria influence particular working principles. These criteria should then be subdivided, limited or generalised using further headings.
- To prepare for the selection process, the important properties of the working principles should be noted.

Sections 6.6. and 11.1 provide further examples that illustrate the search for working principles.

6.4.2 Combining Working Principles

To fulfil the overall function, it is now necessary to elaborate overall solutions from the combination of principles, that is system synthesis. The basis of such combinations is the established function structure which reflects logically and physically possible or useful associations of the sub-functions.

The methods we have been describing, and particularly those with an intuitive bias, are intended to lead to the discovery of suitable combinations, but there are

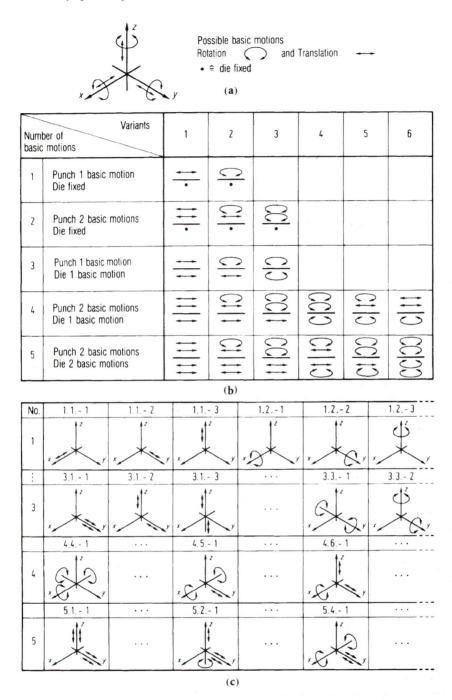

Figure 6.16. Variation possibilities for motions of tools for forming support wires for concrete reinforcing rods, after [6.6]. (a) Possible basic motions; (b) Classification scheme for possible motions of punch and die; (c) Some of the practicable combinations of punch and die motions.

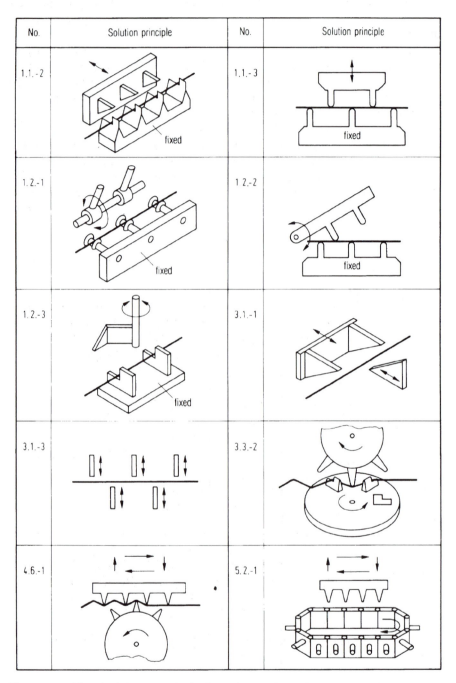

Figure 6.17. Selected solution principles for forming support wires for concrete reinforcing rods on the basis of tool motions in accordance with Figure 6.16c, after [6.6].

also special methods for arriving at such syntheses more directly. In principle, they must permit a clear combination of working principles with the help of the associated physical quantities and the appropriate form design features.

The main problem with such combinations is ensuring the physical and geometrical compatibility of the working principles to be combined, which in turn ensures the smooth flow of energy, material and signals. A further problem is the selection of technically and economically favourable combinations of principles from the large field of theoretically possible combinations.

In 4.1.4 the classification scheme of Zwicky (morphological matrix) has been proposed as particularly suitable for systematically combining solutions (see Fig. 4.18). In this classification scheme the sub-functions and the appropriate solutions (working principles) are entered in the rows of the scheme. By combining a working principle fulfilling a specific sub-function with the working principle for a neighbouring sub-function one obtains an overall solution in the form of a possible

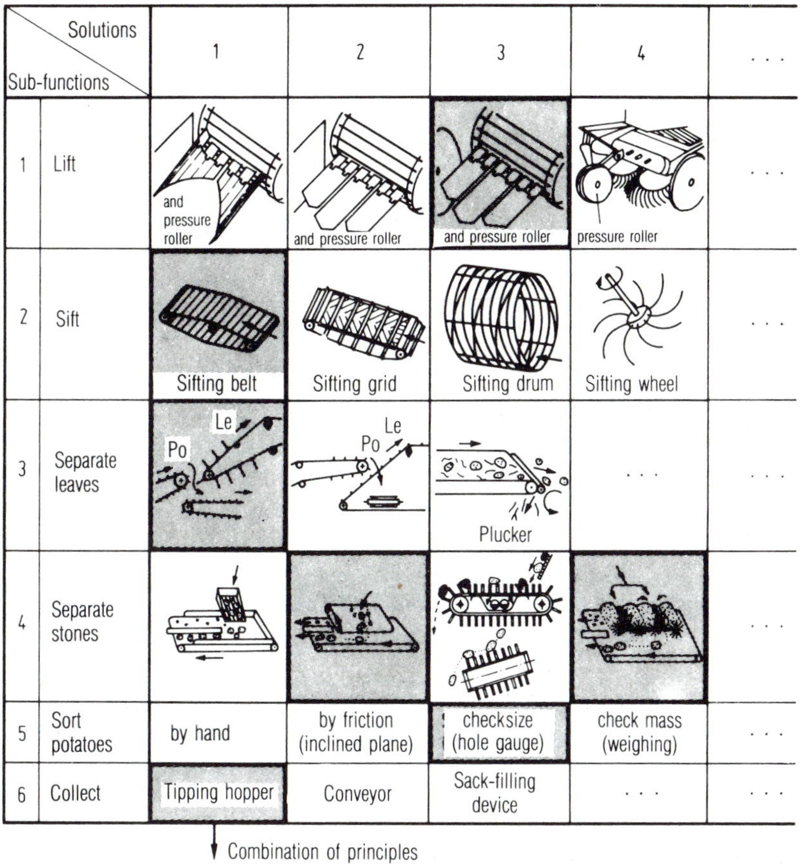

Solutions / Sub-functions	1	2	3	4	. . .
1 Lift	and pressure roller	and pressure roller	and pressure roller	pressure roller	. . .
2 Sift	Sifting belt	Sifting grid	Sifting drum	Sifting wheel	. . .
3 Separate leaves	Le Po	Le Po	Plucker	. . .	. . .
4 Separate stones					. . .
5 Sort potatoes	by hand	by friction (inclined plane)	checksize (hole gauge)	check mass (weighing)	. . .
6 Collect	Tipping hopper	Conveyor	Sack-filling device	. . .	. . .

Combination of principles

Figure 6.18. Combinations of principles for designing a potato harvesting machine in accordance with the overall function structure shown in Fig. 6.9, after [6.1].

working structure. In this process only those working principles that are compatible should be combined.

Figure 6.18 shows an example of combinations for a potato harvesting machine. It consists of working principles that are suitable for the sub-functions in the function structure shown in Fig. 6.9. These have been made more concrete through rough sketches so that the assessment of their compatibility is facilitated. The actual harvesting machine based on this working structure is shown in Fig. 6.19.

A further example is the development of a test rig. The classification scheme shown in Fig. 6.14 is the basis for a combination of working principles using the function structure variants shown in Fig. 6.20, with which a test rig for the loading of shaft-hub connections with impulsive torques can be developed [6.11]. Figure 6.21 shows the combinations of working principles that result in seven different working structures. As can be seen, the order of combining the sub-functions has been modified.

In Fig. 6.14 several working principles have been eliminated because they did not seem appropriate for fulfilling the specified requirements. This early selection is important to restrict the subsequent effort of combining and making the solutions more concrete.

Combining solutions using mathematical methods (see 4.1.4) is only possible for working principles whose properties can be quantified. However, this is seldom possible at this early stage. Examples where it is possible are variant designs and control system designs, such as those using electronic or hydraulic components.

To sum up:

- Only combine compatible sub-functions (useful tool – the compatibility matrix shown in Fig. 4.19).

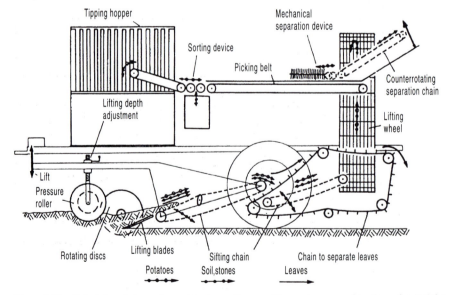

Figure 6.19. Principle solution of a potato harvesting machine, using a combination of principles from Figure 6.18.

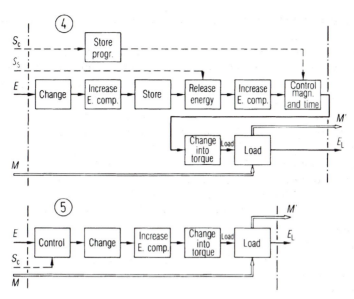

Figure 6.20. Function structure variants for impulse-loading test machine, after [6.11].

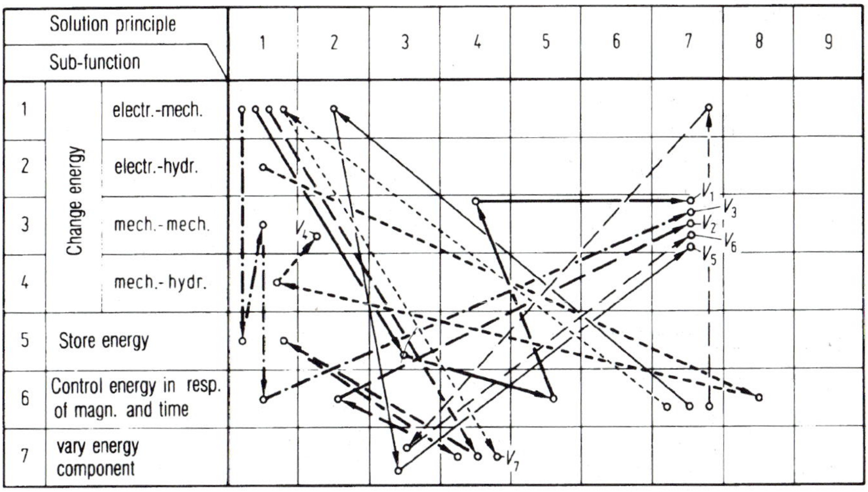

Solution principle Sub-function		1	2	3	4	5	6	7	8	9
1	electr.-mech.									
2	electr.-hydr.									
3	mech.-mech.									
4	mech.-hydr.									
5	Store energy									
6	Control energy in resp. of magn. and time									
7	vary energy component									

Figure 6.21. Combination scheme showing seven combinations of solution principles in accordance with Fig. 6.14.

Variant 1: 1.1–5.3–6.5–3.4–3.7
Variant 2: 1.1–7.4–5.1–7.4–6.2–3.7
Variant 3: 1.1–5.1–3.1–6.1–3.7
Variant 4: 2.1–6.8–4.1–3.2

Variant 5: 6.7–1.2–7.3–3.7
Variant 6: 6.7–1.7–7.3–3.7
Variant 7: 6.7–1.1–7.4

- Only pursue such solutions as meet the demands of the requirements list and look like falling within the proposed budget (see selection procedures in 4.2.1 and 6.4.3).
- Concentrate on promising combinations and establish why these should be preferred above the rest.

6.4.3   Selecting Suitable Working Structures

Because working structures are in general not very concrete and the properties are only known qualitatively, the most suitable selection procedure is the one described in 4.2.1. This procedure is characterised by the activities of *selecting* and *indicating preferences*, and makes use of a schematic *selection chart* that provides a clear overview and can be checked. Figure 6.22 shows an example of such a selection chart for the different working structures that resulted from the combination scheme in Fig. 6.21. Only four solution variants have been selected. For a more detailed assessment it is necessary to make the solution proposals more concrete.

Another possibility for making a rapid selection is the application of two-dimensional classification schemes, similar to the compatibility matrices shown in Fig. 4.19. Figure 6.24 provides an example.

The specification of a test rig for gear couplings (see Fig. 6.23) demanded an axial displacement in the test coupling so that the forces which then appear could be measured. The possible position of the displacement (classifying criterion of the rows) and the axial force input (classifying criterion of the columns) were combined into the classification scheme shown in Fig. 6.24. The various combinations were checked and unsuitable variants were eliminated for a number of immediately obvious reasons.

6.4.4   Practical Uses of Working Structures

The development of working structures is the most important stage for original designs. This stage makes most demands on the creativity of designers. This creativity is influenced by cognitive psychological processes associated with problem solving, by the use of a general working methodology and by generally applicable solution finding and evaluation methods. As a consequence, the approach in this stage is very diverse and depends on the novelty of the task: that is the number of new problems to be solved; on the mentality, ability and experience of the designers; and on the product ideas from product planning or clients.

The procedure suggested in 6.4.1 to 6.4.3 provides the basis for an expedient stepwise design process. The actual process can vary considerably. For *original designs without precedents*, the initial search for solutions should focus on the *main function* that appears to be *crucial* for the overall function (see Fig. 6.12). This guideline is only to be applied when working principles have to be found for several sub-functions that make up a complex overall function. For the solution-determining main function, one must first select some preliminary physical effects

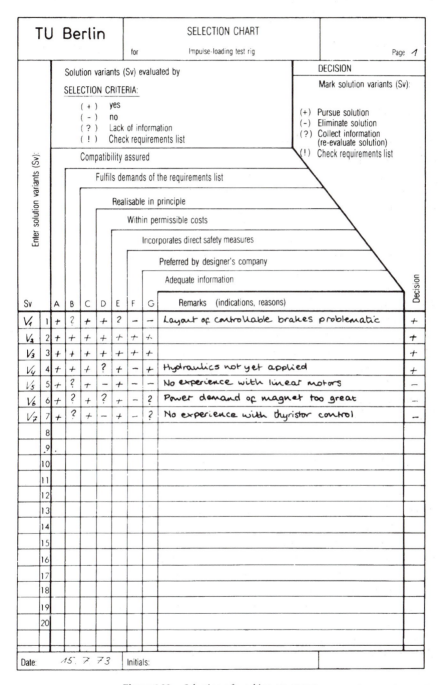

Figure 6.22. Selection of working structures.

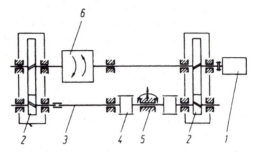

Figure 6.23. Sketch showing the principle of a test rig for gear couplings. *1* drive; *2* gearbox; *3* high-speed shaft; *4* test gear coupling; *5* adjustable bearing block for setting the alignment; *6* device for applying torque.

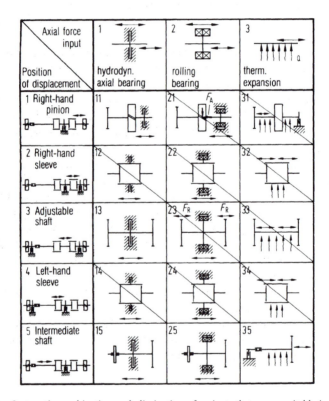

Figure 6.24. Systematic combination and elimination of variants that are unsuitable in principle.
Combination 12, 14: Disturbance of coupling kinematics.
Combination 21: F_A too great (life of rolling bearings too short).
Combination 23: $2 \cdot F_R$, hence life of rolling bearings too short.
Combination 22, 24: Peripheral speed too great (life of rolling bearings too short).
Combination 31–34: Thermal length too small.

or working principles using intuition-based methods, literature and patent searches or previous products. The relationship between the functions in these solutions must be analysed to identify other important sub-functions for which physical effects and working principles have to be found. These working principles are selected from those that are compatible with the other working principles selected to fulfil the main functions. A simultaneous, independent search for working principles for all sub-functions will, in general, be too elaborate and will result in several working principles that would have to be eliminated later from the overall combination, thus resulting in unnecessary effort.

An important strategy for the creation of solution fields is the systematic variation of the physical effects and form design features that were recognised as essential in the initial solutions. *Classification schemes* are very useful, but usually need several trials, based on variation and correction of the classifying criteria, before an optimum scheme can be arrived at. This requires some experience.

When *concrete solution ideas* are available from product planning or other sources, these have to be analysed to identify their essential solution-determining characteristics. These are then systematically varied and combined to arrive at a solution field.

In the case of *evolutionary developments*, the known working principles and working structures should be checked to see if they still meet current technological standards or the latest requirements.

When an approach is strongly based on *intuition* or when previous experience is applicable, working structures that fulfil the overall function will often be found directly without first searching for solutions for the individual sub-functions (working principles).

In particular the *stepwise* generation of working principles, through the search for physical effects and the subsequent form design features, is often mentally integrated by producing *sketches of solutions*. This is because designers think more in configurations and representation of principles than in physical equations.

The use of intuition-based and discursive-systematic methods can lead quickly to extensive solution fields. To limit subsequent design effort, these should be reduced as soon as *feasible working principles* emerge by checking against the demands in the requirements list.

At this stage it is often not possible to assess the characteristics of a principle solution with quantitative data, particularly with regard to production and cost. Therefore the selection of suitable working principles requires an *interdisciplinary team* discussion (similar to a value analysis team, see 10.4) to base the qualitative decision on a broad spectrum of experience.

6.5   Developing Concepts

6.5.1   Firming up into Principle Solution Variants

The principles elaborated in 6.4 are usually not concrete enough to lead to the adoption of a definite concept. This is because, as the search for a solution is based on the function structure, it is aimed, first and foremost, at the fulfilment of a technical function. A concept must, however, also satisfy the conditions laid down in 2.1.8, at least in essence, for only then is it possible to evaluate it. Before concept variants can he evaluated, they must he firmed up which as experience has shown, almost invariably involves a considerable effort.

The selection process may already have revealed gaps in information about very important properties, sometimes to such an extent that not even a rough and ready decision is possible, let alone a reliable evaluation. The most important properties of the proposed combination of principles must first be given a much more concrete *qualitative*, and often also a rough *quantitative*, definition.

Important aspects of the working principle (such as performance and susceptibility to faults) and also of the embodiment (such as space requirements, weight and service life) and finally of important task-specific constraints must be known, at least approximately. More detailed information need only be gathered for promising combinations. If necessary, a second or third selection must follow the collection of further information.

The necessary data are essentially obtained with the help of such proven methods as:

- rough calculations based on simplified assumptions;
- rough sketches or rough scale-drawings of possible layouts, forms, space requirements, compatibility etc;
- preliminary experiments or model tests to determine the main properties, or approximate quantitative statements about the performance and scope for optimisation;
- construction of models to aid analysis and visualisation (for example, kinematic models):
- analogue modelling and systems simulation, often with the help of computers, for example stability and loss analyses of hydraulic systems using electrical analogies;
- further searches of patents and the literature with narrower objectives; and
- market research of proposed technologies, materials, bought-out parts etc.

With these fresh data it is possible to firm up the most promising combinations of principles to the point at which they can be evaluated (see 6.5.2). The properties of the variants must reveal technical as well as economic features so as to permit the most accurate evaluation possible. It is therefore advisable when firming up into principle solutions to bear possible later evaluation criteria (see 4.2.2) in mind, the better to elaborate the information in a purposeful way.

An example will show how it is possible to firm up working principles into principle solutions. To that end, we return once more to our fuel gauge.

Figure 6.25 shows the working principle of the first proposal. Estimates of the weights and inertia forces form the basis of the firming up procedure.

Total force of 20 to 160 litres of the liquid (static):

$$F_{tot} = \rho \cdot g \cdot V = 0.75 \times 10 \times (20 \ldots 160) = (150 \ldots 1200) \text{ N (fuel)}$$

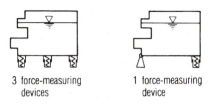

3 force-measuring   1 force-measuring
devices   device

Figure 6.25.   Solution principle 1 (Table 4.2) measure weight of liquid. (Signal = force.)

Additional forces due to acceleration $\pm$ 30 m/s^2 (only the liquid is taken into consideration):

$$F_{add} = m \cdot a = (15 \ldots 120) \times \pm\, 30 = \pm\, (450 \ldots 3600) \text{ N}$$

If the force is converted into movement it can be detected, for instance with the help of a potentiometer. (The suppression of movements resulting from acceleration forces calls for considerable damping.)

It is possible to obtain the total force, and hence the quantity of liquid, statically, either directly by measuring three bearing forces or indirectly by measuring just one bearing force (see Fig. 6.25).

Result: develop solution further, provide damping, seek appropriate solutions and firm them up by means of rough scale drawings. Figure 6.26 shows the result. Once the necessary parts and their arrangement are drawn, the proposal can be evaluated. This confirms the indication in Fig. 4.20 that the effort to complete solution variant 1 will be too high.

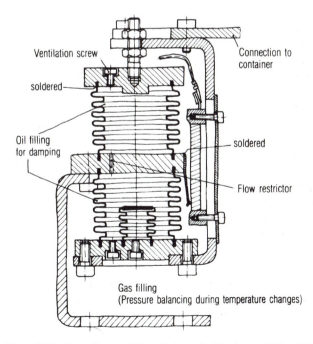

Ventilation screw

Connection to
container

soldered

Oil filling
for damping

soldered

Flow restrictor

Gas filling
(Pressure balancing during temperature changes)

Figure 6.26.   Embodiment of the solution principle shown in Fig. 6.25.

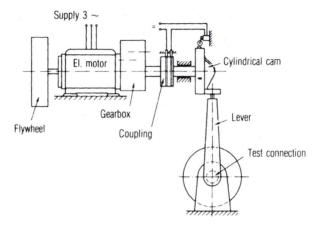

Figure 6.27.   Concept variant V_2.

A further example is the working structure variant V_2 of the impulse loading test rig shown in Fig. 6.21 with the combined working principles given in Fig. 6.14. This variant was considered feasible in the first selection round (see Fig. 6.22), but had to be firmed up before a proper evaluation could take place. Figure 6.27 shows a sketch of the principle of this solution. The essential features are the cylindrical cam to control the impulsive torque and the moment of inertia of the flywheel (energy storage).

The following calculations were necessary to evaluate the solution against technical and economic criteria.

Can the cylindrical cam shown in Fig. 6.28 produce the required torque increase of $dT/dt = 125 \times 10^3$ Nm/s and a maximum torque of $T_{max} = 15 \times 10^3$ Nm?

Calculation steps:

- Time needed to reach the maximum torque at the required rate:

$$\Delta t = \frac{15 \times 10^3}{125 \times 10^3} = 0.12 \text{ s}$$

- Force at the end of the loading lever:

$$F_{max} = \frac{T_{max}}{l} = \frac{15 \times 10^3}{0.85} = 17.6 \times 10^3 \text{ N}$$

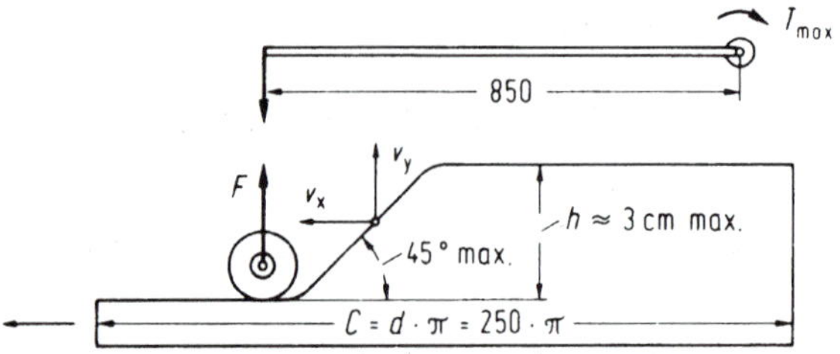

Figure 6.28.   Development of cylindrical cam.

The loading lever is treated as a weak cantilever spring with the end moving through a distance of $h = 30$ mm with a force of F_{max} in such a way that the permissible bending stress is not exceeded.

- Tangential velocity of the cylindrical cam:

$$v_x = v_y = \frac{h}{\Delta t} = \frac{30}{0.12} = 250 \text{ mm/s}$$

- Angular velocity and rpm of cylindrical cam:

$$\omega = \frac{0.25}{0.125} = 2.0 \text{ rad/s}; \; n = \frac{60\omega}{2\pi} = 19 \text{ rev/min}$$

- Period of revolution:

$$t_r = \frac{2\pi}{\omega} = 3.14 \text{ s}$$

Since the switching time of electromagnetically operated clutches for connecting and disconnecting the cam drive is in the region of a few tenths of a second, there should be no problems in applying this principle. The magnitude of, and rate of increase in, the impulse torque loading can be altered by means of interchangeable cams and also by varying the period of revolution.

Steps for estimating flywheel moment of inertia:
- Estimate of the energy needed for the impulse (and hence of the energy to be stored) on the assumption that all load carrying parts are elastically deformed. Stored energy at maximum impulse torque loading:

$$E_{max} = \tfrac{1}{2}F_{max} \cdot h = 260 \text{ J}$$

This amount of energy is needed in the time interval $\Delta t = 0.12$ s.
- Flywheel dimensions:
Select: Maximum rpm, $n_{max} = 1200$ rev/min; $\omega \cong 126$ rad/s
For flywheel dimensions $r = 0.2$ m and $w = 0.1$ m, the flywheel mass $m_f = 100$ kg, and moment of inertia $J_f = \tfrac{1}{2}m_f \cdot r^2 = 2$ kgm^2.
Stored energy of flywheel:

$$E_f = \tfrac{1}{2}J_f \cdot \omega^2 = 159 \times 10^2 \text{ J}$$

- Rotational speed after the impulse:

$$E_{after} = E_f - E_{max} = 15\,640 \text{ J}$$

$$\omega_{after} = \sqrt{\frac{2E_{after}}{J_f}} = 125 \text{ rad/s}; \; n_{after} = 1190 \text{ rev/min}$$

The drop in rpm is thus seen to be very low. Hence all that is needed is a motor with a small output.

6.5.2   Evaluating Principle Solution Variants

In section 4.2.2 we explained generally applicable evaluation methods, in particular Use-Value Analysis and the VDI 2225 procedure [6.18].

For the evaluation of principle solution variants the following steps are recommended:

Identifying Evaluation Criteria

This step is based, first of all, on the *requirements list*. During a previous selection procedure (see 6.4.3) unfulfilled demands may have led to the elimination of variants that were found to be unsuitable in principle. Further information was gathered subsequently by firming up into principle solutions. Hence it is advisable, with all the newly acquired information, to establish first of all whether all the proposals to be evaluated still satisfy the demands of the requirements list. This can involve a new yes/no decision – that is a new selection.

It is only to be expected that, even at the present more concrete stage, this decision cannot be made with certainty for all the variants unless much further effort is applied, which the designers may not be able to provide at this stage. At the given state of information, it may only be possible to decide how likely it is that certain requirements can be fulfilled. In that case, the requirements in question may become evaluation criteria.

A number of requirements are minimum requirements. It has to be established whether or not these should be exceeded. If they should, further evaluation criteria may be needed.

For evaluation during the conceptual phase, both the *technical* and also the *economic characteristics* should be considered as early as possible [6.8]. At the firming-up stage, however, it is not usually possible to give the costs in figures. Nevertheless, the economic aspects must be taken into consideration, at least qualitatively, and so must industrial and environmental safety requirements.

Hence it is necessary to consider technical, economic and safety criteria at the same time. It is suggested [6.11] that the evaluation criteria are derived from the main headings in Fig. 6.29. These are in accordance with the embodiment design checklist (see 7.6).

Every heading in the checklist relevant for the task must be assigned at least one evaluation criterion. The criteria must, moreover, be independent of one another in terms of the overall objective, so as to avoid multiple evaluations. Consumer criteria are essentially contained in the first five and last four headings; producer criteria in the headings: embodiment, production, quality control, assembly, transport and costs.

Evaluation criteria are accordingly derived from:

1. Requirements from the requirements list
    *   Probability of satisfying the demands (how probable, despite what difficulties?).

Main headings	Examples
Function	Characteristics of essential auxiliary function carriers that follow of necessity from the chosen solution principle or from the concept variant
Working principle	Characteristics of the selected principle or principles in respect of simple and clear-cut functioning, adequate effect, few disturbing factors
Embodiment	Small number of components, low complexity, low space requirement, no special problems with layout or form design
Safety	Preferential treatment of direct safety techniques (inherently safe), no additional safety measures needed, industrial and environmental safety guaranteed
Ergonomics	Satisfactory man-machine relationship, no strain or impairment of health, good aesthetics
Production	Few and established production methods, no expensive equipment, small number of simple components
Quality control	Few tests and checks needed, simple and reliable procedures
Assembly	Easy, convenient and quick, no special aids needed
Transport	Normal means of transport, no risks
Operation	Simple operation, long service life, low wear, easy and simple handling
Maintenance	Little and simple upkeep and cleaning, easy inspection, easy repair
Recycling	Easy recovery of parts, safe disposal
Costs	No special running or other associated costs, no scheduling risks

Figure 6.29. Checklist with main headings for design evaluation during the conceptual phase.

- Desirability of exceeding minimum requirements (how far exceeded?).
- Wishes (satisfied, not satisfied, how well satisfied?).

2. General technical and economic characteristics (to what extent present, how satisfied?).

The checklist headings for design evaluation during the conceptual phase in Fig. 6.29 should be referred to.

During the conceptual phase, the total number of evaluation criteria should not be too high: 8 to 15 criteria are usually enough (see Fig. 6.51).

Weighting the Evaluation Criteria

The evaluation criteria adopted may differ markedly in importance. During the conceptual phase, in which the level of information is fairly low because of the relative lack of embodiment, weighting is not generally advisable.

It is much more advantageous in the selection of evaluation criteria to strive for an approximate balance, ignoring low-weighted characteristics for the time being. As a result, evaluation will be concentrated on the main characteristics and hence be clear at a glance. Extremely important requirements, however, which cannot be ignored until later, must be introduced with the help of weighting factors.

Compiling Parameters

It has proved useful to list the identified evaluation criteria in the sequence of the checklist headings and to assign the parameters of the variants to them. Whatever quantitative information is available at this stage should also be included. Such quantitative data generally result from the step we have called "firming up into principle solution variants". However, since it is impossible to quantify all the characteristics during the conceptual phase, the qualitative aspects should be put into words and correlated with the value scale.

Assessing Values

Though the attribution of points raises problems, it is not advisable to evaluate too timidly during the conceptual phase.

Those using the 0-4 scale proposed in VDI Guideline 2225 may feel the need to assign intermediate values, particularly when there are many variants, or when the evaluation team cannot agree on a precise point. It may prove helpful in such cases to attach a tendency sign (↑ or ↓) to the point in question (see Fig. 6.51). Identifiable tendencies can then be taken into account when estimating the evaluation uncertainties. The 0-10 scale, again, may suggest a degree of accuracy that does not really exist. Here, arguments about a point are often superfluous. If there is absolute uncertainty in the attribution of points, which happens quite often during the evaluation of concept variants, the point under consideration should be indicated with a question mark (see Fig. 6.51).

During the conceptual phase it may prove difficult to put actual figures to the costs. It is not therefore generally possible to establish an *economic rating* R_e with respect to the manufacturing costs. Nevertheless, the technical and economic aspects can be identified and separated qualitatively, to a greater or lesser extent. The *strength diagram* (see Fig. 4.28) can be used to much the same effect (see also Figs. 6.30 to 6.32 which are for the test rig shown in Fig. 6.23).

In a similar way a classification based on consumers' and producers' criteria often proves useful. Since the consumers' criteria usually involve technical ratings while the producers' criteria involve economic ratings, it is possible to proceed to a similar classification to the one mentioned above. There are three possible forms of representation, namely:

- technical rating with implicit economic aspects (see Figs. 6.34 and 6.51); or
- separate technical and economic ratings (see Figs. 6.30 to 6.32); or
- *additional* comparison of consumers' and producers' criteria.

Which one is chosen depends on the problem and the amount of information available.

Determining Overall Value

The determination of the overall value is a matter of simple addition, once points have been assigned to the evaluation criteria and the variants. If, because of the

| Variant | 11 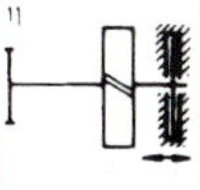 | 13 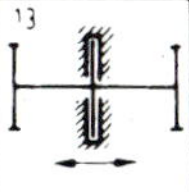 | 15 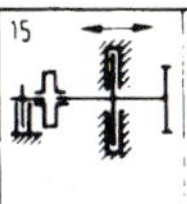 | 25 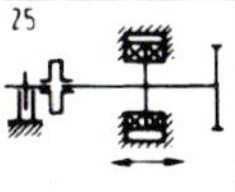 | 35 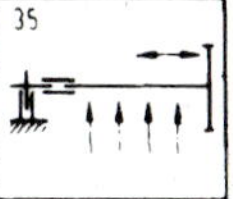 |
techn. criteria					
1) Small disturbance of coupling kinematics	(1) 3	4	4	4	3
2) Simple operation	3	4	4	4	3
3) Easy exchange of coupling	4	3	4	4	4
4) Functional safety	2	4	3	3	3
5) Simple construction	(1) 2	2	2	2	3
Total	14	17	17	17	16
$R_t = \dfrac{\text{Total}}{20}$	0.7	0.85	0.85	0.85	0.80

(1) Torque changes with axial displacement of pinion

Figure 6.30.   Technical evaluation of the remaining principle solution variants, see Fig. 6.24.

| Variant | 11 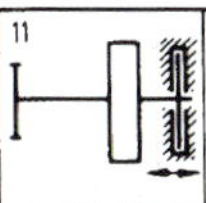 | 13 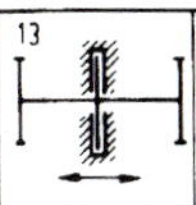 | 15 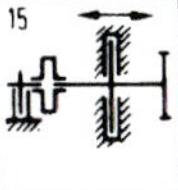 | 25 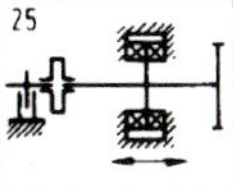 | 35 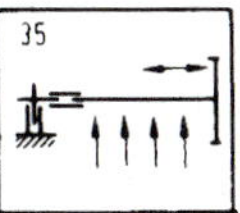 |
econ. criteria					
1) low material costs	2	3	4	4	(1) 2
2) low reassembly costs	2	(2) 1	3	3	3
3) Short testing time	2	4	3	3	2
4) Possibility of manufacturing in own workshop	3	3	3	3	2
Total	9	11	13	13	9
$R_e = \dfrac{\text{Total}}{16}$	0.56	0.69	0.81	0.81	0.56

(1) Austenitic shaft   (2) Torque measuring shaft must be moved

Figure 6.31.   Economic evaluation of the remaining principle solution variants, see Fig. 6.24.

evaluation uncertainty, it is only possible to assign a range of points to individual variants, or if tendency signs are used, one can additionally determine the possible minimum or maximum overall point number and so obtain the probable overall value range (see Fig. 6.51).

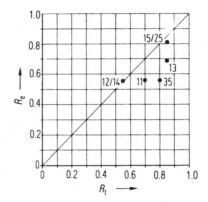

Figure 6.32. Comparison of the technical and economic ratings of the principle solution variants in Figs. 6.30 and 6.31.

Comparing Concept Variants

A relative value scale is generally more suitable for purposes of comparison. In particular, it makes it fairly simple to tell whether particular variants are relatively close to or far from the target. Concept variants that are some 60 per cent below the target are not worth further development.

Variants with ratings above 80 per cent and a balanced value profile – that is, without extremely bad individual characteristics – can generally be moved on to the embodiment design phase without further improvement.

Intermediate variants, too, may, after the elimination of weak spots or an improved combination, be released for embodiment design.

It often happens that two or more variants are found to be practically equivalent. It is a very grave mistake, in that case, to base the final decision on such slight differences. Instead, evaluation uncertainties, weak spots and the value profile should be looked at more closely. It may also be necessary to firm up on such variants during a further step. Schedules, trends, company policy and so on must be assessed separately and taken into account [6.8].

Estimating Evaluation Uncertainties

This step is very important, especially during the conceptual phase, and must not be omitted. Evaluation methods are mere tools, not automatic decision mechanisms. Uncertainties must be determined as indicated earlier. At this point, however, only such informational gaps need be closed as bear on the best concept variants (for example, variant B in Fig. 6.51).

Searching for Weak Spots

During the conceptual phase, the value profile plays an important role. Variants with a high rating but definite weak spots (unbalanced value profile) may prove extremely troublesome during subsequent development. If, because of an unidentified evaluation uncertainty, which is more likely to occur in the conceptual than in the embodiment phase, a weak spot should make itself felt later, then the whole concept may be put in doubt and all the development work may prove to have been in vain.

In such cases it is very much less risky to select a variant with a slightly lower rating but a more balanced value profile.

Weak spots in favourite variants can often be eliminated by the transfer of better sub-solutions from other variants. Moreover, with better information, it is possible to search for a replacement for the unsatisfactory sub-solution. Thus the criteria we have listed played an essential role in the selection of the best variant in the problem discussed in 6.6 (see Fig. 6.51). When estimating evaluation uncertainties and also in the search for weak spots it is advisable to assess the probability and magnitude of the possible risk, especially in the case of important decisions.

We will continue using the test rig shown in Figs. 6.14, 6.20, 6.21, 6.22 and 6.27 to illustrate the application of an objectives tree (see Fig. 4.22), an evaluation chart (see Fig. 4.27) and a value profile (see Fig. 4.31).

Important wishes in the requirements list provide a series of evaluation criteria of varying complexity. These are assessed and elaborated with the help of the checklist shown in Fig. 6.29. Next, a hierarchical classification (objectives tree) is drawn up to facilitate closer identification, and better assignment, of the weighting factors and the parameters of the variants (see Fig. 6.33).

In Fig. 6.34 parameter magnitudes and values assigned to the variants have been set against the evaluation criteria.

It appears that variant V_2 has the highest overall value and the best overall rating. However, variant V_3 follows close behind.

For the detection of weak spots, it is advisable to draw a value profile (see Fig. 6.35). The profile shows that variant V_2 is well balanced in respect of all the important evaluation criteria. With a weighted rating of 68%, variant V_2 thus represents a good *principle solution (concept)* with which to start the embodiment design phase.

6.5.3 Practical Approach to Finding Concepts

The selection of the concept, or the principle solution, provides the *basis for starting the embodiment design phase* (see Fig. 6.1). This often indicates a need for changes in organisation and personnel because the nature of the work alters. Thus firming up of suitable working structures into principle solution variants and the subsequent evaluation at the end of the conceptual design phase are of major importance for product development. The large number of variants have to be reduced to one concept, or just a few, to be pursued further. This decision incurs a heavy responsibility and can only be made when the principle solutions are in a state suitable for evaluation. This needs representations of the solutions which in some cases may be rough scale layouts, involving preliminary calculations and sometimes

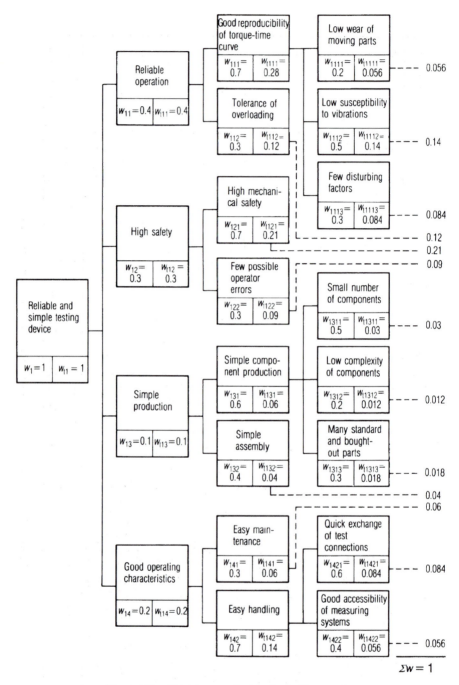

Figure 6.33. Objectives tree for impulse-loading test rig.

No.	Evaluation criteria	Wt.	Parameters	Unit	Variant V_1 Magn. m_{i1}	Value v_{i1}	Weighted value WV_{i1}	Variant V_2 Magn. m_{i2}	Value v_{i2}	Weighted value WV_{i2}	Variant V_3 Magn. m_{i3}	Value v_{i3}	Weighted value WV_{i3}	Variant V_4 Magn. m_{i4}	Value v_{i4}	Weighted value WV_{i4}
1	Low wear of moving parts	0.056	Amount of wear	-	high	3	0.168	low	6	0.336	average	4	0.224	low	6	0.336
2	Low susceptibility to vibrations	0.14	Natural frequency	s^{-1}	410	3	0.420	2370	7	0.980	2370	7	0.980	< 410	2	0.280
3	Few disturbing factors	0.084	Disturbing factors	-	high	2	0.168	low	7	0.588	low	6	0.504	(average)	4	0.336
4	Tolerance of overloading	0.12	Overload reserve	%	5	5	0.600	10	7	0.840	10	7	0.840	20	8	0.960
5	High mechanical safety	0.21	Expected mechan. safety	-	average	4	0.840	high	7	1.470	high	7	1.470	very high	8	1.680
6	Few possible operator errors	0.09	Possibilities of operator errors	-	high	3	0.270	low	7	0.630	low	6	0.540	average	4	0.360
7	Small number of components	0.03	No. of components	-	average	5	0.150	average	4	0.120	average	4	0.120	low	6	0.180
8	Low complexity of components	0.012	Complexity of components	-	low	6	0.072	low	7	0.084	average	5	0.060	high	3	0.036
9	Many standard and bought-out parts	0.018	Proportion of standard and bought-out components	-	low	2	0.036	average	6	0.108	average	6	0.108	high	8	0.144
10	Simple assembly	0.04	Simplicity of assembly	-	low	3	0.120	average	5	0.200	average	5	0.200	high	7	0.280
11	Easy maintenance	0.06	Time and cost or maintenance	-	average	4	0.240	low	8	0.480	low	7	0.420	high	3	0.180
12	Quick exchange of test connections	0.084	Estimated time needed to exchange test connections	min	180	4	0.336	120	7	0.588	120	7	0.588	180	4	0.336
13	Good accessibility of measuring systems	0.056	Accessibility of measuring systems	-	good	7	0.392	good	7	0.392	good	7	0.392	average	5	0.280
		$\Sigma w_i =$ 1.0				$OV_1 = 51$ $R_1 = 0.39$		$OWV_1 =$ 3.812 $WR_1 = 0.38$	$OV_2 = 85$ $R_2 = 0.65$		$OWV_2 =$ 6.816 $WR_2 = 0.68$	$OV_3 = 78$ $R_3 = 0.60$		$OWV_3 =$ 6.446 $WR_3 = 0.64$	$OV_4 = 68$ $R_4 = 0.52$		$OWV_4 =$ 5.388 $WR_4 = 0.54$

Figure 6.34. Evaluation of the four concept variants for the impulse-loading test rig.

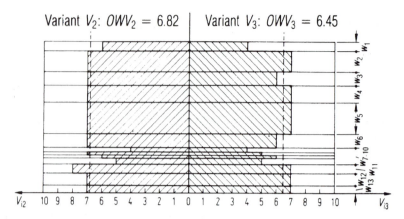

Figure 6.35. Value profile for detection of weak spots.

tests. From research in industry and universities [6.11], it is known that calculating and drawing add up to 60% of the total time spent on conceptual design.

The *representation* of working principles and working structures in the form of simple sketches is likely to remain the domain of conventional sketching. Rough layouts and in particular the more important details of solutions will increasingly be represented using CAD [6.12]. Sketching working structures by hand has the advantage that one does not need to consider the formalities of CAD user interfaces during this highly creative stage. Firming up solution principles using CAD is useful, despite the effort needed to enter the product model into the system, because making variations to the layout and individual components becomes very efficient. For dynamic systems it is also possible to do initial simulations using the CAD model.

In any case it is expedient for *reasons of efficiency* and for identifying essential characteristics, not to firm up the whole working structure to the same level of detail. The aim should be to focus on those working principles, components or parts of the structure that are essential for the evaluation of the working principles and selection of the one that will be transferred to the embodiment stage. Richter provides proposals for this task [6.13].

It has to be emphasised again that *iterations* often occur in the steps mentioned in 6.4 and 6.5. On the one hand it might be necessary to detail working principles in order to combine and select them, on the other a completely new idea for a working principle might emerge while making a rough layout of a principle solution.

It must be stressed that principle solutions or concepts have to be *unambiguously documented*. It must also be clear which parts of the working structure or function carriers can be realised by existing and standard components, and which ones will need to be specially designed.

6.6 Examples of Conceptual Design

6.6.1 Mechanical System

This section illustrates how the approach can be applied to a material flow problem. An example of signal flow has been used throughout the previous sections in this chapter. An example of energy flow can be found in Figs. 6.14, 6.20 to 6.22, 6.27, 6.28 and 6.33 to 6.35 (conceptual design) and in 7.7 (embodiment design).

One-handed Household Water Mixing Tap

A one-handed mixing tap is a device for regulating water temperature and through-flow with one hand. This task was sent to the design department by the planning department in the form shown in Fig. 6.36.

Step 1: Clarifying the Task and Elaborating the Requirements List

New data on fittings, standards, safety regulations and ergonomic factors led to the replacement of the original requirements list by the revised version shown in Fig. 6.37.

Step 2: Abstracting to Identify the Essential Problems

The basis of abstraction is the requirements list, from which it is possible to arrive at Fig. 6.38. Simple household solutions for mixing taps suggested that the chosen solution principle must be based on metering out the water through a diaphragm or valve. Such alternatives as heating and cooling by the introduction of external

---

One-handed water mixing tap

Required: one-handed household water mixing tap with the following characteristics:

Throughput	10 l/min
Max. pressure	6 bar
Normal pressure	2 bar
Hot water temperature	60°C
Connector size	10 mm

Attention to be paid to appearance. The firm's trade mark to be prominently displayed. Finished product to be marketed in two years' time Manufacturing costs not to exceed DM 30 each at a production rate of 3000 taps per month.

Figure 6.36. One-handed mixing tap. Example of an assignment suggested by the product planning department.

TH Darmstadt		REQUIREMENTS LIST for One-handed mixing tap			Page 1
Changes	D W	Requirements			Resp.
	D	1 Throughput (mixed flow) max 10 l/min at 2 bar			
	D	2 Max. pressure 10 bar (test pressure 15 bar as per DIN 2401)			
	D	3 Temp. of water: standard 60°C, 100°C (short-time)			
	D	4 Temperature setting independent of throughput and pressure			
	W	5 Permissible temp fluctuation ±5°C at a pressure diff. of ±5 bar between hot and cold supply			
	D	6 Connection: 2 × Cu pipes, 10 × 1 mm, l = 400 mm			
	D	7 Single-hole attachment Ø 35$^{+2}_{-1}$ mm, basin thickness 0–18 mm (Observe basin dimensions DIN EN 31, DIN EN 32, DIN 1368)			
	D	8 Outflow above upper edge of basin: 50 mm			
	D	9 To fit household basin			
	W	10 Convertible into wall fitting			
	D	11 Light operation (children)			
	D	12 No external energy			
	D	13 Hard water supply (drinking water)			
	D	14 Clear identification of temperature setting			
	D	15 Trade mark prominently displayed			
	D	16 No connection of the two supplies when valve shut			
	W	17 No connection when water drawn off			
	D	18 Handle not to heat above 35°C			
	W	19 No burns from touching the fittings			
	W	20 Provide scalding protection if extra costs small			
	D	21 Obvious operation, simple and convenient handling			
	D	22 Smooth, easily cleaned contours, no sharp edges			
	D	23 Noiseless operation (≤20 dB as per DIN 52218)			
	W	24 Service life 10 years at about 300 000 operations			
	D	25 Easy maintenance and simple repairs. Use standard spare parts			
	D	26 Max. manuf. costs DM 30 (3000 units per month)			
	D	27 Schedules from inception of development			
		conceptual embodiment design detail design prototype design			
		after 2 4 6 9 months			
		Replaces 1st issue of 12.6.1973			

Figure 6.37.   Requirements list for one-handed mixing tap.

energy through heat exchangers etc could be dismissed: they were more expensive and involved a time lag. Selecting sound solution principles without further investigation, because they have proved their worth in previous company products, is a common and justified approach.

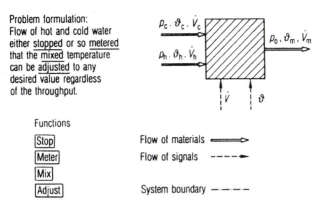

Problem formulation:
Flow of hot and cold water either stopped or so metered that the mixed temperature can be adjusted to any desired value regardless of the throughput.

Functions

Stop
Meter
Mix
Adjust

Flow of materials ⟹

Flow of signals ----→

System boundary —·—·—

Figure 6.38. Problem formulation and overall function as per requirements list, Figure 6.37. V = volume flow rate; p = pressure; ϑ = temperature. Index: c = cold, h = hot, m = mixed, o = atmosphere.

Next, the physical relationships for the diaphragm or valve flow rate and the temperature of mixed flows of similar fluids were determined (see Fig. 6.39).

Temperature and flow rate adjustments are based on the same physical principle—valve or diaphragm.

On *changing the flow rate* $\dot{V}_m$, the flows must be changed linearly and in the same sense as the signal setting $s_{\dot{v}}$. The temperature ϑ_m, however, must remain unchanged—that is, the relationship $\dot{V}_c / \dot{V}_h$ must remain constant and independent of the signal positions $s_{\dot{v}}$.

On *changing the temperature* ϑ_m, the volume flow rate $\dot{V}_m$ must remain unchanged—that is, the sum of $\dot{V}_c + \dot{V}_h = \dot{V}_m$ must remain constant. To that end the component flows $\dot{V}_c$ and $\dot{V}_h$ must be changed linearly and in the opposite sense to the signal setting s_ϑ.

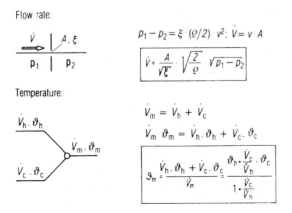

Figure 6.39. Physical relationships for flow rate and temperature of mixed flows of the same fluid.

Step 3: Establishing Function Structures

The function structures were derived from the sub-functions:

- Stop – meter – mix,
- Adjust flow rate,
- Adjust temperature.

The physical principle being well known, the function structures could be varied and developed to determine the best system and its behaviour (see Figs. 6.40 to 6.42). From the results the function structure shown in Fig. 6.42 was chosen as the most satisfactory.

Step 4: Searching for Solution Principles to Fulfil the Sub-functions

Brainstorming was used as a first attempt to solve the problem: "vary two flow areas, simultaneously or successively, in one sense by one movement and in the

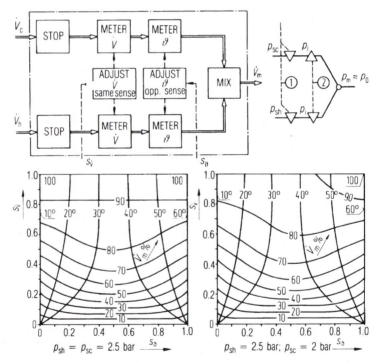

Figure 6.40. Function structure for a one-handed water mixing tap based on Fig. 6.38. Metering flow ① and adjusting temperature ② separately before mixing. In the graphs, lines of constant temperature and constant percentage flow rate have been plotted for given temperature settings (s_ϑ) and flow rate settings ($s_{\dot{v}}$). Through mutual effects of the pressures on the inlets at ① and ② the temperature and flow characteristics are not linear except for the setting $s_{\dot{v}} = 0.825$, and hence are unsuitable for small flow rates. At a pressure difference between the cold and hot water supplies (in this case $p_{sh} - p_{sc} = 0.5$ bar) the lines shift. The settings are no longer independent of each other even for the setting $s_{\dot{v}} = 0.825$ (diagram on right).

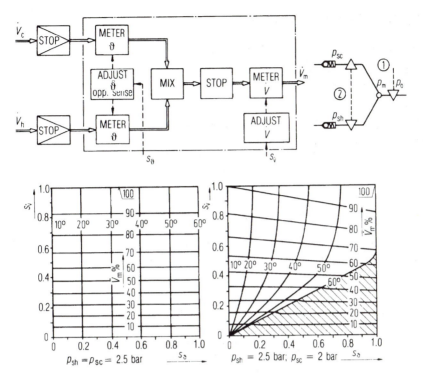

Figure 6.41. Function structure based on Fig. 6.38 in which the temperature is set before, and the flow metered after, mixing. With equal pressures in the supply pipes, the flow and temperature settings are independent of each other due to equal pressure differences across each temperature-flow-metering valve. The behaviour is linear. With different supply pressures, however, the characteristic ceases to be linear and is strongly displaced, especially with small quantities, when the pressure in the mixing chamber approximates the smaller supply pressure. If it is exceeded, then cold or (here) hot water only will run out regardless of the temperature setting.

opposite sense by a second, independent, movement". This resulted from the sub-function: "meter flow rate and temperature" in accordance with the function structure shown in Fig. 6.42. The results are shown in Fig. 6.43.

Analysis of Brainstorming Results

The solutions suggested during the brainstorming session were checked to establish whether the $\dot{V}$ and ϑ settings were independent. An analysis of the combined movements suggested the following solution principles:

1. *Solutions with separate movements for $\dot{V}$ and ϑ tangential to the valve seat face*
   - The independence of the $\dot{V}$ and ϑ setting is only guaranteed if each of the flow areas of the valves is bounded by two edges running parallel to the corresponding movements. This implies that the movements must proceed at an angle to each other and in a straight line. Every valve setting thus has two pairs of straight and parallel bounding edges (see Fig. 6.44). This

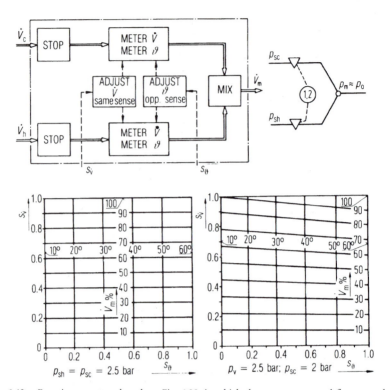

Figure 6.42. Function structure based on Fig. 6.38, in which the temperature and flow at each inlet is metered out independently and then mixed. Linear temperature and flow characteristics. No serious changes even at different supply pressures.

ensures that when one setting is adjusted the other setting is not simultaneously adjusted.

- Distribution of bounding edges. Each of the components producing the valve flow areas must have at least two edges that face each other and lie in the direction of the movement.
- With the $\dot{V}$ setting, both valve areas must approach zero simultaneously.
- With the ϑ setting, one area must approach zero as the other approaches its maximum $\dot{V}_{max}$.
- This implies, with $\dot{V}$ settings, that the bounding edges on both valve areas must move towards each other or away from each other in the same sense. With ϑ settings, the bounding edges on the two valve areas must move in the opposite sense to each other.
- The seat face may be plane, cylindrical or spherical.
- Solutions of this type can be effected with a single valve element, and seem simple to design.

Figure 6.43. Result of a brainstorming session to discover solution principles for the assignment "vary two flow areas, simultaneously or successively, in one sense by one movement and in the opposite sense by a second, independent movement".

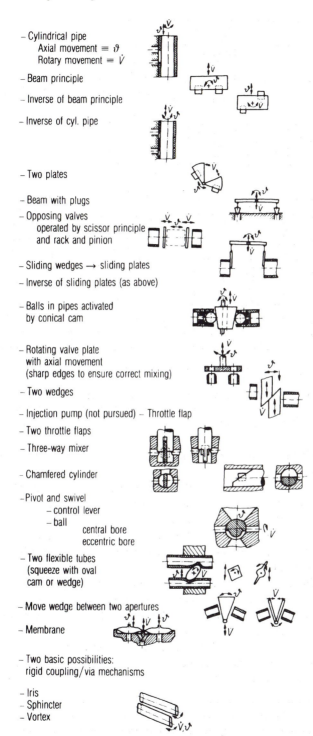

- Cylindrical pipe
  Axial movement ≡ ϑ
  Rotary movement ≡ $\dot{V}$

- Beam principle

- Inverse of beam principle

- Inverse of cyl. pipe

- Two plates

- Beam with plugs

- Opposing valves
  operated by scissor principle
  and rack and pinion

- Sliding wedges → sliding plates

- Inverse of sliding plates (as above)

- Balls in pipes activated
  by conical cam

- Rotating valve plate
  with axial movement
  (sharp edges to ensure correct mixing)

- Two wedges

- Injection pump (not pursued) – Throttle flap

- Two throttle flaps

- Three-way mixer

- Chamfered cylinder

- Pivot and swivel
      – control lever
      – ball
              central bore
              eccentric bore

- Two flexible tubes
  (squeeze with oval
  cam or wedge)

- Move wedge between two apertures

- Membrane

- Two basic possibilities:
  rigid coupling/via mechanisms

- Iris
- Sphincter
- Vortex

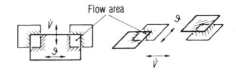

Figure 6.44.   Movements and bounding edges of valve positions.

2. *Solutions with separate movements for $\dot{V}$ and ϑ normal to the valve seat face*
    - This group includes all movements which involve lifting a valve from its seat face. However, only a movement at right angles to the seat face is possible in practice.
    - The independent setting of $\dot{V}$ and ϑ can only be achieved with additional control elements (coupling mechanism).
    - The design seems to require greater effort.
3. *Solutions with one type of movement for $\dot{V}$ and ϑ tangential to the seat face*
    - To guarantee the independence of the $\dot{V}$ and ϑ settings, additional coupling elements are needed.
    - The solutions are similar to those listed under 2. They only differ in the shape of the seat face and the resulting movement.
4. *Solutions with one movement for $\dot{V}$ normal to, and one movement for ϑ tangential to, the seat face and vice versa*
    - These solutions do not, even with the help of coupling mechanisms, satisfy the demand for independent $\dot{V}$ and ϑ settings. The overall function is not achieved.

The solutions of the first group (movements for $\dot{V}$ and ϑ tangential to the valve seat face) have an unambiguous behaviour and seem to be less complex. Therefore they will be pursued, a formal selection procedure not being necessary. On the other hand useful working parts and types of movement still have to be analysed. This analysis results in the classification criteria shown in Fig. 6.45, with the least suitable characteristics indicated with (−). Figure 6.46 shows a classification scheme of possible working principles based on different working parts and movements.

Step 5: Selecting Suitable Working Principles

All working principles fulfil the demands of the requirements list and promise to be economic. Hence all three working principles are firmed up into principle solutions.

Step 6: Firming up into Principle Solution Variants

With the help of further research into possible setting or operating elements which we have not discussed here, the working principles could then be firmed up into principle solution variants which can be evaluated (see Figs. 6.47 to 6.50).

	Classifying Criteria	Parameters
Rows	Form of working elements	Flat plate Wedge (–) Cylinder Cone (–) Ball Special elastic body (–)
Columns	Coupling of Movements	Direct (one part) Indirect (mechanism) (–)
	Movement	$\left.\begin{array}{l}\dot V\\ \dot\vartheta\end{array}\right\}$ One element $\left.\begin{array}{l}\dot V\\ \dot\vartheta\end{array}\right\}$ Several elements
	Direction of Movement $\dot V$ and $\dot\vartheta$ Type of Movement $\dot V$ and $\dot\vartheta$	Normal to seat face ($\perp$) (–) Tangential to seat face ($\longrightarrow$) Transitional Rotational

Figure 6.45. Classifying criteria and parameters for working principles of one-handed water mixing tap.

Form of valve \ Type of movement		trans./trans. 1	trans./rot. 2	rot./rot. 3
plane plate	A	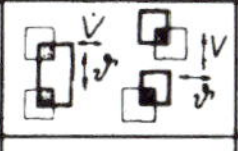	o	o
cylinder	B	o		o
cone	C	o	o	o
sphere	D	o	o	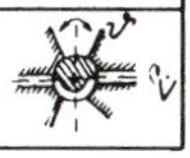

Figure 6.46. Classification scheme for solutions of the one-handed mixing tap problem. Movement tangential to the seat face. Two independent movements at an angle for V and ϑ.

Step 7: Evaluating the Principle Solutions

In accordance with VDI 2225, this step was taken with the help of an evaluation chart. In addition, evaluation uncertainties and weak spots were examined (see Fig. 6.51).

Thanks to the balanced profile and the discernible improvement possibilities, Solution B (see Fig. 6.48) was found to be preferable to all others. The ball Solution D

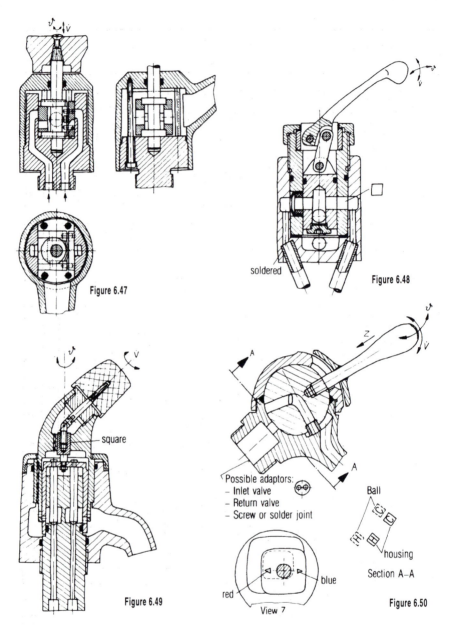

Figure 6.47

Figure 6.48

soldered

square

Possible adaptors:
– Inlet valve
– Return valve
– Screw or solder joint

Ball

housing

Section A–A

Figure 6.49

Figure 6.50

View 7

red

blue

Figure 6.47. One-handed mixing tap, solution variant A: "plate solution with eccentric and pull-and-turn grip".
Figure 6.48. One-handed mixing tap, solution variant B: "cylinder solution with lever".
Figure 6.49. One-handed mixing tap, solution variant C: "cylinder solution with end valves and additional sealing".
Figure 6.50. One-handed mixing tap, solution variant D: "ball solution".

TH Darmstadt		EVALUATION CHART													
		for One-handed mixing tap									Page *1*				
In the order of the checklist headings		P: present variant (P): possible after improvement		**A**		**B**		**C**		**D**		**E**		**F**	
	No.	Evaluation criterion	W	P	(P)	P	(P)	P	(P)	P	(P)	P	(P)	P	(P)
Funct.	1	Reliability of stopping flow without drips	1	*1*		3		3 I	4	*1*					
Work. Princ.	2	Reliable, reproducible setting (calcium-resistant, few wearing parts)	1	2		3		2 I	3	3					
Embod	3	Low space requirement	1	3↑		2		2		4					
Prod.	4	Few parts	1	1		2 I		1 W		4					
	5	Simple manufacture	1	1		3		2		1↕	4				
Assy.	6	Easy assembly	1	2		3		2		2↑ J	3				
Operation	7	Convenient operation, sensitive setting	1	1		3		4		2					
	8	Easy upkeep (easy to clean)	1	4↑		2 I		3		2					
Maint.	9	Simple maintenance (with standard tools. fittings need not be dismantled)	1	1		3		2 W		1↕ J	3				
	10														
	11														
	12														
	13														
	14														

? Evaluation uncertain ↑ Tendency: better ↓ Tendency: worse	$P_{max} = 4$	Σ	16		24	(26)	21	(23)	20	(26)				
	R_t		0,45		0,67		0,58		0,56					
	Ranking		4		1	(1)	2	(3)	3	(2)				

Justification (J), Weak spot (W), Improvement (I) of variant / criterion

C1	Provide rubber seal
B4	Simplify lever mechanism
D6 D9	Indeterminate position of ball during assembly
B8	Improve with B4
D9	Attachment of lever

Decision	Develop solution B with improvement of control elements Solution D: Examine production possibilities, present results in 2 months

Date: *11. 10. 73*	Initials: Dhz

Figure 6.51. One-handed mixing tap: Evaluation of concept variants A, B, C, D.

(see Fig. 6.50) would only have been considered if further studies into production and assembly problems had been undertaken and led to positive results.

Step 8: Result

Drawings of Solution B with improvements to the operating lever in respect of space requirements, easier cleaning and number of parts were produced. The level of information for Solution D was improved with a view to re-examining it for final evaluation.

7 Embodiment Design

Embodiment design is that part of the design process in which, starting from the working structure or concept of a technical product, the design is developed, in accordance with technical and economic criteria and in the light of further information, to the point where subsequent detail design can lead directly to production (see 3.2).

7.1 Steps of Embodiment Design

Having elaborated the principle solution during the conceptual phase, the underlying ideas can now be firmed up. During the embodiment phase, at the latest, designers must determine the overall layout design (general arrangement and spatial compatibility), the preliminary form designs (component shapes and materials) and the production processes, and provide solutions for any auxiliary functions. In all this, technological and economic considerations are of paramount importance. The design is developed with the help of scale drawings, critically reviewed, and subjected to a technical and economic evaluation.

In many cases several embodiment designs are needed before a definitive design appropriate to the desired solution can emerge.

In other words, the *definitive layout* must be developed to the point where a clear check of function, durability, production, assembly, operation and costs can be carried out. Only when this has been done is it possible to prepare the final production documents.

Unlike conceptual design, embodiment design involves a large number of corrective steps in which analysis and synthesis constantly alternate and complement each other. This explains why the familiar methods underlying the *search for solutions* and *evaluation* must be complemented with methods facilitating the *identification of errors* (design faults) and *optimisation*. The *collection of information* on materials, production processes, repeat parts and standards involves a considerable effort.

The embodiment process is complex in that:

- many actions have to be performed simultaneously;

- some steps have to be repeated at a higher level of information; and
- additions and alterations in one area have repercussions on the existing design in other areas.

Because of this, it is not always possible to draw up a strict plan for the embodiment design phase. However, it is possible to suggest a general approach. Particular problems may demand deviations and subsidiary steps, which can rarely be predicted precisely.

It is always advisable to proceed from the qualitative to the quantitative, from the abstract to the concrete, and from rough to detailed designs, and to make provision for checks and, if necessary, for corrections (see Fig. 7.1).

1. Using the requirements list, the first step is to identify those *requirements that have a crucial bearing* on the embodiment design:
- Size-determining requirements such as output, through-put, size of connectors etc;
- arrangement-determining requirements such as direction of flow, motion, position etc; and
- material-determining requirements such as resistance to corrosion, service life, specified materials etc.

Requirements based on safety, ergonomics, production and assembly involve special design considerations which may affect the size, arrangement and selection of materials (see 7.2).

2. Next, scale drawings of the *spatial constraints* determining or restricting the embodiment design must be produced (for instance drawings showing clearances, axle positions, installation requirements etc).
3. Once the embodiment-determining requirements and spatial constraints have been established, a rough layout, derived from the concept, is produced with the emphasis on the embodiment-determining *main function carriers*, that is, the assemblies and components fulfilling the main functions.

The following subsidiary problems must be settled, due regard being paid to the principles of embodiment design (see 7.4):

- Which main functions and function carriers determine the size, arrangement and component shapes of the overall layout? (For instance, the blade profiles in turbo-machines or the flow area of valves)
- What main functions must be fulfilled by which function carriers jointly or separately? (For instance, transmitting torque and allowing for radial movement by means of a flexible shaft or by means of a stiff shaft plus a special coupling)

This step is similar to the division into realisable modules as shown in Fig. 1.10.

4. Preliminary layouts and form designs for the embodiment-determining main function carriers must be developed; that is, the general arrangement, component shapes and materials must be determined provisionally. To that end, it is advisable to work systematically through the first three headings of the checklist in Fig. 7.2. The result must meet the overall spatial constraints and

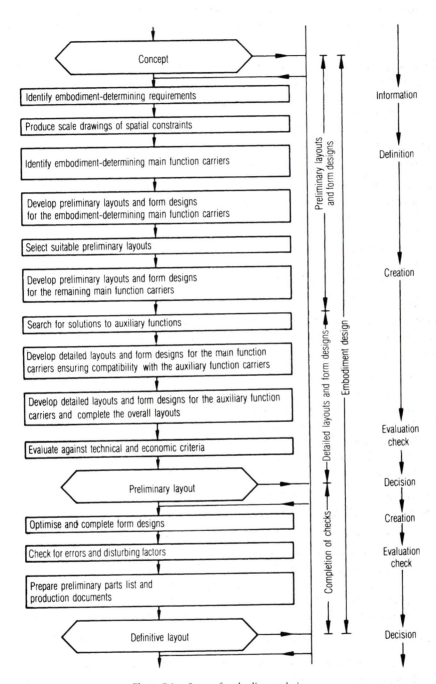

Figure 7.1.   Steps of embodiment design.

then be completed so that all the relevant main functions are fulfilled (for instance by specifying the minimum diameters of drive shafts, provisional gear ratios, minimum wall thicknesses etc). Known solutions or existing components (repeat parts, standard parts etc) must be shown in simplified form. It may be useful to start working on selected areas only, combining these later into *preliminary layouts.*

5. One or more suitable preliminary layouts must be selected in accordance with the procedures described in 4.2.1 (modified if necessary) and with the checklist in 7.2.

6. Preliminary layouts and form designs must now be developed for the remaining main function carriers that have not yet been considered because known solutions exist or they are not embodiment-determining until this stage.

7. Next, determine what essential *auxiliary functions* (such as support, retention, sealing and cooling) are needed and, where possible, *exploit known solutions* (such as repeat parts, standard parts, catalogue solutions). If this proves impossible, *search for special solutions,* using the procedures already described in Chapter 6.

8. Detailed layouts and form designs for the main function carriers must now be developed in accordance with the embodiment design rules and guidelines (see 7.3 to 7.5), with due attention to standards, regulations, detailed calculations and experimental findings, and also to the problem of compatibility with those auxiliary functions that have now been solved. If necessary, divide into assemblies or areas that can be elaborated individually.

9. Proceed to develop the detailed layouts and form designs for the auxiliary function carriers, adding standard and bought-out parts. If necessary, refine the design of the main function carriers and combine all function carriers into overall layouts.

10. Evaluate the layouts against technical and economic criteria (see 4.2.2). If a particular project requires several concepts to be put in more concrete form prior to evaluation, then the embodiment process must not, of course, be pursued beyond what the evaluation of the variants demands. Depending on the circumstances, it is thus possible, in some cases, to take a decision just as soon as the main function carriers have reached the preliminary layout stage, while in other cases the decision will have to be deferred until after a great deal of detail design. In either event, all the designs to be compared must be on one and the same level of embodiment, since otherwise no reliable evaluation is possible.

11. Fix the preliminary overall layout.

12. Optimise and complete the form designs for the selected layout by elimination of the *weak points* that have been identified in the course of the evaluation. If it should prove advantageous, repeat the previous steps and adopt suitable sub-solutions from less favoured variants.

13. Check this layout design for errors (design faults) in function, spatial compatibility etc (see checklist in 7.2) and for the effects of disturbing factors. Make what improvements may be needed. The achievement of the objectives with

respect to cost (see Chapter 10) and quality (see Chapter 9) must be established at this point at the latest.

14. Conclude the embodiment design phase by preparing a preliminary *parts list* and preliminary *production and assembly documents*.

15. Fix the *definitive layout* design and pass on to the detail design phase.

In the embodiment phase, unlike the conceptual phase, it is not necessary to lay down special methods for every individual step.

The *representation* of the layout and form designs may be based on standard drawing conventions or, if necessary, on simplified scale drawings, as suggested by Lupertz [7.186].

The *search for solutions* for auxiliary functions and other subsidiary problems is based either on the procedure described in Chapter 4, but simplified as far as possible, or else directly on catalogues. Requirements, functions and solutions with appropriate classifying criteria have already been elaborated.

The *embodiment* (layout and form designs) of the function carriers is based on the checklist and involves reference to the principles of mechanics and to structures and materials technology. It calls for calculations ranging from the simplest to complex differential equations or the method of finite elements applied with the help of computers. For these calculations, the reader is referred to the literature listed in 7.5.1, and for even more complex calculations to the domain specific literature. In some cases it might be necessary to build prototypes or undertake specific tests.

In addition, fixed rules and principles for this phase, to be elaborated later, must be followed.

Because of the fundamental importance of the *identification of errors* (design faults) in several of the steps, the reader is especially referred to Chapter 9.

In the elaboration of embodiment designs, many details have to be clarified, confirmed or optimised. The more closely they are examined, the more obvious it becomes whether the right solution concept has been chosen. It may appear that this or that requirement cannot be met, or that certain characteristics of the chosen concept are unsuitable. If this is discovered during the embodiment phase, it is advisable to re-examine the procedure adopted in the conceptual phase, for no embodiment design, however perfect, can hope to correct a poor solution concept. This is equally true of the working principles applicable to the various sub-functions.

However, even the most promising solution concept can cause difficulties in detail design. This often happens because various features were originally treated as subordinate or as not in need of further clarification. Attempts to solve these sub-problems compel designers to reiterate the appropriate steps while retaining the chosen solution concept.

Experiences with the proposed approach for embodiment design have confirmed its validity but have also revealed the following important points [7.226]:

- If prior research has been undertaken or embodiment variants already exist, the step of producing preliminary embodiments can often be left out.

- Preliminary embodiments can always be left out when only detailed improvements are required.
- The solutions for auxiliary functions usually influence the preliminary embodiment of the main function carriers.
- Many products are not developed from scratch, but, based on new requirements and experiences, are developed further or improved. Experience has shown that it is useful to start by analysing the *failures* and *disturbing factors* for the existing solution (see 9.2 and 9.3) and based on that analysis to develop a new requirements list (see Fig. 7.2). The result of the clarified task will show whether a new working structure, that is a new principle solution, is required or whether it is sufficient to modify the existing embodiment. It is possible to start at many different places within the overall approach. In some cases a new product can be produced by making improvements to the details. In other cases, tests of the existing or modified modules may be necessary. The required steps in the overall approach must be selected appropriately.
- A characteristic of successful designers is that they continuously check and monitor their actions to identify the direct and indirect effects.

To sum up, embodiment design involves a flexible approach with many iterations and changes of focus. The individual steps have to be selected and adapted to the particular situation. While paying due regard to the fundamental links between the steps and the recommendations we have provided, the ability to organise one's own approach is important (see 2.2.1 to 2.2.3).

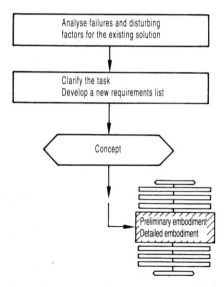

Figure 7.2. Embodiment design phase based on the development of an existing solution. Which of the steps shown in Fig. 7.1 has to be completed follows from an analysis of failures and disturbing factors.

7.2 Checklist for Embodiment Design

Embodiment design is characterised by *repeated deliberation and verification* (see 7.1).

Every embodiment design is an attempt to fulfil a given function with appropriate layout, component shapes and materials. The process starts with preliminary scale layout drawings based on spatial requirements and a rough analysis, and proceeds to consider safety, ergonomics, production, assembly, operation, maintenance, recycling and costs.

In dealing with these factors, designers will discover a large number of interrelationships, so that their approach must be progressive as well as iterative (verification and correction). Notwithstanding this double character, however, the approach must always be such as to allow the speedy identification of those problems that must be solved first.

Though individual factors may be closely interrelated, designers can derive important checklist headings from the general objectives and constraints (2.1.8) which, moreover, provide them with a useful procedural order and a systematic check on each step. The checklist thus not only provides a strong mental impetus, but also ensures that nothing essential is forgotten in the embodiment phase (see Fig. 7.3).

All in all, reference to the headings will help designers to develop and test their progress in a systematic and time-saving way. Each heading should be examined in turn, regardless of its interrelationship with the rest.

The actual sequence is no indication of the relative importance of the various headings, but ensures a systematic approach. For instance, it would be futile to deal with assembly problems before ascertaining if the required performance or minimum durability is ensured. The checklist thus provides a consistent scrutiny of embodiment design and one that is easily memorised.

7.3 Basic Rules of Embodiment Design

The following basic rules apply to all embodiment designs. If they are ignored problems are introduced and breakdowns or accidents may occur. They underlie nearly all the steps listed in 7.1. When used in conjunction with the checklist (see Fig. 7.3) and with the design fault identification method (see Chapter 9) they also help with selection and evaluation.

The basic rules of *clarity*, *simplicity* and *safety* are derived from the general objectives, that is:

- fulfilment of the technical function;
- economic feasibility; and
- individual and environmental safety.

Headings	Examples
Function	Is the stipulated function fulfilled? What auxiliary functions are needed?
Working principle	Do the chosen working principles produce the desired effects and advantages? What disturbing factors may be expected?
Layout	Do the chosen overall layout, component shapes, materials and dimensions provide: adequate durability (strength) permissible deformation (stiffness) adequate stability freedom from resonance unimpeded expansion acceptable corrosion and wear with the stipulated service life and loads?
Safety	Have all the factors affecting the safety of the components, of the function, of the operation and of the environment been taken into account?
Ergonomics	Have the human–machine relationships been taken into account? Have unnecessary human stress or injurious factors been avoided? Has attention been paid to aesthetics?
Production	Has there been a technological and economic analysis of the production processes?
Quality control	Can the necessary checks be applied during and after production or at any other required time, and have they been specified?
Assembly	Can all the internal and external assembly processes be performed simply and in the correct order?
Transport	Have the internal and external transport conditions and risks been examined and taken into account?
Operation	Have all the factors influencing the operation, such as noise, vibration, handling, etc been considered?
Recycling	Can the product be reused or recycled?
Maintenance	Can maintenance, inspection and overhaul be easily performed and checked?
Costs	Have the stipulated cost limits been observed? Will additional operational or subsidiary costs arise?
Schedules	Can the delivery dates be met? Are there design modifications that might improve the delivery situation?

Figure 7.3.   Checklist for embodiment design.

The literature contains numerous rules of, and guidelines for, embodiment design [7.183, 7.193, 7.212, 7.220, 7.241]. On closer analysis it appears that clarity, simplicity and safety are fundamental to all of them.

Clarity, that is clarity of function or the lack of ambiguity of a design, facilitates reliable prediction of the performance of the end product and in many cases saves time and costly analyses.

Simplicity generally guarantees economic feasibility. A smaller number of components and simple shapes are produced more quickly and easily.

Safety imposes a consistent approach to the problems of strength, reliability, accident prevention and protection of the environment.

In short, by observing the three basic rules, designers can increase their chances of success because they focus attention on, and help to combine, functional efficiency, economy and safety. Without this combination no satisfactory solution is likely to emerge.

7.3.1 Clarity

In what follows we shall be applying the basic rule of clarity to the various headings of the checklist (see Fig. 7.3):

Function

Within a given function structure, an unambiguous interrelationship between the various sub-functions and the appropriate inputs and outputs must be guaranteed.

Working Principle

The chosen working principle must, in respect of the physical effects:

- reveal a clear relationship between cause and effect, thus ensuring an appropriate and economical layout.

The chosen working structure must:

- guarantee an orderly flow of energy, material and signals.

If it does not, undesirable and unpredictable effects such as excessive forces, deformations and wear may ensue. For this reason alone, it is advisable to avoid the so-called "double restraints", the more so as they can cause further difficulties during production and assembly.

By paying attention to the deformations associated with a given loading, and also to thermal expansion, designers can make the necessary allowances for possible expansion in a given direction.

The widely used bearing pairs, with a locating and a non-locating bearing (see Fig. 7.4a), avoid "double restraints" and have a clearly defined behaviour. The stepped bearing pair (see Fig. 7.4b), on the other hand, should be specified only when the expected changes in length are negligible or when the resulting play is permissible. By contrast, a spring-loaded arrangement, in which the operating axial force F_a must not exceed the pre-load F_p, will permit a clear definition of the force transmission path (see Fig. 7.4c).

Combined bearing arrangements often present problems. The combination shown in Fig. 7.5a consists of a needle roller bearing which is intended to transmit the radial forces and a ball bearing which is meant to transmit the axial forces. However, this particular arrangement does not clearly define the transmission path for the radial forces, because the inner and outer races of both bearings are restrained radially. As a result the service life cannot be predicted accurately. The arrangement shown in Fig. 7.5b, on the other hand, satisfies the clarity rule with the same elements, provided the designer ensures, during assembly, that the right-hand race has enough radial play in the housing, thus making certain that the ball bearing transmits axial forces only.

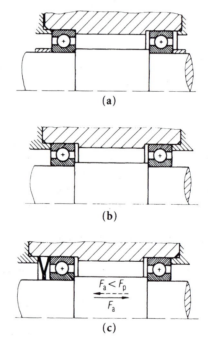

Figure 7.4. Basic bearing arrangements:

(a) Locating and non-locating arrangement: left-hand locating bearing takes up all the axial forces, right-hand sliding bearing permits unimpeded axial movement due to thermal expansion; accurate calculations possible.

(b) Stepped bearing arrangement: the axial loading of the bearings depends on the preload and thermal expansion and cannot be clearly determined; a modification is the "floating arrangement" in which the bearings are provided with axial clearance; in that case, thermal expansion is possible to a limited extent but there is no precise shaft location.

(c) Spring-loaded bearing arrangement: here the disadvantages of the stepped bearing arrangement are largely eliminated, though the constantly-applied axial load may reduce the bearing life; forces resulting from thermal expansion can be determined by spring force deflection diagrams; shaft is precisely located provided axial force F_a acts only towards the right or does not exceed the pre-loading F_p.

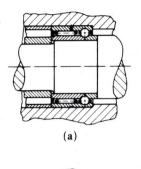

(a)

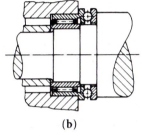

(b)

Figure 7.5. Combined rolling-element bearing:
(a) Transmission path of radial forces not clear; (b) combined rolling bearing with the same elements as in (a), but clear identification of the transmission paths of the radial and axial forces.

Layout

The layout (general arrangement) and form design (shapes and materials) require a clear definition of the magnitude, type, frequency and duration of loads.

If these data are not available, the implementation must be based on reasonable assumptions and the expected service life specified accordingly. In any case, the embodiment must be such that the loads can be defined and calculated under all operating conditions.

No impairment of the function or the durability of a component must be allowed to arise.

Similarly, behaviour in respect of stability, resonance, wear and corrosion must be clearly established.

Very often one comes across double restraints "for safety's sake". Thus a shaft-hub connection designed as a shrink fit will not have a better load-carrying capacity if it is also provided with a key as in Fig. 7.7. The extra element merely ensures correct positioning in the circumferential sense, but because of the reduction in the area at A, the resulting stress concentration at B and the presence of complicated and almost incalculable stresses at C, it decreases the strength in a drastic and fairly unpredictable manner.

Schmid [7.259] has shown that an axially pre-loaded taper joint for the transmission of torque requires a spiralling motion when the hub is assembled on the shaft in order to ensure a reliable shrink fit, and the use of a key would prevent this.

Figure 7.8 shows a housing adapter for a centrifugal pump which can be used to provide various annulus profiles to fit different blade shapes so that new housings need not be constructed for each case. Unless the intermediate pressure in

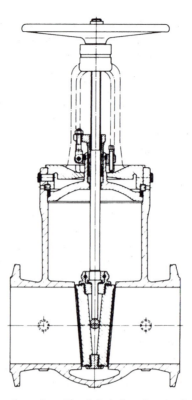

Figure 7.6.   Gate valve with relatively large lower collecting area.

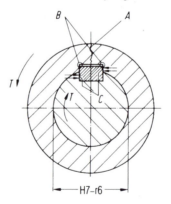

Figure 7.7.   Combined shaft-hub connection by means of shrink fit and key. An example of not applying the principle of clarity.

the gap between the adapter and the housing can be clearly regulated, or some other means of attachment is used, the adapter might travel upwards and damage the blades by rubbing against them.

   This is particularly true when similar fits (H7-j6) are chosen for the two locating diameters which are approximately the same size. This is because, depending on manufacturing tolerances and working temperature, gaps may appear, the relative

sizes of which are unpredictable and which produce unknown intermediate pressures in the space between the adapter and the housing. The solution shown in Fig.
7.8 (detail) ensures, by means of the specially designed connecting passage A
(which must have a flow area roughly four to five times greater than the maximum
gap area that might appear at the upper locating diameter), a clearly definable
intermediate pressure corresponding to the lower inlet pressure of the pump. As
a result the housing adapter is always pressed downwards when the pump is in
operation, and attachments are only needed as locating aids for assembly and to
prevent any tendency of the adapter to rotate.

Serious damage has been reported in gate valves whose operational or loading
conditions were not clearly defined [7.134, 7.135]. When closed, gate valves separate, say, two steam pipes and at the same time close off the inside of the valve
housing. The result is a small, self-contained pressure chamber as in Fig. 7.6. If
condensate has collected in the lower part of the valve housing, and steam appears
on the inlet side with the valve closed so that the valve is heated, then the enclosed
condensate may evaporate and produce an unpredictable increase in pressure
inside the valve housing. The result is either a ruptured housing or serious damage
to the housing cover connection. If the latter is self-sealing, serious accidents may
ensue since, in contrast to what happens with overloaded bolted flange connections, there is no preliminary leakage and hence no warning.

The danger lies in the failure to specify clear operational and loading conditions.
Possible remedies are as follows:

- Connect the inner chamber of the gate valve housing to an appropriate steam
  pipe, operational conditions permitting ($p_{valve} = p_{pipe}$)
- Protect the valve housing against excess pressure (p_{valve} restricted)
- Drain the valve housing, thus avoiding collection of condensate ($p_{valve} \approx p_{external}$)

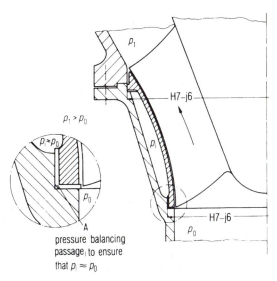

Figure 7.8. Housing adapter in a cooling-water pump.

- Design valves in such a way as to minimise the housing volume (collection of condensate kept low).

Similar phenomena in welded membrane seals are discussed in [7.221].

Safety

See basic rule in 7.3.3.

Ergonomics

In human-machine relationships, correct operation must be ensured by the logical layout of equipment and controls.

Production and Quality Control

These must be facilitated by:

- clear and comprehensive data in the form of drawings, parts lists and instructions;
- insistence on adherence to the prescribed production and organisational procedures.

Assembly and Transport

Much the same is true of assembly and transport. A planned assembly sequence preventing mistakes should be incorporated in the design (see 7.5.8).

Operation and Maintenance

Clear installation instructions and the appropriate embodiment design must ensure that:

- the performance is easily checked;
- inspection and maintenance involve the smallest possible variety of tools and equipment;
- the scope and schedules of inspection and maintenance are defined; and
- inspection and maintenance can be checked after they have been carried out (see 7.5.10).

Recycling

Designers should provide (see 7.5.11):

- clear separation of materials that are incompatible with regard to recycling; and
- clear sequences of assembly and disassembly.

7.3.2 Simplicity

For technical applications the word "simple" means "not complex", "easily understood" and "easily done".

A solution seems simpler if it can be effected with fewer components, because the probability of lower production costs, less wear and lower maintenance is then greater. However, this is only true if the layout and shapes of the components are kept simple. Hence designers should always aim at the minimum number of components with the simplest shapes [7.183, 7.212, 7.221].

As a rule, however, a compromise has to be made: fulfilment of the function always demands a certain number of components. Cost-efficiency often imposes a decision between numerous components with simple shapes but with greater overall costs, and a single and cheaper cast component with the greater uncertainty it may entail in delivery.

Returning to the checklist:

Function

In principle, only a minimum number and a clear and consistent combination of sub-functions will be pursued during consideration of the function structure.

Working Principle

In selecting working principles, only those involving a small number of processes and components, having obvious validity, and involving low costs are taken into consideration.

What constitutes "simpler" in individual cases depends on the problem and the constraints.

In the development of the one-handed mixing tap (see 6.6.1) several solution principles were proposed. One group (see Fig. 6.46) involved two independent adjustments in directions tangential to the valve seat face (types of motion: translation and rotation). The other group (see Fig. 6.43), though involving only movements in one direction (normal or tangential to the seat face) required an additional coupling mechanism to convert the two single adjustments into one direction of movement. Quite apart from the fact that, in the second group, the pre-set temperature is often lost when the tap is shut off, all solutions represented in Fig. 6.43 involve a greater design effort than does the first group. Hence designers should always begin with a group such as that depicted in Fig. 6.46.

Layout

Here the simplicity rule requires:

- geometrical shapes which can be analysed simply for strength and stiffness; and
- symmetrical shapes which provide clearer identification of deformations during production and under mechanical or thermal loads.

In many cases, designers can reduce the work of calculation and experimentation significantly if they try, by means of a simple design, to facilitate the application of basic mathematical principles.

Safety

See under 7.3.3.

Ergonomics

The human-machine relationship should also be simple (see 7.5.5) and can be significantly improved by means of:

- sensible operating procedures;
- clear physical layout; and
- easily comprehensible signals.

Production and Quality Control

Production and quality control can be simplified, that is speeded up and improved, if:

- geometrical shapes permit the use of well-established, time-saving methods;
- the production processes involve short setting up and waiting times; and
- shapes are chosen to facilitate the inspection process.

Leyer, discussing changes in production methods [7.181] uses the example of a sliding control valve approximately 100 mm long to demonstrate how the replacement of a complicated casting by a brazed product made of geometrically simple turned parts helped to overcome difficulties and paved the way for more economical production.

Pursuing his line of approach, we discover that further simplifications are possible (see Fig. 7.9). Step 3 helps to simplify the geometrical shape of the central, tubular part. Step 4 (fewer parts) can be taken when the surface areas at right angles to the valve axis need not be retained.

A further example is provided by the one-handed mixing tap discussed earlier. The design of the lever arrangement shown in Fig. 7.10 is expensive to make and difficult to clean (slits, open recesses). The one shown in Fig. 7.11 is much simpler and also more suitable for longer production runs. The lever, whose end can slide and rotate in a circumferential groove, requires a smaller number of parts and avoids wear in areas that are difficult to readjust. All in all, therefore, this solution is by far the better because it is more economic, easier to clean and looks nicer.

Assembly and Transport

Assembly is simplified, that is facilitated, speeded up, and rendered more reliable, if:

Redundancy layouts cannot, however, replace the safe-life or fail-safe principles. Two cable cars operating in parallel will, admittedly, increase the reliability of passenger transport, but contribute nothing to the passengers' safety. The redundant layout of aeroplane engines will not increase safety if any of the engines has a tendency to explode and hence to endanger the system.

In short, an increase in safety can only be guaranteed if the redundant element satisfies the safe-life or the fail-safe principle.

Adherence to all the principles we have mentioned – that is the attainment of safety in general – is greatly facilitated by the principle of the division of tasks (see 7.4.2) and by the two basic rules of clarity and simplicity, as we shall now try to show with the help of an example.

The principle of the division of tasks and the clarity rule have been applied with great consistency to the construction of a helicopter rotor head (see Fig. 7.17), and help the designers to come up with a particularly safe construction based on the safe-life principle. Each of the four rotor blades exerts a radial force on the rotor head due to the centrifugal inertia force, and a bending moment due to the aerodynamic loading. The rotor blades must also be able to swivel so that their angles of incidence can be changed. A high safety level is achieved by the following measures:

- A completely symmetrical layout so that the external bending moments and the radial forces at the rotor head cancel out.
- The radial forces are transmitted exclusively by the torsionally flexible member Z to the main central component where they cancel each other out.
- The bending moment is only transmitted through part B and is taken by the roller bearings in the rotor head.

As a result, every component can be optimally designed in accordance with its task. Complicated joints and shapes are avoided and a high safety level is attained.

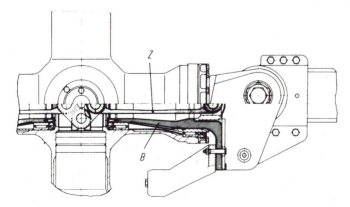

Figure 7.17.   Rotor blade attachment of a helicopter based on the principle of the division of tasks (Messerschmitt-Bölkar system).

3 Indirect Safety Principles

Indirect safety methods involve the use of special *protective systems* and *protective equipment*. They are applied whenever direct safety methods prove inadequate. A detailed discussion of indirect safety methods for technical systems can be found in [7.229]. In what follows, the most important elements of these methods are described.

Protective systems react when danger occurs. To that end their function structure includes a *signal transformation* with an input that captures the danger and an output that removes it.

The working structure of a protective system is based on a function structure with the following main functions: capture – process – act. Examples are the multiple redundant monitoring of temperatures in a nuclear reactor, the monitoring of robots in inaccessible workplaces, the sealing of areas which are subject to X-rays, and the automatic checking of the locking of centrifuge covers prior to operation. The required actions can involve removing, limiting or separating.

Protective devices fulfil protective functions without transforming signals.

Examples are a pressure safety valve (see Fig. 7.21), a shaft coupling that slips with torque overload, a pin that shears to limit excessive forces, and safety belts in cars. Their main action is removing or limiting. They form part of a protective system.

Protective barriers fulfil protective functions without acting.

These barriers are passive, and not able to act on their own. They do not transform signals and therefore do not require a function structure with an input and an output. They protect by separating, that is, by keeping persons and equipment at a distance from danger using physical barriers, covers, fences etc. They are described in DIN 31001, Parts 1 and 2 [7.62, 7.63]. Locking devices, according to Part 5 of this standard [7.64], are regarded as protective systems.

Basic Requirements

All safety measures have to fulfil the following basic requirements:

- reliable operation;
- protection throughout danger period; and
- resistance to tampering.

Reliable Operation

Reliable operation of protective systems means that: the working principle and the embodiment allow unambiguous operation; the layout follows the established rules; production and assembly are quality controlled; and the protective systems and devices are rigorously tested. The safety modules and their functional links should be based on direct safety principles and demonstrate safe-life or fail-safe behaviour.

Protection Throughout Danger Period

This requirement means that:

- the protective function has to be available from the start of the dangerous situation and last throughout the period of danger; and
- the protective function does not cease or the protective device is not removed before the dangerous situation has completely ended.

Figure 7.18 shows example layouts of safety fence contacts for a machine guard. Closed contacts signal safety fence in position. Layout *a* has severe deficiencies because the contact movement relies upon the spring force alone. If the spring breaks or the contacts stick together, the contact will not be broken, that is the machine can be started with the safety fence still open. Layout *b* will provide protection throughout the danger period. Sticking contacts will be opened because the effect relies on form rather than spring force, and in the event of parts getting broken they will not fall onto the contacts. Layout *c* also makes use of form for activation, but adds spring force and bi-stable behaviour (see 7.4.4). Further examples can be found in [7.229].

Resistance to Tampering

Resistance to tampering means that the protection cannot be reduced or removed by unintended or intended actions. With respect to the safety fence contact in Fig. 7.18, it should be designed such that actions that prevent its correct operation are not possible. The best way to achieve this is by using a cover that cannot be opened without tools or without stopping the machine.

In the following paragraphs the requirements of protective systems and devices are listed.

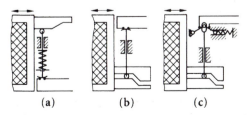

Figure 7.18. Layouts of safety fence contacts for a machine guard. (a) Protection not guaranteed because contact movement relies on a spring force alone. (b) Protection guaranteed because activation relies on form fit. (c) Bi-stable behaviour added to form fit activation in *b*.

Protective Systems

Protective systems render endangered plant or machinery automatically safe, with
the aim of preventing danger to persons and machinery. In principle, the following
possibilities are available:

When danger occurs, *prevent the consequences* by:

• putting the plant or machinery out of action; and
• preventing any plant or machinery in a dangerous state from being put into
  operation.

When there is a continuous danger, *avoid its effects* by:

• introducing protective measures.

The basic requirements "reliable operation", "protection throughout danger period", and "resistance to tampering" are supported by the following requirements.

Warning

When a protective system produces changes in the working conditions, a warning
must be given that indicates the change and the cause of the warning. Examples
are "oil level too low", "temperature too high", and "safety fence open".

There are recommended acoustic and optical signals [7.71], colours for warning
lights and push buttons [7.79], and special safety symbols [7.49].

If the dangerous situation emerges so *slowly* that action by the operators can
reduce the danger, then a warning should be given before a protective action is
initiated (two-stage action).

Between the two stages, there should be a sufficiently large and clearly defined
step of the danger variable. For example, if pressure is the monitored danger variable, a warning could be given at 1.05 p_{normal} and shut down initiated at 1.1 p_{normal}.

If the dangerous situation emerges too *fast*, the protective system should react
immediately and signal its response clearly. The terms slow and fast have to be
interpreted in the context of the cycle time of the technical process and the necessary reaction time [7.261].

Self-monitoring

A protective system must be self-monitoring, that is, it must not only be triggered
when the system breaks down, but also by faults of its own. This requirement is
best satisfied by the *stored energy principle* because, when this is applied, the energy
needed for activating the safety device is stored within the system and any disturbance of, or fault in, the protective system will release that energy and switch off
the plant or machinery. This principle can be used not only in electronic protective
systems but also in systems using other types of energy.

This principle has been used in the valve shown in Fig. 7.19. When the valve

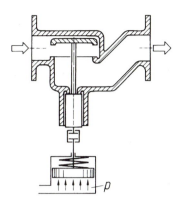

Figure 7.19. Layout of a quick-action valve. In the event of a drop in oil pressure p, the spring force, the flow pressure on the valve face and weight of the valve act together to guarantee the rapid closure of the valve.

opens the spring is compressed by the operating oil pressure. When the oil pressure reduces, the spring extends and the valve closes. Failure of the spring will not inhibit the closure of the valve because of the particular configuration. The selected flow direction and the suspended configuration support the requirement of providing protection throughout the danger period.

A further example of the use of the stored energy principle in a hydraulic system is shown in Fig. 7.20. In this protective system, a pump *1* with a pressure-regulating valve *2* ensures a constant pre-pressure p_p. The protective system with the pressure p_s is connected to the pre-pressure system by means of an orifice *3*. Under normal conditions, all outlets are closed, so that the quick-action stop valve *4* is held open by the pressure p_s to admit the energy supply of the machine. In case of a faulty axial shaft position, piston valve *5* at the end of the shaft opens, pressure p_s drops, and further energy supplies are cut off by the quick-action stop valve *4*. The same effect is produced by damage to the pre-pressure or protective system, for example by pipe fracture, lack of oil or pump failure. The system is self-monitoring.

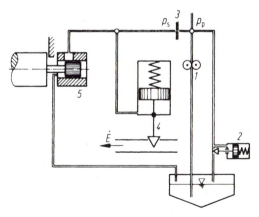

Figure 7.20. Hydraulic protection system to prevent incorrect axial shaft positions based on the stored energy principle.

A system that provides energy only in case of danger cannot detect a failure in its own system. Therefore this approach should only be used to provide the warning signals of a protective system when a monitoring system is also available and the system is checked regularly.

The possibility that a protective system, based on the stored energy principle, can cause interruptions that are not caused by a dangerous situation but by the protective system itself, should be met by increasing the reliability of the system elements.

Redundancy

The failure of a protective system should be seen as a real possibility. Because a single protective system may break down, its mere doubling or replication ensures greater safety: it is unlikely that all the systems will fail at once. A solution that is often applied is redundancy based on *2 from 3 selection*. Three sensors are used to detect the same danger signal (see Fig. 7.16). Only when at least two sensors signal the critical value is the protective action such as switching off the machine initiated. Thus the failure of a single sensor does not reduce the protective cover and its failure will not trigger an unnecessary protective action [7.192].

This is, however, only true provided that the replicated protective systems do not all fail due to a common fault. Safety is considerably increased if the double or multiple systems work independently of one another and are, moreover, based on different working principles. In that case, common faults, for instance due to corrosion, will not have catastrophic consequences: the simultaneous breakdown of all such systems is highly improbable.

Figure 7.21 illustrates protective devices to prevent excessive pressure in pressure vessels. The mere doubling would not protect against common failures such as corrosion or inappropriate materials. The use of different working principles, however, reduces the possibility of simultaneous failure.

When redundant configurations are linked in parallel or series, the values at which they are triggered should be carefully staggered within an appropriate range. In this manner a primary and secondary protection is established. In many cases the primary protection can receive its signals from an existing control system, if it has the characteristics of a protective system.

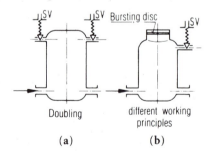

Figure 7.21. Protective system against excessive pressure build up in pressure vessels: (a) two safety valves (not safe against common faults); (b) safety valve and shear plate, double principle protective system.

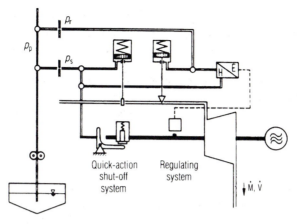

Figure 7.22.   Double principle stored energy protective system against overspeeding.

This requirement is met in the control of steam turbines shown in Fig. 7.22 [7.291]. In the case of overspeeding, the energy supply is cut off by two systems differing in principle. Increases in speed first bring in the regulating system whose speed measurement and regulating valve are independent of, and different in principle from, the quick-action shut-off system.

Speed is measured with three identical, but independent, magnetic sensors. They take their measurements from a gear wheel on the turbine shaft (see Fig. 7.23). Their primary purpose is the electronic-hydraulic speed control of the machine. In addition, every signal is compared with a reference signal to prevent excess speed. This comparison is based on the 2 from 3 principle. Each measurement circuit is monitored separately and any failures signalled. If two fail, the system is shut down immediately.

The measurement and the activation of the quick-action system, however, are based on a mechanical principle. Figure 7.24 shows quick-action pins that, in case of excess speed, move out rapidly against the retaining springs and strike a trigger. This in turn activates the quick-action shut-off system hydraulically. The turbine is provided with two such bi-stable devices that trigger at 110% and 112% excess speed respectively (see 7.4.4).

A common hydraulic supply of the control and quick-action shut-off system based on the stored energy principle is acceptable because both are based on a common self-monitoring principle.

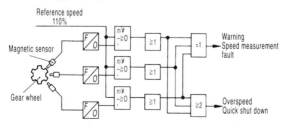

Figure 7.23.   Electronic speed control and speed monitoring using a redundant layout based on the 2 from 3 principle (Brown Boveri).

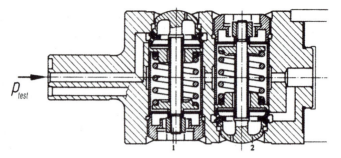

Figure 7.24. Quick-action shut-off system, after [7.291]. Two quick-action pins are provided that trigger at 110% and 112% respectively. p_{test} indicates the connection for oil pressure testing.

Bi-stability

Protective systems and devices must be designed with a clearly defined triggering value. When this value is attained, the protective reaction must be initiated immediately and unambiguously. This can be achieved by using the bi-stable principle (see 7.4.4). Below the triggering value, the system is in a stable state. When the triggering value is attained, an unstable condition is created deliberately. This avoids intermediate states and transfers the system rapidly into its second stable state.

Preventing System Restarts

After a protective system has been activated, that system should not automatically return a machine to normal operation, even if the danger recedes. The activation of a protective system is always triggered by an unusual situation. After shutdown, the situation should be checked and evaluated, and subsequent restarting should follow a clearly structured procedure. For example, the safety regulations covering automatic protection systems [7.278], and other machines used in manufacturing [7.361], prescribe procedures for restarting.

Testability

A protective system should allow its functioning to be tested without having to create a real dangerous situation. However, it might be necessary to simulate the dangerous situation in order to trigger the protective system. During a simulation the effects used must be similar to the real danger and all possible danger conditions checked.

In our example of the speed control system, this means a planned increase in speed up to the excess speed at which the protective system triggers. If this is not possible or cannot be done frequently, it is possible to simulate the centrifugal inertia force by using oil pressure to trigger the system. For this simulation the machine does not have to be shut down. Figure 7.24 shows the oil channel. The oil simulates an increase in the centrifugal inertial force on the quick-action shut-off pins so that they are triggered and their action tested without attaining an excess speed.

With redundant protective systems it is possible to isolate individual systems from the machine to test them. Any other redundant protective systems can remain active and continue to monitor safety during the test.

When designing a protective system care must be taken to ensure that after test procedures which only check part of the system, the protective system automatically returns into its fully operational state.

From the previous paragraphs the following points emerge:

- protection must be retained during testing;
- testing must not introduce new dangers; and
- after testing the parts tested should return automatically to their fully operational state.

Often a *start-up check* is useful or even prescribed. This check allows the operation of a machine only after its functions have been tested by activating the protective system. Safety regulations, for example, often prescribe this type of start-up check for power tools with automatic safety devices [7.278].

Protective systems must be *tested regularly*, that is:

- before the first operation;
- at regular predetermined intervals; and
- after every service, repair or modification.

The procedures should be described in operating manuals and the results documented.

Relaxing Requirements

The question may be asked whether it is necessary to meet the testability requirement as well as that of self-monitoring. Even protective systems based on the stored energy principle include elements whose full functionality can only be assessed through testing. Examples include the operation of the quick-action pins in Fig. 7.24 and sticking contacts in an electric switch.

Relaxation of the safety system requirements is only permissible when the probability of failure is so small and the consequences of any failure are so limited that the overall risk is acceptable. This will only be the case with redundancy requirements when system tests are easy and carried out regularly. This occurs when these tests are part of normal operation, for example when start-up checks are implemented. This often applies to protective systems associated with safety at work.

If human life is endangered or large scale damage may occur, leaving out redundancy is neither justified nor economic. Which redundancy is applied, for example

2 from 3 selection, replication of the same principle or principle redundancy, depends on the specific context and the level of risk.

Protective Barriers

The purpose of a protective barrier is to isolate people and objects from the source of danger, and to protect them from a variety of dangerous effects. DIN 31001 Part 1 [7.62] and Part 2 [7.63] deal mainly with protection against physical contact with dangerous static and moving parts, and against objects and particles that break away. Elaborate illustrations and examples are given in [7.229].

   The desired principle solution (see Fig. 7.25) is to prevent contact by providing:

- a full *enclosure*; or
- a *cover* for a particular side; or
- a *fence* to maintain a safe distance.

Safety distances play an essential role when it is possible to reach through or around fences or barriers. These distances are determined by body dimensions and ranges of reach. DIN 31001 Part 1 [7.62] gives clear safety distances depending on body dimensions and posture.

   With respect to contact protection and protection against objects and particles that break away, DIN 31001 Part 2 [7.63] only permits the use of those materials that can fulfil their protective function on the basis of their durability, shape stability, temperature resistance, corrosion resistance, resistance to aggressive substances, and their permeability to those aggressive substances.

4 Designing for Safety

Here, too, the checklist (see 7.2) can prove a great help. Safety criteria must be scrutinised in respect of all the headings listed [7.330].

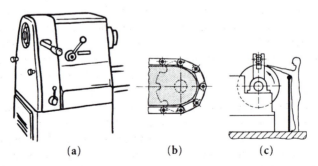

(a)   **(b)**   **(c)**

Figure 7.25. Examples of protective barriers: (a) full enclosure; (b) cover for a particular side; (c) fence to maintain safe distance.

Function and Working Principle

It is important to establish whether or not the function is fulfilled safely and reliably by the chosen solution. Likely *faults and disturbing factors* must be taken into account as well. It is not always clear, however, to what extent allowance must be made for exceptional, purely hypothetical, circumstances that could affect the function.

The correct estimation of the *scope and likelihood of a risk* should be based on the successive negation of each of the functions to be fulfilled and on an analysis of the likely consequences (see 9.2). Sabotage need not necessarily be considered in this context, although general safety methods are likely to decrease its effects.

What we have to consider first and foremost are failures due to possible disturbances of the structure, operation and environment of a machine, and what preventive steps should be taken. Harmful effects not due to technological factors (such as operator ignorance) cannot be eliminated by technical systems but must be considered and if possible limited.

A further question is whether the direct safety method we have been discussing is adequate, or whether safety should be increased by additional protective systems. Finally, we might also ask whether, should it be impossible to make adequate safety provisions in a particular case, the whole project should be abandoned.

The answer depends on the *degree of safety* that has been attained, on the *probability of unpreventable damage or accident* and on the *magnitude of the possible consequences*. Objective standards are often lacking, particularly in the case of new methods and their application. It has been argued that technical risks must be no greater than the risks humans must expect from natural causes [7.151]. However, this must be a matter for discretion. The final decision should, in any case, reflect a responsible attitude towards the human race.

Layout

External loads produce stresses in components. By analysis we determine their magnitude and frequency (steady and/or alternating loads). The various types of stress produced can be determined by calculation or experiment.

Materials technology provides the designer with limiting values of stress for particular conditions (tension, compression, bending, shear and torsion), beyond which the material will fail. Since these values are usually obtained from tests on specimens and not from tests on the components themselves, the stresses to which the latter are subjected should be kept well within the limits if adequate *durability* is to be guaranteed.

The ratio of the limiting stress σ_L of the material to the acceptable working stress σ_W in a component is called the *safety factor, $SF = \sigma_L / \sigma_W$*.

The value of a safety factor depends on uncertainties in the determination of the limiting stress; on uncertainties in the load assumptions; on the calculation methods; on the production processes; on the (uncertain) influence of shape, size and environment; and also on the probability and importance of possible failures.

The determination of safety factors still lacks generally valid criteria. An investigation by the authors has shown that published recommended safety factors cannot be classified by type of product, branch of engineering or such other criteria as toughness of material, size of component, probability of failure etc. Tradition, figures based on one-off and often inadequately explained failures, hunches and experiences are often the basis of numerical data from which no generally valid statements can be derived.

What figures are given in the literature must therefore be treated with circumspection. Their application usually calls for knowledge of the individual circumstances and of the special practices or regulations of the branch of engineering in question.

Toughness, that is the ability to undergo plastic deformation before failure and thus relieve stress concentrations caused by unevenly distributed loads, is one of the most important safety features any material can have. The usual overspeed spinning tests of rotors with the correspondingly high stresses they set up, and also the required overpressure tests of pressure vessels – provided that they are built of tough materials – are good examples of the direct safety method aimed at reducing stress concentrations in finished components.

Because toughness is a crucial safety-enhancing property of materials, it is not enough simply to aim at greater yield strength. Since, in general, the toughness of materials decreases with increasing yield strength, it is essential to ensure minimum toughness, otherwise the benefits of plastic deformation are no longer guaranteed.

Dangerous too are those cases in which the material turns brittle with time or for other reasons (for instance, due to radiation, corrosion, heat, or surface coatings). This is particularly true of synthetic materials.

If the safety of a component is calculated merely by the difference between the computed stress and the maximum permissible stress, a vital point is missed.

Of the utmost importance is the loading condition and the effect on the properties of the material due to ageing, heat, radiation, weathering, operating conditions and production processes, for instance welding and heat treatment. Residual stresses must not be underestimated either: brittle (fast) fractures without plastic deformation can occur suddenly and without warning.

The avoidance of a build-up of additive stresses, of brittle materials, and of production processes that encourage brittle fractures, is therefore an essential requirement of the direct safety method.

If plastic deformation is monitored at a critical point, or can be used to impede the function in such a way that the danger can be noticed before humans or machines are endangered, it becomes a fail-safe method [7.221].

Elastic deformations must not be allowed to disturb the smooth functioning of a machine, for instance through loss of clearance. If this happens the force transmission paths or the expansions can no longer be determined with certainty and overloading or fracture may ensue. This is true of stationary no less than of moving parts (see 7.4.1).

By *stability* we refer not only to the basic stability of a machine but also to its stable operation. Disturbances should be counteracted by stabilising effects, that is by automatic return to the initial or normal position. Designers must ensure neutral equilibrium or that potentially unstable states do not lead to a build-up of disturbances that might get out of control (see 7.4.4).

Resonances produce increased stresses that cannot be accurately determined. They must be avoided unless the vibrations can be sufficiently damped. This applies not only to the stability problem, but also to such associated phenomena as noise and vibration which impair the efficiency and health of operators.

Thermal expansions must be taken into account under all operating conditions if overloading and impairment of the function are to be avoided (see 7.5.2).

Inefficient *seals* are a common cause of breakdown or trouble. Careful choice of seals, provision for pressure relief at critical sealing points and careful attention to fluid dynamics help to overcome these problems.

Wear and the resulting particles can also impede operational safety, and must therefore be kept within tolerable limits. In particular, designers should ensure that such particles do not damage or interfere with other components. They should be removed as near as possible to their point of origin.

Uniform *corrosion* reduces the designed thickness of components. Local corrosion, particularly of components subject to dynamic loading, may appreciably increase stress concentrations and lead to fast fractures with little deformation. There is no such thing as permanent stability under corrosion – the load capacity of components decreases with time. Apart from fretting corrosion and fatigue corrosion, stress corrosion can also be very serious for certain materials subject to tensile stresses in the presence of corrosive media. Finally, corrosion products can impede the functioning of machines, for instance by jamming valve spindles, control mechanisms etc (see 7.5.4).

Ergonomics and Industrial Safety

The application of ergonomic principles to industrial safety involves the careful scrutiny of safety at work and of the human-machine relationship. A great many books and papers have been devoted to this subject [7.35, 7.67, 7.201, 7.273, 7.330]. In addition DIN 31000 [7.61] specifies the basic requirements of design for safety, and DIN 31001, Parts 1, 2 and 10 [7.62, 7.63] deal with protective equipment. Regulations by various professional bodies, factory inspectorates etc must be scrupulously observed in all branches of engineering, and so must a great deal of special legislation [7.115] (see also [7.361]). In a book of this kind it is impossible to examine every aspect of industrial safety, but operator ignorance and fatigue are two factors that should always be taken into consideration. For that reason alone, machines must be designed on ergonomic principles (see 7.5.5). Tables 7.1 and 7.2 list the minimum requirements for industrial safety.

Production and Quality Control

Components must be designed in such a way that their qualities are maintained during production (see 9). To that end special quality controls must be instituted,

Table 7.1. Harmful effects associated with various types of energy

Protect humans and environment against harmful effects

Headings	Examples
Mechanical	Relative movement of human and machine, mechanical vibrations, dust
Acoustic	Noise
Hydraulic	Jets of liquid
Pneumatic	Jets of gas, pressure waves
Electrical	Passage of current through body, electrostatic discharges
Optical	Dazzle, ultra-violet radiation, arcs
Thermal	Hot and cold parts, radiation, inflammation
Chemical	Acids, alkalis, poisons, gases, vapours
Radioactive	Nuclear radiation, X-rays

Table 7.2. Minimum industrial safety requirements in mechanical devices

In mechanical devices protruding or moving parts should be avoided in areas where human contacts might occur

Protective equipment regardless of the operational speed is required:

— for gear, belt, chain and rope drives
— for all rotating parts longer than 50 mm, even if they are completely smooth
— for all couplings
— in case of danger from flying parts
— for potential traps (slides coming up against stops; components pushing, or rotating against, each other)
— descending components (weights, counter-weights)
— for slots, for example, at material inputs. The gap between parts must not exceed 8 mm; in the case of rollers, the geometrical relationship must be examined and, if necessary, special guards must be installed

Electrical installation must always be planned in collaboration with electrical experts. In the case of *acoustic, chemical* and *radioactive* danger, expert advice must be sought for the requisite protection

if necessary by special regulations. Designers must help to avoid the emergence of dangerous weak spots in the course of production processes (see 7.3.1, 7.3.2 and 7.5.7).

Assembly and Transport

The loads to which a product will be subjected during assembly and transport must be taken into consideration during the embodiment design phase. Welds carried out during assembly must be tested and, where necessary, heat treated. All major assembly processes should, whenever possible, be concluded by functional checks.

Firm bases and support points should always be provided and marked clearly. The weights of parts heavier than 100 kg should be marked where they can be seen easily. If frequent dismantling is called for, the appropriate lifting points must be incorporated. Suitable handling points must be provided for transport and marked clearly.

Operation

Operation and handling must be safe [7.61, 7.62]. The failure of any automatic device must be indicated at once so that the requisite actions can be taken.

Maintenance

Maintenance and repair work must only be undertaken when the machine is shut down. Particular care is needed to ensure that assembly or adjusting tools are not left behind in the machine. Safety switches must ensure that the machinery is not started unintentionally. Centrally placed, easily accessible and simple service and adjustment points should be provided. During inspection or repair, safe access should be possible through the provision of handrails, steps, non-slip surfaces etc.

Costs and Schedules

Cost and schedule requirements must not affect safety. Cost limits and delivery dates are ensured by careful planning, and by implementing the correct concepts and methods, not by cutting corners. The consequences of accidents and failures are generally much greater and graver than the effort needed to prevent them.

7.4 Principles of Embodiment Design

The general principles of embodiment design have been discussed at some length in the literature. Kesselring [7.161] set out principles of minimum manufacturing costs, minimum space requirements, minimum weight, minimum losses, and optimum handling (see 1.2.2). Leyer [7.182] discussed the principle of lightweight construction. It is obviously neither possible nor desirable that all these principles should be implemented in every technical solution—one of them might be crucial, the rest merely desirable. Which principle must predominate in a given case can only be deduced from the task and the company's general objectives. By proceeding systematically, elaborating a requirements list, abstracting to identify the crux of the problem, and also by following the checklist given in Fig. 5.7, designers transform these principles into a concrete proposal that enables them to determine manufacturing costs, space requirements, weights etc. These have to be consistent with the requirements list.

The systematic approach also highlights the question of how, with a given problem and a fixed solution principle, a function can be best fulfilled and by what type of function carrier. Embodiment design principles facilitate this part of the design process. In particular, they help with Steps 3 and 4, but also with Steps 7 to 9 as listed in 7.1 above.

For the relatively common task of transmitting forces or moments, it seems advisable to establish special "principles of force transmission".

Tasks requiring changes in the type or variations in the magnitude of a force are primarily fulfilled by the appropriate physical effects, but designers must also apply the "principle of minimum losses" [7.161] for energy-conservation or economic reasons, which they do by adopting a small number of highly efficient steps. This principle also applies to the efficient conversion of one type of energy into another, whenever this should be required. In that case the design problem, in terms of generally valid functions, reduces essentially to one of channelling, connecting and storing.

Energy storage problems involve the accumulation of potential and kinetic energy, be it directly or indirectly through the collection of material. The storage of energy, however, raises the question of the stability of the system, and the consequent application of the "principles of stability and bi-stability".

Often, several functions have to be fulfilled by one or several function carriers. Here the "principle of the division of tasks" may be useful to designers. Its application involves a careful analysis of the functions and their assignment to function carriers. This analysis of functions is also helpful for the application of the "principle of self-help", when supplementary effects must be identified and exploited.

In applying embodiment design principles, designers may find that they run counter to certain requirements. Thus the principle of uniform strength may conflict with the demand for minimum costs; the principle of self-help may conflict with fail-safe behaviour (see 7.3.3); and the principle of equal wall thickness chosen for the purpose of simplifying the production process [7.183] may conflict with the demand for lightweight construction.

These principles represent many strategies that are only applicable under certain conditions. In using them, designers must strike a balance between competing demands. To that end, the present authors have developed what they consider to be important embodiment design principles, which will now be presented. Most are based on energy-flow considerations and, by analogy, they apply equally well to the flow of material and of signals.

7.4.1 Principles of Force Transmission

1 Flowlines of Force and the Principle of Uniform Strength

The problems solved in mechanical engineering generally involve forces and/or motions and their connection, change, variation or channelling, and involve the conversion of energy, material and signals. The generally applicable function "channel forces" includes the application of loads to, the transfer of forces between, and the transmission of forces through components and devices [7.36, 7.183, 7.298].

In general, designers should try to avoid all sudden changes of direction in the flowlines of force, that is force transmission path, caused by sharp deflections and abrupt changes of cross section. The idea of "flowlines of force" aids the visualisation of the force transmission path (load path) through components and devices, and is analogous to flowlines in fluid mechanics.

Leyer [7.182, 7.183] has dealt with the transmission of forces at some length, so that we can dispense with a detailed discussion of the problem. Designers are advised to consult these important texts. Leyer, moreover, emphasises the complex interaction between the functional, technological and production aspects.

Force transmission must be understood in a broad sense, that is, it must include the application, transfer and transmission of bending and twisting moments.

The external loads applied to a component produce axial and transverse forces plus bending and twisting moments at every section. These set up stresses, direct and shear, and produce longitudinal, lateral (Poisson) and shear strains (elastic or plastic deformations).

The section dimensions transmitting the forces are obtained by "mental dissection" of the components at the point under consideration.

The sum of the stresses over these sections produces internal forces and moments which must be in equilibrium with the external loads.

The stresses, determined from the section dimensions, are then compared with the material properties of tensile strength, yield strength, fatigue strength, creep strength etc, due regard being paid to stress concentrations, surface finish and size effects.

The *principle of uniform strength* [7.5, 7.298] aims, with the help of appropriate materials and shapes, to achieve uniform strength throughout a mechanical device over its anticipated operational life. Like the principle of lightweight construction [7.182], it should be applied whenever economic circumstances allow.

This important consideration often misleads designers into neglecting the deformations (strains) associated with the stresses. It is, however, these very deformations that often throw light on the behaviour of components and tell us what we need to know about their functional efficiency.

2 Principle of Direct and Short Force Transmission Path

In agreement with Leyer [7.183] we consider the following principle of great importance:

- If a force or moment is to be transmitted from one place to another with the *minimum possible deformation*, then the *shortest and most direct* force transmission path is the best.

This principle, which leads to the minimum number of loaded areas, ensures:

- minimum use of materials (volume, weight); and
- minimum deformation.

This is particularly true if it is possible to solve a problem using tensile or compressive stresses alone, because these stresses, unlike bending and torsional stresses, produce smaller deformations. When a component is in compression, however, special attention must be paid to the danger of buckling.

If, on the other hand, we require a flexible component capable of *considerable elastic deformation*, then a design using *bending or torsional stresses* is generally the more economical.

The principle is illustrated in Fig. 7.26 using the mounting of a machine frame on a concrete foundation – where different requirements demand supports with different stiffnesses. This, in turn, has repercussions on the operational behaviour of the machine: different natural and resonant frequencies, modified response to additional loads etc. The more rigid solutions are obtained with minimum material and space requirements by means of a short support under compression; the most flexible solution by means of a spring, which transmits the force in torsion. If we look at other design solutions, we find many examples of the same principle: for

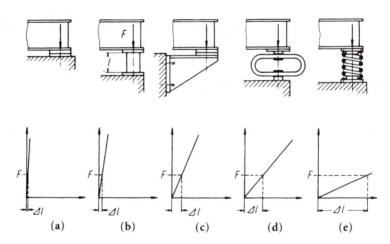

Figure 7.26.   Supporting a machine frame on a concrete foundation:
(a) very rigid support due to short force transmission path and low stress on the baseplates;
(b) longer force transmission path, but still a rigid support with tubes or box sections under compression;
(c) less rigid support with pronounced bending deformation (a stiffer construction would involve a greater use of materials);
(d) more flexible support under bending stresses;
(e) very flexible support using a spring, which transmits the load in torsion. This can be used for altering the resonance characteristics.

example, in the torsion bar springs of motor cars or in flexible pipes that rely on bending or torsional deformations.

The choice of means thus depends primarily on the nature of the task – that is on whether the force transmission path must be designed for stability with maximum stiffness, or whether certain force-deformation relationships must be satisfied first and stability can be treated as a subsidiary problem.

If the *yield point is exceeded*, then the following have to be taken into consideration (see Fig. 7.27):

- When a component is loaded by a force, it is invariably subjected to deformation. If the yield point is exceeded, then the linear-elastic relationship between the force and the deformation no longer holds. Relatively small changes in the force near the peak of the force-deformation curve may produce unstable conditions leading to fracture, because the load-bearing cross-sections may be reduced more rapidly than the strength is increased due to strain hardening. Examples are tie rods, centrifugal inertia forces on a disc and weights on a rope. The necessary safety precautions must always be taken.

- When a component is deformed, then a reaction force is set up. So long as the impressed deformation does not change, the force and the stress remain unchanged as well. If the peak is not reached, the component remains stable so that the yield point can be exceeded without danger. Beyond the yield point, a large change in deformation will lead to only a small change in the force. Admittedly, any preload must not be augmented with further operational loads in the same sense, since otherwise the conditions described above will prevail. Further requirements are the use of tough materials and the avoidance of a build-up of multi-axial stresses in the same sense. Examples are highly-distorted shrink fits, preloaded bolts and clamps.

3 Principle of Matched Deformations

Designs matched to the flowlines of force avoid sharp deflections of the transmission path and sudden changes in cross section, thus preventing the uneven

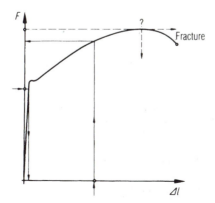

Figure 7.27. Force deformation diagram of tough materials. Arrows indicate cause-effect relationship.

distribution of stresses with high stress concentrations. A visualisation of the flow-lines of force, though very graphic, does not always reveal the decisive factors involved. Here, too, the key is the deformation of the affected components.

The principle of matched deformations states that related components must be designed in such a way that, under load, they will deform *in the same sense* and, if possible, *by the same amount*.

As an example let us take soldered or glued connections in which the solder or adhesive layer has a different modulus of elasticity from that of the material to be joined. Figure 7.28a illustrates the resulting deformation [7.194]. The deformations and the thickness of the solder or adhesive layers have been greatly exaggerated. The load F, which is transmitted across the junction of parts 1 and 2, produces distinct deformations in the overlapping parts, the adhesive layer being subjected to particularly marked deformation near the edges due to differences in the relative deformation of parts 1 and 2. While part 1 bears the full load F at the upper edge of the adhesive layer and is therefore stretched, part 2 does not yet bear a load. The relative shift in the adhesive layer sets up a local shear stress that exceeds the calculated mean value.

A particularly unsatisfactory result is shown in Fig. 7.28b where, as a result of opposite and unmatched deformations of parts 1 and 2, the deformation in the adhesive layer is considerably increased. This example makes it clear why provision should be made for deformations to take place in the same sense and, if possible, to be equal in magnitude. Magyar [7.190] has made a mathematical study of the relationships between load and shear stress: the result is shown qualitatively in Fig. 7.29.

The same phenomenon also occurs between nuts and bolts in bolted joints [7.354]. The nut (see Fig. 7.30a) is in compression and the bolt is in tension, that is they are deformed in the opposite sense. In the modified nut (see Fig. 7.30b) a

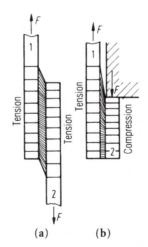

(a)   (b)

Figure 7.28. Overlapping adhesive or solder joint with strongly exaggerated deformation from [7.194]:
(a) Parts 1 and 2 deformed in the same sense.
(b) Parts 1 and 2 deformed in the opposite sense.

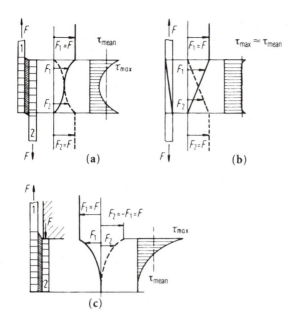

Figure 7.29. Distribution of forces and shear stresses in overlapping joints with layer of adhesive or solder, from [7.190]:
(a) overlapped on one side (bending stress neglected);
(b) spliced with linearly decreasing thickness;
(c) pronounced "deflection of the flowlines of force" with deformations in the opposite sense (bending stress neglected).

deformation in the same sense is set up in the leading threads, which gives rise to a smaller relative deformation and hence a more even distribution of the load borne by individual threads. Wiegand [7.354] has been able to demonstrate this effect by showing that such nuts have a longer service life. Paland [7.228] has shown more recently that standard nuts are not as unsatisfactory as Maduschka [7.188] has suggested, because the moment $F \cdot h$ produces additional outward deformations of the nut and thus relieves the leading threads of their load. The load-relieving deformation of the nut due to this moment and also to the bending of the threads can be considerably increased by using material with a lower modulus of elasticity. If, on the other hand, the load-relieving deformations are resisted by a very stiff nut or a very small lever arm h, then the type of load distribution described by Maduschka would ensue.

As a further example, let us take a shaft-hub connection formed by a shrink fit. In essence, this too involves the deformation of two components [7.129]. In transmitting the torque, the shaft experiences a torsional deformation that decreases as the torque is transferred to the hub. The hub, for its part, is deformed in accordance with the transmitted torque.

Figure 7.31a shows that the maximum relative deformation occurs at A. In the case of alternating torques, this may lead to fretting corrosion; moreover the right-hand end, to all intents and purposes, contributes nothing to the transfer of the torque.

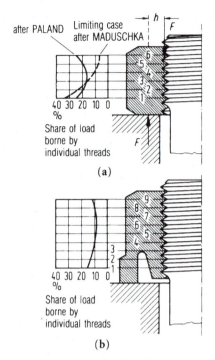

Figure 7.30. Nut shapes and load distribution after [7.354]:
(a) Standard nut: limiting case after Maduschka [7.188]; Paland [7.228] allowing for deformation due to moment $F \cdot h$.
(b) Modified nut with matched deformations in the tension part.

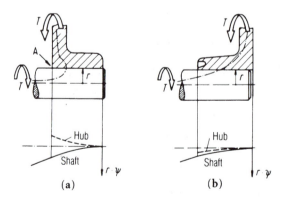

Figure 7.31. (a) Shaft-hub connection with strong "force flowline deflection". Torsional deformations of shaft and hub in opposite sense (ψ = angle of twist).
(b) Shaft-hub connection with gradual "force flowline deflection". Torsional deformations of shaft and hub in the same sense.

The solution shown in Fig. 7.31b is much better because the resulting deformations are in the same sense. The best solution appears when the torsional stiffness of the hub is matched to that of the shaft. The transfer of torque then takes place along the whole length of the connection and high stress concentrations are avoided.

Even if the shrink fit were replaced with a keyed connection, the layout depicted in Fig. 7.31a would, because the torsional deformations are in the opposite sense, set up very high contact stresses in the neighbourhood of A. The layout depicted in Fig. 7.31b, on the other hand, will, because the deformations are in the same sense, ensure an even stress distribution [7.200].

The principle of matched deformations can also be applied to bearings as in Fig. 7.32.

Mention must also be made of welded joints. Here the residual stresses which occur on cooling and the stress concentrations caused by deflections of the force transmission path can be reduced by careful design [7.6, 7.13].

The principle of matched deformations must be taken into account not only in the transfer of forces from one component to another, but also in the division or combination of forces or moments. A well known problem is the simultaneous propulsion of wheels that have to be placed at a considerable distance from one another, for instance in crane drive assemblies. In the layout shown in Fig. 7.33a, the left side has a relatively high torsional stiffness due to the short force transmission path, and the right side a relatively low torsional stiffness because of its greater path length. When the torque is first applied, the left wheel will be set in motion while the right wheel remains stationary until the right hand part of the shaft has twisted sufficiently to transmit the torque. The drive assembly has a tendency to run skew.

It is essential to provide the same torsional stiffness to both parts of the shaft so as to ensure an appropriate division of the initial torque. This can be achieved in two distinct ways if the input torque is taken in one position only: either by

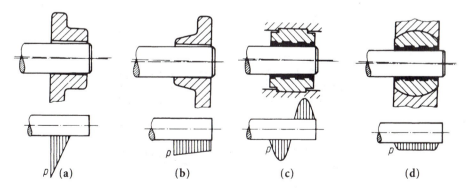

Figure 7.32. Force transmission in bearings:
(a) edge compression because of insufficient adaptation of the bearing to the deformed shaft;
(b) more even bearing pressure because of matched deformations;
(c) lacking adjustment to shaft deformation;
(d) more even bearing pressure because of adaptability of bearing bush.

symmetrical layout (see Fig. 7.33b); or by adaptation of the torsional stiffness of the appropriate parts of the shaft (see Fig. 7.33c).

4 Principle of Balanced Forces

Those forces and moments that serve the function directly, such as the driving torque, the tangential tooth force and the load torque in a gearbox, can, in accordance with the definition of a main function, be described as *functionally determined main forces*.

In addition, there are many forces or moments that do not serve the function directly but cannot be ignored, for instance:

- the axial force produced by a helical gear;
- the force resulting from a pressure difference, for instance across the blades of a turbine or across a control valve;
- tensile forces for producing a friction connection;
- inertia forces due to linear acceleration or rotation of components; and
- fluid flow forces inasmuch as they are not the main forces.

Such forces and moments accompanying the main ones are called *associated forces* and may either produce an auxiliary effect (see auxiliary function) or else appear merely as invariable concomitants.

Associated forces place additional loads on the components and require an appropriate layout or must be taken up by further surfaces and elements such as stiffening members, collars, bearings etc. As a result, weights are increased and further frictional losses may be incurred. For that reason, the associated forces

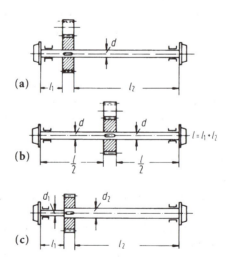

Figure 7.33. Application of the principle of matched, here equal, deformations in crane drives:
(a) unequal torsional deformation of lengths l_1 and l_2;
(b) symmetrical layout ensures equal torsional deformation;
(c) asymmetrical layout with equal torsional deformation due to adaptation of torsional stiffnesses.

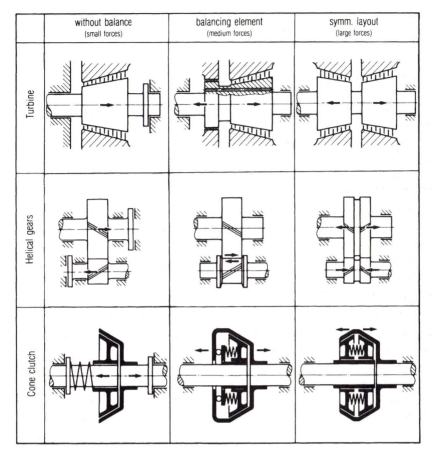

	without balance (small forces)	balancing element (medium forces)	symm. layout (large forces)
Turbine			
Helical gears			
Cone clutch			

Figure 7.34. Fundamental solutions for balancing associated forces illustrated by means of a turbine, helical gears and a cone clutch.

must, whenever possible, be balanced out at their place of origin, thus obviating the need for a heavier construction or for reinforced bearing and transfer elements.

As has been shown in [7.219], this balance of forces is essentially ensured by two types of solution:

- balancing elements; or
- symmetrical layout.

Figure 7.34 shows how the associated forces can be balanced in a turbine, helical gears and a cone clutch, with the help of the principle of direct and short force transmission path. As a result, no bearing position is additionally loaded and the designs are highly economical.

When it comes to the balancing of inertia forces, we find that a rotationally symmetrical layout is inherently balanced. The same solution principle is applied for reciprocating masses, as we know from automobile engineering. If the number of cylinders is too small to ensure a perfect balance, either special balancing elements, weights or shafts [7.239] are introduced, or cylinders are arranged symmetrically, as for instance in opposed cylinder engines.

As a general rule (which, however, can be ignored if there are overriding reasons for doing so) balancing elements should be chosen for relatively small or medium forces, and a symmetrical layout for relatively large forces.

All in all, we can say of the transmission of forces (in the discussion of which the physically undefinable but descriptive idea of the flowlines of force is most helpful) that:

- the flowlines of force must always be closed; and
- sharp deflections of the flowlines of force and changes in the "density" of the lines resulting from sudden changes in cross section must be avoided.

The concept of the flowlines of force should be considered in conjunction with the following principles:

The *principle of uniform strength* which ensures, through the careful selection of materials and shapes, that each component is of uniform strength and contributes equally to the overall strength of a device throughout its service life.

The *principle of direct and short force transmission path* which ensures minimum volume, weight and deformation, and which should be applied particularly if a rigid component is needed.

The *principle of matched deformations* which ensures the matching of deformations of related components so that stress concentrations are avoided and the function can be reliably fulfilled.

The *principle of balanced forces* which ensures, with the help of balancing elements or a symmetrical layout, that the associated forces accompanying the main ones are reacted as close as possible to their place of origin, so that material quantities and losses can be kept to a minimum.

7.4.2 Principle of the Division of Tasks

1 Assignment of Sub-Functions

Even during the setting up and variation of the function structure, it is important to determine to what extent several functions can be replaced by a single one, or whether one function can be subdivided into several sub-functions (see 6.3).

These questions reappear in the embodiment phase, when the problem is to fulfil the requisite functions with the choice and assignment of suitable function carriers. We ask:

- What sub-functions can be fulfilled with one function carrier only?
- What sub-functions must be fulfilled with the help of several, distinct function carriers?

So far as the number of components and the space and weight requirements are concerned, a single function carrier fulfilling several functions would, of course, be the best. In respect of the production and assembly processes, however, this may prove disadvantageous, if only because of the complicated shape of the resulting component. Nevertheless, for economic reasons, the attempt should always be made to fulfil several functions with a single function carrier.

Numerous assemblies and components can fulfil several functions simultaneously or successively.

Thus a shaft on which a gearwheel has been mounted transfers the torque and the rotating motion simultaneously, and, at the same time, takes up the bending moments and shear forces resulting from the normal tooth force. It also locates the gears axially and, in the case of helical gears, carries the axial force components from the teeth. In conjunction with the body of the gearwheel, it provides sufficient stiffness to ensure correct mating of the teeth.

A pipe flange connection makes possible the connection and separation of the pipes, ensures the sealing of the joint and transmits all forces and moments in the pipe resulting from residual tension, from thermal expansion or from unbalanced pipe loads.

A turbine casing provides the appropriate inlet and outlet flow areas for the fluid, provides a mounting for the stationary blades, transmits the reaction forces to the foundation, and ensures a tight seal.

The wall of a pressure tank in a chemical plant must combine a retaining with a sealing function and stave off corrosion, while not interfering with the chemical process.

A deep groove ball bearing, apart from its centring task, transmits both radial and axial forces and occupies a relatively small volume, for which reasons it is a popular machine element.

The combination of several functions in a single function carrier may often prove economically advantageous, but may have certain drawbacks. These do not usually appear unless:

- the capacity of the function carrier has to be increased to the limit in respect of one or several functions; or
- the behaviour of the function carrier must be kept absolutely constant in one important respect.

As a rule, it is impossible to optimise the carrier of several combined functions. Instead, designers have recourse to the *principle of the division of tasks* [7.222], by which a special function carrier is assigned to every function. Moreover, in borderline cases, it may even be useful to distribute a single function over several function carriers.

The principle of the division of tasks:

- allows very much better exploitation of the component concerned;
- provides for greater load capacity; and
- ensures unambiguous behaviour, and hence fosters the basic rule of clarity (see 7.3.1).

This is because the separation of tasks facilitates optimum design in respect of every sub-function and leads to more accurate calculations. In general, however, the constructional effort becomes correspondingly greater.

To determine whether the principle of the division of tasks can be usefully applied, the functions must be analysed with a view to determining if the simultaneous fulfilment of several functions in one carrier introduces constraints or mutual interference. If it does, then it is best to settle for individual function carriers.

2 Division of Tasks for Distinct Functions

Examples from various fields illustrate the advantage of the division of tasks for distinct functions.

In large gearboxes, as found for instance between a turbine and a generator, it is advisable, because of thermal expansion of the foundations and bearings and also because of the torsional oscillations, to use a radially and torsionally flexible shaft whilst maintaining the shortest possible axial length on the output side [7.218]. However, because of the forces between the gear teeth, the transmission shaft must be as rigid as possible. Here the principle of the division of tasks leads to the following arrangement: the gearwheel is fitted to a stiff hollow outer shaft with the shortest possible distance between the bearings; while the radially and torsionally flexible component takes the form of an inner torsion shaft (see Fig. 7.35).

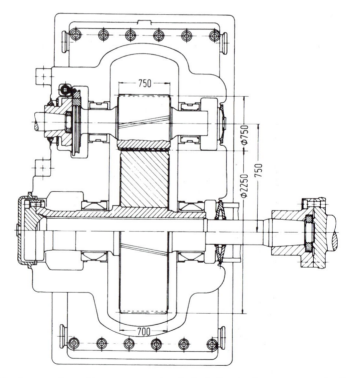

Figure 7.35. Large gearbox with an output torsion shaft; the bearing forces are transmitted over a stiff hollow shaft; the inner torsion shaft is radially and torsionally flexible, from [7.218] (Siemens-Maag).

Modern pressure-fed boilers are built with a membrane wall, as shown in Fig. 7.36. The furnace must be gas-tight. Moreover, optimum heat transfer to the water demands thin walls with large surface areas. Beyond that, thermal expansion and pressure differences between the furnace and its environment must also be taken into consideration, and so must the weight of the walls. This complex problem is solved with the help of the principle of the division of tasks. The tubular walls with their welded lips constitute the sealed furnace. The forces resulting from the pressure differences are transferred to the special supports outside the heated area which also carry the weight of the, usually suspended, walls. Articulated arms between the tubular wall and the supports allow for unimpeded thermal expansion. Thus every part can be designed in accordance with its special task.

The clamp connection in a superheated steam pipe shown in Fig. 7.37 has also been designed on the principle of the division of tasks. The sealing and load carrying functions are assigned to different function carriers: the sealing function is performed by the welded membrane seal, which is axially loaded by the tension in the clamp. Tensile forces or bending moments should not be carried by the seal, whose function and durability would thereby be destroyed, so the load-carrying function is performed by the clamp which, in its turn, is designed on the principle of the division of tasks. The clamp is made up of segments, which transmit forces and bending moments by means of a close tolerance fit, and shrink rings hold the clamp segments together by friction in a simple and effective manner. Every part can be optimally designed for its particular task and is easily analysed.

The casings of turbines must ensure a tight seal under all operational and ther-

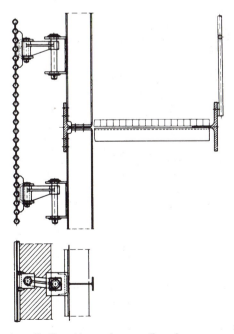

Figure 7.36. Section of boiler with membrane walls and separate supports (Babcock).

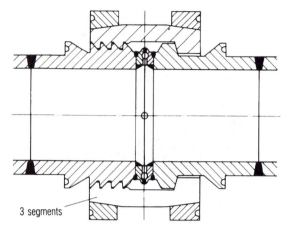

3 segments

Figure 7.37.   Clamp connection in a superheated steam pipe (Zikesch).

mal conditions if they are to conduct the working fluid with minimum loss and turbulence. They must also provide an annular area and a support for the stationary blades. During temperature changes, sectioned casings with an axial flange have a particular tendency to distort and to lose sealing power due to marked changes in shape at the inlet and outlet [7.236].

This effect can be offset by a separate blade carrier, that is, by a division of tasks. The annular area and stationary blade attachment can be designed regardless of the larger casing with its inlet and outlet sections. The outer casing can then be designed exclusively for durability and sealing power (see Fig. 7.38).

A further example is provided by the synthesis of ammonia, which involves feeding nitrogen and hydrogen into a container under high pressures and temperatures. If the hydrogen were allowed to come into direct contact with a ferritic steel container, it would penetrate into and decarbonise the latter, producing decomposition at the grain boundaries with the formation of methane [7.119]. The solution is again based on the division of tasks. The sealing function is provided by an inner casing of austenitic steel which is resistant to hydrogen, while support and

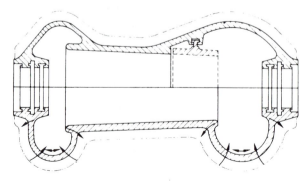

Figure 7.38.   Axially divided turbine housing from [7.236]; lower half conventional; upper half with separate blade carrier.

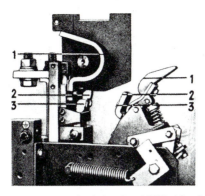

Figure 7.39. Arrangement of contacts in circuit breaker (AEG) *1* breaker contacts; *2* intermediate contacts; *3* main contacts.

strength are provided by a surrounding pressure chamber constructed of high-tensile ferritic steel, not resistant to hydrogen.

In the electrical circuit-breaker illustrated in Fig. 7.39, two or even three contact systems are provided. The breaker contacts *1* take the arcing current during the closing or opening of the switch, and the main contacts *3* carry the current under normal conditions. The breaker contacts *1* are subject to burning, that is to wear and tear, and must be designed accordingly, while the main contacts must be designed to carry the full working current.

The division of tasks is also illustrated in Fig. 7.40: the Ringfeder connector carries the torque while the corresponding cylindrical surfaces ensure the central location and seating of the pulley, something the Ringfeder connector cannot provide by itself.

A further example is provided by the design of rolling element bearings in which the service life of the locating bearing is increased by the clear separation of the

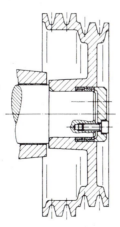

Figure 7.40. Ringfeder connector plus centralising surfaces.

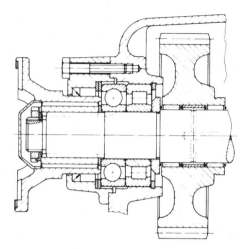

Figure 7.41. Locating bearing with separate transmission paths for radial and axial forces.

transmission paths of radial and axial forces (see Fig. 7.41). The outer race of the deep-groove ball bearing is not supported radially and hence transmits axial forces only, while the roller bearing transmits radial forces only.

The principle of the division of tasks has been applied consistently to the construction of composite flat belts. They are made up, on the one hand, of a synthetic material capable of carrying high tensile loads and, on the other hand, of a chrome leather layer on the contact surface which provides a high coefficient of friction for the transfer of the load.

Yet another example is provided by the rotor blade attachment in a helicopter (see Fig. 7.17).

3 Division of Tasks for Identical Functions

If increases in load or size reach a limit, a single function can be assigned to several, identical function carriers. In other words, the *load can be divided* and then recombined later. There are numerous examples.

The load capacity of a V-belt cannot be increased at will by increases in its cross section (number of load-carrying strands per belt) because, for a given pulley diameter, an increase in the belt height h (see Fig. 7.42) leads to an increase in

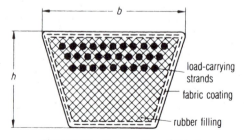

Figure 7.42. Cross section of V-belt.

the bending stress. As a result of the ensuing deformation, the rubber, which has hysteresis properties and is also a poor conductor of heat, becomes overheated and this reduces its life. A disproportionally wide belt, on the other hand, loses the stiffness needed to take up the normal forces acting on the wedge-shaped surfaces of the pulley. An increase in load-carrying capacity can, however, be obtained by dividing the overall load into part loads each appropriate to the load limit and normal life of the individual belt (multiple arrangement of parallel V-belts).

The coefficient of thermal expansion of superheated steam pipes made of austenitic steel is approximately 50 per cent higher than that of pipes made of the usual ferritic steel. Such pipes, moreover, are particularly stiff. At constant inner pressures and fixed material property limits, the ratio of outer to inner pipe diameter remains constant if the inner diameter is changed. However, while the throughput at constant flow velocities varies as the square of the inner diameter, the bending and torsional stiffnesses vary as its fourth power. The substitution of z pipelines for a single large pipe would admittedly lead to increased pressure and heat losses for the same flow area, but would reduce by $1/z$ the stiffness resisting thermal expansion. With four or eight pipelines the individual reaction forces would then be no more than 1/4 or 1/8 of that present in a single pipe [7.40, 7.299]. In addition, the reduction in wall thickness leads to a reduction in thermal stresses.

Gearboxes, and epicyclic gearboxes in particular, make use of the principle of the division of tasks, or rather of forces, in the form of multiple meshing, which will increase the transmission capacity of the gearbox, provided that the thermal effects can be kept within reasonable limits. In the symmetrical layout of epicyclic gearboxes based on the principle of balanced forces (see 7.4.1 - 4) even the bending moment in the shaft is eliminated because the forces produced by the gears cancel out. However, the torsional deformation is increased because of the greater load capacity (see Fig. 7.43). In large gearboxes, this principle is applied to advantage in the form of multiple drives equipped with spur gears, which have external teeth only and hence are more easily manufactured. As Ehrlenspiel [7.97] has shown, it is possible to increase the load capacity with the number of force transmission

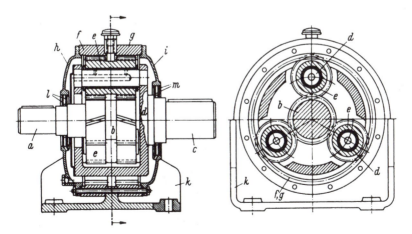

Figure 7.43.   Epicyclic gearbox with balanced forces, from [7.98].

paths, though not in direct proportion because each step introduces a different flank geometry with a slightly greater flank loading. Basic arrangements are depicted in Fig. 7.44.

One problem of the principle of the division of tasks is the uniform participation of all the elements in the fulfilment of the function, that is the provision of a uniform distribution of forces or loads. In general, this can only be achieved if:

- the participating elements adjust themselves automatically to balance out the forces; or
- appropriate flexibility is specially provided in the force transmission paths.

In the case of multiple V-belt drives, the tangential forces produce slight extensions of the belts which help to offset any dimensional errors in the lengths of the belts or in the pulleys or any lack of parallelism in the shaft, and thus ensure equal load sharing.

In the case of the multiple pipeline discussed above, the individual pipe loss coefficients, the relationships between inflow and outflow, and also the geometry of the pipe layouts must be kept similar, or else the individual loss coefficients must be small and not greatly affected by the flow speeds.

In the case of multiple gears, either a strictly symmetrical arrangement must ensure equal stiffnesses and temperature distributions throughout the gearbox or special flexible or adjusting elements [7.98] must ensure the equal participation of all the components.

Figure 7.45 illustrates a flexible arrangement. Further balancing components, such as elastic and articulated joints, are described in [7.98].

All in all, the principle of the division of tasks provides for increases in the maximum load capacity or for wider applications. By spreading tasks over several function carriers, we also gain a clearer picture of the relationship between forces and their effects, and, what is more, can increase the output, provided only that a balanced division of forces is maintained by adjustable or self-regulating elements.

In supporting structures (such as bearing supports) where force transmission is divided, a more balanced load distribution can be achieved by adjusting the stiff-

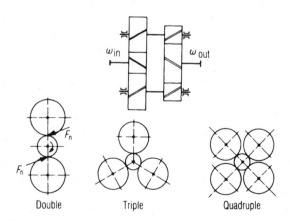

Double   Triple   Quadruple

Figure 7.44. Basic arrangements of multiple gears, from [7.218].

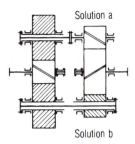

Figure 7.45. Balanced forces in multiple gears by means of flexible torsion shafts, from [7.218].

ness. During the stiffness analysis, the location and direction of the external forces have to be considered carefully because they influence the deformation behaviour. This analysis can be facilitated by the use of Finite Element (FE) methods (see the principle of matched deformation 7.4.1 - 3).

In general, the application of the principle of the division of tasks calls for greater effort on the part of designers, which must be offset by greater overall economy or safety.

7.4.3 Principle of Self-Help

1 Concepts and Definitions

In the last section we discussed the principle of the division of tasks and showed how it could help to increase load capacity and to provide a clearer definition of the behaviour of the components. To that end, we analysed the various sub-functions and assigned them to such function carriers as neither influence nor interfere with one another.

The same analysis can also be used in conjunction with the *principle of self-help* to achieve, through the appropriate choice of system elements and their arrangement, a mutual supportive interaction that improves the fulfilment of the function. Under normal conditions (normal loading), "self-help" provides for greater effect or relief; in emergency situations (overloading), it provides for greater safety.

In a self-helping design, the *overall effect* is made up of an initial effect and a supplementary effect.

The *initial effect* sets off the physical process required by the solution but is insufficient on its own.

The *supplementary effect* is obtained from the functionally determined main forces (gearbox torque, sealing force etc) and/or from the associated forces (axial force produced by helical gears, centrifugal inertia force, force due to thermal expansion etc), provided, of course, that the two sets of forces are clearly correlated. A supplementary effect may also be obtained from appropriate changes in force transmission paths.

The idea of formulating the self-help principle was first suggested by the Bredtschneider-Uhde self-sealing cover, which is particularly suitable for pressure vessels [7.253]. Figure 7.46 shows how it works. A relatively small force provided by the central bolt *2*, suffices to press the cover *1* against the metal seal *5*. The initial effect of this force ensures that the parts make the proper contact. With increasing operational pressure a supplementary effect is produced, thanks to which the sealing force between cover and tank is increased appropriately. The internal pressure thus provides the required sealing force automatically.

Inspired by this self-sealing solution, the principle of self-help was formulated in [7.221] and further analysed and elaborated by Kühnpast [7.177].

It may be useful to specify the quantitative contribution of the supplementary effect *S* to the overall effect *O* in producing the degree of self-help:

$$\chi = S/O = 0. . .1.$$

The gain from self-help solutions can be expressed in terms of one or several technical characteristics: efficiency, service life, use of materials, technical limit etc. The self-help gain is defined as:

$$\gamma = \frac{\text{technical characteristic with self-help}}{\text{technical characteristic without self-help}}$$

Whenever the application of the self-help principle calls for a greater effort on the part of designers, then it must bring clear technical or economic advantages.

Identical design approaches may turn out to be *self-helping* or *self-damaging*, depending on the layout. Take the case of an inspection cover (see Fig. 7.47). So long as the pressure inside the tank is greater than the pressure outside, the layout shown on the left is self-helping, because the pressure on the cover (supplementary effect) increases the sealing effect (overall effect) of the initial tension-screw force (initial effect).

The layout shown on the right, by contrast, is self-damaging because the pressure on the cover decreases the sealing effect (*O*) of the initial tension-screw force (*I*). If, however, the tank were kept at below atmospheric pressure, the left layout would be self-damaging, the right layout self-helping (see also Fig. 7.48).

This example shows that the degree of self-help depends on the resultant effect: in the present case the effect on the sealing force resulting from the elastic forces,

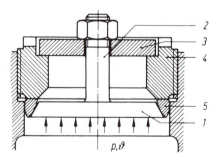

Figure 7.46.   Self-sealing cover:
1 cover; *2* central bolt; *3* cross member; *4* element with saw tooth thread; *5* metal sealing ring; *p* = internal pressure; ϑ = temperature.

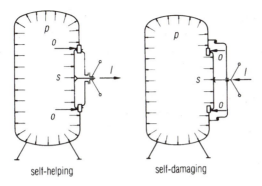

Figure 7.47.   Layout of an inspection cover.
I = initial effect; S = supplementary effect; O = overall effect; p = internal pressure.

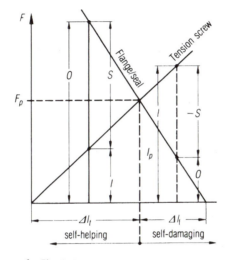

Figure 7.48.   Force diagram for Fig. 7.47:
F = forces; F_p = preload; Δl = change in length; subscript t = tension screw; subscript f = flange/seal.

and not on the simple addition of the force exerted by the screw and the force acting on the cover.

Figure 7.48 can also be considered as a force-deformation diagram of a bolted connection with a preload and a working load. The conventional bolted flange connection may be called self-damaging inasmuch as, under operational conditions, the overall effect—that is the flange sealing—becomes smaller than the preload. Also the loading of the bolts is increased at the same time. If possible, therefore, only such self-helping arrangements should be chosen as increase the overall effect, while reducing the loading of the bolts (Figs. 7.49 a-d illustrate such arrangements).

For practical purposes, it is useful to classify self-helping solutions in accordance with Table 7.3.

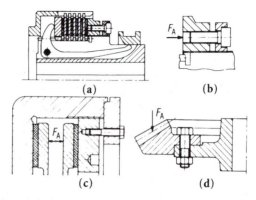

Figure 7.49.   Self-helping bolted connections:
(a) Multiple disc clutch with adjustment ring; (b) force acting on the adjustment ring; (c) adjustable
disc of two-disc friction clutch; (d) crown wheel attachment, symmetrical take-up of forces.

Table 7.3.   Summary of self-help solutions

		Normal load	Overload
Type of self-help	Self-reinforcing	Self-balancing	Self-protecting
Supplementary effect due to	Main and associated forces	Associated forces	Altered force transmission path
Important features	Main or associated forces act in the same sense as other main forces	Associated forces act in the opposite sense to main forces	Force transmission path altered by elastic deformation; limitation of function permissible

2 Self-Reinforcing Solutions

In self-reinforcing solutions, the supplementary effect is obtained directly from a main or associated force and adds to the initial effect to produce a greater overall effect.

This group of self-helping solutions is the most common. Under part-load conditions, it ensures greater service life, less wear, higher efficiency etc, because the components are only loaded to an extent needed to fulfil the function at any particular moment.

As a first example, let us consider a continuously adjustable friction drive (see Fig. 7.50).

The preload spring a presses the freely movable cup wheel c on the drive shaft b against the cone wheel d, thus providing the initial effect. Once a torque is applied, the roller follower e attached to shaft b is pressed against the cam f formed on the cup wheel c where it produces a normal force F_n that can be resolved into a tangential force F_t and an axial force F_a which, for its part, increases the contact force F_c applied to the cone wheel in a fixed proportion to the applied torque T:

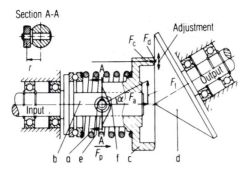

Figure 7.50. Continuously adjustable friction drive:
a: preload spring; *b*: drive shaft; *c*: cup wheel; *d*: cone wheel; *e*: roller follower; *f*: cam formed on the cup wheel; *r*: radius on which F_t and F_a act.

$$F_a = T/(r \cdot \tan \alpha).$$

The force F_a represents the supplementary effect gained from the torque. The overall effect is obtained from the spring preload force F_p, plus the axial force F_a which varies as the torque T (see Fig. 7.51). The tangential driving force F_d on the cone, which determines the transmittable torque, is therefore:

$$F_d = (F_p + F_a) \cdot \mu$$

and the degree of self-help:

$$\chi = S/O = F_a/(F_p + F_a).$$

It is obvious that the contact pressure between the wheels, which helps to determine the wear and the service life of the drive, must not exceed what is strictly necessary. A conventional solution (no self-reinforcement) would have demanded an axial force produced exclusively by the spring preload corresponding to the maximum torque, and therefore maximum pressure being applied to the contact

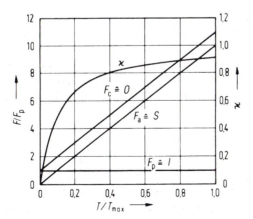

Figure 7.51. Degree of self-help (χ) and initial (I), supplementary (S) and overall (O) effect against the relative torque T/T_{max} for the friction drive (Fig. 7.50).

area under all loads. As a result the bearings, too, would have had to carry a considerably greater load, which would have led to a reduced service life or demanded a much heavier construction.

A rough calculation shows that if the actual loading is, say, 75 per cent of the nominal maximum load, then the bearing load would be reduced by about 20 per cent which, because of the exponential relationship of service life to load, can lead to a doubling of the life of the bearings. In that case, the self-help gain in respect of the service life becomes:

$$\gamma_L = \frac{\text{Life with self-help}}{\text{Life without self-help}} = \left(\frac{C/0.8P}{C/P}\right)^n = 1.25^3 = 2$$

Figure 7.49 shows various self-reinforcing layouts of contact surfaces loaded by bolts, in which the frictional forces are increased by the operational forces while the bolts themselves are off-loaded.

The application of the principle of self-help in the design of self-reinforcing brakes has been described by Kühnpast [7.177] and Roth [7.247]. Depending on the application, even self-damaging, and in this case self-weakening, solutions can prove interesting, inasmuch as they reduce the effect of variations of the coefficient of friction on the braking moment [7.107, 7.247].

Self-reinforcing seals (see Fig. 7.52) provide us with further examples. In them,

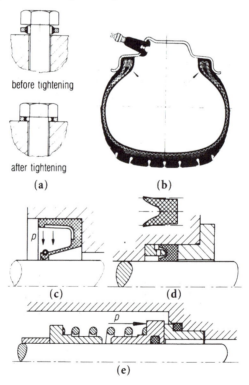

Figure 7.52. Self-reinforcing seals; (a) self-sealing washer; (b) tubeless tyre; (c) radial-shaft seal; (d) sleeve seal; (e) sliding-ring seal.

the operating pressure against which the seal has to be applied is used to produce the supplementary effect.

Finally we must mention one case in which the supplementary effect is produced by an associated force. In hydrostatic axial bearings, the centrifugal inertia effect leads to an increase in oil pressure which, at high revolutions, will help to improve the load-carrying capacity, provided the heat can be removed (see Fig. 7.53). The supplementary effect leads to an improvement in the load-carrying capacity due to the increased oil pressure resulting from the centrifugal effect alone; the overall effect is due to the load-carrying capacity of the combined static and dynamic pressures. According to Kühnpast [7.177] it should be possible at, say, 166 rev/s and $\chi = 0.38$, to obtain a gain in self-help of $\gamma = 1.6$.

The supplementary effect of another associated force, namely that caused by the effect of temperature on the shrink-fitted rings of a turbine, is discussed in [7.221].

3 Self-Balancing Solutions

In self-balancing solutions, the supplementary effect is obtained from an associated force, and offsets the initial effect to produce an improved overall effect.

A simple example is provided by turbo-machines. A blade attached to a rotor is subject to a bending stress due to the tangential force acting upon it and also to an axial tensile stress due to the centrifugal inertia force. The two are additive and, because a certain stress must not be exceeded, the transferable tangential force is reduced (see Fig. 7.54). If, however, the blade is attached at an angle, a supplementary effect is produced: an additional bending stress due to the centrifugal inertia force acting on the offset centre of gravity of the blade opposes the original bending stress and thus allows the application of a larger tangential force — that is a greater overall effect. How far this balancing process can be carried depends on the aerodynamic and mechanical conditions.

A self-balancing effect can also be produced by allowing thermally induced forces

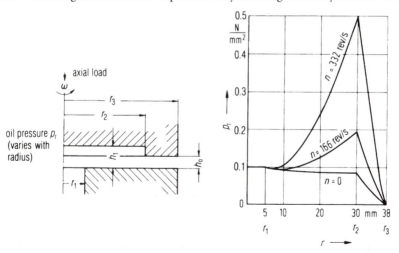

Figure 7.53. Self-help effect in hydrostatic axial bearings, from [7.177].

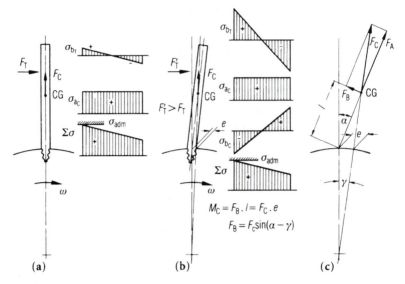

CG = Centre of gravity
F_T = Tangential force; F_C = Centrifugal force; σ_{b_T} = Bending stress due to F_T; σ_{a_C} = Axial stress due to F_C;
σ_{b_C} = Bending stress due to F_C; F_A = Axial component of F_C; F_B Bending component of F_C

Figure 7.54. Self-balancing solution for turbine blades; (a) conventional solution; (b) leaning of the blade produces a balancing supplementary effect due to the additional bending stresses produced by the centrifugal force (σ_{b_C}) which oppose the bending stresses caused by the tangential force (σ_{b_T}); (c) diagram of forces.

(stresses) to oppose other forces (stresses); for instance those resulting from excess or other mechanical loads (see Fig. 7.55).

All the examples we have given are intended to encourage the design of technical systems where:

- forces and moments with their resulting loads cancel out as far as possible; or
- additional forces or moments are produced in a clearly defined way so that it is possible to balance them out.

4 Self-Protecting Solutions

In general, in the event of an overload, we do not want components to be destroyed, unless, of course, they have been deliberately designed as weak links.

In particular, we try to protect components that are frequently subject to slight overloads. If special safety arrangements, for instance to limit the load, are not essential, then a self-protecting solution may prove advantageous. It will sometimes be simplicity itself.

Self-protecting solutions derive their supplementary effect from an additional force transmission path that, in case of excess loading, is generally created after a given elastic deformation has taken place. As a result, the distribution of the flow-lines of force is altered and the load-carrying capacity increased. Admittedly, in

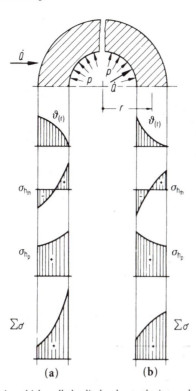

Figure 7.55.   Hoop stresses in a thick-walled cylinder due to the internal pressure σ_{h_p} and temperature differences at nearly steady heat flow $\sigma_{h_{th}}$:
(a) non-balancing solution, thermal stress is added to the maximum mechanical stress on the inner surface; (b) self-balancing solution, thermal stress opposes maximum mechanical stress on the inner surface.

that case, the functional properties associated with normal conditions may become altered, limited or suspended.

The springs shown in Fig. 7.56 have such self-protecting properties. In case of excess loading, the spring elements which are normally subject to torsional or bending stresses will transmit the additional force directly. The same effect may also be produced if the springs are shock-loaded (see Fig. 7.56b).

Figure 7.57 shows the layout of elastic couplings in which restriction of the spring movements provides additional force transmission paths with consequent loss of flexibility but with increased load-carrying capacity. The original springs are removed from the force transmission path. In Fig. 7.57a, the load-carrying capacity of the bar springs is altered inasmuch as, besides the normal bending, a powerful shear force between the two halves of the coupling appears with overloads.

Figure 7.57b shows a coupling that, strictly speaking, may be considered a borderline case between a division of tasks and a self-protecting solution. The buffers will only take up forces in cases of overloading. The characteristic of the spring elements remains unchanged. However, the force transmission path is altered after a given elastic deformation has taken place.

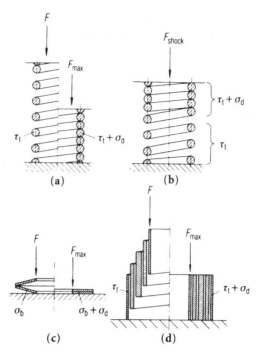

Figure 7.56.   Self-protecting solution in springs; (a) to (d) force transmission path changed, the normal function is suspended or limited in case of excess loading.

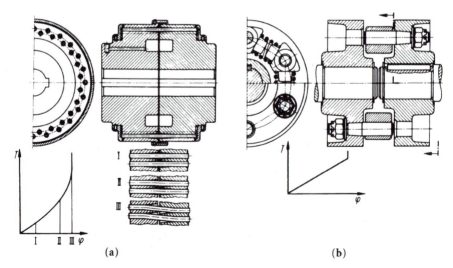

Figure 7.57.   Self-protecting solution in couplings; change of force transmission paths with loss of elastic properties in case of overloading; (a) bar spring coupling; (b) elastic coupling with coil springs and special buffers to take up the force in case of overloading.

Kühnpast [7.177] also mentions cases in which there is an uneven stress distribution over a cross-section and where plastic deformation can then be used for purposes of self-protection. In such cases, however, sufficiently tough materials and adequate dimensional stability are needed.

It is hoped that the principle of self-help based on self-reinforcing, self-balancing and self-protecting solutions will encourage designers to examine every conceivable arrangement in an effort to arrive at an effective and economical solution.

7.4.4 Principles of Stability and Bi-Stability

From mechanics, we know the concepts of stable, neutral and unstable equilibrium, as illustrated in Fig. 7.58.

In elaborating solutions, designers must always consider the effect of disturbances and try to keep the system stable by devising means whereby the disturbances can be made to cancel out, or at least to mitigate one another.

If disturbances are self-reinforcing, we have unstable or bi-stable behaviour. This effect is desirable in certain solutions, in which case we speak of planned instability.

1 Principle of Stability

By applying this principle, designers try either to ensure that disturbances cancel out or else to reduce their particular effects.

Reuter [7.237] has discussed this subject at length and we shall now look at some of his examples.

In the design of pistons for pumps or regulating devices, the main objective is to achieve stable behaviour and minimum friction.

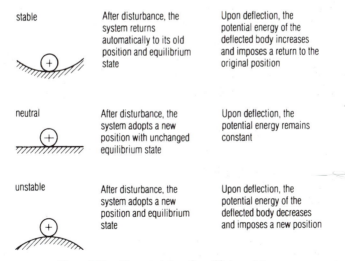

stable	After disturbance, the system returns automatically to its old position and equilibrium state	Upon deflection, the potential energy of the deflected body increases and imposes a return to the original position
neutral	After disturbance, the system adopts a new position with unchanged equilibrium state	Upon deflection, the potential energy remains constant
unstable	After disturbance, the system adopts a new position and equilibrium state	Upon deflection, the potential energy of the deflected body decreases and imposes a new position

Figure 7.58.   Characteristics of equilibrium states.

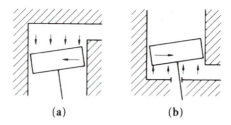

Figure 7.59.   Piston in cylinder, tilted due to a disturbance, from [7.237]:
(a) resulting pressure distribution produces an effect that increases the disturbance (unstable behaviour);
(b) resulting pressure distribution produces an effect that opposes the disturbance (stable behaviour).

Figure 7.59a shows the layout of a piston with unstable characteristics. Disturbances due, say, to inaccuracies in the cylinder bore can tilt the piston slightly and produce pressure distributions over the piston that encourage further tilting (unstable behaviour). Stable behaviour is ensured by the layout shown in Fig. 7.59b, which, however, has a disadvantage: the piston rod inlet has to be sealed off on the pressure side.

According to [7.237], the layout shown in Fig. 7.59a can be stabilised by the measures shown in Fig. 7.60 a-d. They ensure that a disturbance will itself initiate such pressure distributions as tend to correct the misalignment.

Another example is the well known case of hydrostatic bearings with oil pockets distributed around the periphery. When the bearing is loaded, the leakage path below the load is reduced with the result that pressure builds up in the affected oil pocket and decreases in the opposite one. Thanks to the combined effect, the bearing can take up the load with very small shaft displacement.

The stuffing boxes and seals of turbo-machinery must always be designed for thermo-stable behaviour [7.237]. The seal of a turbo-charger shown in Fig. 7.61 is a case in point. In the thermo-unstable layout (see Fig. 7.61a) most of the frictional heat generated by contact forces will flow into the rotor which will heat up further, expand, and hence increase the contact forces. In the stable arrangement (see Fig. 7.61b), by contrast, the frictional heat will cause the contact forces to be reduced. A disturbance thus produces a self-limiting effect.

A similar approach is used in the design of taper roller bearings. Thus, in the

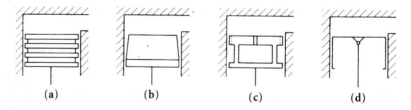

Figure 7.60.   Measures for improving the resulting pressure distribution, from [7.237]:
(a) unstable behaviour mitigated by pressure-equalising grooves;
(b) stable behaviour through conical piston;
(c) through pressure pockets;
(d) through joint fitted above centre of gravity of the piston.

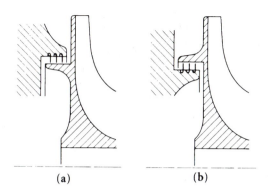

Figure 7.61.   Seal in turbocharger, from [7.237].

layout shown in Fig. 7.62a, heating of the shaft, by excessive loading for instance, will tend to increase the load even further because of the expansion of the shaft due to the increased frictional heat. The arrangement shown in Fig. 7.62b, by contrast, will lead to a load reduction. In the case under consideration, this reduction must not, however, be allowed to reach the point where one of the bearings becomes unloaded, because the shaft at that point would then not be located radially and the bearings easily damaged.

Another interesting example of thermo-stable behaviour is provided by the double-helical gears used in marine gearboxes [7.348].

2 Principle of Bi-Stability

In some cases, unstable or bi-stable behaviour is positively welcome. This happens when, on reaching a limit, a clearly distinct state or position is required and no intermediate state is acceptable. The requisite instability is initiated when a selected physical quantity reaches a limiting value and then introduces self-reinforcing effects which cause the system to spring into a second stable state. This bi-stable behaviour is required for switches and protective systems (see 7.3.3).

A well known application is in the design of safety and alarm valves [7.237] which, on reaching a limiting pressure, will spring from a completely closed to a completely open position. This avoids undesirable settings with a low flow rate or flutter and wear of the valve seat.

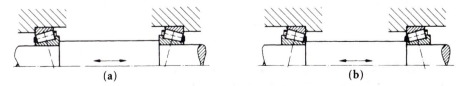

Figure 7.62.   Taper roller bearings in which the shaft heats up more than the housing:
(a) thermal expansion leads to increased loading and hence to unstable behaviour;
(b) thermal expansion leads to reduced loading and hence to stable behaviour.

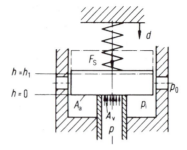

Figure 7.63. Solution principle for a valve with an unstable opening mechanism: d = precompression of spring; s = stiffness of spring; F_s = spring force; h = lift of valve head; p = pressure on valve; p_l = limiting pressure just sufficient to open the valve; p_i = intermediate pressure on opening of valve; p' = pressure after opening of valve; p_0 = atmospheric pressure; A_v = valve-opening surface area; A_a = additional surface area.

Valve closed: $F_s = s \cdot d > p \cdot A_v$, $h = 0$

Valve just open: $F_s = s \cdot d \leq p_l \cdot A_v$, $h \approx 0$

Valve opening fully: $F_s = s(d + h) < p \cdot A_v + p_i \cdot A_a$, $h \rightarrow h_1$

Valve fully open: $F_s = s(d + h_1) = p'(A_v + A_a)$, $h = h_1$ (new equilibrium position).

Figure 7.63 illustrates the solution principle.

Up to the limiting pressure $p = p_l$, the valve remains closed under the preload of the spring. If this pressure is exceeded, then the valve head will lift off very slightly. The result is an intermediate pressure p_i, the valve head throttling the outlet. This intermediate pressure acts on the additional surface A_a of the valve head and produces a supplementary opening force that offsets the elastic force of the spring F_s to such an extent that the valve head lifts rapidly. In the open state, a different intermediate pressure p' is set up and keeps the valve open. To close the valve, the pressure must be reduced considerably below the limiting opening pressure, because, in the open state, the pressure is applied to a greater surface area.

One application is the pressure switch for monitoring bearing oil pressure shown in Fig. 7.64. If the bearing oil pressure drops below a certain value, the piston jumps open and the pressure inside the safety system is reduced with consequent shut-off of the endangered machinery.

The principle of bi-stability is also applied to the design of quick shut-off devices in which a striker pin under a spring preload has its centre of gravity slightly offset from the centre of rotation (see Fig. 7.65). Once a limiting angular speed is reached,

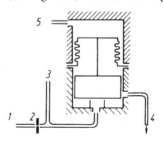

Figure 7.64. Diagrammatic sketch of a pressure switch to monitor bearing oil pressure, from [7.237]. *1* main oil system pressure; *2* orifice; *3* safety system activating quick shut-off valves; *4* drainage (no pressure); *5* bearing oil pressure.

the striker pin begins to move against the spring preload. The resulting increase in the eccentricity of the centre of gravity leads to an increase in the centrifugal inertia force acting on the pin, which is flung out even without any further increase in the angular speed. For this to happen, however, the rate of increase in the centrifugal force with x must be greater than that of the opposing spring force when the centre of gravity of the pin begins to move. The forces must be equal in the limiting state ($\omega = \omega_1$), this can be achieved provided that:

$$dF_c/dx > F_s/dx \text{ or } m \cdot \omega^2_1 > s$$

Once it has been displaced to the outside, the pin strikes a catch which, in turn, activates the quick shut-off mechanism.

7.4.5 Principles for Fault Free Design

In high precision products, in particular, but also for other technical systems, an embodiment should be sought in which the number of potential faults is minimised. This can be achieved by:

- designing a simple structure with simple components which have few close tolerances;
- adopting specific design measures to minimise the causes of faults;
- selecting working principles and working structures whose functions are largely independent of any disturbing effects, or which only have a low interdependency (see 7.3.1 – basic rule of clarity); and
- ensuring that any potential disturbing factors influence two parameters that compensate each other at the same time (see 7.4.1 – principle of balanced forces).

Examples of this important principle [7.174, 7.257, 7.340] that result in simpler production and assembly and maintain product quality are: elastic and adjustable configuration used in multi-gear gearboxes to balance out tooth tolerances (see

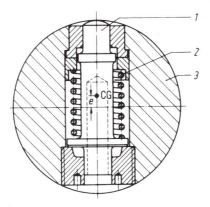

Figure 7.65. Quick shut-off pin *1* in shaft *3* with centre of gravity CG offset by *e* and spring *2* holding the pin in the normal position, from [7.237].

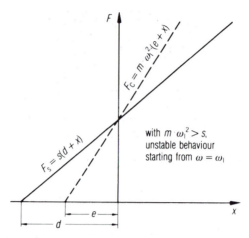

Figure 7.66. Graph of spring force and centrifugal force against the displacement x of the centre of gravity of the quick shut-off pin (Fig. 7.65). e = eccentricity of centre of gravity; d = spring precompression; ω_1 = limiting angular speed beyond which the pin lifts off.

Figs. 7.43 and 7.45); low stiffness of screws and springs to reduce the manufacturing tolerances in prestressed screwed connections and suspension systems; simple structures with few parts, low tolerances, and few toleranced joints; the possibility to adjust and reset to allow lower tolerances on individual components; the principle of stability (see 7.4.4).

Figure 7.67 shows a simple example: a link that is independent of play for the precise transfer of position. By making the ends of the link dome shaped based on a shared spherical surface, the distance between the driving component and receiving component remains the same despite tilting of the plunger caused by any play in the guides [7.174].

The example in Fig. 7.68 illustrates how continuous adjustments can be incorporated to make it easier to maintain a volume with very tight tolerances in, for example, a split mould.

Figure 7.69 shows a further example. In a micro-fiche reader it is important to keep the objective lens perpendicular to the micro-fiche, which is held between glass sheets. The usual solution is to mount the lens in a cylindrical body with tight tolerances with its axis perpendicular to the glass surface. The solution in Fig. 7.69, however, locates the cylindrical body directly on the glass sheet and therefore automatically maintains it perpendicular to the surface of the glass.

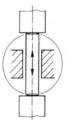

Figure 7.67. Link that is independent of play for the precise transfer of position [7.174].

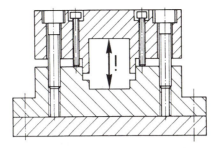

Figure 7.68. Continuous adjustment provided to maintain tight tolerances.

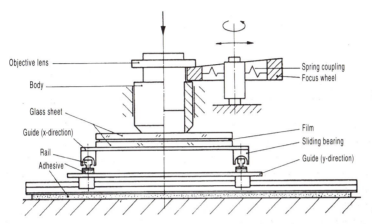

Figure 7.69. Automatically adjusting function chain in a micro-fiche reader.

7.5 Guidelines for Embodiment Design

7.5.1 General Considerations

In addition to the three basic rules of clarity, simplicity and safety derived from the general objectives (see 7.3), designers should also follow a number of embodiment design guidelines based on the general constraints set out in 2.1.8 and the checklist in Fig. 7.2.

In what follows we cover the most important guidelines, without making any claims to completeness. Detailed discussions are dispensed with whenever summaries or special accounts have been published, to which the reader is referred.

When *designing for durability* (stress requirements), designers should refer to the literature covering machine elements [7.127, 7.170, 7.180, 7.187, 7.212, 7.294, 7.301]. Special attention should be paid to changes in loading conditions with time and to the correct estimates of the level and type of the resulting stresses. Damage-accumulation criteria help to improve service-life predictions [7.113, 7.116, 7.123, 7.130, 7.204, 7.264, 7.267, 7.268, 7.302]. When determining stresses, stress concentrations and/or multi-axial stress conditions should be taken into account [7.34, 7.208, 7.209, 7.295, 7.309, 7.349]. Assessments of durability should be based on the material properties and the appropriate failure criteria [7.5, 7.28, 7.123, 7.140, 7.206, 7.293, 7.295, 7.310, 7.325, 7.326].

When *designing to allow for deformation, stability and resonance,* designers should refer to the appropriate calculations in mechanics and machine dynamics: mechanics and strength problems [7.23, 7.38, 7.180, 7.269]; vibration problems [7.168, 7.189]; stability problems [7.230]; and method of finite elements [7.362]. In 7.4.1 we dealt briefly with the problems of designing with due allowance for the deformation caused by the transmission of forces.

Designing to allow for *expansion* and *creep,* that is temperature phenomena, are discussed in 7.5.2 and 7.5.3, and designing against *corrosion* in 7.5.4.

Wear poses an extraordinarily complex problem that is currently being examined from many sides [7.39, 7.124, 7.166, 7.173, 7.232, 7.280, 7.339].

Safety problems were treated at some length in 7.3.3.

General *ergonomic* problems are discussed in 7.5.5.

The *aesthetics* of technical products involves the special rules set out in 7.5.6.

Design for *production* and *assembly,* including *quality control* and *transport,* is dealt with at some length in 7.5.7 and 7.5.8.

Designing to *standards* (see 7.5.9) helps with this aspect and also in reducing *costs* and improving *schedules.*

The problems involved in design for *operation* and *maintenance* depend very much on the product and its use. The reader is referred to 7.5.10 and [7.66, 7.344] also to the ergonomics literature listed in 7.5.5.

The *recycling* aspects of embodiment design are discussed in 7.5.11.

7.5.2 Design to Allow for Expansion

Materials used in technical systems tend to expand when they are heated. The resulting problems must be taken into consideration not only in the design of thermal devices in which higher temperatures must be expected as a matter of course, but also in high-performance engines and devices in which frictional heating can occur and special cooling is employed. As a result, several areas will be affected by local heating. Moreover, devices whose environmental temperature fluctuates significantly will only work properly if the physical effects of thermal expansion have been allowed for in the design [7.221].

Apart from the thermal effects of linear expansion, designers must also consider the purely mechanical expansion of parts subjected to heavy loading.

1 Expansion

Expansion has been the subject of a host of special studies. For solid bodies the coefficient of linear expansion is defined as:

$$\alpha = \Delta l/(l \cdot \Delta \vartheta_{\mathrm{m}})$$

where Δl = change in length (expansion) due to a temperature rise of $\Delta \vartheta_{\mathrm{m}}$, $l =$

the length of the component under consideration, and $\Delta\vartheta_m$= mean temperature difference to which the body is subjected.

The coefficient of linear expansion defines the expansion of a solid along one co-ordinate axis only, while the coefficient of cubical expansion defines the relative change of volume per degree of temperature rise. For homogeneous solids its value is three times that of the coefficient of linear expansion.

Coefficients of expansion should be understood as mean values over the particular temperature range $\Delta\vartheta_m$; they depend not only on the material but also on the temperature. At higher temperatures, the coefficient usually increases.

Figure 7.70 gives the coefficients of linear expansion of distinct groups of engineering materials. It shows that with commonly used combinations of metals, for

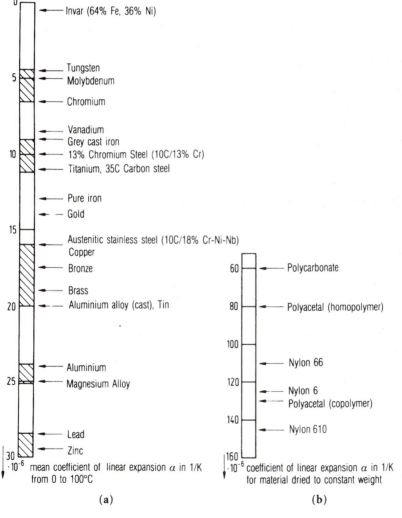

(a)

(b)

Figure 7.70. Mean coefficient of linear expansion for various materials; (a) metallic (b) synthetic.

example of 35C carbon steel with austenitic (10C/18% Cr-Ni-Nb) steel, or of grey cast iron with bronze or aluminium, great care must be taken to allow for relative expansions because of the significant differences in the coefficients of thermal expansion. With large dimensions, even the relatively small differences between, say, 35C carbon steel and 13 per cent chromium steel (10C/13% Cr) can cause serious problems.

Metals with a low melting point, such as aluminium and magnesium, have greater coefficients of thermal expansion than metals with a high melting point such as tungsten, molybdenum and chromium.

Nickel alloys have different coefficients depending on their nickel content. Very low values occur in the range of 32-40 per cent by weight, with 36% Ni-64% Fe, known as "Invar", having the lowest coefficient.

Synthetic materials have significantly higher coefficients of expansion than metals.

2 Expansion of Components

To calculate changes in length, Δl, designers must know the temperature distribution (position and time) in the component and hence the mean temperature change with respect to the initial value.

If the temperature distribution does not change with time, we speak of a *steady* or fixed expansion. If the temperature distribution changes with time, we speak of an *unsteady* or fluctuating expansion.

In the case of steady expansion, the physical quantities on which the expansion of the components depends are obtained from the basic equations:

$$\Delta l = \alpha \cdot l \cdot \Delta \vartheta_m \qquad \Delta \vartheta_m = \frac{1}{l} \int_0^l \Delta \vartheta(x)\, dx$$

The change in length Δl is therefore dependent on:

- the coefficient of linear expansion α ;
- the length l of the component; and
- the mean temperature change $\Delta \vartheta_m$ over this length, and can be determined accordingly.

The value thus determined has a direct bearing on the design: every component must be clearly located and must only have as many degrees of freedom as are necessary for its proper functioning. In general, a point is fixed and the requisite translational and rotational movements are set by appropriate guides, for example slides, bearings etc. A body in space (a satellite or helicopter) has three translational degrees of freedom in the x, y and z directions and three rotational degrees of freedom about the x, y and z axes. A sliding pivot (for example the non-locating bearing of a shaft) provides two degrees of freedom – one translational and one rotational. A body clamped at one point (for example a built-in beam), on the other hand, has no degrees of freedom.

Layouts based on these considerations alone do not, however, allow for expansion automatically, as we shall now demonstrate.

Figure 7.71a shows a body clamped at one point with no degrees of freedom. On thermal expansion it can expand freely from this point along the various axes. Figure 7.71b shows a plate that can be rotated about the z axis and thus has one degree of freedom. As shown in Fig. 7.71c, this single degree of freedom can be simply removed by means of a slide. Were this plate to expand under uniform temperature increases, it would have to rotate about the z axis, for the slide does not lie in the direction of the expansion that results from the change of length in the x and y directions. If the slide allowed only translational movement and did not also act as a pivot, then jamming would occur. By fitting the slide in the direction of one of the coordinates (see Fig. 7.71d) it is possible to avoid the rotation of the component.

After deformation due to thermal expansion, geometric similarity will only be maintained if the following conditions are met:

- The coefficient of expansion α must be constant throughout the component (isotropy), which can be taken for granted in practice provided that only one kind of material is used and that the temperature differences are not too great.

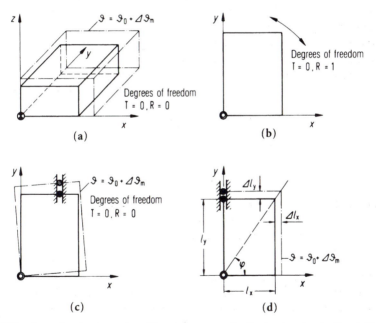

Figure 7.71. Expansion due to steady uniform temperature distribution; continuous line: initial state; broken line: higher temperature state:
(a) body attached to a fixed point;
(b) plate can rotate about the z axis; that is, one degree of freedom;
(c) plate as in (b) but with single degree of freedom removed by an additional sliding pivot;
(d) plate as in (b) but allowing for expansion without rotation. It would also be possible to use simple slides which might equally well be arranged along the x axis as along a line through the z axis inclined at $\tan \varphi = l_y/l_x$.

- The thermal strains ε along the x, y, z axes must be such that:

$$\varepsilon_x = \varepsilon_y = \varepsilon_z = \alpha \cdot \Delta\vartheta_m \quad [7.196]$$

If α is constant throughout a component, then the mean temperature increase must be the same for all three axes, so that we have:

$$\Delta l_x = l_x \cdot \alpha \cdot \Delta\vartheta_m$$
$$\Delta l_y = l_y \cdot \alpha \cdot \Delta\vartheta_m$$
$$\Delta l_z = l_z \cdot \alpha \cdot \Delta\vartheta_m$$

and for the x and y axes:

$$tan\ \varphi = \frac{\Delta l_y}{\Delta l_x} = \frac{l_y}{l_x} \qquad \text{(See Fig. 7.71d)}$$

- The component must not be subjected to additional thermal loads, which will not happen if, for instance, it is completely surrounded by a source of heat [7.196].

As a rule, however, different temperatures are measured in a single component. Even in the simplest case, with the temperature distribution changing linearly along the x axis (see Fig. 7.72a), a change in angle is produced which, again, can only be taken up by a guide with a sliding as well as a pivoting movement. A simple slide, which allows translational movement with one degree of freedom, can only be used if the guide lies along the line of symmetry of the deformation (see Fig. 7.72b).

If this condition is not fulfilled, a further degree of freedom must be allowed. Hence we obtain the rule that guides that take up thermal expansion and have one degree of freedom only must lie on a line through the fixed point, and this line must be the symmetry line of the deformed state.

The deformed state can be caused by load-dependent and temperature-dependent stresses, in addition to the expansion itself. Since the stress and temperature distribution also depend on the shape of the component, the required symmetry

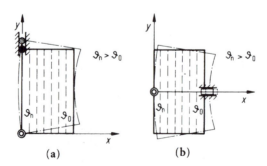

Figure 7.72. Expansion under non-uniform temperature distribution, here decreasing linearly along the x axis:
(a) Plate corresponding to Fig. 7.71d; non-uniform temperature distribution produces deformation shown by broken line; sliding pivot required;
(b) Guide placed on symmetry line of deformed state so that a simple slide can be used.

line of the deformed state should, in the first instance, be sought both along the symmetry line of the component and also along that of the superimposed temperature field. However, as Fig. 7.72b shows, this symmetry line may not be easily identifiable from the component shape and temperature distribution, so that the ultimate state of deformation must also be taken into account. That state, as we said earlier, may also be caused by external loads. To that extent, our remarks also apply to guides of components subject to large mechanical deformations. An example will be found in [7.13].

The following examples serve as further illustrations.

Figure 7.73 is the plan view of a device whose temperature decreases from the centre to the periphery. It is supported on four feet. In Fig. 7.73a one of the feet was chosen as the fixed point. If the device is not to rotate or jam, the guide may only be placed along the symmetry line of the temperature field, that is on the opposite foot. Figure 7.73b shows a method of providing guides along symmetry lines, without a designated fixed point. The intersection of the lines through the guides constitutes an imaginary fixed point from which the device can expand evenly in all directions. In that case, two guides, for example *1* and *2*, could be omitted.

Figure 7.74 shows the location of inner casings in outer casings when a common centre must be maintained as, for instance, in turbines. If the deformed shape of these components is not completely rotationally symmetrical, then the guides must be placed on the symmetry lines to prevent jamming of guides due to, say, oval deformation of the casings (see Fig. 7.74b). Such oval deformation is caused by temperature differences, especially during the warm-up phase. The imaginary fixed point lies on the longitudinal axis of the casing or shaft.

Figure 7.75 shows an austenitic steel high-temperature steam inlet pipe *a* which must be fitted into a ferritic steel outer casing *b* while protruding into a ferritic steel inner casing *c*. Because of marked differences in the two coefficients of expansion and also because of the considerable temperature differences between the components, particular attention must be paid to relative expansion. An imaginary fixed point is provided by the rotationally symmetrical guides *d*, an arrangement ensuring the unimpeded expansion of the austenitic component along any line

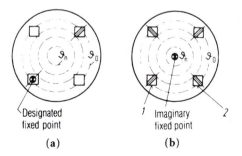

Figure 7.73.   Plan view of a device, whose temperature decreases from the centre to the periphery, mounted on four feet:
(a) Designated fixed point on one foot; simple slide along a line that is also the symmetry line of the temperature field;
(b) Imaginary fixed point in the centre of the device formed by the intersection of the lines of expansion.

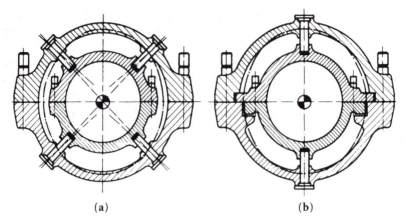

Figure 7.74.   Location of inner casings in outer casings:
(a) Arrangement of guides does not allow for expansion; oval deformation of the housings can cause guides to jam;
(b) Arrangement allowing for expansion; guides lie along symmetry lines; no jamming with oval deformation.

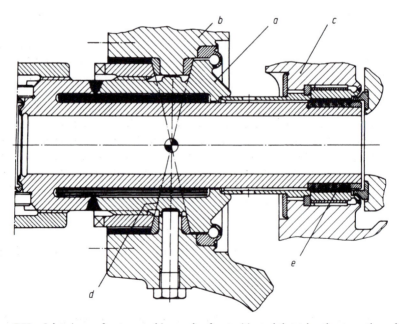

Figure 7.75.   Inlet pipe *a* of a steam turbine made of austenitic steel that takes the steam through the ferritic steel outer casing *b* to the inner casing *c*. Expansion planes through guideways *d* determine an imaginary fixed point. Piston ring seals at *e* permit the axial and radial expansion of the end of the inlet pipe (BBC).

through the imaginary fixed point. Because the temperature distribution at that point is fairly uniform, the respective radial and axial expansions produce a resulting expansion along the indicated lines.

By contrast, the insertion of the inlet pipe into the inner casing must allow independent expansion along two axes, because the fixed point of the inlet pipe and the fixed point of the inner casing are not identical and no definite temperature distributions can be assigned to the components. The double degree of freedom is obtained with the help of the piston-ring seal e which permits the independent axial and radial movements of the inlet pipe.

3 Relative Expansion of Components

So far, we have been considering expansion in a relatively stable environment. Very often, however, the relative expansion of two (or more) components has to be taken into account, especially in the case of mutual loadings or when certain clearances must be maintained. If in addition the temperature varies with time, then designers are faced with a very difficult problem. The relative expansion of the two components is:

$$\delta_{rel} = \alpha_1 \cdot l_1 \cdot \Delta\vartheta_{m1(t)} - \alpha_2 \cdot l_2 \cdot \Delta\vartheta_{m2(t)}$$

Steady-State Relative Expansion

If the relative mean temperature difference does not vary with time, and if the coefficients of linear expansion are identical, then all that has to be done to minimise the relative expansion is to even out the temperature or else to select materials with different coefficients of expansion. Often both are necessary.

This can be seen in the case of a flanged connection consisting of a steel stud and an aluminium flange [7.214]. Because the aluminium has a higher coefficient of expansion, a temperature rise will increase the load on the stud, which may lead to failure (see Fig. 7.76a). This can be prevented, on the one hand, by increasing the length of the stud and using a sleeve and, on the other hand, by using components with appropriate coefficients of expansion (see Fig. 7.76b). If relative expansion is to be avoided altogether, then we must have:

$$\delta_{rel} = 0 = \alpha_1 \cdot l_1 \cdot \Delta\vartheta_{m1} - \alpha_2 \cdot l_2 \cdot \Delta\vartheta_{m2} - \alpha_3 \cdot l_3 \cdot \Delta\vartheta_{m3}$$

With $l_1 = l_2 + l_3$ and $\lambda = l_2 / l_3$ the relative length of sleeve to flange becomes:

$$\lambda = \frac{\alpha_3 \cdot \Delta\vartheta_{m3} - \alpha_1 \cdot \Delta\vartheta_{m1}}{\alpha_1 \cdot \Delta\vartheta_{m1} - \alpha_2 \cdot \Delta\vartheta_{m2}}$$

With steady-state expansion, $\Delta\vartheta_{m1} = \Delta\vartheta_{m2} = \Delta\vartheta_{m3}$ and with steel ($\alpha_1 = 11 \times 10^{-6}$), Invar ($\alpha_2 = 1 \times 10^{-6}$) and aluminium alloy ($\alpha_3 = 20 \times 10^{-6}$) as the chosen materials (see Fig. 7.76b), we have $\lambda = l_2 / l_3 = 0.9$.

Designers will be familiar with the complicated expansion problems associated

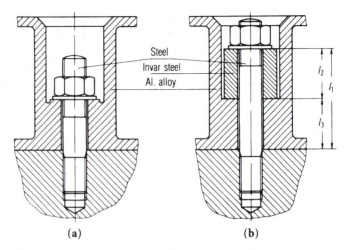

Figure 7.76. Connection by means of a steel stud and aluminium flange [7.214]:
(a) stud endangered because aluminium flange has greater expansion;
(b) incorporation of Invar expansion sleeve with a coefficient of expansion close to 0 helps to balance the relative expansion of flange and stud.

with the pistons of internal combustion engines. Here, the temperature distribution over and along the piston differs even in the near-steady state and, what is more, differences in the coefficients of expansion of piston and cylinder must also be taken into account. One solution is the use of an aluminium-silicon alloy with a relatively small coefficient of expansion (smaller than 20×10^{-6}), of expansion-inhibiting inserts that are also good heat conductors, and of a flexible piston skirt. The bimetal effect provided by steel inserts also helps to match the shape of the piston skirt to that of the cylinder [7.191] (see Fig. 7.77). A further possibility is to make the piston oval shaped.

If, on the other hand, the choice of materials is restricted in practice, then

Figure 7.77. Piston of internal combustion engine made of aluminium-silicon alloy with steel inserts which inhibit circumferential expansion; moreover the bimetal effect ensures optimum adaptation of the piston skirt to the cylinder (Mahle) from [7.191].

designers must rely on temperature adjustments. In high-power generators, for instance, large lengths of insulated copper rod must be embedded in the steel rotors. For insulation purposes alone the absolute and relative expansions must be kept as small as possible. Here the only solution is to keep the temperature level to a minimum by cooling [7.179, 7.342]. Moreover, if these fast-running rotors have large dimensions, thermal imbalances may occur; even though the temperature distribution is relatively uniform, the rotor, because of its complicated structure and the various materials that have gone into it, may not always and everywhere display the same temperature-dependent properties. This can only be remedied if the expansions are kept under control by the carefully planned introduction of appropriate cooling or heating.

Unsteady Relative Expansion

If the temperature changes with time, for instance during heating or cooling processes, we often find a relative expansion much greater than that which is found in the steady, final state. This is because the temperature of the individual components can differ considerably. In the common case of components of equal length and equal coefficients of expansion, we have:

$$\alpha_1 = \alpha_2 = \alpha \text{ and } l_1 = l_2 = l$$

$$\delta_{rel} = \alpha \cdot l \left(\Delta\vartheta_{ml(t)} - \Delta\vartheta_{m2(t)} \right)$$

The heating of components has been examined by, among others, Endres and Salm [7.99, 7.252]. No matter whether we assume a step or linear temperature change in the heating medium, the heating curve will be characterised by a time constant. If, for instance, we consider the temperature change $\Delta\vartheta_m$ of a component during a sudden temperature increase $\Delta\vartheta^*$ of the heating medium, then, under the admittedly approximate assumption that the surface and mean temperatures of the components are equal – which, in practice, is approximately true only for relatively thin walls and high thermal conductivity – we obtain the curve shown in Fig. 7.78, with:

$$\Delta\vartheta_m = \Delta\vartheta^* (1 - e^{-t/T})$$

Here t is the time and T is the time constant such that:

$$T = \frac{c \cdot m}{h \cdot A}$$

where:
   c = specific heat of the component;
   m = mass of the component;
   h = heat transfer coefficient of the heated surface of the component; and
   A = heated area of the component.
Despite the simplification involved, this approach may be considered fundamental.
   With two components having different time constants, we obtain temperature

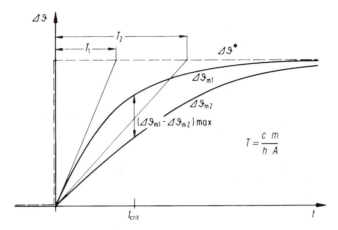

Figure 7.78. The effects on two components with different time constants of a step temperature change, $\Delta\vartheta^*$, in the heating medium.

curves that, at a given critical time, will have a maximum difference. At this point we have maximum relative expansion, and must provide clearances to accept the expansion or run the risk of excessive stresses beyond the yield point.

Two identical temperature curves appear if the time constants of the two components can be equalised. In that case, there is no relative expansion. This objective cannot always be achieved, but in order to render the time constants approximately equal, that is, to reduce the relative expansion, the following relationship:

$$T = c \cdot \rho \cdot \frac{V}{A} \cdot \frac{1}{h},$$

where V = volume of the component and ρ = density of the component, can be used by designers to:

- adapt the ratio of the volume V to the heated surface area A; or
- adjust the heat transfer coefficient h by means of, say, lagging.

Figure 7.79 gives the relationship V/A for a number of simple but representative bodies.

An example is shown in Fig. 7.80. Here, the problem is to ensure adequate clearance for a valve spindle so that it can move safely and smoothly in its sleeve, even during temperature changes. In Fig. 7.80a, the sleeve has been incorporated in the housing. When heated, the spindle will quickly expand radially, while the sleeve, which transfers its heat readily to the housing, remains cooler for a longer time. As a result, the clearance between the spindle and the sleeve will diminish dangerously.

In Fig. 7.80b, the sleeves are sealed axially but can expand freely radially. Moreover, their volume to area ratio is such that spindle and sleeves have approximately equal time constants. As a result, the clearance remains more or less uniform at all temperatures and can therefore be kept small. The surface of the valve spindle and the inner surface of the sleeves are heated by steam leaks, so that we have:

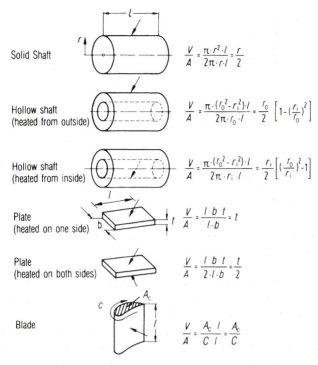

Solid Shaft

$$\frac{V}{A} = \frac{\pi \cdot r^2 \cdot l}{2\pi \cdot r \cdot l} = \frac{r}{2}$$

Hollow shaft
(heated from outside)

$$\frac{V}{A} = \frac{\pi \cdot (r_0^2 - r_i^2) \cdot l}{2\pi \cdot r_0 \cdot l} = \frac{r_0}{2}\left[1 - \left(\frac{r_i}{r_0}\right)^2\right]$$

Hollow shaft
(heated from inside)

$$\frac{V}{A} = \frac{\pi \cdot (r_0^2 - r_i^2) \cdot l}{2\pi \cdot r_i \cdot l} = \frac{r_i}{2}\left[\left(\frac{r_0}{r_i}\right)^2 - 1\right]$$

Plate
(heated on one side)

$$\frac{V}{A} = \frac{l \cdot b \cdot t}{l \cdot b} = t$$

Plate
(heated on both sides)

$$\frac{V}{A} = \frac{l \cdot b \cdot t}{2 \cdot l \cdot b} = \frac{t}{2}$$

Blade

$$\frac{V}{A} = \frac{A_c \cdot l}{C \cdot l} = \frac{A_c}{C}$$

Figure 7.79. Volume-surface area relationship of various geometrical bodies, arrows point to heated surfaces.

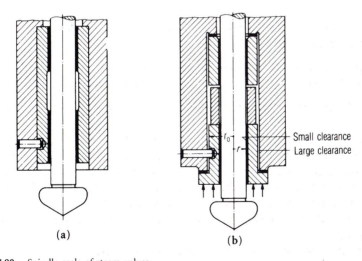

Small clearance
Large clearance

(a)   (b)

Figure 7.80. Spindle seals of steam valves:
(a) fixed sleeve requires relatively large spindle clearance because it has not been designed to allow for expansion;
(b) radially free and axially sealed sleeve permits small spindle clearance because spindle and sleeve have been designed to have the same time constant.

$$(V/A)_{spindle} = r/2$$
$$(V/A)_{sleeve} = (r_0^2 - r_i^2)/2r_i;$$

with $r_i = r$ and $(V/A)_{spindle} = (V/A)_{sleeve}$, we have

$$r/2 = (r_0^2 - r^2)/2r$$
$$r_0 = r \cdot \sqrt{2}$$

Figure 7.81 shows various steam turbine housings. With appropriate design it is possible to adapt the volume to area ratio of the housing and also the heat transfer coefficient and size of the heated surface to the time constants of the shaft and thus keep the blade clearances approximately constant when starting (heating) the turbine.

There are several well known methods for reducing the heat transfer coefficient of a component (for example by insulation) and thus for slowing down the heating and reducing the relative expansion.

The ideas we have just put forward are applicable wherever temperatures change with time, and particularly wherever relative expansion goes hand in hand with clearance reductions that are likely to endanger the functioning of turbines, piston engines and machines operating in hot environments.

7.5.3 Design to Allow for Creep and Relaxation

1 Behaviour of Materials Subject to Temperature Changes

When designing components subject to temperature changes we must take into account not only the expansion effect, but also the creep properties of the materials. The temperatures involved need not necessarily be very high, though they usually are. However, there are some materials that will, even at temperatures well below 100°C, behave in much the same way as metals do at very high temperatures.

Beelich [7.7] has examined this subject at some length and in what follows we shall base ourselves largely on his findings.

Materials in common use, pure metals no less than alloys, have a polycrystalline structure and a temperature-dependent behaviour. Below a *critical temperature*, the

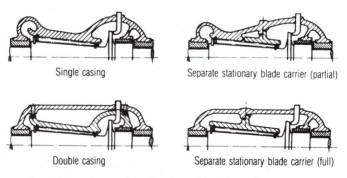

Single casing	Separate stationary blade carrier (partial)
Double casing	Separate stationary blade carrier (full)

Figure 7.81. Steam turbine housings with different time constants.

stability of the inter-crystalline bonds is largely independent of time, and the yield point can be used to determine the strength of components. Components at temperatures above the critical temperature are strongly influenced by the time-dependent behaviour of the material. In this temperature range, materials will, under the influence of load, temperature and time, experience a gradual plastic deformation that, after a given period, may lead to fracture. The ensuing time-dependent fracture stress is much lower than the 0.2 per cent proof stress at the same temperature determined by short-term experiments (see Fig. 7.82). Critical temperature and creep strength depend largely on the materials used and must both be taken into consideration. With steels, the critical temperature lies between 300°C and 400°C.

When working with synthetic materials, designers must allow for their viscoelastic behaviour even at temperatures below 100°C.

In general, the modulus of elasticity changes inversely with the temperature (see Fig. 7.83a). The smallest changes occur with nickel alloys.

As the modulus of elasticity drops, so does the stiffness of the components, and of synthetic components in particular (see Fig. 7.83b). In their case, designers must know the temperature at which the modulus of elasticity drops suddenly to relatively low values.

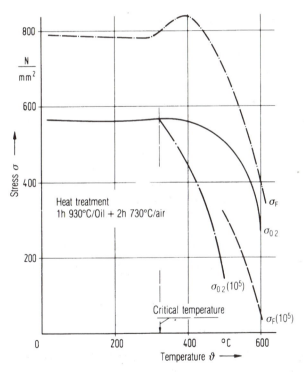

Figure 7.82.   Characteristic values determined by high-temperature tensile strength and creep experiments with 21C/1.5% Cr-Mo-V steel at various temperatures; critical temperature is the intersection of the curves of 0.2% proof stress and the stress for 0.2% creep strain in 10^5 hours.

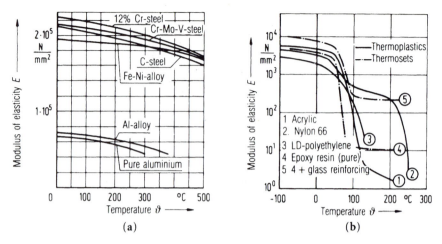

Figure 7.83.   Relationship of modulus of elasticity of various materials to temperature; (a) metals, (b) synthetic materials.

2 Creep

Components that are put under loads for long periods at high temperatures will, in addition to the strain given by Hooke's Law ($\varepsilon = \sigma / E$) also experience plastic deformation (ε_{plast}) with time. This property of materials, which is known as *creep*, depends on stress, the effective temperature ϑ and time.

We say a material creeps if the strain of the components increases under constant stress [7.7]. The creep curves of various materials are well known [7.109, 7.147].

Creep at Room Temperature

Before we can design components loaded to near the yield stress, we must know how they react in the transition region between the elastic and the plastic states [7.147]. With persistent, static loads in this transition region, we can expect primary creep in metals even at room temperature (see Fig. 7.84). The resulting plastic deformations are small and merely affect the dimensional stability of a particular component. In general, steels show little creep when subject to stress $\leq 0.75 \cdot \sigma_{0.2}$ or $\leq 0.55 \cdot \sigma_F$, whereas, in the case of synthetic materials, a reliable assessment of the mechanical behaviour can only be made by consideration of the temperature and time-dependent characteristics.

Creep Below the Critical Temperature

Previous studies [7.147, 7.160] of metals have shown that the customary calculations, based on high-temperature yield strength as the maximum permissible stress for short-term loads, additional thermal loads and load variations, suffice up to the critical temperature.

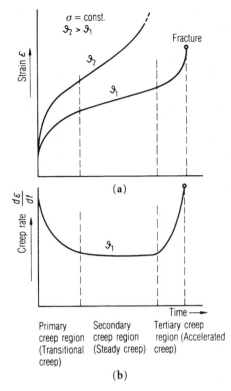

Figure 7.84. Strain (a) and creep rate (b) with duration of load (schematic representation); character-istics of the various creep phases.

With components that must have high dimensional stability, however, the characteristics of the material determined by creep experiments must also be taken into consideration, even at moderately high temperatures. Unalloyed and low-alloy boiler-making steels and even austenitic steels show varying degrees of creep depending on length of operation and working temperature.

Synthetic materials experience structural changes even at slightly elevated temperatures. These transformations may lead to a marked temperature and time-dependence of the properties of the materials, which is not the case with metals. In specific cases these changes are irreversible and referred to as thermal ageing [7.169, 7.197].

Creep Above the Critical Temperature

In this temperature region, mechanical loads will cause deformations in metals at far below the appropriate yield strength, that is the materials will creep. This creep leads to gradual deformation of components and can lead to loss of function. In general, this process can be divided into three phases [7.147, 7.160] (see Fig. 7.84). For components affected by temperature changes, the beginning of the tertiary creep phase must be considered dangerous. This region begins at approximately 1 per cent permanent strain. Figure 7.85 shows the 10^5 hour creep strengths $\sigma_{1\%/10^5}$ at 500°C of various steels.

3 Relaxation

In loaded systems (springs, bolts, tension wires, shrink fits), the necessary preload produces an overall strain ε (elongation Δl). Because of creep and settling of the material due to plastic flow at the bearing surfaces and split lines, the ratio of plastic to elastic deformation gradually increases. The phenomenon of decreasing elastic strain at constant overall strain is called *relaxation* [7.100, 7.351, 7.352].

Loaded components are usually preloaded at room temperature. Because the modulus of elasticity varies with the temperature (see Fig. 7.83), the preload decreases at higher temperatures even without a change in length of the loaded system.

The preload remaining at operational conditions, though reduced, will lead to creep at high temperatures and hence to a further drop in the preload (relaxation). The residual clamping force is also affected by production and operation-determined factors, for instance by the assembly preload, the design of the loaded system, the nature of the contact surfaces, and the influence of superimposed stresses (normal or tangential to the surface). Studies of the relaxation of bolted flanges [7.100, 7.351, 7.352] have shown that plastic deformation also occurs at the split lines and bearing surfaces (settlement) and in the threads (creep and settlement).

To sum up, we can say that, with metallic components:

- The drop in preload depends on the relative stiffness of the parts loaded against

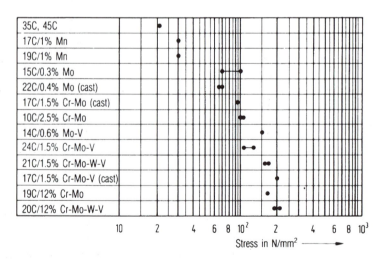

Figure 7.85. Stresses corresponding to a 1% permanent strain of various steels after 10^5 hours at 500°C [7.213].

each other. The more rigid the connection, the greater the drop in the preload due to plastic deformation (creep and settlement).

- Although settlement can be appreciably offset during the tightening of bolted flanges or the assembling of shrink fits, designers should, where possible, provide for few but accurately machined surfaces (split lines, bearing surfaces).
- There is a temperature limit beyond which the material cannot be properly used. In addition, designers should always choose materials in which the appropriate yield point is not reached even with superimposed operational stresses.
- In the short term, high initial preloads (initial clamping forces) give rise to higher residual clamping forces. In the longer term, the residual clamping forces become relatively independent of the initial preload.
- Joints that have already undergone relaxation can be tightened up if the toughness of the material permits. As a rule, creep of about 1 per cent, which leads to the tertiary creep region, must not be exceeded.
- If joints are subjected to an alternating load in addition to the static preload (see Fig. 7.86), then, as experiments have shown, the amplitudes tolerated during relaxation-dependent decreases in the mean stress are considerably greater than those tolerated at constant mean stress. However, relaxation-dependent decreases in the mean stress will often lead to a loosening of the joint.

When using bolted joints made of synthetic materials, designers try to take advantage of their small electrical and thermal conductivities, their resistance to corrosion, their high mechanical damping, their small specific weights etc. In addition, such joints must, of course, have the appropriate strength and toughness.

Special attention must also be paid to preload decay, lest the functioning of the joints be seriously impaired.

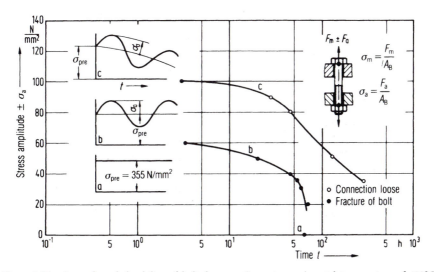

Figure 7.86. Strength and durability of bolted connections at experimental temperature of 450°C [7.351]; size M12; bolt 34C/1% Cr steel; nut 35C steel; static preload stress 355 N/mm²; (a) creep experiment (b) alternating fatigue (Wöhler) experiment (c) fatigue relaxation experiment.

Special studies [7.202, 7.203] have shown that in synthetic, unlike metallic, materials:

- the preload remaining after a given time and at room temperature is determined by the material itself and its tendency to absorb moisture; and
- continual changes in the absorption and release of moisture have a particularly deleterious effect.

4 Design Features

In order to increase the potential life of components subject to long-term loads, designers must familiarise themselves with the behaviour with time of the material involved. According to [7.147], it is dangerous to use short-term values to predict load responses for periods of 10^5 hours or longer.

It is impossible to avoid thermal stresses in all components by specifying the use of highly alloyed materials. Appropriate design features are often more useful than changes in the material.

The design must be such as to keep creep within permissible limits, which can be done by means of:

- a high elastic strain reserve, which helps to keep down additional loads due to temperature fluctuations (see Fig. 7.87);
- insulation or cooling of the component as in steam turbines and in gas turbines (see Fig. 7.88);
- the avoidance of mass concentrations which, in unsteady processes, may lead to increased thermal loading; and
- the prevention of creep in unwanted directions which can cause functional

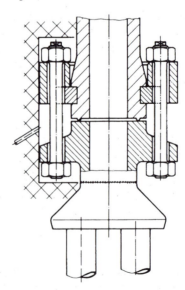

Figure 7.87.   Austenitic-ferritic steel flanged joint for operating temperatures of 600°C [7.285].

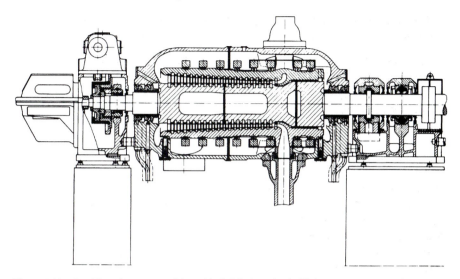

Figure 7.88.   Double casing steam turbine with shrink rings that hold the inner casing together. Relaxation of the shrink rings is reduced by cooling with exhaust steam. As the machine increases its output, the shrink rings exert an increasing pressure thanks to growing temperature differences between the steam inlet and outlet. The shrink rings are seated on heat-inhibiting segments which, with the help of shims, permit the original shrink fit to be restored after relaxation (BBC).

failure (for instance the jamming of valve spindles) or dismantling problems (see Fig. 7.89).

In Fig. 7.89a the material of the cover creeps into the relief groove. The cover, which heats up more quickly, presses against the centring surface and also creeps at point y. The cover shown in Fig. 7.89b is a better design since, despite the creep, it can be dismantled easily. In addition the cover has been made hollow so that it cannot exert a significant radial force on the centring surface.

In other words, the part which is moved during dismantling should not project axially beyond the fixed part [7.221].

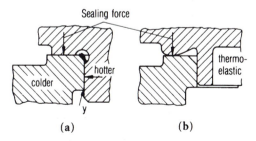

Figure 7.89.   Centring and sealing of a cover [7.221].
(a) Dismantling is impeded because the material creeps into the relief groove and at y.
(b) Convex sealing edge provides a better seal with smaller clamping forces. Creep does not impede dismantling thanks to improved design.

7.5.4 Design Against Corrosion Damage

It often happens that corrosion can only be reduced, not completely avoided; moreover the use of corrosion-proof materials may not be economical. Rubo [7.251] emphasises the use of components with the same corrosion resistance in a machine. He also points out that from an economic point of view it is not always viable to use corrosion resistant materials, and that suitable embodiments can be used that will retain the functionality despite corrosion. This suggests a shift from focusing on corrosion protection to designing machines and their components to be corrosion tolerant. It follows that designers must tackle corrosion with appropriate concepts or special embodiment design features. The measures they take will depend on the type of corrosion anticipated. An extensive description of the types of corrosion and many useful design features are provided in the guidelines on designing against corrosion [7.172]. Spähn, Rubo and Pahl [7.227, 7.283] describe various types of corrosion and the appropriate remedies. The following remarks are largely based on their findings. In order to provide a systematic approach, these are set out slightly differently from [7.81, 7.82].

1 Causes and Effects of Corrosion

While the formation of metal oxide layers in dry environments and at higher temperatures tends to increase chemical resistance to corrosion, relatively weak electrolytes are formed in conditions below the dew point and these generally lead to electrochemical corrosion [7.282]. Corrosion is also fostered by the fact that different components have contacting surfaces with different properties, for instance due to the inclusion of various noble or base metals, to differences in crystalline structure, and to residual stresses set up, for instance, by heat treatment and welding. In addition, wherever the design calls for slits or holes, local differences in electrolyte concentration appear even in the absence of differences in electric potential resulting from the use of different materials.

According to [7.81, 7.227] (see Fig. 7.90) we must distinguish between:

- free surface corrosion;
- contact corrosion;
- stress corrosion; and
- selective corrosion within the material.

The preventive measures depend on the respective causes and effects. Various examples are given in 7.5.4-5.

2 Free Surface Corrosion

The corrosion of free surfaces can be uniform or locally concentrated. The latter is particularly dangerous because, in contrast to uniform corrosion, it leads to a high stress concentration and is often difficult to predict. It is, therefore, necessary to pay particular attention right from the start to potential danger zones.

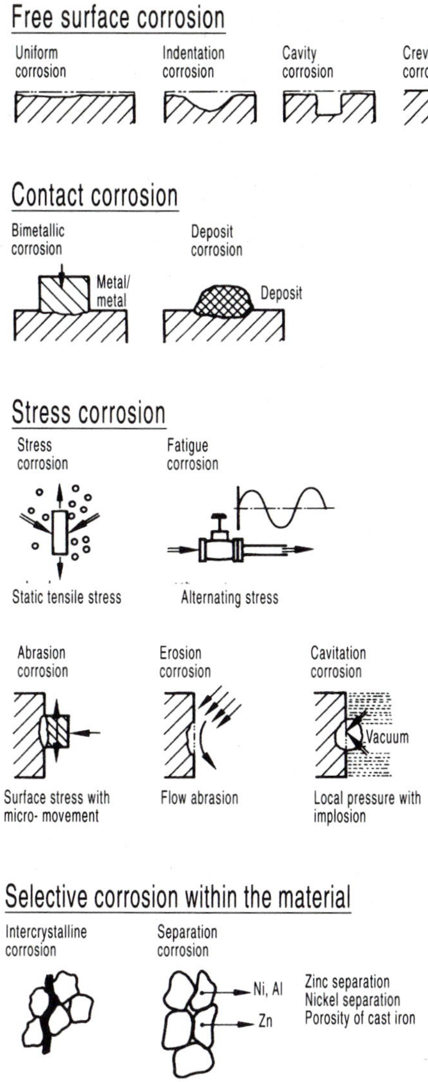

Free surface corrosion

Uniform corrosion	Indentation corrosion	Cavity corrosion	Crevice corrosion

Contact corrosion

Bimetallic corrosion

Metal/metal

Deposit corrosion

Deposit

Stress corrosion

Stress corrosion

Static tensile stress

Fatigue corrosion

Alternating stress

Abrasion corrosion

Surface stress with micro- movement

Erosion corrosion

Flow abrasion

Cavitation corrosion

Vacuum

Local pressure with implosion

Selective corrosion within the material

Intercrystalline corrosion

Separation corrosion

Ni, Al

Zn

Zinc separation
Nickel separation
Porosity of cast iron

Figure 7.90. Types of corrosion.

Uniform Corrosion

Cause:

The presence of moisture (weak electrolytes) combined with oxygen from the environment, particularly below the dew point.

Effects:

Extensive, uniform corrosion of the surface; in steel, for instance, approximately 0.1 mm per annum in a normal atmosphere. Sometimes more pronounced locally, especially in zones kept frequently below the dew point and hence subject to moisture concentration. Uniform corrosion is fostered by greater activity of the medium, higher flow velocity, and intensive heat transmission.

Remedies:

- Provision of uniform service life by means of appropriate wall thicknesses and materials.
- Design based on a concept that obviates corrosion or makes it economically acceptable (see Example 1 below).
- The use of small and smooth surfaces involving geometrical shapes with a maximum volume to surface area ratio (see Example 2 below).
- The avoidance of moisture traps (see Fig. 7.91).
- The avoidance of temperatures below the dew point by good insulation and the prevention of hot or cold bridges (see Example 3 below).
- The avoidance of flow rates greater than 2 m/s.
- The avoidance of areas of high and differing thermal loads on heated surfaces.
- The application of a protective coating, possibly in conjunction with cathodic protection.

Indentation Corrosion

This type of corrosion is not uniform over the surface.

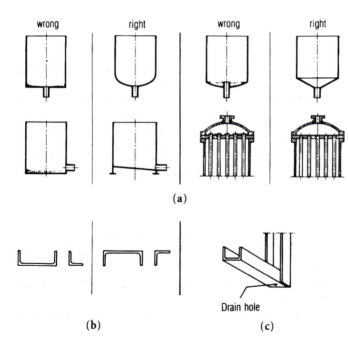

(a)

(b)   (c)

Figure 7.91. Drainage of components susceptible to corrosion: (a) design of floors encouraging and impeding corrosion; (b) wrong and right arrangement of steel sections; (c) brackets made of channel section with drain hole.

Cause:

There are components [7.82] with anodic and cathodic areas that cause differences in the rate of corrosion. These differences are usually caused by inhomogeneous material, by a medium with varying concentrations or by local influences such as temperature and radiation.

Remedies:

- Removal on inhomogeneity and varying influences.
- Covering the surface with a protective coating. Damage to this coating, however, will cause strong local corrosion (see cavity corrosion).

Cavity Corrosion

Cavity corrosion is concentrated on small surfaces with relatively deep indentations, with the depth being at least as great as the width. A clear distinction between indentation and cavity corrosion is not always possible.

Cause:

Similar to indentation corrosion, but its occurrence is more localised.

Remedy:

- Basically the same as for indentation corrosion, however particular attention should be paid to reduction or prevention.

Crevice Corrosion

Cause:

Most often, the accumulation of acidic electrolytes (moisture, aqueous medium), following the hydrolysis of corrosion products in crevices etc. In rust and acid-proof steels, there is a breakdown of passivity due to depletion of oxygen in a crevice. Typically this type of corrosion is caused by insufficient ventilation.

Effects:

Increased corrosion in hidden areas. Increased stress concentration in areas that are, in any case, under greater stress. Danger of fracture or separation without prior warning.

Remedies:

- The provision of smooth, crevice-free surfaces and connections.
- The provision of weld seams without permanent crevices; the use of butt seams or through-welded fillet seams (see Fig. 7.92).
- The sealing of crevices, for instance by providing protruding parts with moisture-proof sleeves or coatings.
- The enlargement of crevices so that throughflow prevents the accumulation of moisture.

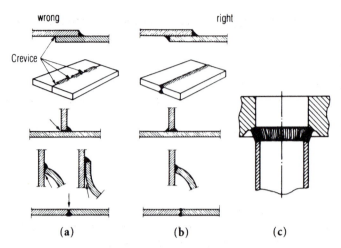

Figure 7.92. Examples of welded joints: (a) susceptible to corrosion in crevices; (b) correct design according to [7.282]; (c) crevice-free welding of pipes; also improves resistance to stress corrosion cracking.

3 Contact Corrosion

Bimetallic Corrosion

Cause:

The contact of two metals with different potentials in the presence of an electrolyte, that is a conductive fluid or vapour [7.281].

Effects:

The baser of the two metals will corrode more rapidly than the nobler round the contact area, and the faster the smaller its surface area (galvanic corrosion). Once again, the stress concentration is increased and corrosion products may be deposited. Such deposits have secondary effects of various kinds, for instance the production of sludge, contamination of the medium etc.

Remedies:

- Use combinations of metals with small potential differences and hence a small contact current.
- Prevent action of electrolytes on the contact area by providing local insulation between the two metals.
- Avoid electrolytes altogether.
- If necessary, resort to planned corrosion by introducing still baser materials in the form of sacrificial anodes.

Deposit Corrosion

Cause:

Unwanted materials become deposited on the surface or in crevices and cause potential difference at particular locations. These deposits can come from existing corrosion, the surrounding medium, vaporisation residues, excess sealing material etc.

Remedies:

- Avoid, filter or collect the deposits.
- Prevent water traps, aim at smooth flow, maintain reasonable speed and self-drainage (see Fig. 7.91a).
- Rinse or clean the components.

Transition Zone Corrosion

Cause:

Changes of state of the medium or its components from the liquid to the gaseous phase and vice versa tend to increase the danger of corrosion of metallic surfaces in the transition zone. That danger may be increased further by encrustations [7.282].

Effects:

This type of corrosion is concentrated in the transition zone and is the more pronounced the more sudden the change of state and the more aggressive the medium [7.248].

Remedies:

- Gradual input and removal of heat using a heating or cooling element.
- Reduction of turbulence, and hence of heat transfer coefficients at the inlet of the affected medium, for instance by means of guide plates.
- Provision of corrosion-resisting jackets at critical points (see Examples 3 and 4).
- Avoidance of transition zone problems by appropriate design features (see Fig. 7.93).
- Continuous change of fluid level, for example by stirring.

4 Stress Corrosion

Components susceptible to corrosion are often mechanically loaded, either stati-

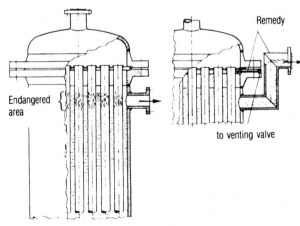

Figure 7.93. Increased corrosion at the transition from the gaseous to the liquid state [7.282] due to concentration of the medium in the region of the water line of a vertically arranged condenser. This can be remedied by raising the water level.

cally or dynamically. The mechanical stresses produced by these loads can cause several serious corrosion phenomena.

Fatigue Corrosion

Cause:

Corrosive attacks on a component subjected to mechanical fatigue loading appreciably reduce its strength. The greater the loading, the more intense the corrosion and the shorter the life of the component.

Effects:

Fracture without distortion, as in fatigue failure. Because the corrosion products, especially in slightly corrosive media, can only be seen under a microscope, this type of corrosion is often mistaken for normal fatigue failure.

Remedies:

- Minimisation of alternating mechanical or thermal stresses and especially avoidance of oscillatory stresses due to resonance phenomena.
- Avoidance of stress concentrations.
- Provision of compressive stresses on the surface by shotblasting, roller burnishing, nitriding, etc to increase the working life.
- Avoidance of contact with corrosive media (electrolytes).
- Provision of surface coating (for example rubber, baked enamel, hot dip galvanisation, aluminium etc).

Stress Corrosion

Cause:

Certain sensitive materials tend to develop trans- or inter-crystalline cracks if static tensile stresses are combined with a specific trigger.

Effects:

Depending on the medium [7.282], various very fine and rapidly developing trans- or inter-crystalline cracks appear in the component. Adjacent parts are not affected.

Remedies:

- Avoidance of sensitive materials, which may not, however, be possible because of other requirements. These materials are: unalloyed carbon steels, austenitic steels, brass, magnesium, aluminium alloys and titanium alloys.
- Substantial reduction or complete avoidance of tensile stresses on the attacked surfaces.
- Introduction of compressive stresses on the surface, for instance by shrink fits, by pre-loaded multi-layer materials or by shot-blasting.
- Reduction of residual tensile stresses by annealing.
- Application of cathodic coatings.
- Avoidance of corrosive influences by lowering the concentration and temperature.

Strain Induced Corrosion

Cause:

Under repetitive large extensions or compressions any protective outer layer cracks and opens repeatedly. This removes the protection and local corrosion will occur.

Remedy:

- Reduce the magnitude of any extensions and compressions

Erosion, Cavitation and Abrasion Corrosion

Corrosion may accompany erosion and cavitation, in which case the breakdown of the material is accelerated. The basic remedy is the avoidance or reduction of erosion and cavitation by hydrodynamic means or special design features. Only when this is not possible should such hard surface treatments as metal spraying or hard chrome coating be considered. Abrasive corrosion can be caused by relatively small movements between two surfaces subject to surface stresses (see also 7.4.1-3). The effected surfaces form hard oxidation products (so called abrasion rust) that speed up the process. At the same time stress concentrations increase. The most effective remedy is the removal of the abrasive movement, for example through elastic suspensions or hydrostatic bearings.

Abrasion spots have to be avoided. They can appear, for instance, as a result of thermal expansion, or of pipes vibrating against their guides etc. In either case, the oxidic protection layer on the surfaces of the rubbing parts may become damaged. Exposed metallic areas have a more negative electrochemical potential than those covered with a protective layer. If the fluid medium is an electrolyte, these relatively small exposed areas will be broken down electrochemically unless the protective layer can be regenerated.

Remedies:

- Reduce the vibration of the pipes by reducing the flow velocity inside them and/or change the distances between the guides.
- Increase gaps between the pipes and their guides so that no rubbing contact takes place.
- Increase the wall thickness of the pipes, thus increasing their stiffness and the tolerable corrosion rate.
- Use pipe materials that readily accept protective coatings.

5 Selective Corrosion within a Material

In the case of selective corrosion only certain interfaces in the material matrix are effected. Of importance are:

- intercrystalline corrosion of stainless steels and aluminium alloys;
- the so-called porosity or graphite corrosion of cast iron when iron particles separate out; and
- de-zincification of brass (zinc separation).

Cause:

Many material constituents or intercrystalline areas are less corrosion resistant than the bulk material matrix.

Remedy:

- Suitable selection of materials and their processing, such as adopting welding procedures which avoid producing a corrosion-sensitive material structure. Designers need to consult a materials expert when this type of corrosion is thought likely.

In general, designers should aim at ensuring the maximum and uniform life of all components [7.248, 7.251]. If it should prove economically impossible to meet these requirements with the appropriate choice of materials and layout, then designers must provide for the regular monitoring of all areas and components particularly prone to corrosion, for instance by visual inspection and regular measurements of wall thicknesses, directly by mechanical or ultrasonic methods and/or indirectly by means of corrosion probes that can be scrutinised and replaced at regular intervals.

Corrosion should never be allowed to proceed to the point where it threatens safety (see 7.3.3-4).

Finally, the reader is referred back to the principle of the division of tasks (see 7.4.2), with the help of which even difficult corrosion problems can be solved. Thus, one component might provide protection against corrosion and provide a seal, while another provides support or transmits forces. As a result, the combination of high mechanical stresses with corrosion stresses is avoided, and the choice of materials for any one component becomes easier [7.222].

6 Examples of Designing Against Corrosion Damage

Example 1

Lye is used to absorb CO_2 from a gaseous mixture under pressure, and the CO_2-enriched lye is then forced to surrender much of its CO_2 by expansion (regeneration). The position of the expansion chamber in a gas-washing plant is determined by the following factors.

If the lye were expanded immediately behind the washing tower (see Fig. 7.94, point A) the pipework to B would have to withstand lower pressures and would accordingly allow a saving in wall thickness. However, because of the release of CO_2, the aggressiveness of the lye permeated with CO_2 bubbles would increase to such an extent that the cheap unalloyed pipe steel commonly used would prove inadequate and hence have to be replaced with a more expensive rust and acid-proof material. For that reason, it is far better to keep the CO_2-enriched lye under pressure until it enters the regeneration tower (point B).

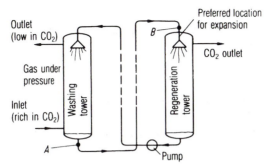

Figure 7.94. Influence of the point chosen for the expansion of CO_2-enriched lye on the choice of material for the pipework from A to B.

Example 2

Designers have to choose between two methods of storing compressed gases (see Fig. 7.95): (a) 30 cylindrical containers, each with a capacity of 50 litres and a wall thickness of 6 mm; and (b) 1 spherical container with a capacity of 1.5 m³ and a wall thickness of 30 mm; solution (b) is less prone to corrosion for two reasons:

- The surface exposed to corrosive attack is approximately 6.4 m², and is about five times smaller than it is in (a). In other words, less material is lost through corrosion to the same depth.
- For an anticipated corrosion depth of 2 mm in 10 years, the loss of strength in (a) is such that the wall of the container must be increased to a thickness of 8 mm, while corrosion to a depth of 2 mm in the 30 mm wall of container (b) is relatively insignificant. The spherical container can be dimensioned by considering strength requirements only and is therefore the better design of the two.

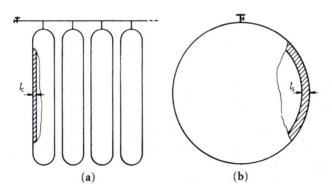

Figure 7.95. Influence of container shape on corrosion [7.248] in the case of gases stored at 200 bar: (a) in 30 cylinders with a capacity of 50 litres each; (b) in a sphere with a capacity of 1.5 m³.

Example 3

Figure 7.96a shows a container holding a mixture of superheated steam and CO_2 [7.248]. The outlet is not insulated and cooling leads to the formation of a condensate with strong electrolytic properties. Corrosion will attack at the transition zone between the condensate and the gases with the result that the outlet may break away.

Figure 7.96b shows a solution using insulation and Fig. 7.96c one using separate components made of more durable materials.

Example 4

In a heated pipe carrying moist gases, the inlet to the heated area is particularly prone to corrosion (see Fig. 7.97a). A less sudden transition (see Fig. 7.97b) or an extra protective sleeve (see Fig. 7.97c) offer remedies.

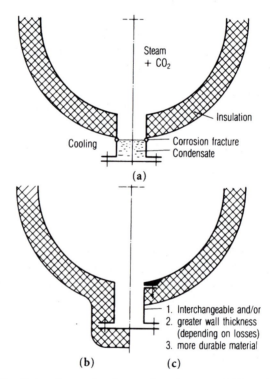

Figure 7.96. Outlet of a container for superheated steam and CO_2 under pressure:
(a) original design;
(b) insulated outlet avoiding condensation;
(c) other corrosion-resistant variants with separate components.

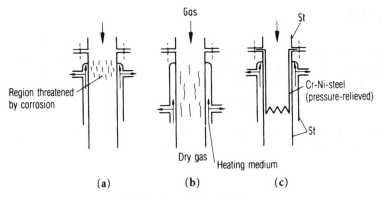

Figure 7.97. Corrosion in a heated pipe [7.248]:
(a) severe corrosion at the inlet due to sudden transition;
(b) sudden transition avoided;
(c) protective sleeve covers critical zone and mitigates sudden transition.

7.5.5 Design for Ergonomics

Ergonomics deals with the characteristics, abilities and needs of humans and, in particular, the interfaces between humans and technical products. A knowledge of ergonomics can lead to an embodiment that:

- adapts technical products to humans; and
- matches humans to products or activities by selection based on education and experience [7.327].

The range of technical products includes domestic products and those used for hobbies and leisure.

1 Fundamentals

The starting point is the human being in as far as he or she is the operator, user or recipient. Humans can work with or be affected by technical systems in many different ways (see 2.1.6). In this context it is helpful to address biomedical, physiological and psychological issues.

Biomechanical Issues

The operation and use of products requires specific body postures and movements. These result from the spatial situation resulting from the embodiment of a product (for example the position and movement of controls) combined with the body dimensions [7.69]. This relationship can be represented and evaluated using templates of body dimensions [7.72] (see Fig. 7.98).

The maximum forces humans can exert are given in [7.73]. To find the acceptable forces for a particular situation, however, also requires knowledge of frequency, duration, age, gender, experience and fitness, as well as knowledge of the methods used to calculate these influences [7.32. 7.131].

Physiological Issues

Body postures and movements resulting from the operation and use of technical products involve static and dynamic muscle actions. Muscle action requires the circulatory system to supply blood to the muscles based on the external loading. For static muscle action (for example when supporting a load) the blood throughput is throttled and recovery of the muscle is postponed. For this reason large loads can only be sustained for short periods.

From an ergonomic point of view, it is important to distinguish between loads, stresses and fatigue. Loads are external influences. Loads produce stresses related to individual characteristics such as age, gender, fitness, health, and training. The

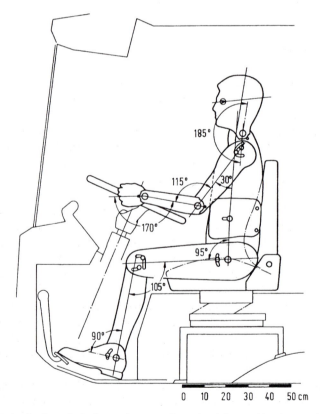

Figure 7.98. Application of a body template to evaluate the sitting position in a truck, after [7.72].

result of stress is fatigue, which depends on the intensity and duration of the stress. Recovery is achieved through relaxation. Fatigue-like situations such as monotony, however, are not recoverable through relaxation but require a change of activity [7.120].

A further physiological requirement for human life and work is a normal body temperature between 36°C and 38°C. Despite external hot and cold situations, and continuous heat generation within the body (increased during heavy work), the body temperature in the brain and other parts remains nearly constant because of the heat transferred by the blood. Working requirements and climatic influences have to be matched through technological measures, for example ventilation or organisational changes such as work breaks [7.70].

The senses also play an important role in work and leisure activities. Physiological variables involved in vision, for example, are minimum, optimum and maximum light intensity and light contrast [7.53, 7.55, 7.71]. The variables related to hearing are noise level and noise differential [7.331], and these must be taken into account when designing acoustic warning signals in a noisy environment [7.71]. The relevant signals must be based on the sensor characteristics of human beings. For the processing of these signals in the nervous system and the brain, no clear models exist. However it is known that humans filter inputs to each of the sensors according to experience, interest etc.

Psychological Issues

Several psychological issues have to be considered in the design of technical artefacts. The use of sensors, for example, implies that the processing of the signals involves a series of steps that can be influenced in a variety of ways. Examples include optical illusions, not hearing or seeing unimportant things, and different interpretations. Guiding attention therefore is an important embodiment design principle. This is true for the embodiment of control rooms as well as for the placing of indicators and signs on products.

The process of sensing, deciding and acting usually proceeds undisturbed. When this process, which is partially unconscious, is disturbed, conscious thinking is used to bring certainty back into the process of perceiving, deciding and acting. In products where the structure and functionality cannot be seen from the outside, the cause of and remedy for unusual phenomena, such as disturbances, cannot be clarified by thinking. It is therefore necessary to convey the required information through sufficient and clear signs and through operating manuals. A well designed product should minimise the thinking required for its operation so that thinking capacity can focus on the actual task. The requirement is for an obvious configuration, that is, one which during operation avoids thought processes that can be easily disturbed or are susceptible to errors. For example the relation between the movement of a control and the resulting response should be obvious and simple.

Perception and thought focus on the actual action. Learning is defined as storing successful actions and knowledge for later use. For the operation and use of products, for example, one has to take into account that a series of actions learnt earlier may be reintroduced out of habit. Subsequent versions of a similar product should, therefore, avoid introducing unnecessary changes in operation or use, in particular opposite movements or different positions for similar control actions should be avoided. Such changes must never be introduced if the consequences of an error could lead to direct or indirect safety risks.

Directing and constraining human activities excessively using technical or organisational systems can have a negative effect on motivation and behaviour, especially over a long period. All such activities should therefore leave space for free actions.

2 Human Activities and Ergonomic Constraints

Humans can be involved or affected by technical processes either actively or passively. In an active relation they can act and are deliberately involved in the technical system. That is they execute certain functions such as activating, controlling, monitoring, loading, removing, registering etc. In general, the following repetitive activities are undertaken in an activity cycle:

- Preparing for the activity, eg going to work.
- Gathering and processing information, eg observing and orienting, drawing conclusions, deciding on an action.
- Undertaking the activity, eg activating, connecting, separating, writing, drawing, talking, giving signs, thinking.
- Checking results, eg identifying status, checking measured values.
- Stopping the activity or starting a new one, eg cleaning, closing, going away, starting new activity cycle.

When the involvement of human beings is functional, that is deliberate, then this involvement should be planned carefully and suitable arrangements made. This should start early in the design process, even when clarifying the task (see 5.2). It is often necessary to represent this involvement in the function structure (see 6.3).

Active Human Involvement

Whether it is sensible and useful to involve humans in technical systems has to be assessed from the point of effectiveness, efficiency and humanity (dignity and appropriateness). This initial and basic consideration influences and determines, to a large extent, the involvement of humans and thereby the solution principle.

The following ergonomic aspects can be useful in the generation of solutions and as evaluation criteria [7.327] (see Table 7.4):

- Is human involvement necessary or desirable?
- Will the involvement be effective?
- Is involvement easy to achieve?
- Can the involvement be sufficiently precise and reliable?
- Is the activity clear and sensible?

Table 7.4. Ergonomic aspects for the requirements list and the evaluation criteria [7.327]

Active human involvement in a system to fulfil a task:

— necessary, desired
— effective
— simple
— fast
— precise
— reliable
— error free
— clear, sensible
— learnable

Active or passive involvement through disturbing effects and side effects on humans:

— tolerable stress
— low fatigue
— low annoyance
— no physical danger, safe
— no health risk or loss
— stimulation, change, holding attention, no monotony
— personal development

- Can the activity be learnt?

Only when the answers to these questions are positive, should the involvement of people in technical systems be considered.

Passive Human Involvement

Not only those actively involved but also those passively involved will experience disturbing effects and side effects from technical systems (see terminology in 2.1.6). The effects of energy, material and signal flows and the environment, such as vibrations [7.320], light [7.52-54], climate [7.70], and noise [7.331] are very important. These effects have to be identified early so that they can considered during the selection of the working principle and the development of the embodiment.

The following questions can be useful and can also serve as evaluation criteria (see Table 7.4):

- Are the stresses tolerable and is the emerging fatigue recoverable?
- Has monotony been avoided and is stimulation, change and attention ensured?
- Are annoyances or disturbances few or non-existent?
- Has physical danger been avoided?
- Has health risk or loss been excluded?
- Does the work allow for personal development?

When these questions cannot be answered satisfactorily, then another solution should be selected or the existing solution considerably improved.

3 Identifying Ergonomic Requirements

In general it is not easy for designers to find immediately satisfactory answers to the questions listed above. As described in Guideline VDI 2242 [7.327], the problem of identifying the most important influences and suitable measures can be approached in two ways.

Object-Based Approach

In many cases the technical object that has to be ergonomically embodied is known and documented, eg a control panel, a driver's seat, a piece of office equipment, or an item of protective clothing. In such cases it is useful to apply the checklist for objects in Part 2 of Guideline VDI 2242 [7.328]. It is also important to be sensitive to the particular requirements of the object under consideration and to make use of the guidelines in Table 7.5. Just reading the guidelines can be very instructive and can help clarify the issues. Concrete design features can be based on the insights acquired or obtained from the literature listed below.

Table 7.5.   Guidelines with characteristics to identify ergonomic requirements [7.327]

Characteristics	Examples
Function	Division of functions, type of functions, type of activities
Working principle	Type and intensity of the physical or chemical effects, consequences such as vibration, noise, radiation, heat
Embodiment	
— Type	Type of elements, configuration, type of operation
— Form	Ergonomic overall form and elements, division based on symmetry and proportion, aesthetically pleasing
— Position	Configuration, arrangement, distance, direction of effect and visibility
— Size	Dimensions, working area, contact surfaces
— Number	Amount, division
Energy	Adjustment force, adjustment direction, resistance, damping, pressure, temperature, moisture
Material	Colour and surface finish, contact properties such as safe to touch, easy surface to hold
Signal	Labelling, text, symbols
Safety	Free of danger, avoiding danger sources and spots, inhibit dangerous movements, protective measures

Effect-Based Approach

In new situations, that is when no object has been defined, it is useful to adopt the following approach. The effects related to existing, and thus known, energy, material and signal flows of technical systems are identified and compared with the ergonomic requirements. When there are limitations, intolerable loads or even dangers to safety, other solutions must be sought. Effects such as mechanical forces, heat and radiation are derived from the individual types of energy and the form in which they appear. The material flow has to be checked to identify whether the suggested materials are flammable, easy to ignite, poisonous, cancer causing etc. For this purpose Guideline VDI 2242 Part 2 [7.328] provides a checklist with effects that gives an indication of existing problems.

The following literature is also useful:

Design of work space 7.67, 7.74, 7.120, 7.131, 7.132, 7.207, 7.263, 7.327, 7.328, 7.357
Work physiology 7.245
Work psychology 7.211
Illumination 7.26, 7.52-54
Computers at work 7.84, 7.85
Climate 7.70, 7.263
Operation and handling 7.31, 7.67-69, 7.72, 7.73, 7.80, 7.152, 7.210, 7.242-244
Vibration, noise 7.320, 7.331, 7.335
Monitoring and control 7.71, 7.75, 7.76.

7.5.6 Design for Aesthetics

1 Aims

Technical products should not only fulfil the required technical function as defined by the function structure (see 6.3) but also be aesthetically pleasing to their users. A considerable change has occurred recently in user expectations and the way products are judged.

VDI Guideline 2224 [7.323] focuses on the aesthetics of products. Starting with a technical solution, the guideline provides rules for the external form or shape such as: compact, clear, simple, unified, in line with function, and compatible with materials and with production processes. The form chosen can give a strong static impression, eg calm, or a light moving impression, eg floating. Through the use of colour and gloss, and the use of light and shade, the surface structure can enhance or diminish impression and expression. The guideline provides examples of good and bad embodiments that combine both functional and pleasing forms.

In many products nowadays aesthetics is as important as technical function. This is particularly true for products aimed at large markets and used directly by users in their daily lives. In such cases the emphasis is not only on aesthetics and use, but also on feelings such as prestige, fashion and lifestyle. The forms, or better the embodiments, of consumer products are determined primarily by industrial designers, artists and psychologists. While ensuring the technical function, they select the forms, shapes, colours and graphics, that is, the overall appearance, based on human feelings and values.

Expression and style play an important role, for example a military appearance in radio products, a space age impression in lights, a safari image in cars, or a nostalgic feel in telephones. The body of a car, for example, is strongly based on artistic and psychological criteria and not only on technical criteria such as low air resistance and transportation efficiency.

It is clear that all the requirements regarding function, safety, use and economy have to be fulfilled. The aim of designers, however, is to create products that appeal to customers. Given this aim, industrial design lies between engineering and art, and has to address ergonomic and visual issues in the same way as engineering design has to address function and safety issues. In addition, the company image has to be promoted to underline the individuality of its products. It is clear that with such complex requirements, the involvement of industrial designers should not be left to the end of the design process. They should be part of the design team and involved from the beginning of the task clarification phase. In special circumstances they can even help formulate the task or undertake preliminary design studies.

The result of this approach is a design process that proceeds from "outside to inside". Continuous collaboration between industrial designers and engineering designers is required to ensure that the requirements of appearance, expression and impression still allow the technical functions to be fulfilled within the forms and shapes created.

In this collaboration engineering designers should not try to replace industrial designers, but should focus on developing the technical and economical aspects of the product. In the same way that technical solutions are developed, visual variants have to be proposed and evaluated, and models and prototypes made to decide on the final appearance of the product. When searching for solutions, the same methods as described for the engineering design process can be used, such as brainstorming, stepwise development of variants through sketches, and a systematic variation of configuration, form and colour. Tjalve [7.300] gives a very clear example of such a development (see Fig. 7.99). This clarity is evident throughout his book and illustrates the way in which form and embodiment can be varied. He emphasises that the following factors influence each other and determine the appearance of the product:

- engineering (purpose, function, construction structure);
- production (process, assembly, cost);
- sales and distribution (packaging, transport, storage, company image);
- use (handling, ergonomics); and
- disposal (recycling).

Seeger [7.274] underlines the close link between embodiments that focus on ergonomics and are in line with the use. Klöcker [7.165] focuses more on physiological and psychological aspects. In his latest work Seeger [7.275, 7.276] discusses the

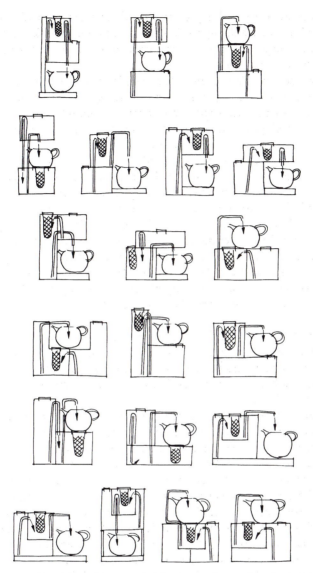

Figure 7.99.   Systematic variation of the structure of an automatic tea maker after [7.300], investigating the configuration of the water kettle, the tea container and the tea pot.

basic knowledge used for the development and embodiment of industrial products. Their appearance is developed from the structure, form, colour and graphics. Of crucial importance are the impressions experienced by observers. Knowledge on this topic can be found in the literature of the partially overlapping areas of physiology, psychology and ergonomics.

2 Visual Information

In general, the technical function and the selected technical solution, together with its construction structure, determine the configuration and form and hence the appearance of assemblies and components. This results in a *functional embodiment* that is often difficult to change. An example of a simple functional embodiment is a spanner (lever arm and shape of bolt head), and a complex one is a dredger (kinematic requirements, shape of dredging buckets, power train, location of operator).

Human beings not only see this functional embodiment but also other visual impressions such as stability, compactness, and a modern or striking appearance and they also expect information on operational procedures, safe areas, potential dangers etc which together form the *information presentation*. In the embodiment design phase, the functional embodiment and the necessary or desired information presentation should be integrated. Based on Seeger [7.274], we list the essential information presentation areas and some related rules.

Information about Market and User

When determining this type of information presentation, it is important to consider the type of user being addressed such as the technical expert, prestige seeker, nostalgia lover, and avant-garde. In general the overall appearance should be:

- simple, uniform, pure, and embody style;
- structured and well proportioned; and
- identifiable and definable.

Information about Purpose

This information presentation should enable the purpose of the product to be easily recognised and understood. The outer shape, colouring and graphics should support the identification of the function and the actions involved, such as where a tool has to be inserted and which parts exert forces.

Information about Operation

Information presentation about the correct operation and intended use should:

- be centrally located and recognisable, eg control elements with a function related layout;
- be ergonomically appropriate in accordance with the action space of human limbs;
- be labelled clearly, eg gripping and stepping areas;

- identify the operational status; and
- use safety signs and colours [7.49, 7.50].

Information about Manufacturer and Distributor

This information presentation expresses origin or house style. It contributes to continuity, confidence in known quality, involvement in the further development of successful products, and membership of a group. This can be achieved by easily recognisable and repeated elements, though the style and expression can be adapted to current fashion.

3 Guidelines for Achieving Good Aesthetics

Information presentation is achieved by a specific and intended *expression* such as lightness, compactness and stability, and by related *structure, form (shape), colour* and *graphics*. The following recommendations have to be considered (see Figs. 7.100 to 7.102).

Select an Expression

- Provide a recognisable and uniform expression that creates an impression in the observer that is in accordance with the aim, for example stable, light and compact.

Structure the Overall Form

- Order in an identifiable way, such as block shape, tower shape, L-shape, C-shape etc.
- Divide into clearly distinguishable areas with identical, similar or adapted form elements.

Unify Form

- Minimise variations in form and position, for example use only circular shapes with horizontal orientation along the main axis, or only rectangular forms with vertical orientation.
- Introduce form elements and alignments appropriate to the basic form selected, for example use the split lines of assemblies. Arrange the form by bringing several edges to one point or by running them parallel to one another. Support the intended expression with form elements and appropriate lines, such as horizontal lines to emphasise length. Keep an eye on the overall profile.

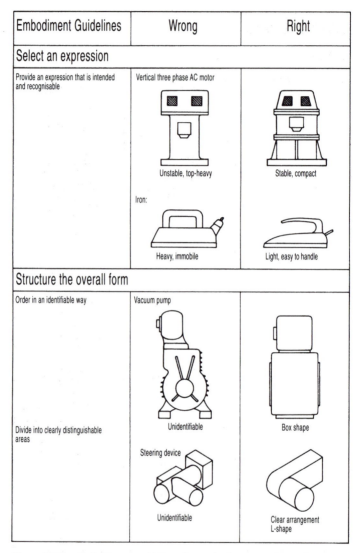

Figure 7.100.   Embodiment guidelines for aesthetics: expression and structure.

Support Using Colour

- Match colours to the form.
- Aim at few colours and material differences.
- In case of several colours, choose one main colour supported by complementary colours. For contrast use black and white, for example use a black colour to contrast yellow, a white colour to contrast red, green, blue etc (see also safety colours).

Complement with Graphics

- Use uniform styles for fonts and graphic symbols.

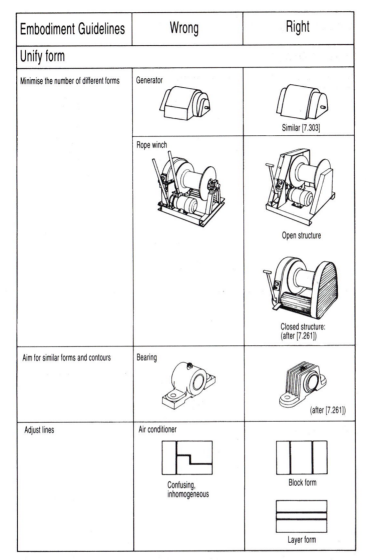

Figure 7.101.   Embodiment guidelines for aethetics: unifying form.

- Unify expression by using the same processes for the graphics, for example etching, painting or embossing.
- Adjust size, form and colour of the graphics to the other forms and colours.

7.5.7 Design for Production

1 Relationship between Design and Production

The crucial influence of design decisions on *production costs*, *production times* and the *quality of the product* is described in [7.29, 7.171, 7.332, 7.337]. *Design for*

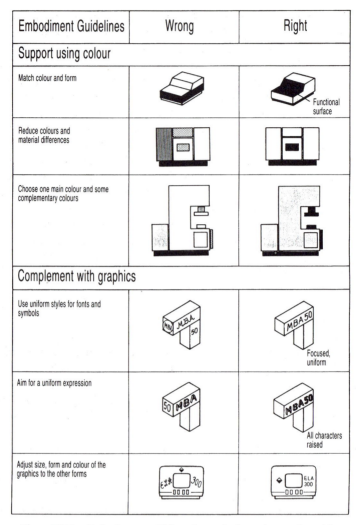

Figure 7.102. Embodiment guidelines for aesthetics: colour and graphics.

production means designing for the minimisation of production costs and times while maintaining the required quality of the product.

By *production*, we usually refer to:

- producing components in the narrow sense by accepted processes [7.58] (primary forming, secondary forming, material removal, joining, finishing, changing material properties);
- assembly, including transport of components; quality control; materials handling; and operations planning.

Designers will accordingly do well to consult the checklist (7.2) under the headings "production", "quality control", "assembly" and "transport".

In what follows we shall first concentrate on the design of components or assemblies in the narrower sense, while paying due regard to quality control and improvement of the overall production procedure. In 7.5.8 we shall then examine design features for improved assembly and transport.

Design for production is greatly facilitated if, from the earliest possible stage, decisions are backed up with data compiled by the standards department, the planning and estimating department, the purchasing department and the production manager. Figure 1.5 shows how the flow of information can be improved by systematic means, appropriate organisational measures and the use of computer systems (CAD/CAM, CIM). In original designs of mass-produced articles, systematic procedures and exchanges of views are essential. Their use, however, should also be encouraged in one-off and small batch production, especially in the case of adaptive designs, where designers are often forced to make decisions that influence production without consultation, if only to meet delivery dates.

Design for production has become an increasingly important activity in the wake of growing automation. By observing the basic rules of simplicity and clarity (see 7.3), designers are already proceeding along the correct lines. The principles of embodiment design (see 7.4), too, can lead them to a better and safer fulfilment of a given function and to the best solution from a production point of view. Another step in the same direction is the application of general and company standards (see 7.5.9).

For an overview of the influence designers have on production rationalisation, and of the interdependence of design and production, the reader is referred to Table 7.6. That table highlights several groups of problems with a direct bearing on the rationalisation of production procedure [7.15], namely:

Appropriate overall layout design, which determines the production procedure by the breakdown of the product into assemblies and individual components (in-house or bought-out, new, repeat or standard).

Appropriate form design of components, which determines the production procedure and processes, and the quality of components.

Appropriate selection of materials, which determines the production procedure and processes, the materials handling and quality control.

Appropriate use of standard and bought-out components, which influence the production capacity, the storage and the costs.

Appropriate documentation, which must be adapted to the production procedure and processes, and to quality control.

2 Appropriate Overall Layout Design

The overall layout design, developed from the function structure, determines the division of a product into assemblies and components. With the overall layout designers:

- determine the source of the components, that is whether they are in-house, bought-out, standard or repeat parts;
- determine the production procedure, for instance whether the parallel manufacture of individual components or assemblies is possible;

Table 7.6. Relationship between design and production

Design	Production
Overall layout design:	
— Assemblies	— Production process
— Components	— Assembly and transport possibilities
— Bought-out parts	— Batch size of similar components
— Standard parts	— Proportion of in-house and bought-out
— Joining and assembly	items
— Transport aids	— Quality control
— Quality control	
Component form design:	
— Shape and dimensions	— Production procedure
— Surface finishes	— Means of production, machine tools
— Tolerances	— Measuring instruments
— Limits and fits	— In-house and bought-out components
	— Quality control
Materials selection:	
— Type of material	— Production procedure
— Treatment	— Means of production, machine tools
— Quality control	— Materials handling (purchase, storage)
— Semi-finished materials	— In-house and bought-out parts
— Availability	— Quality control
Standard and bought-out components:	
— Repeat parts	— Purchases
— Standard parts	— Storage
— Bought-out parts	— Stock control
Production documentation:	
— Workshop drawings	— Execution of orders
— Parts lists	— Production planning
— Databases	— Production control
— Assembly instructions	— Quality control
— Testing instructions	— CAM, CAP, CAQ, CIM

- determine the dimensions and the approximate batch sizes of similar components, and also the means of joining and assembly;
- select suitable fits; and
- influence quality control procedures.

Conversely, such production limitations as the capacity of machines, assembly and transport facilities etc, naturally have repercussions on the choice of the overall layout.

The appropriate sub-division of the overall layout can give rise to differential, integral, composite and/or building-block methods of construction.

Differential Construction Method

By differential construction we refer to the breakdown of a component (a carrier of one or several functions) into several easily produced parts. This idea comes from lightweight engineering [7.141, 7.350], where that approach was introduced for the purpose of optimising load-carrying capacity. In both cases, we are entitled to speak of the "principle of sub-division for production".

As an example of the differential method let us take the rotor of a synchronous generator (see Fig. 7.103).

The large forging shown at the top *a* is divided into several rotor discs consisting of simple forged parts and two considerably smaller flanged shafts *b*. Each of the latter could also be subdivided into shaft, disc support flange, and coupling flange in the form of a welded construction *c*. The reason for this differential construction might be the market situation (price, delivery date) of large forgings, and the easier adaptation of the generator to various output requirements (rotor sizes) and types of coupling. A further advantage is that the parts can be produced as stock and not necessarily to a specific order. However, the illustration also demonstrates the limitations of the differential approach – beyond a certain rotor length and diameter, the machining costs become too great and the stiffness of the joints too problematical.

Figure 7.104 shows the magnet support of a large-scale DC motor which can either be cast in one piece or else be built up from sheet metal and welded. The production costs of the second design are some 25 per cent lower than those of the first, and this despite the fact that the differential construction involves several processes. However, the cost reduction is not constant but depends on the relative market situation of castings, sheet metal and semi-finished material.

Another differential construction is shown in Fig. 7.105. In the winding machine *a*, the winding head is integrated with the drive unit on a common shaft. The differential solution *b* was developed to facilitate the parallel production of drive units and winding heads to meet various customer requirements. In this way, a

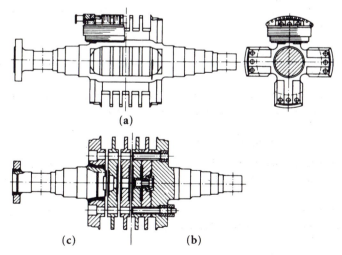

Figure 7.103. Rotor of synchronous generator after [7.13] (AEG-Telefunken):
(a) as forged part;
(b) as disc construction with forged flanges;
(c) and with welded flanges.

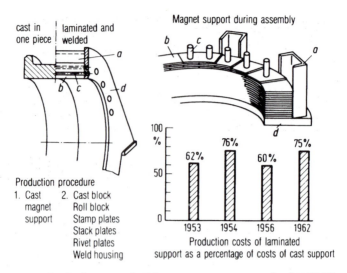

Figure 7.104. Production costs of a DC motor magnet support, after [7.167] (Siemens).

small number of standard drive units can be combined with a large number of winding heads.

The replacement of forged and cast constructions with welded constructions incorporating suitable semi-finished parts provides a further example of this method.

The differential construction methods also influence the production time. Figure 7.106 shows an example of the production sequence for a medium-powered electric motor. The times spent on acquiring the material and on producing the components and assemblies are indicated by the lengths of the horizontal lines. The diagram not only makes clear where improvements can be made by the choice of more quickly procurable raw and semi-finished materials or by keeping these materials in stock, but also where different production steps could be taken in parallel. Thus by allowing the stator laminations to be built up in parallel with the construction of the housing (two time-consuming operations), a significant reduction in the overall production schedule is possible in comparison to older designs in which the stator laminations could only be inserted followed by the windings after the casing had been welded.

All in all, differential designs have the following advantages, disadvantages and limitations:

Advantages:

- use of easily available and favourably priced semi-finished materials or standard parts;
- easier acquisition of forged and cast parts;
- easier adaptation to existing factory layout (dimensions, weight);
- increase in component batch sizes;
- reduction in component dimensions allowing easier assembly and transport;
- simpler inspection (smaller components and larger batch sizes);

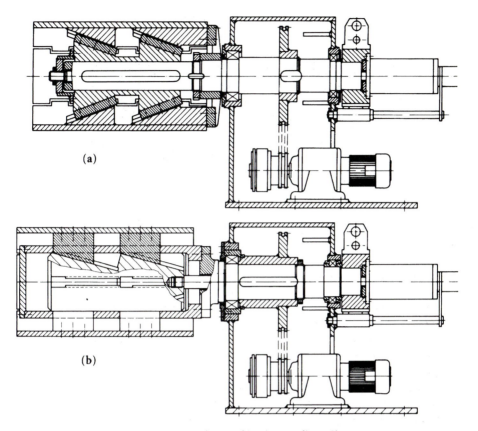

Figure 7.105.   Winding machine (Ernst Julius KG):
(a) winding head with integrated drive unit;
(b) winding head with separate drive unit.

- easier maintenance, for instance by simple replacement of worn parts;
- easier adaptation to special requirements; and
- reduced risk of missing delivery dates.

Disadvantages or limitations:

- greater machining outlay;
- greater assembly costs;
- greater need for quality control (smaller tolerances, necessary fits etc); and
- limitations of function by joints (stiffness, vibration, sealing).

Integral Construction Method

By integral construction we refer to the combination of several parts into a single component. Typical examples are cast constructions instead of welded constructions, extrusions instead of connected sections, welded instead of bolted joints etc. In lightweight engineering this type of construction is often used to avoid stress concentrations and to save weight [7.141, 7.350].

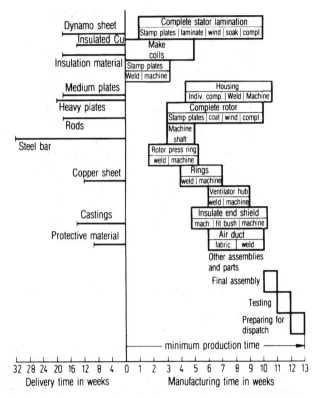

Figure 7.106. Production procedure of an electric motor from the series shown in Fig. 8.17 (AEG Telefunken).

This method is often used for product optimisation because of the economic benefits of integrating several functions into one component. This method can indeed be an advantage for specific technical, production and procurement situations, particularly for labour intensive production.

Figure 7.107 shows an example chosen from electrical engineering. Here, a cast and welded construction has been replaced with a single cast component. Though the casting is fairly complicated, it leads to a cost reduction of 36.5 per cent. Naturally, this percentage will vary with the size of the batch and with market conditions.

Another example is the rotor of a hydroelectric generator (see Fig. 7.108). Four different constructions with the same generator output and identical radial loads were investigated. Variant *a* has numerous individual support discs and may therefore be considered a differential construction. In Variant *b* the degree of division is reduced by the use of cast-steel hollow shafts, two support rings and end discs. Variant *c* is an integral construction in that two cast hollow bodies have been bolted together. In Variant *d* the cast construction is split up again (a cast central part, two forged shafts and two support rings). Weight comparisons show that the integral method saves material. In the end, however, Variant *d* was chosen because of difficulties in procuring large castings.

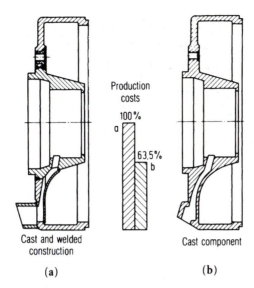

Figure 7.107. End cover of electric motor after [7.167] (Siemens):
(a) composite construction;
(b) integral construction.

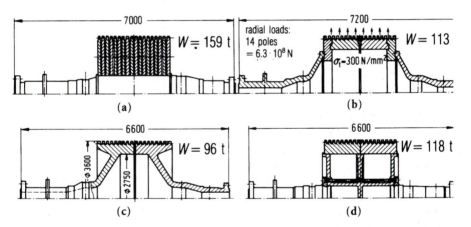

Figure 7.108. Rotor construction for a large-scale hydroelectric generator (Siemens).

The advantages and disadvantages of the integral construction method are easily determined by a reversal of the advantages and disadvantages of the differential method.

Composite Construction Method

By composite construction we refer to:

• the inseparable connection of several, differently made parts into a single component needing further work, for instance, the combination of cast and forged parts;

- the simultaneous application of several joining methods for the combination of components [7.234]; or
- the combination of various materials for optimal exploitation of their properties [7.318].

Figure 7.109 gives an example of the first method: the combination of cast steel components and rolled steel sheet into a welded construction.

Further examples are bogies with cast centres and welded arms, and also the welding of cast bar joints used in steel structures.

Examples of the second method are combinations of adhesives and rivets or of adhesives and bolts.

The combination of several materials into a single part is exemplified by synthetic components with cast-in thread inserts; by composite sound-absorption panels which have two plates separated by a plastic core; and also by rubber/metal components.

Another economical design of the composite type is the combination of steel and pre-stressed concrete [7.121].

Building-Block Construction Method

If the differential method is used to split a component in such a way that the resulting parts and/or assemblies can also be used in other products, then they can

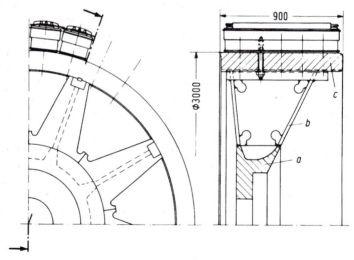

Figure 7.109. Magnet wheel of a hydro-electric generator of composite construction from [7.22] (AEG Telefunken):
(a) Hub – cast steel;
(b) Spoke – rolled steel sheet;
(c) Support – cast steel.

be considered as building-blocks. These are particularly useful if they are economical to produce. In a sense, the utilisation of repeat parts from stock may also be considered a building-block construction method (see 8.2).

3 Appropriate Form Design of Components

In the form design of components designers exert a great influence on production costs, production times and the quality of the product. Thus their choice of shapes, dimensions, surface finishes, tolerances and joints affects:

- the production procedure;
- the machine tools, including bench tools and measuring instruments;
- the choice between in-house components and bought-out components;
- the selection of materials and semi-finished materials; and
- the quality control procedures.

Conversely, production facilities influence the design features. Thus, the available machine tools might limit the dimensions of components and necessitate a split-up into several connected parts or the acquisition of bought-out components.

Many guidelines are available for the appropriate form design of components [7.25, 7.27, 7.33, 7.126, 7.193, 7.212, 7.216, 7.241, 7.284, 7.306, 7.308, 7.313, 7.315, 7.316, 7.319, 7.345, 7.358-360]. Because of the importance of tolerances (geometry, dimension, position and surface) for the production and assembly of components we specifically suggest the following literature [7.46-48, 7.56, 7.156, 7.157]. Important is the use of a *tolerancing basis* appropriate for the specific requirements [7.156]. A distinction is made between the *independent basis* where every tolerance is specified individually, and the *envelope basis* where every geometrical element (such as a circle or pair of parallel surfaces) has an enveloping tolerance zone (maximum metal condition). The latter cannot control deviations in position. For both tolerancing bases, deviations of position are independent of dimensional tolerances. The difference is whether deviations of geometry should be within the envelope. A fit has to remain within the envelope and, using the independent basis, this is indicated on the drawing with a fit specification, for example H7-j8. When the independent basis is used, blanket tolerances for geometry and position should be indicated. The envelope basis only requires a blanket tolerance for position [7.156].

In keeping with the aims of this book, we shall present only essential design suggestions arranged systematically in the form of charts. Our classifying criteria will be the *process steps* (PS) used in the production of the component. In addition, we shall be assigning objectives - *reduction of costs* (C) and *improvement of quality* (Q) - to the various design guidelines. When designing components, designers should always bear these process steps and objectives in mind.

Form Design for Primary Shaping Processes

The form design of components to be shaped by primary processes, for example casting and sintering, must satisfy the demands and characteristics of the processes used.

In cast components (primary shapes obtained from the fluid state) designers must allow for the following process steps: *pattern* (Pa), *casting* (Ca) and *machining* (Ma). Figure 7.110 lists the most important design guidelines. The literature cited contains further information.

In designing *sintered* components (primary shapes obtained from the powder state), designers must allow for *tooling* (To) and *sintering* (Si). In particular, they must be guided by the latest findings of powder technology. The essential guidelines are shown in Fig. 7.111.

Form Design for Secondary Shaping Processes

The form design of components to be shaped by secondary processes (hammer (free) forging, drop forging, cold extrusion, drawing and bending) must adhere to the guidelines listed below. Special consideration for the design of ferrous materials can be found in [7.55] and of non-ferrous metals in [7.60].

With *hammer forging*, designers need only allow for the actual forging process, since no complicated devices (for instance, dies) are involved. The following design guidelines should be observed:

- Aim at simple shapes, if possible with parallel surfaces (conical transitions are difficult) and with large curvatures (avoid sharp edges). *Objectives:* reduction of costs, improvement of quality.
- Aim at light forgings, perhaps by separation and subsequent combination. *Objective:* reduction of costs.
- Avoid excessive deformations or excessive differences in cross-sections due, for instance, to the presence of excessively high and fine ribs or of excessively narrow indentations. *Objective:* improvement of quality.
- Try to place bosses and indentations on one side only. *Objective:* reduction of costs.

Design guidelines for *drop forging* have been collated in Fig. 7.112. They allow for the process steps of: *tooling* (To), *forging* (Fo) and *machining* (Ma).

Figure 7.113 lists design guidelines for the cold extrusion of simple rotationally symmetrical solid and hollow bodies. They allow for the process steps of: *tooling* (To) and *extrusion* (Ex). It must be stressed that only certain types of steel can be used economically. Like all other cold forming methods, cold extrusion gives rise to work hardening, in which the yield strength is raised while the toughness of the material drops significantly. Designers must take this factor into consideration. The best materials for cold extrusion are case-hardening and heat-treatable steels.

For *drawing*, the following design guidelines are recommended in [7.241]:

- Allow for tooling (To): Choose the dimensions in such a way that the smallest possible number of drawing steps is needed. *Objective:* reduction of costs.
- Allow for tooling and drawing (To/Dr): Aim at rotationally symmetrical hollow bodies; producing the corners of rectangular hollow bodies leads to a high loading of the materials and tools. *Objectives:* improvement of quality, reduction of costs.

PS	Guidelines	Objective	Wrong	Right
Pa	Choose simple shapes for patterns and cores (straight lines, rectangles).	C	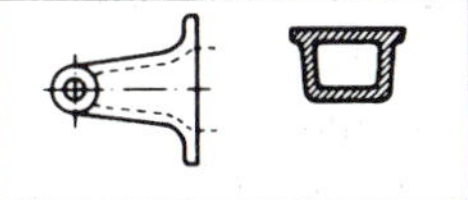	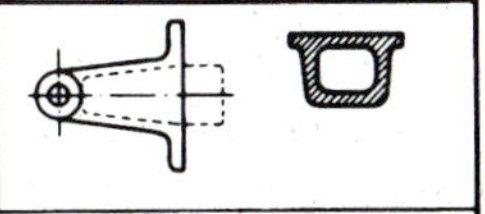
Pa	Aim at undivided patterns, if possible without cores (e.g. by means of open cross sections)	C		
Pa	Provide tapers from the split-line.	Q	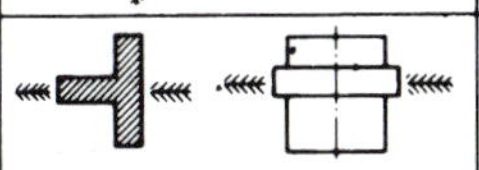	
Pa	Arrange ribs so that pattern can be removed; avoid undercuts.	Q	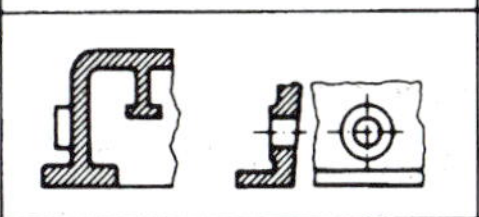	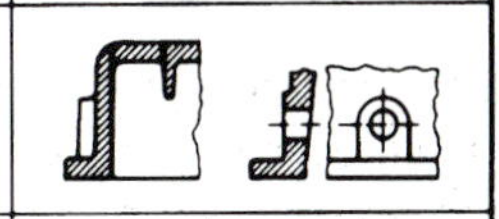
Pa	Ensure accurate location of cores.	Q	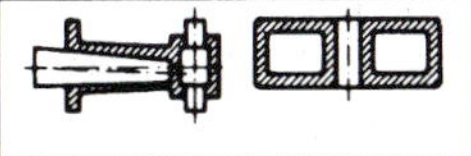	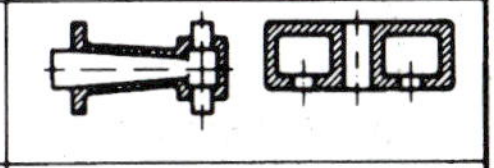
Ca	Avoid vertical sections (bubbles, blowholes) and reduced cross-sections to the risers.	Q	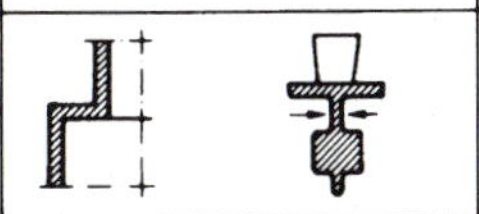	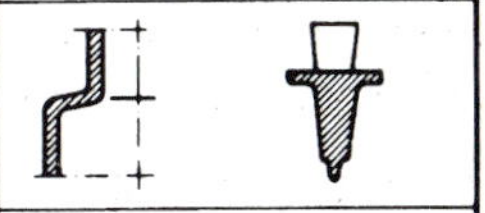
Ca	Aim at uniform wall thicknesses and cross-sections and at gradual changes of cross-section; select material allowing of adequate wall thicknesses and component sizes.	Q	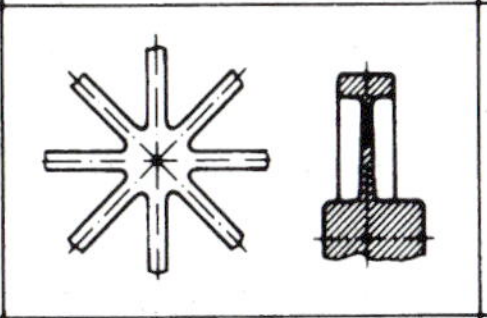	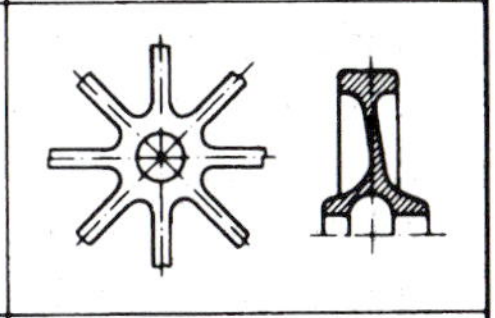
Ma	Set split-lines to avoid misalignment and to permit easy removal of the flash.	C Q	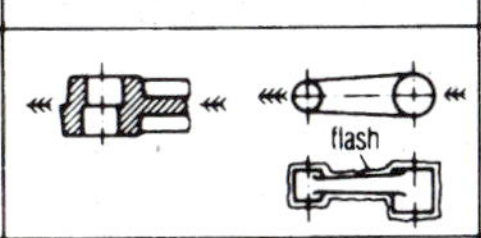	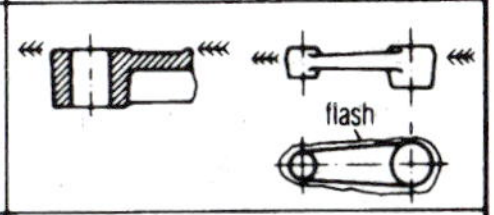
Ma	Arrange castings to ease machining.	C Q	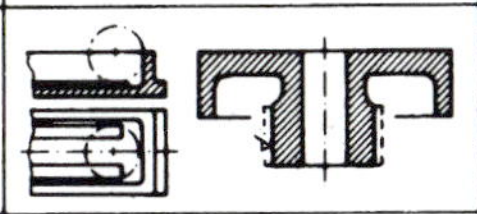	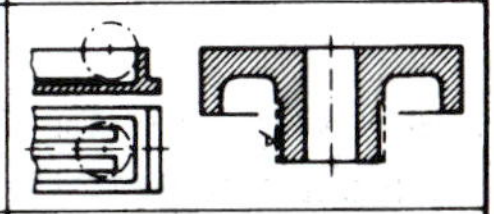
Ma	Provide adequate support surfaces.	Q C	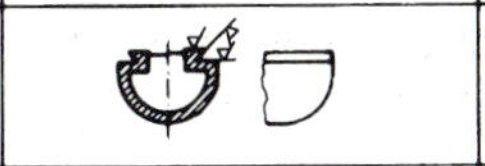	
Ma	Avoid sloping machining and boring surfaces.	C Q	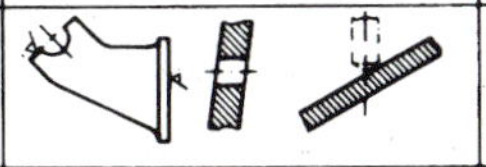	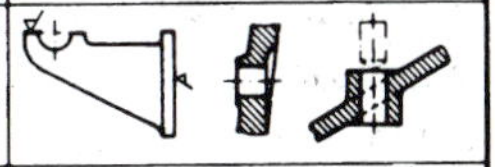
Ma	Combine machining processes by appropriate arrangement of machining and boring surfaces.	C	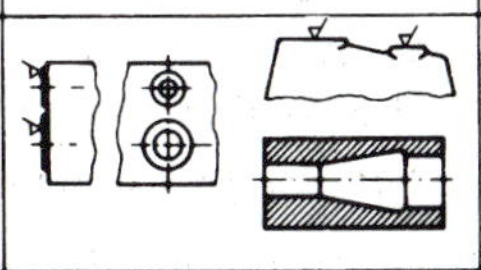	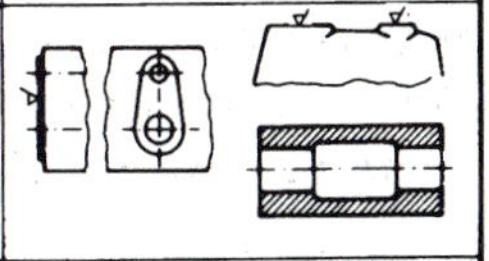
Ma	Avoid unnecessary machining by breaking up large surfaces.	C		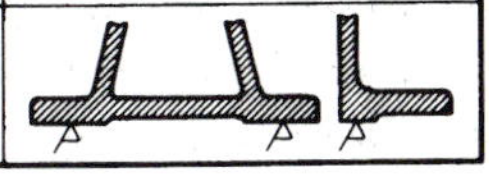

Figure 7.110. Design guidelines with examples for cast components, in accordance with [7.126, 7.193, 7.212, 7.241, 7.358, 7.359].

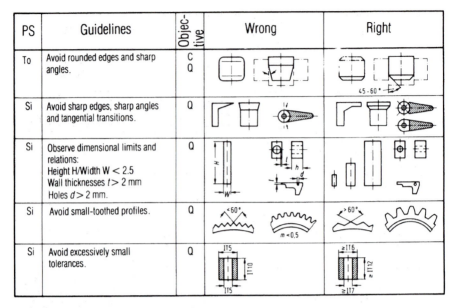

PS	Guidelines	Objective	Wrong	Right
To	Avoid rounded edges and sharp angles.	C Q		45-60°
Si	Avoid sharp edges, sharp angles and tangential transitions.	Q		
Si	Observe dimensional limits and relations: Height H/Width W < 2.5 Wall thicknesses t > 2 mm Holes d > 2 mm.	Q		
Si	Avoid small-toothed profiles.	Q	< 60° m < 0.5	> 60°
Si	Avoid excessively small tolerances.	Q	IT5 IT10 IT5	≥ IT6 ≈ IT12 ≥ IT7

Figure 7.111. Design guidelines with examples for sintered components, after [7.106].

- Allow for drawing (Dr): Choose tough materials. *Objective:* improvement of quality.
- Allow for drawing (Dr): For the design of flanges see [7.216]. *Objective:* improvement of quality.

Bending (cold bending) as it is used for the manufacture of sheet metal components in precision and electrical engineering, and also for casings, claddings and air ducts in general mechanical engineering involves two separate steps, namely *cutting* (Cu) and *bending* (Be). Designers must accordingly allow for both. The design guidelines shown in Fig. 7.114 apply to the bending process alone; cutting is covered under the next heading.

Form Design for Separation

Of the separating procedures mentioned in [7.57, 7.58], we shall only consider "machining with geometrically defined cuts" (turning, boring, milling), "machining with geometrically undefined cuts" (grinding), and "separating" (cutting). In all separating processes designers must allow for *tooling* (To), including clamping, and *machining* (Ma).

Design for tooling involves:

- The provision of adequate clamping facilities. *Objective:* improvement of quality.
- A preferential sequence of operations that does not necessitate the reclamping of components. *Objectives:* reduction of costs, improvement of quality.

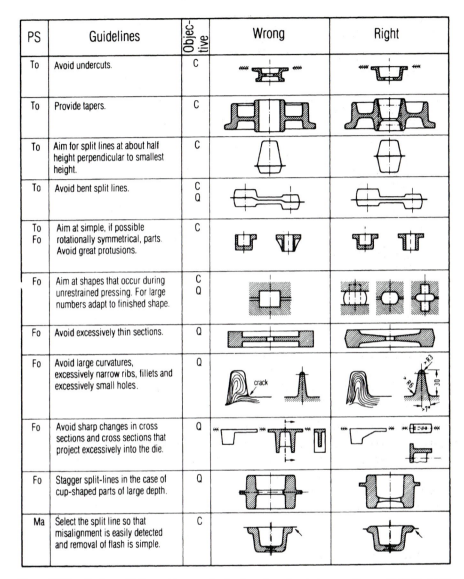

PS	Guidelines	Objec- tive	Wrong	Right
To	Avoid undercuts.	C		
To	Provide tapers.	C		
To	Aim for split lines at about half height perpendicular to smallest height.	C		
To	Avoid bent split lines.	C Q		
To Fo	Aim at simple, if possible rotationally symmetrical, parts. Avoid great protusions.	C		
Fo	Aim at shapes that occur during unrestrained pressing. For large numbers adapt to finished shape.	C Q		
Fo	Avoid excessively thin sections.	Q		
Fo	Avoid large curvatures, excessively narrow ribs, fillets and excessively small holes.	Q		
Fo	Avoid sharp changes in cross sections and cross sections that project excessively into the die.	Q		
Fo	Stagger split-lines in the case of cup-shaped parts of large depth.	Q		
Ma	Select the split line so that misalignment is easily detected and removal of flash is simple.	C		

Figure 7.112. Design guidelines with examples for drop-forged parts in accordance with [7.25, 7.158, 7.241, 7.308, 7.363].

- The provision of adequate tool clearances. *Objective:* improvement of quality.

  Design for machining in all separating processes involves:

- The avoidance of unnecessary machining, that is, reduction of machined areas, fine surface finishes and close tolerances to the absolute minimum (protruding bosses and cut-outs placed at same height or depth are advantageous). *Objective:* reduction of costs.

PS	Guidelines	Objec-tive	Wrong	Right
To Ex	Avoid undercuts.	Q C	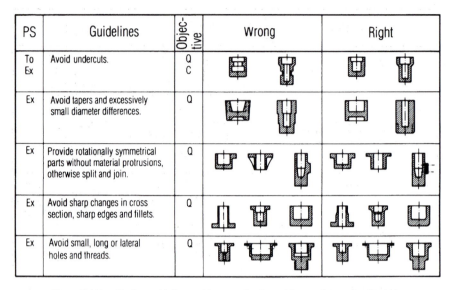	
Ex	Avoid tapers and excessively small diameter differences.	Q		
Ex	Provide rotationally symmetrical parts without material protrusions, otherwise split and join.	Q		
Ex	Avoid sharp changes in cross section, sharp edges and fillets.	Q		
Ex	Avoid small, long or lateral holes and threads.	Q		

Figure 7.113.   Design guidelines with examples for cold extrusions, after [7.108].

- The location of machined surfaces parallel or perpendicular to the clamping surfaces. *Objectives:* reduction of costs, improvement of quality.
- The choice of turning and boring in preference to milling and shaping. *Objective:* reduction of costs.

Figure 7.115 represents the design guidelines for components machined by turning; Fig. 7.116 for components machined by boring; Fig. 7.117 for components machined by milling; and Fig. 7.118 for components machined by grinding. In the design of cut-out components, the characteristics of the *tools* (To) and of the *cutting method* (Cu) [7.138] must be taken into consideration (see Fig. 7.119).

Form Design for Joining

Of the joining methods discussed in [7.59] we shall only consider welding under the above heading. For separable joints the reader is referred to 7.5.8.

Welding involves three process steps, namely *preparation* (Pr), *welding* (We) and *finishing* (Fi). The following design guidelines apply:

- Pr, We, Fi: avoid the imitation of cast designs; preferably select standard, easily obtainable or prefabricated plates, sections or other semi-finished materials; make use of composite constructions (cast/forged components). *Objective:* reduction of costs.
- We: adapt the material, welding quality and welding sequence to the required strength, sealing and shape. *Objectives:* reduction of costs, improvement of quality.
- We: aim for small welding seams and small dimensions to reduce damage

PS	Guidelines	Objective	Wrong	Right
Be	Avoid complex bent parts (material watse); rather split and join.	C	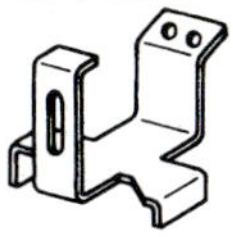	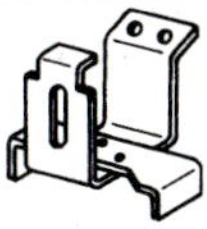
Be	Allow for minimum values of bending radii (bulging in the compression area and overstretching in the tension area) flange height and tolerances.	Q	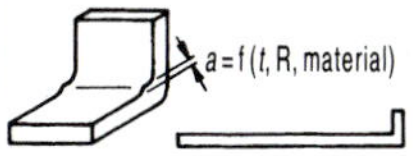 $a = f(t, R, \text{material})$	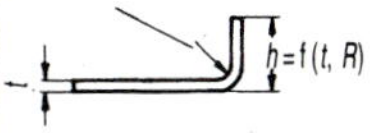 $R = f(t, \text{material})$ $h = f(t, R)$
Be	Provide sufficient distance between pre-pierced holes and bend.	Q		
Be	Aim at holes and notches to cross the bend when it is not possible to provide the minimum gap.	Q		
Be	Avoid sloping edges and tapers in the region of the bend.	Q	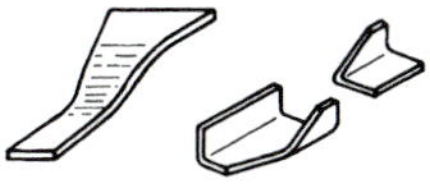	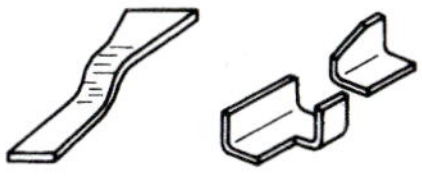
Be	Provide clearances at the corners when all sides are to be bent up.	Q	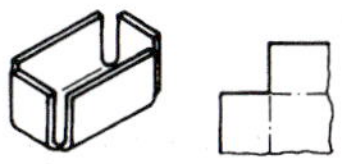	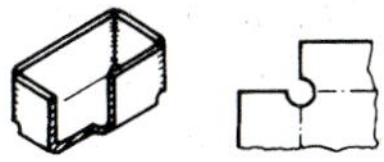
Be	Provide Folded seam of sufficient width.	Q		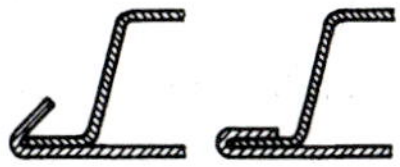
Be	Aim at large access openings for hollow shapes and undercut bends.	Q C	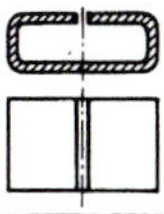	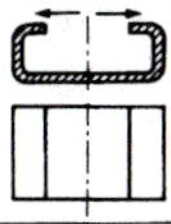
Be	Provide stiffening at sheet edges.	A		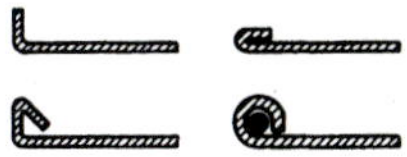
Be	Aim at identical indentation forms.	A	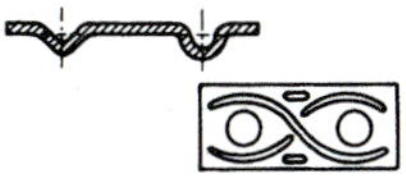	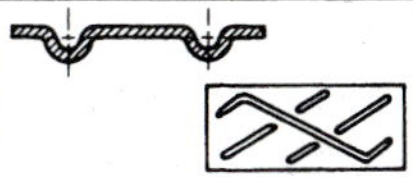

Figure 7.114. Design guidelines with examples for bent parts, after [7.1, 7.25].

through heating and to simplify handling. *Objectives:* improvement of quality, reduction of costs.

- We, Fi: minimise the amount of welding (heat input) to avoid or reduce distortion and corrective work. *Objectives:* improvement of quality, reduction of costs.

Further guidelines are given in Fig. 7.120.

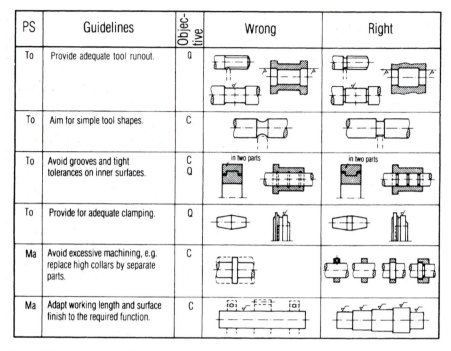

PS	Guidelines	Objective	Wrong	Right
To	Provide adequate tool runout.	Q		
To	Aim for simple tool shapes.	C		
To	Avoid grooves and tight tolerances on inner surfaces.	C Q	in two parts	in two parts
To	Provide for adequate clamping.	Q		
Ma	Avoid excessive machining, e.g. replace high collars by separate parts.	C		
Ma	Adapt working length and surface finish to the required function.	C		

Figure 7.115. Design guidelines with examples for components machined by turning, in accordance with [7.193, 7.241].

PS	Guidelines	Objective	Wrong	Right
To Ma	Where possible, use boring tools on blind holes.	C Q		
To Ma	Provide starting and finishing flats for holes breaking through angled surfaces.	Q		
To	Aim for continuous holes, avoiding blind holes.	C		

Figure 7.116. Design guidelines with examples for components machined by boring, in accordance with [7.193, 7.212, 7.241].

4 Appropriate Selection of Materials and of Semi-Finished Materials

An optimum choice of materials and semi-finished materials is difficult to make because of interactions between characteristics of the function, working principle, layout and form design, safety, ergonomics, production, quality control, assembly, transport, operation, maintenance, costs, schedules and recycling. When expensive

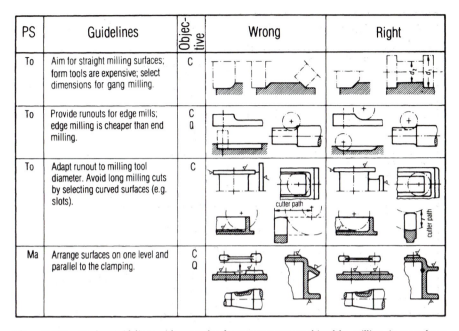

PS	Guidelines	Objec-tive	Wrong	Right
To	Aim for straight milling surfaces; form tools are expensive; select dimensions for gang milling.	C		
To	Provide runouts for edge mills; edge milling is cheaper than end milling.	C Q		
To	Adapt runout to milling tool diameter. Avoid long milling cuts by selecting curved surfaces (e.g. slots).	C	cutter path	cutter path
Ma	Arrange surfaces on one level and parallel to the clamping.	C Q		

Figure 7.117. Design guidelines with examples for components machined by milling, in accordance with [7.193, 7.241].

PS	Guidelines	Objec-tive	Wrong	Right
To	Avoid edge limitations.	Q C		
To	Provide runouts for grinding wheels.	Q		
To	Aim for unimpeded grinding by appropriate selection of surfaces.	C Q		
To Ma	Give preference to equal blend radii (if no runout possible) and to equal tapers.	C Q		

Figure 7.118. Design guidelines with examples for components machined by grinding, in accordance with [7.241].

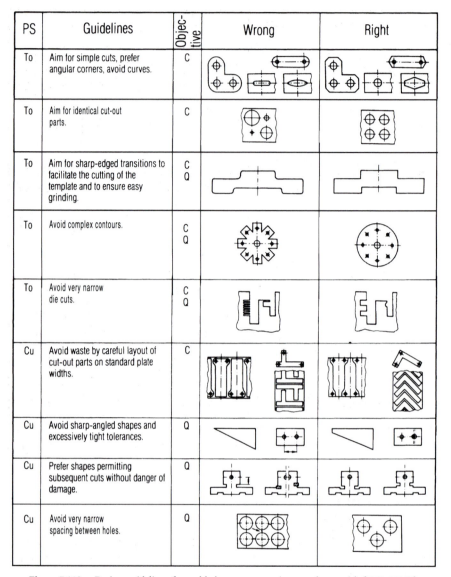

PS	Guidelines	Objec-tive	Wrong	Right
To	Aim for simple cuts, prefer angular corners, avoid curves.	C		
To	Aim for identical cut-out parts.	C		
To	Aim for sharp-edged transitions to facilitate the cutting of the template and to ensure easy grinding.	C Q		
To	Avoid complex contours.	C Q		
To	Avoid very narrow die cuts.	C Q		
Cu	Avoid waste by careful layout of cut-out parts on standard plate widths.	C		
Cu	Avoid sharp-angled shapes and excessively tight tolerances.	Q		
Cu	Prefer shapes permitting subsequent cuts without danger of damage.	Q		
Cu	Avoid very narrow spacing between holes.	Q		

Figure 7.119. Design guidelines for welded components, in accordance with [7.25, 7.241].

materials are involved, their careful selection is nevertheless of the utmost economic importance (see 10). In general, designers are advised to consult the checklist (see Fig. 7.3) and to evaluate the materials accordingly. The selected material and the resulting processing and machining of the components, their quality and the market conditions influence:

- the production procedure;
- the choice of machine tools, including bench tools and measuring instruments;
- materials handling, for example, purchasing and storage;

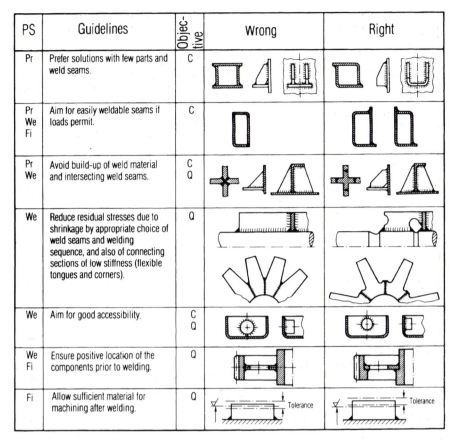

PS	Guidelines	Objective	Wrong	Right
Pr	Prefer solutions with few parts and weld seams.	C		
Pr We Fi	Aim for easily weldable seams if loads permit.	C		
Pr We	Avoid build-up of weld material and intersecting weld seams.	C Q		
We	Reduce residual stresses due to shrinkage by appropriate choice of weld seams and welding sequence, and also of connecting sections of low stiffness (flexible tongues and corners).	Q		
We	Aim for good accessibility.	C Q		
We Fi	Ensure positive location of the components prior to welding.	Q		
Fi	Allow sufficient material for machining after welding.	Q		

Figure 7.120. Design guidelines for welded components, in accordance with [7.25, 7.212, 7.241, 7.301, 7.306].

- quality control; and
- the choice between in-house and bought-out parts.

The close relationship between design, production procedures and materials technology calls for cooperation between designers, production engineers, materials experts and buyers.

The most important recommendations for the selection of materials for primary shaping processes (for example casting and sintering) and secondary shaping processes (for example forging, extrusion etc) have been set out by Illgner [7.148]. Few designers are completely familiar with the selection of materials needed for such new manufacturing methods as ultrasonic welding, electron-beam welding, laser technology, plasma cutting, spark erosion and electrochemical processes. These topics are discussed in [7.37, 7.96, 7.142, 7.195, 7.256, 7.270, 7.284].

Closely connected with the selection of materials is the choice of semi-finished materials (for example tube, standard extrusions etc). Because of the common method of costing by weight, designers tend to think that cost reduction invariably

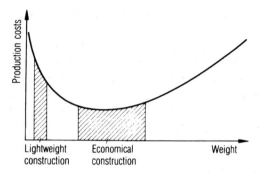

Figure 7.121. Cost areas for lightweight and economical constructions, from [7.324].

goes hand in hand with weight reduction. However, as Fig. 7.121 makes clear, that belief is often mistaken.

The following example throws further light on this problem. Figure 7.122 shows a welded electric motor housing. The old layout involved eight different plate thicknesses to achieve the required stiffness with a minimisation of weight. In the modified design, however, the number of plate thicknesses was deliberately reduced although this increased the weight. This change in the design involved the replacement of standard flame-cutting by numerically controlled machines. The extra outlay was to be justified by keeping the programming and re-equipment costs low and by maximum utilisation of the plate material through stacking before cutting [7.8]. A cost analysis showed that, despite an increase in weight due to oversizing of some of the housing parts, the new design was cheaper than the old thanks to lower labour costs and lower production overheads. Admittedly, the actual saving was not very great, but this example serves to show that minimisation of weight, which often involves a great deal of design and technical effort, does not necessarily lead to minimisation of costs. Moreover, even when the calculated cost reductions due to the incorporation of semi-finished materials and simplification in manufacturing methods are not great, the actual savings may be much

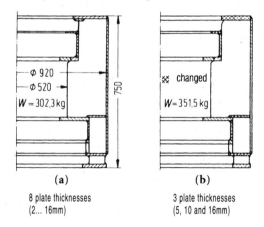

Figure 7.122. Electric motor housing of welded construction (Siemens):
(a) current design;
(b) proposed design.

greater because of the consequent reduction in idle time and time spent on operations scheduling (see 10).

A further example of the economic use of semi-finished materials is given in Fig. 7.123 which shows the plate-cutting plan for a welded motor housing. To allow the use of circular blanks for the end wall bearing shields *d*, the end walls are made from four parts *b*, which are then welded together. The resulting aperture, even after machining, is smaller than the bearing shield made from the blank. In addition, this arrangement provides the support feet *c*.

5 Appropriate Use of Standard and Bought-Out Components

Designers should always try to use components that do not have to be specially manufactured but are readily available as *repeat*, *standard*, or *bought-out* parts. In that way, they can help to create favourable supply and storage conditions. Easily available bought-out parts are often cheaper than parts made in-house.

The importance of standard parts has been stressed on several occasions.

The decision whether components are to be made in-house or bought-out depends on the following considerations:

- number (one-off, batch or mass production);
- whether production is for a specific order or for the general market;
- the market situation (costs, delivery dates of materials and bought-out parts);
- the possibility of using existing production facilities;
- the labour situation; and
- the available or desired degree of automation.

These factors influence not only the decision whether in-house production is to be preferred to sub-contract production, but also the design approach. Unfortunately, most of the factors vary with time. This means that a particular decision may be justified at the time when it was made but may no longer be right if the market or labour situation and the production capacity have changed. Particularly in the case of one-off or batch products of the heavy engineering industry, the production and market situation have to be re-examined at regular intervals.

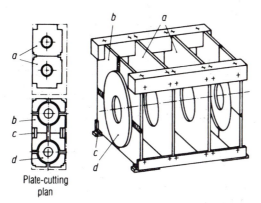

Plate-cutting
plan

Figure 7.123. Electric motor housing. Welded construction with plate plan, from [7.178] (Siemens).

6 Appropriate Documentation

The effect of production documents (in the form of drawings, parts lists and assembly instructions) on costs, delivery dates, product quality etc is often underestimated. The layout, clarity and comprehensiveness of such documents have a particularly marked influence on highly mechanised and automated production methods. They determine the execution of the order, production planning, production control and quality control.

7.5.8 Design for Ease of Assembly

1 Types of Assembly

Designers do not only have a major influence on the costs (see 10) and the quality of the production of components, but also on the costs and quality of assembly [7.356].

By assembly we refer to the combination of components into a product and to the auxiliary work needed during and after production. The cost and quality of an assembly depend on the type and number of operations and on their execution. The type and number, in their turn, depend on the layout design of the product and on the type of production (one-off or batch production).

The following guidelines for designing for ease of assembly can therefore be no more than general hints [7.2, 7.42, 7.101, 7.102, 7.341, 7.343, 7.356]. The aims of the guidelines are to simplify, standardise, automate and ensure quality. In individual cases, they may be influenced or overridden by reference to the following headings of the checklist (see Fig. 7.3): function, working principle, layout, safety, ergonomics, production, quality control, transport, operation, maintenance and recycling. The particularities of specific cases have to be checked [7.139, 7.184, 7.185, 7.235, 7.317].

According to [7.3, 7.288, 7.334] the following essential operations are involved:

- *Storing* (St) of parts to be assembled, if possible in a systematic manner. Automatic assembly further necessitates the programmed supply of parts and connecting elements.
- *Handling* (Ha) of components, including:
  - *identifying* the part by fitter or robot;
  - *picking-up* the part, if necessary in conjunction with individual selection and dispensing; and
  - *moving* the part to the assembly point, if necessary in conjunction with
  - separation, manipulation etc.
- *Positioning* (Po) (placing the part correctly for assembly), and aligning (final adjustment of the position of the part before and possibly after joining).
- *Joining* (Jo) parts by the provision of appropriate connections. According to [7.59], the following operations must also be included here:
  - bringing together, for example by inserting, superposing, suspending or folding;
  - filling, for example by soaking;

- pressing together, for example by bolting, clamping or shrink fitting;
- joining by primary processes, for example by fusing, casting and vulcanising;
- joining by secondary processes, for example by bending or auxiliary components; and
- joining by the combination of materials, for instance by welding, soldering or gluing.

- *Adjusting* (Ad) to equalise tolerances, to restore the required play etc [7.289].
- *Securing* (Se) the assembled parts against unwanted movements under operational loads.
- *Inspecting* (In). Depending on the degree of automation, various testing and measuring operations must be performed, possibly between individual assembly operations.

These operations are involved in every assembly process, their importance, sequence and frequency depending on the number of units (one-off assembly, batch assembly) and the degree of automation (manual, part automatic or fully automatic assembly).

The linking of assembly operations or assembly cells can be divided into the following types: unbranched, branched, single level and multi-level assembly. The assembly process can be stationary or flowing. It is also important whether it takes place within the company or on site, by experts or less well trained personnel or customers.

In general, the improvements one can make to automate assembly will also simplify manual assembly and vice versa. The selected type of assembly and the embodiment are closely related, that is they influence one another.

2 Designing the Layout

In accordance with the steps of embodiment design (see 7.1) it seems useful to start considering assembly even while working on the layout of the working structure and the construction structure. An easy to assemble layout can be achieved by:

- structuring,
- reducing,
- standardising, and
- simplifying

the assembly operations. This will lead to a reduction of expenditure because the assembly process is improved and ensure product quality because assembly is clearer and easier to control [7.95, 7.105, 7.279]. A layout which has been selected for these reasons could also lead to a reduction in the number of components or at least the standardisation of components.

The embodiment guidelines that focus on ease of assembly are classified in Fig. 7.124. The column "operation" contains the assembly operations that are primarily affected by the specific embodiment guidelines. The third column indicates whether the guideline leads to an improvement of *manual assembly* (MA) or *automated assembly* (AA), or both. This classification should ease the use and selection of embodiment guidelines for specific assembly situations.

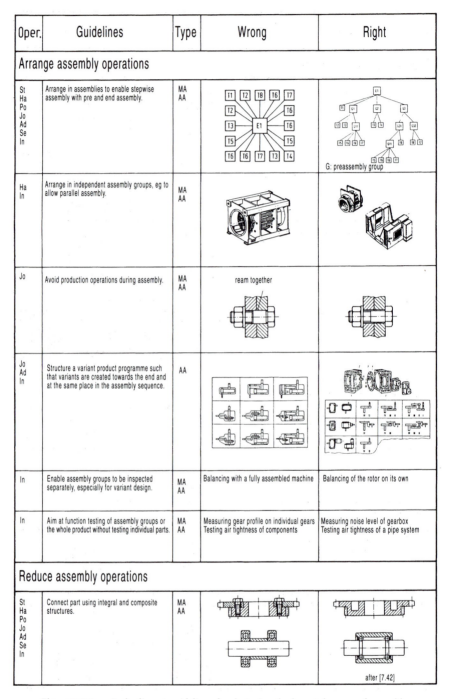

Oper.	Guidelines	Type	Wrong	Right

Arrange assembly operations

St Ha Po Jo Ad Se In	Arrange in assemblies to enable stepwise assembly with pre and end assembly.	MA AA		G: preassembly group
Ha In	Arrange in independent assembly groups, eg to allow parallel assembly.	MA AA		
Jo	Avoid production operations during assembly.	MA AA	ream together	
Jo Ad In	Structure a variant product programme such that variants are created towards the end and at the same place in the assembly sequence.	AA		
In	Enable assembly groups to be inspected separately, especially for variant design.	MA AA	Balancing with a fully assembled machine	Balancing of the rotor on its own
In	Aim at function testing of assembly groups or the whole product without testing individual parts.	MA AA	Measuring gear profile on individual gears Testing air tightness of components	Measuring noise level of gearbox Testing air tightness of a pipe system

Reduce assembly operations

| St Ha Po Jo Ad Se In | Connect part using integral and composite structures. | MA AA | | after [7.42] |

Figure 7.124a. Embodiment guidelines for designing the layout for ease of assembly.

Oper.	Guidelines	Type	Wrong	Right
Reduce assembly operations				
St Ha Po Jo Ad Se In	Use function integration to reduce number of parts.	MA AA	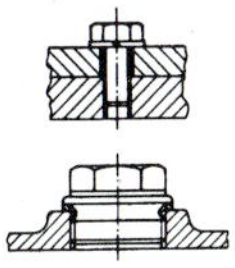	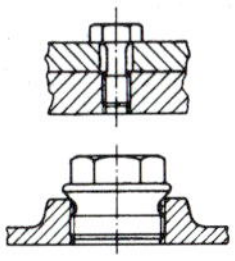 after [7.42]
Jo	Execute assembly operations simultaneously.	AA	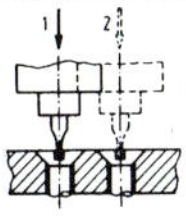	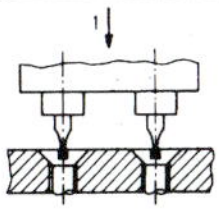
Jo Ad Se	Reduce number of interfaces to be joined.	MA AA	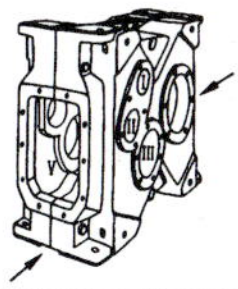	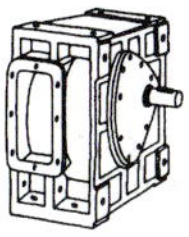
Ad Se In	Avoid disassembly to test functions of assembled groups and products.	MA AA	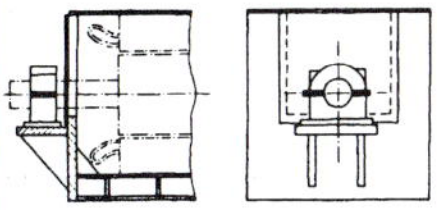 Air gap measuring not possible	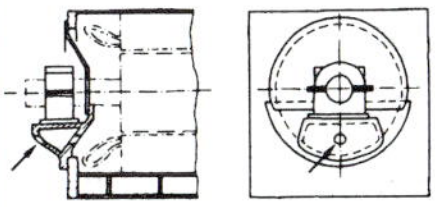 Air gap measuring directly possible
Standardise assembly operations				
Po Jo In	Provide a basic component in every assembly group, eg to allow interlocking constructions.	AA	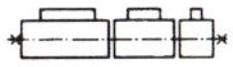	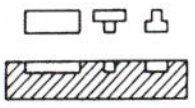 after [7.145]
Jo	Aim for uniform joining directions and procedures within an assembly group.	AA	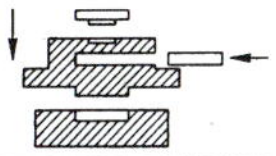	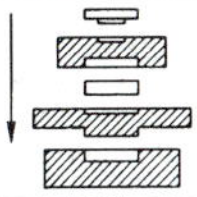 after [7.145]
Simplify assembly operations				
Po Jo Ad Se In	Constrain assembly operations (clear assembly sequence).	MA	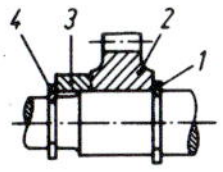 4 3 2 1 or: 4 3 2 1	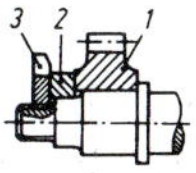 3 2 1
Jo	Combine production and assembly operations.	MA AA	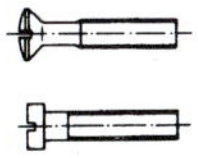	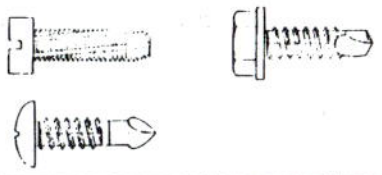
Ad In	Provide access for tests, enable visual inspection.	MA AA	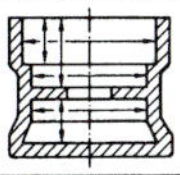	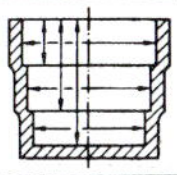

Figure 7.124b.   Embodiment guidelines for designing the layout for ease of assembly.

3 Designing Assembly Interfaces

Another important aspect of improving assembly is the design of the interfaces that are influenced by the layout. Here too the aim is to:

- reduce,
- standardise, and
- simplify

the interfaces in order to reduce the number of connecting elements and assembly operations, and to minimise the quality requirements of the interfacing elements [7.2, 7.112, 7.150, 7.292].

In Fig. 7.125 the embodiment guidelines are again classified according to the aims and the affected assembly operations.

4 Designing Interface Elements

Closely linked to the design of interfaces is the design of the interface elements. The aims are to:

- enable, and
- simplify

the automatic storage and handling, including the identification, ordering, picking up and moving the interface elements. This is particularly important for the application of automatic assembly machines (AA) [7.2, 7.103, 7.292, 7.317]. Figure 7.126 shows the design guidelines.

In summary, the essential guidelines can be derived from the basic guidelines of *simplicity* (simplify, standardise, reduce) and *clarity* (avoiding over and under constraining) (see 7.3.1 and 7.3.2). Further examples in [7.2, 7.104, 7.112, 7.114, 7.265, 7.266, 7.333].

5 Guidelines for Application and Selection

Designing for ease of assembly should, in line with the overall approach (see 7.1), involve the following five steps [7.112, 7.266].

Step 1: Draw up demands and wishes for the requirements list that determine or influence assembly. This list will specify requirements such as:

- individually designed product or variant range;
- number of variants;
- safety and legal requirements;
- production and assembly constraints;
- test and quality requirements;
- transport and packaging requirements;
- assembly and disassembly requirements for maintenance and recycling; and
- requirements related to assembly operations undertaken by the user.

Oper.	Guidelines	Type	Wrong	Right

Reduce interfaces

Oper.	Guidelines	Type	Wrong	Right
St Ha Jo Ad Se	Reduce connecting elements, eg by using clamp and snap connections.	MA AA	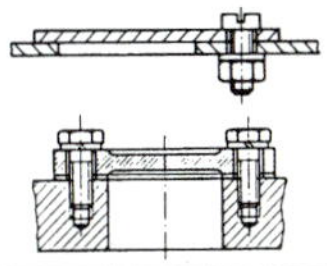	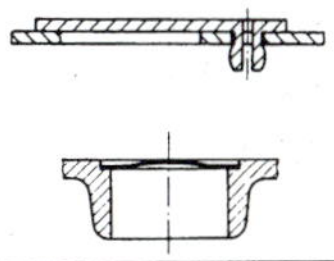
St Ha Jo	Reduce connecting elements by using special connecting elements.	MA AA	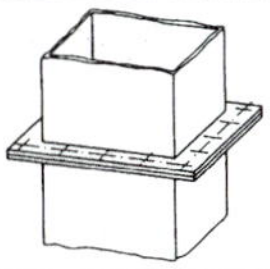	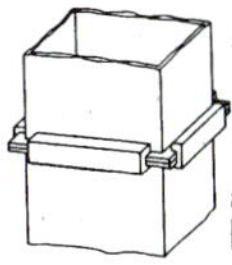 after [7.238, 7.258, 7.289]
St Jo Se	Aim for direct connections without connecting elements.	MA AA	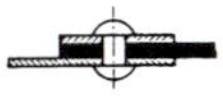	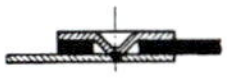 after [7.289]
Po	Aim for self-adjustment and positioning.	AA	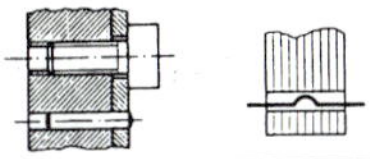	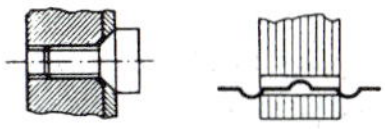
Se	Prefer self-locking connecting elements, eg through elastic-plastic deformation.	AA	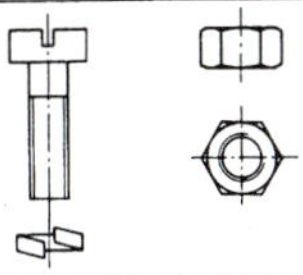	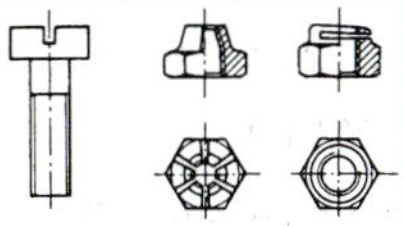 with locking adhesive

Standardise interfaces

Oper.	Guidelines	Type	Wrong	Right
St Ha Jo	Use identical connecting elements, if possible even for different functions.	MA AA	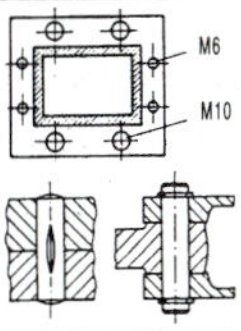 M6 M10	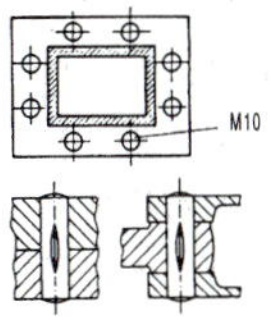 M10

Simplify interfaces

Oper.	Guidelines	Type	Wrong	Right
St Ha	Prefer connecting elements that can be delivered by belt or in a continuous flow.	AA	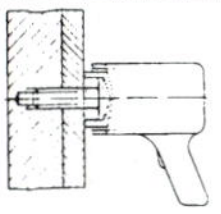	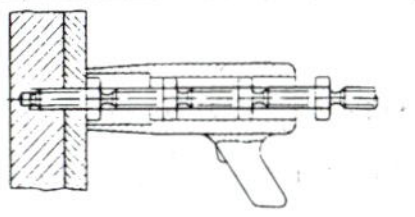
Ha Jo	Ease handling and connecting movements, eg by supporting at the centre of gravity.	MA AA	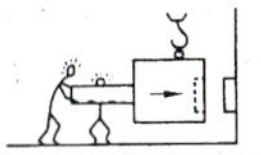	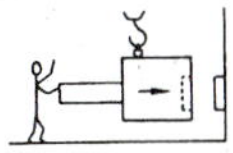 after [7.42]
Po Jo	Avoid dimension chains with tight tolerances by dividing the dimension chain.	MA AA	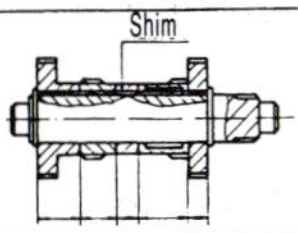 Shim	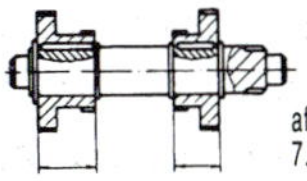 after [7.238 7.258]

Figure 7.125a.   Embodiment guidelines for designing assembly interfaces for ease of assembly.

Oper.	Guidelines	Type	Wrong	Right
Simplify interfaces				
Po Jo	Avoid double fits to enable unambiguous positioning and to reduce tolerances on dimensions.	MA AA	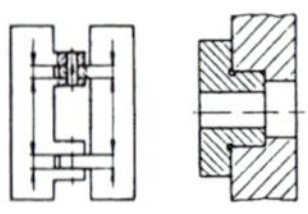	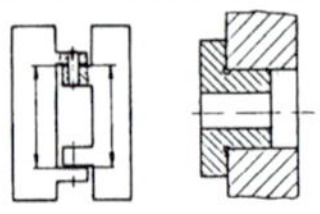
Po Ad	Prefer simple adjustments or provide positioning guides.	MA AA	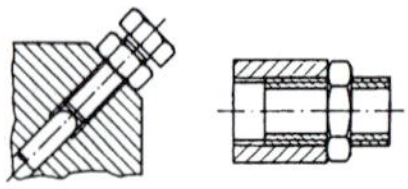	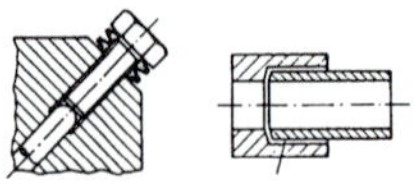 glued
Po Ad	Enable continuous adjustment.	MA AA	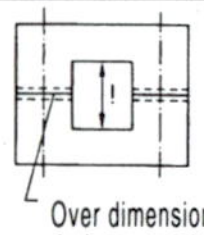 Over dimensioned, adapt during assembly	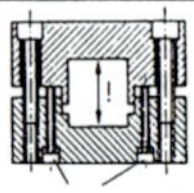 Adjustment screws, adjust during assembly after [7.289]
Po Ad	Aim for accessibility to allow adjustment without disassembling other parts.	MA AA	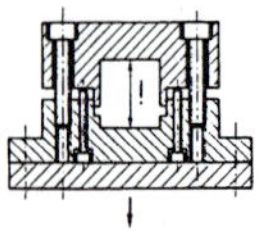	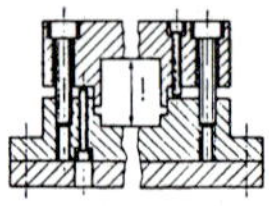 after [7.289]
Po Ad	Compensate tolerances by using compensation components.	MA	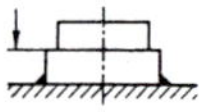	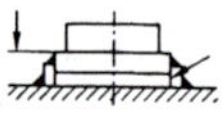
Po Ad In	Provide reference surfaces, edges and points.	MA AA	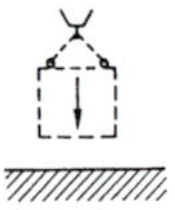	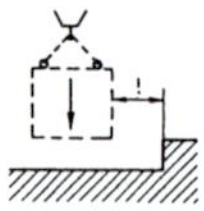 after [7.42]
Po Ad In	Aim for unambiguous and independent adjustment operations.	MA AA	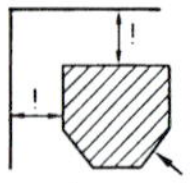	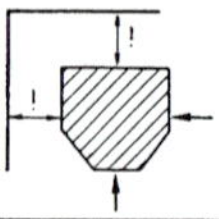
Jo	Prefer translational joining motions.	AA	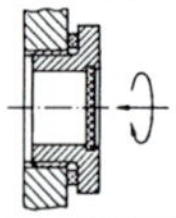	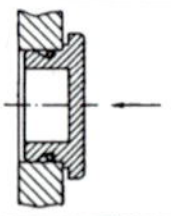
Jo	Avoid joining motions involving multiple axes, in particular curves.	AA	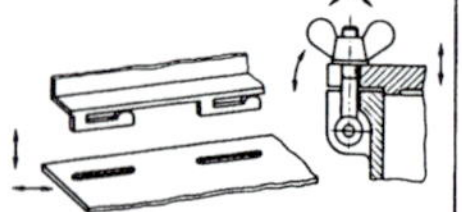	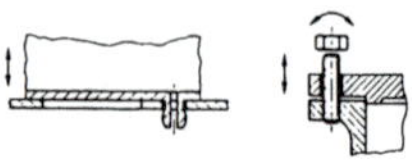
Jo	Avoid long joining paths.	MA AA	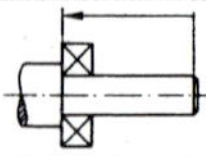	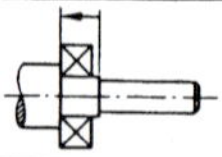 after [7.42]

Figure 7.125b.   Embodiment guidelines for designing assembly interfaces for ease of assembly.

Oper.	Guidelines	Type	Wrong	Right
Simplify interfaces				
Jo	Avoid hindering caused by air cushions.	MA AA	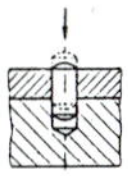	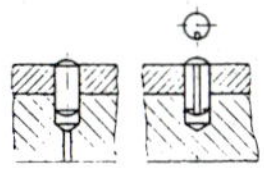
Jo	Provide tapering to ease joining.	MA AA	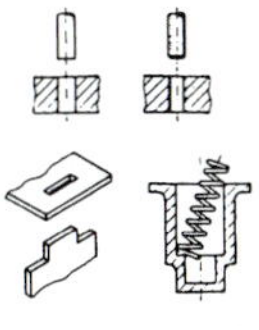 after [7.42, 7.258]	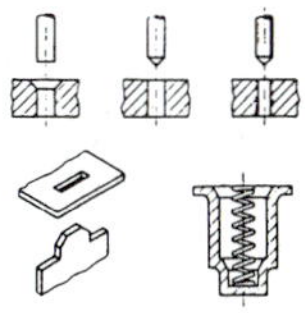
Jo	Divide large interfaces into several smaller ones.	MA AA	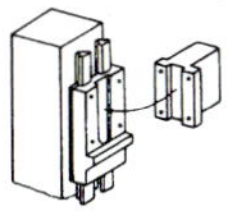	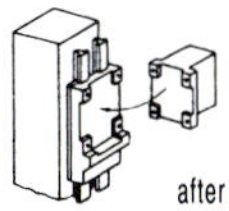 after [7.238, 7.258]
Jo Ad	Avoid simultaneous operations that influence each other.	MA AA	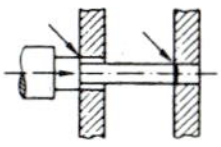	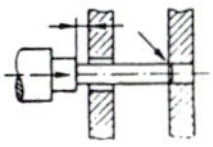 after [7.42]
Jo Ad	Provide access for assembly tools.	MA AA	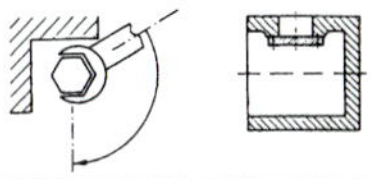	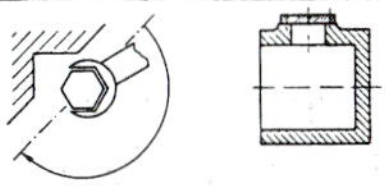
Jo Ad Se	Prefer connecting elements with elastic, elastic-plastic or material tolerance compensation.	MA AA	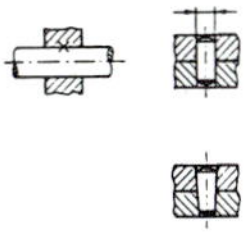	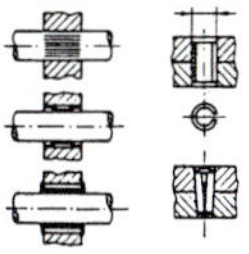 knurled / tolerance ring / cast after [7.42]
Jo Se	Allow for large tolerances through assembly parts that are flexible.	MA AA	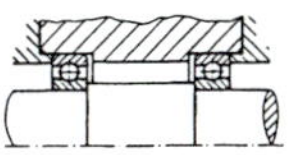	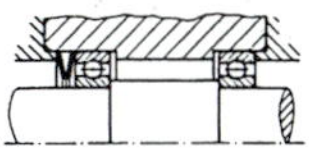
Ad	Adapt using standardised matching parts without disassembling.	MA AA	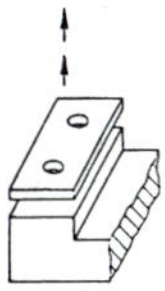	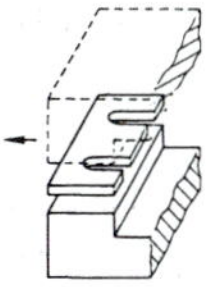
Se	Apply locking elements that are easy to assemble.	AA	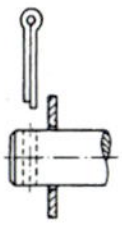	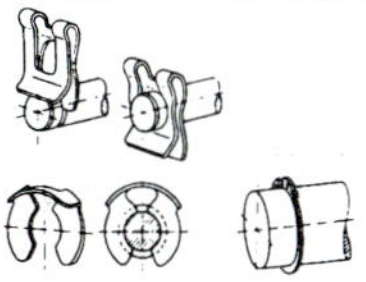

Figure 7.125c.   Embodiment guidelines for designing assembly interfaces for ease of assembly.

Oper.	Guidelines	Type	Wrong	Right
Enable and simplify automatic storage and handling				
St	Prefer interface elements having a stable position.	AA	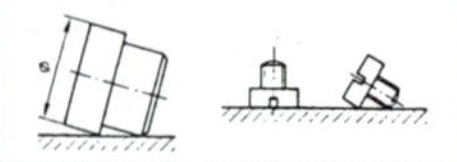	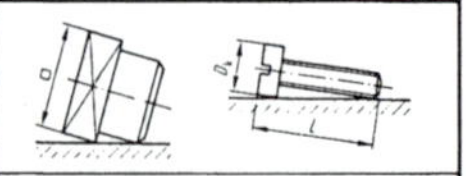
St	Avoid identical interface elements that can interlock.	MA AA	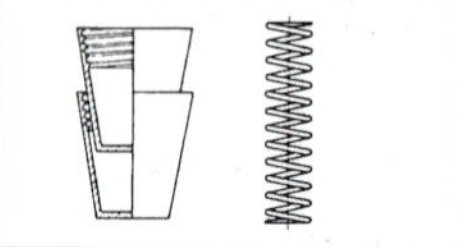	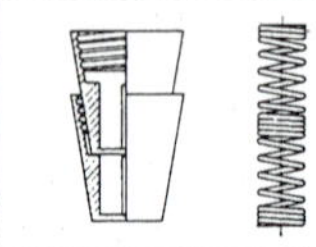 after [7.149]
St Ha	Aim for interface elements that can roll.	AA	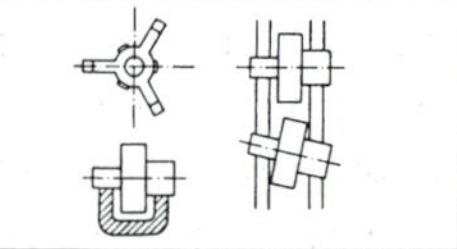	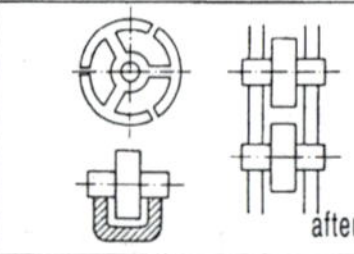 after [7.143]
Ha	Aim for symmetric contours when a specific position is not required.	AA	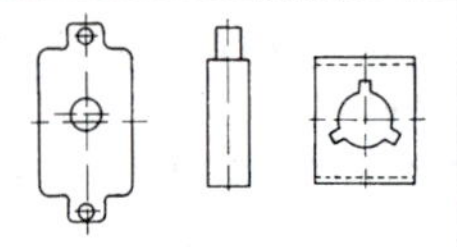	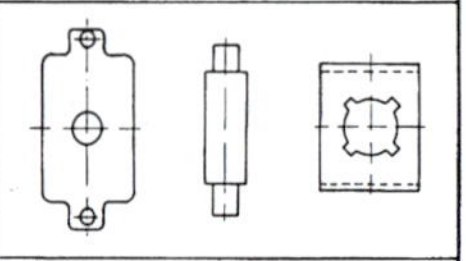
Ha	Aim for geometric identifiers.	AA	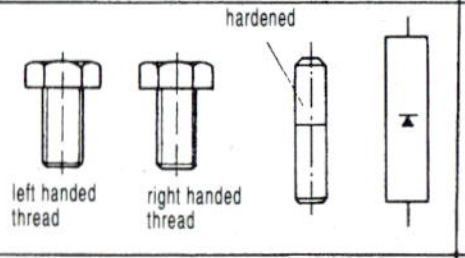	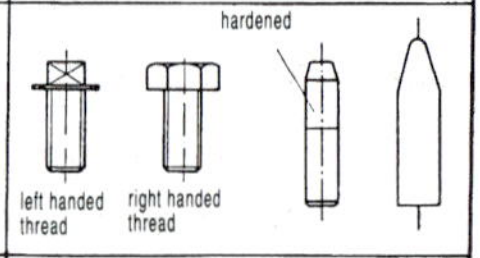
Ha	Prefer indentifiers on the outer contour.	AA	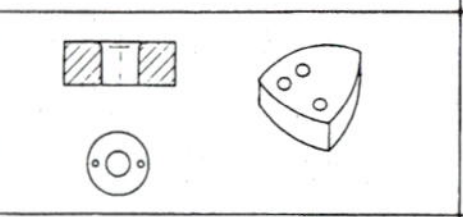	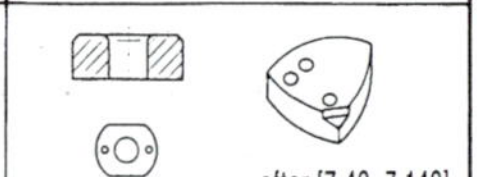 after [7.42, 7.149]
Ha	Avoid near symmetry when a specific position is required.	AA	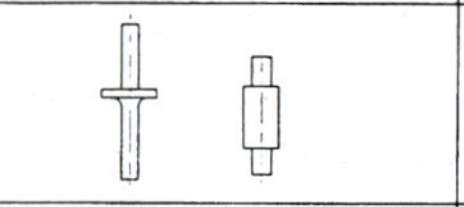	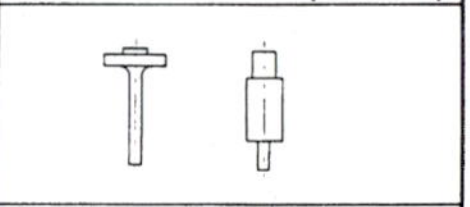
Ha	Ease handling by interface elements that can be suspended and prefer a position based on the centre of gravity.	AA	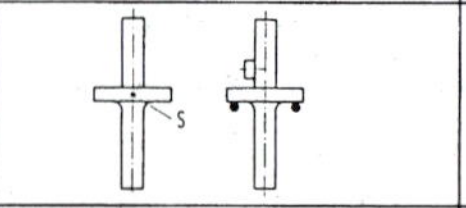	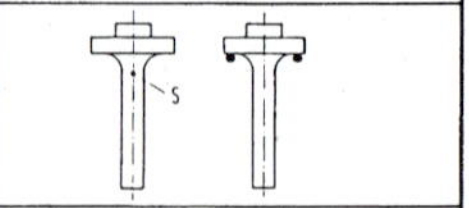
Ha	Provide features and surfaces outside functional surfaces to aid handling.	MA AA	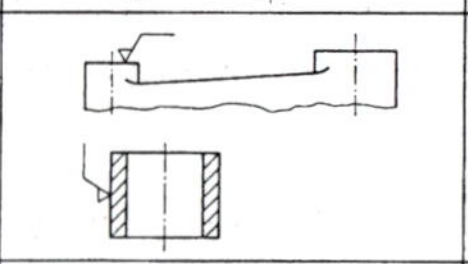	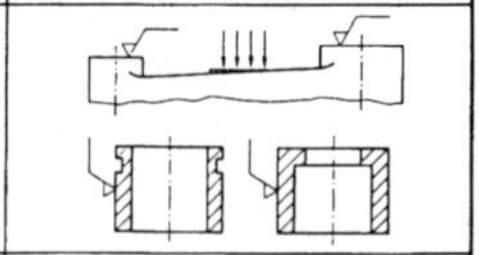
Ha	Position handling surfaces based on the centre of gravity.	MA AA	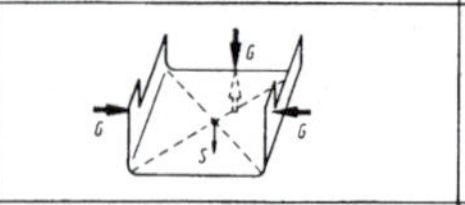	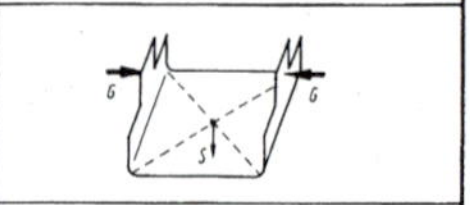
Ha	Aim for interface elements with stable geometry.	MA AA	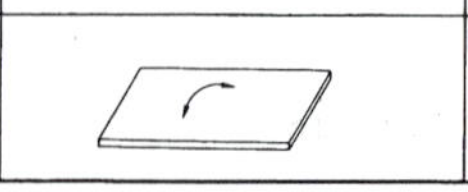	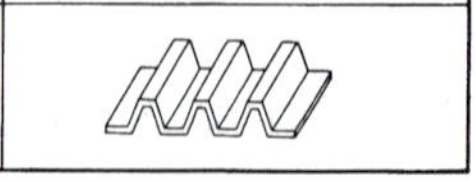

Figure 7.126.   Embodiment guidelines for designing interface elements for ease of assembly.

Step 2: Check for ways of easing assembly by using technical opportunities in the principle solution (working structure) and especially in the overall layout (construction structure).

- Reduce the number of variants in a product range by using series and modular construction (see 8) or by concentrating on a few different types.
- Apply the embodiment guidelines of Fig. 7.124 and use these to select layouts.

Step 3: Embody the assemblies, interfaces and interface elements that determine the assembly process.

- Apply the embodiment guidelines of Figs. 7.125 and 7.126 and use these to select embodiment variants.
- Take into account special production and assembly restrictions (batch size; available machine tools; manual, semi-automatic or automatic assembly).
- Select connecting elements and processes not only based on functional requirements (strength, sealing, or corrosion resistance) but also based on requirements of assembly and disassembly (ease of loosening during disassembly, reuse, potential for automation).
- Consider production and assembly costs together.

Step 4: Evaluate embodiment variants, including the required interfacing procedures, technically and economically.

- To evaluate the ease of assembly of a design, designers should work together with the product planning department because the assembly plan (assembly sequence and structure [7.112]) and the assembly processes and tools, including quality control, cannot be determined by designers alone. An aid for developing an assembly plan is a mental division of the overall layout drawing into its individual elements, that is, to start by drawing up a disassembly plan. The inverse of this can then be the basis of the assembly plan of the product. It can also be useful to simulate the assembly process using computer supported production and assembly planning (CAP) and the production of prototypes.
- The assembly process should be assessed in terms of the supply of sub-contract, bought-out and standard parts.
- Evaluation criteria can be derived from the goals and embodiment guidelines listed in Figs. 7.124 to 7.126, by adapting them to the particular situation and modifying them where necessary.

Step 5: Prepare detailed assembly instructions together with the production documents. This includes overall layout drawings for sub-assemblies and the product (pre-assembly and final assembly), assembly parts-list and other assembly information.

7.5.9 Design to Standards

1 Objectives of Standardisation

An essential feature of the methods we have been discussing is the breakdown of complex into simple problems. Thus, in the conceptual phase, the overall function is broken down into simpler sub-functions to facilitate the search for sub-function carriers or the use of design catalogues. In the embodiment phase, too, it is helpful to work separately on individual areas or assemblies before recombining them into an overall layout design. If we examine this approach in the light of the minimisation of effort, we are bound to ask to what extent generally applicable function carriers can be determined and documented so that designers can have ready access to tested solutions—that is, to known elements and assemblies.

This question has also been raised in connection with standardisation which, according to Kienzle [7.163], can be defined as follows:

"Standardisation lays down the definitive solution of a repetitive technical or organisational problem with the best technical means available at the time. It is therefore a form of technical and economic optimisation limited by the time factor."

Further definitions can be found in [7.86]:

"Standardisation determines the best solution of recurring problems."

or in [7.44]:

"Standardisation is the systematic unification by those concerned of material and immaterial things for the benefit of the community. It should not lead to the economic advantage of an individual case. It stimulates rationalisation and ensures quality in business, technology, science and administration. It serves the safety of people and things as well as the improvement of quality in all areas of life. Furthermore it serves a purposeful ordering of the information in the specific standardisation area. Standardisation takes place on regional, national and international levels."

Standardisation considered as the unification and determination of solutions, for instance in the form of national and international standards (BSI, DIN, ISO), of company standards, or of generally applicable design catalogues, and also of data sheets is becoming of increasing importance in systematic design. Here, the fact that the objectives of standardisation are to limit the range of possible solutions in no way conflicts with the systematic search for a multiplicity of solutions, because standardisation is largely confined to the determination of individual elements, sub-solutions, materials, computation and testing procedures etc, while the search for a multiplicity of solutions and their optimisation is based on the combination or synthesis of known elements and data. Standardisation is therefore not simply an important complement to, but the prerequisite of, the systematic approach, in which various elements are combined as so many building blocks.

It is, however, important to stress the limitations of all types of standardisation. As Kienzle stated: "Standardisation . . . is a form of technical and economic optimisation limited by the time factor".

The data in standards are time-dependent and must be continually updated to reflect technological changes. Standards contain not only the established knowledge of a specific domain but also agreements about topics that have a wider application such as communication standards.

In what follows, we shall be examining the possibilities of, need for and limits of, standards in the design process. In addition, the reader is referred to the comprehensive literature [7.9, 7.14, 7.44, 7.86, 7.90, 7.93, 7.133, 7.164, 7.175].

2 Types of Standard

In technical devices, standards of various origin, content, range of application and complexity are used in the design of even the simplest components [7.144]. Designers averse to standardisation ("Standardisation as a straitjacket" [7.4]) should consider how many standards they use unwittingly in their daily work. If they do so, they will find that standards are the indispensable foundation and prerequisite of all types of design work and can encourage innovation [7.12].

The following discussion of *types of standard* is meant to:

- draw the attention of designers to this important method of acquiring an organised body of information;
- encourage them to make wide use of standards;
- invite them to suggest new standards or, at the very least, to influence the development of standardisation; and
- remind them of the crux of standardisation, namely, the systematic arrangement of facts with a view to their unification and optimisation in the light of functional considerations.

By their *origin* we distinguish between:

- national standards of the BSI (British Standards Institution) or the DIN (Deutsches Institut für Normung – German Standards Institution);
- European standards of the CEN (Comité Européen de Normalisation) and CENELEC (Comité Européen de Normalisation Electrotechnique);
- recommendations of the IEC (International Electrochemical Commission); and
- recommendations and more recently universal standards by the ISO (International Organisation for Standardisation).

By their *content* we distinguish, for instance, between communication standards, classification standards, type standards, planning standards, dimensional standards, material standards, quality standards, procedural standards, operational standards, test standards, delivery standards and safety standards.

By their *scope* we distinguish between basic standards, that is general and interdisciplinary standards; and special standards, that is standards used in specialist fields.

Besides the national and international standards we have mentioned, designers can also have recourse to the rules and regulations published by professional engineering organisations. These are important as they pave the way for further standardisation after initial trials.

Designers can also have recourse to a variety of *internal company standards* and regulations [7.87-89]. These can be classified as follows:

- compilations of representative standards, that is, a *selection* from general standards that is applicable to the special requirements of a particular company, for instance, stock lists and comparisons of old with new standards (synoptic standards);
- catalogues, lists and data sheets on *bought-out parts*, including their storage and

also data on the acquisition (ordering/supply) of raw materials, semi-finished materials, fuels etc;

- catalogues or lists of *in-house parts*, for instance machine elements, repeat parts, assemblies etc;
- information sheets for the purpose of *technical and economic optimisation*, for instance on production capacity, production methods, cost comparisons (see 10.2.2);
- rules and regulations for the *calculation and embodiment design* of machine elements, assemblies, machines and plant, if necessary with a selection of sizes and/or types;
- information sheets on *storage and transport* capacity;
- regulations concerning *quality control*, for example inspection and testing procedures;
- rules and guidelines for the *preparation and processing of information*, for instance of drawings, parts lists, numbering systems and electronic data processing; and
- rules laying down *organisational and working procedures*, for instance the updating of parts lists and drawings.

The relation between company, national, European and international standards is shown in Fig. 7.127. Company standards are developed or selected for specific products or processes and adapted to the actual situation. This implies that their depth and actuality is high. National and international standards require a longer period of development, but are more generally applicable. The variety and depth of these standards is, in general, less. It is more difficult to adapt them to changes and therefore their dissemination and effects are more important.

3 Preparing Standards

Searching for and using standards, regulations and other information during the

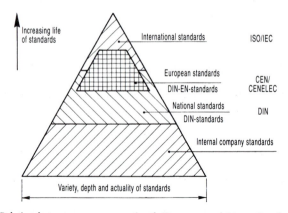

Figure 7.127. Relation between company, national, European and international standards based on DIN.

design process requires considerable effort [7.17, 7.144]. There are various ways in which information is available: folders with standards, BSI or DIN handbooks and guides, micro-fiche and increasingly computer databases. The latter will be widely used in the future as databases are integrated into CAD systems [7.10, 7.12, 7.128, 7.254, 7.255]. This will provide designers not only with the textual information contained in standards, but also with the geometry of the part itself.

4 Using Standards

Though there are no absolutely binding standards in the legal sense at the time of writing, national and international standards are widely treated as regulations, adherence to which is of great advantage in the case of legal disputes. This is particularly true of safety standards [7.30, 7.61, 7.115, 7.272, 7.330].

In addition, all company standards should be considered binding within their sphere of application, not least for economic reasons.

The sphere of application of a given standard is largely set by Kienzle's definition (see above). A standard can only be valid and binding if it does not conflict with technical, economic, safety or even aesthetic demands. Even in the case of such conflicts, however, designers should guard against rejecting or replacing the relevant standards out of hand, without assessing the possible consequences.

Moreover, they should never make such assessments by themselves, but should always consult the standards organisation and the head of department. In what follows, a number of recommendations and hints for the correct use of standards are listed.

First of all, we recommend adherence to national standards since the preferred sizes laid down in them help to determine the dimensions of all components. If these basic standards are ignored, then unpredictable long-term consequences (for instance, in the spare parts service), and grave technical and economic risks may ensue.

The use of standards should be examined against the checklist in 7.2 above.

Function

Can the anticipated overall function or sub-function be fulfilled by the use of a standard solution?

If it cannot, the problem (requirements list) and the chosen function structure should be re-examined before the search for a solution is begun.

Working Principle

Can existing standards help the development of suitable solution principles or concepts?

If they impede this development, then the consequences of ignoring or changing them or of introducing new standards must be subjected to a detailed analysis.

Layout

The basic and special standards – especially constructional, dimensional, material and safety – must be fully taken into account. Testing and inspection procedures also influence the embodiment.

Standards should only be ignored in borderline problems.

Safety

Established component, work and environmental safety standards and regulations must be rigorously observed. Safety standards must always be given precedence over rationalisation procedures and economics.

Ergonomics

The field of ergonomic standards has not yet been adequately developed, so designers would do well to consult the general literature (see 7.5.5) and work in close collaboration with production and safety engineers.

Production

Here, observance of production standards is particularly important and that of factory regulations is binding. This necessitates the continual updating of the relevant standards. Designers should only deviate from production standards after a broad assessment of all the industrial and relevant market (purchase and sales) aspects.

Quality Control

Test standards and inspection rules are essential features of quality control.

Assembly

Good assembly must be ensured by the observation of standard tolerances, limits and fits, and also of test standards and inspection rules.

Transport

Transport, inside as well as outside the factory, is rendered safer, simpler and more economical if the relevant standards are observed.

Operation

The correct operation of engineering products involves the use of various standards, for example, standard symbols and standard operating procedures.

Maintenance

Standard symbols (for instance, circuit diagrams) should be used and service standards should be provided.

Recycling

For reuse and recycling, test, material, quality, dimensional, production and communication standards are particularly important.

Expenditure

Costs and delivery times must be minimised with the help of company standards.

The above list must not be considered exhaustive or universally applicable - the work of designers is much too varied and complex for that, and the range of general and company standards much wider than we have been able to cover in our summary. By working their way down the checklist, designers can tell fairly quickly to what extent a particular standard fits the various headings.

5 Developing Standards

Since designers bear much of the responsibility for the development, production and utilisation of products, they should play a leading role in the revision of existing standards, and the development of new ones. To make a useful contribution to the development of standards, they must first determine whether the revision of an existing standard or the development of a new standard is technically or economically justified. There is rarely a clear-cut answer to this question. In particular, completely reliable assessments of the economic consequences are seldom possible because of the complex effects of in-house costs and market influences, and, in any case, would involve considerable research.

The evaluation criteria set out in Fig. 7.128, once again arranged in accordance with the checklist, can prove of great help in the assessment of existing or newly proposed standards if they are used in conjunction with the usual evaluation procedure. Not all the evaluation criteria we have mentioned apply to the assessment of individual standards. Thus, the evaluation of a drawing standard is influenced by its clarity; by the improvement in communication; by the simplification of the design activity and the overall execution of the order it provides; by the degree to

Headings	Examples
Function	Lack of ambiguity ensured.
Working principle	Market position of the product favourably influenced.
Layout	Material and energy expenditure reduced. Complexity of the product reduced, design work improved and simplified, and use of replacement parts facilitated.
Safety	Safety increased.
Ergonomics	Clarity of instructions improved. Psychological and aesthetic conditions improved.
Production	Materials handling, storekeeping, production and quality control facilitated. Precision and reproducibility ensured. Execution of the orders simplified; planning improved; production capacity increased.
Quality control	Inspection and testing simplified; quality improved.
Assembly	Assembly facilitated.
Transport	Transport and packing simplified.
Operation	Operation clarified.
Maintenance	Replacement of parts improved; spare parts service and maintenance facilitated.
Recycling	Recycling facilitated.
Costs	Costs of, and/or time spent on, design, work preparation, materials handling, production, assembly and quality control reduced. Test costs reduced. Calculations simplified. Electronic data processing reduces costs of standardisation.

Figure 7.128.   Evaluation criteria for the assessment of standards.

which it is generally accepted; and also by the costs its development entails. Before they make an evaluation, standards engineers or designers should therefore grade the importance of the various evaluation criteria and discard those that may not apply. In much the same way as with the recommendations in 4.2.2, there must be an adequate value rating to justify the development of standards.

Finally and by way of summary, the following principles of developing general, and particularly company, standards can be enunciated [7.11, 7.41]. Whether something should be standardised depends on several prerequisites, that is, the envisaged standard must:

- document the state of the art of the technology;
- be accepted by the majority of experts in the field;
- ensure the complete interchangeability of parts, for example if a standardised product is modified in such a way that it can no longer be freely interchanged even in respect of a single feature, its designation (identification number) must be altered;

- only be used if it is economical and useful, that is, there must be a need;
- always support a simple, clear and safe solution;
- not contain any provisions that conflict with the law (for instance, with monopoly restrictions or safety regulations);
- not formulate design and production details;
- not concern topics that are developing rapidly;
- not hinder technical progress;
- not allow subjectiveness or interpretation;
- not standardise fashion and taste;
- not endanger the safety of humans and the environment; and
- not serve a single individual, that is, affected people must be consulted during the development and no standardisation should take place when important groups are opposed.

Moreover the following aspects should be considered:

- Standards must be unambiguous, framed in clear terms and easily understood [7.45].
- Standard dimensions must, as far as possible, agree with preferred number series.
- All standards must be based on SI units.
- Standards should only be altered for technical, not for purely formal, reasons.
- When preparing a standard, drafts should be agreed by all those affected before a final version is drawn up.
- The layout of a standard should support its use and application. In particular the use of computer-based information systems should be facilitated [7.128, 7.254, 7.255].
- Because a standard can be regarded as an artificial system, its preparation should also follow the steps of systematic design (see 3, 6, and 7). This ensures the optimisation of its content and layout, and facilitates an economic development of the standard. Susanto [7.290] proposes a methodology for the development of standards.

7.5.10 Design for Ease of Maintenance

1 Goals and Terminology

Technical systems and products are subject to wear and tear, reduction of useful life, corrosion, contamination and changes in time-dependent material properties, such as embrittlement. After a certain period of time whether in use or not, the actual condition of a system will no longer be the intended one. Deviations from the intended condition cannot always be recognised directly and can cause changes in performance, failures and dangerous situations. This can reduce substantially the functionality, economy and safety. Sudden breakdowns disrupt normal operation, and because they are unexpected involve considerable cost to rectify. Not to check the condition of a system until damage has occurred, possibly involving injury, is not acceptable from a human and economic point of view.

Because systems and products have become more complex, maintenance as a preventive measure has become increasingly important. Designers have a significant influence on maintenance costs and procedures through their selection of the principle solution and embodiment features. We have already emphasised the importance of maintenance in our systematic approach, for example in the guidelines (see 2.1.8, 5.2.2 and 6.5.2) and their application in connection with the basic rules (see 7.3).

Maintenance is related to safety (see 7.3.3), ergonomics (see 7.5.5) and assembly (see 7.5.8). As the sections in this book addressing these topics already include suggestions and rules relevant to maintenance, this section focuses on what is necessary for a general understanding of maintenance and on design for ease of maintenance.

According to [7.66] maintenance involves monitoring and assessing the actual condition of a system and maintaining or recovering the intended condition.

Possible measures are:

- *service* to maintain the intended condition;
- *inspection* to monitor and assess the actual condition; and
- *repair* to recover the intended condition.

The type, extent and duration of service and inspection measures depend obviously on the type of system, its intended function, its required availability, its desired reliability, and on any potential dangers. The selected measures determine whether inspection and service have to take place after a fixed period of time, after a specific number of operating hours or after a particular intensity of load. The maintenance strategy is also influenced by the rate of deterioration of components, for example through wear that reduces operating life. The measures to recover the intended condition have to be taken before components are predicted to fail. Accordingly two types of repair are distinguished:

- *Failure repair* that takes place after a component has failed. This strategy is applied, and is often the only possibility, when failures cannot be predicted accurately. It is important that such failures do not cause danger. The disadvantage is the effect it has on planning. An example is the shattering of a car windscreen. This strategy is not suitable for production plant and in situations where a function has to be fulfilled or where danger is involved.
- *Preventive repair* that takes place before components fail. This can be either determined by *interval* or by *condition*. Interval repair takes place after a fixed period of time, a specific distance or a set number of operations. An example is changing the oil in an vehicle engine after 10 000 km. Condition repair is based on actual performance measures such as the loads or temperatures experienced in operation. When an undesired condition is observed, the service or repair measures must be carried out. An example is changing the oil in a vehicle engine after a certain number of cold starts or the integrated average temperature of the oil. Another example is the exchange of brake linings after a measured amount of wear. Whether the interval or condition strategy is applied

depends on the operating conditions. A combination of the two strategies is also possible. A power station, for example, will use the interval repair strategy using time intervals to safeguard the base load. For components that can last several intervals, the condition repair strategy will be adopted.

More details about maintenance strategies can be found in [7.307]. Predicting the probability of failure and the reliability of components is discussed in [7.246].

2 Design for Ease of Maintenance

Maintenance requirements should have been included in the requirements list (see Fig. 5.7). When solutions have to be selected easily maintained variants should be preferred. Examples are variants that require minimal servicing, include components that can be exchanged easily, and use components with similar life expectancies. During the embodiment phase, it is important to consider accessibility and ease of assembly and disassembly. However, design for ease of maintenance should never compromise safety.

According to [7.307] a technical solution should, in principle, require as few preventive measures as possible. The aim is complete freedom of service by using components of identical life, reliability and safety. The chosen solution should thus incorporate features that make maintenance unnecessary or reduce it substantially.

Only when such features cannot be realised or are too costly, should service and inspection measures be introduced. In principle, the following aims should be met:

- Prevent damage and increase reliability.
- Avoid possibility of errors during disassembly, reassembly and start up.
- Simplify service procedures.
- Make the results of servicing checkable.
- Simplify inspection procedures.

Service measures usually concentrate on refilling, lubricating, conserving and cleaning. These activities should be supported by embodiment features and appropriate labelling based on ergonomic, physiological and psychological principles. Examples are easy access, non-tiring procedures, and clear instructions.

Inspection measures can be reduced to a minimum when the technical solution itself embodies direct safety techniques (see 7.3.3-2) and thus promises high reliability. Overloading, for example, can be avoided using appropriate principles such as self-help that provide protection against failures and disturbing influences (see 7.4.3-4). When service and inspection measures cannot be avoided, the embodiment guidelines discussed earlier should be applied [7.307]. In what follows, we limit ourselves to lists and short explanations.

Technical measures that can reduce the service and inspection effort, and should already have been considered in the conceptual phase, include:

- Prefer self-balancing and self-adjusting solutions.
- Aim at simplicity and few parts.
- Use standard components.

- Allow easy access.
- Provide for easy disassembly.
- Apply modular principles.
- Use few and similar service and inspection tools.

Service, inspection and repair instruction documents have to be prepared, and service and inspection points have to be labelled clearly.

To facilitate the execution of service, inspection and repair measures, the following ergonomic rules, supported by appropriate technical embodiments, should be applied:

- Service, inspection and repair locations should be easily accessible.
- Working environment should follow safety and ergonomic requirements.
- Visibility should be ensured.
- Functional processes and supporting measures should be clear.
- Damage localisation should be possible.
- Exchange of components should be easy.

Instructive examples for each of these requirements can be found in [7.307].

Finally, maintenance should be part of an overall concept. Maintenance procedures must be compatible with functional and operational constraints of the technical system, and must be included in the overall cost along with the purchase and operating costs.

7.5.11 Design for Recycling

1 Aims and Terminology

To save raw materials the following possibilities can be considered [7.154, 7.155, 7.198, 7.347]:

- *reducing material use* through better utilisation (see 7.4.1) and less waste during production (see 7.5.7) [7.16];
- *substituting materials* for those becoming rare and expensive; and
- *recycling materials* by reusing or reprocessing production waste, products and parts of products.

In what follows, possible types of recycling and recycling processes are explained based on [7.329]. They help to understand the embodiment guidelines that support recycling (see Fig. 7.129).

Production waste recycling involves recycling production waste in a new production process, for example offcuts.

Product recycling involves recycling a product or part of it, for example reconditioning a vehicle's engine.

Used material recycling is the recycling of old products and materials in a new production process, for example the reprocessing of materials from scrapped vehicles.

These secondary materials or parts do not necessarily have a lower quality than

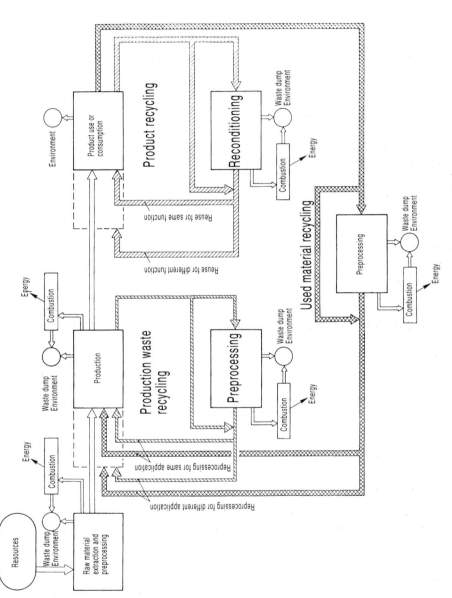

Figure 7.129. Possibilities for recycling, after [7.198, 7.329].

new materials or parts in which case they can be reused. When the quality is significantly reduced, they can only be used for other purposes.

The materials left over from the recycling system end up in waste dumps or in the environment. It is possible that in the future these will also be used as resources.

Within the recycling loops (see Fig. 7.129) various ways of recycling are possible. Basically one can distinguish between *reusing* products and *reprocessing* products, supported by reconditioning and preprocessing.

Reuse is characterised by retaining the product shape whenever possible. This type of recycling represents a high level of utilisation and should therefore be aimed at. Two types of reuse can be distinguished. In the first the product fulfils the same function, eg refillable gas cylinders, in the second a different function, eg reusing car tyres as boat fenders.

Reprocessing destroys the product shape and so this process leads to a lower utilisation value. Two types of reprocessing can be distinguished. In the first reprocessing takes place for application in the same product production process, eg reprocessing the materials from scrapped vehicles, in the second reprocessing takes place for a different application, eg converting old plastic into oil by pyrolysis.

2 Recycling Support Processes

Preprocessing

The reprocessing of production waste and scrap materials is influenced strongly by the necessary preprocessing methods [7.198, 7.296, 7.329].

Compacting of loose scrap by *pressing* eases the process of charging in metal making, but does not allow the separation of materials in mixed scrap. It is therefore only suitable for recycling non-mixed production waste and scrap metals (eg waste food cans).

Cutting heavy or large products can be done by *shears* or *flame cutting*. These methods are particularly suitable when the materials have to be subsequently separated.

Separating can take place in a *shredding plant* based on the principle of a hammer pulveriser in which a rotating hammer tears apart the product. In series with this pulveriser are other processes such as dust removal, magnetic separation, size separation, and manual sorting of materials. Shredded scrap has high quality because of its high density, purity and uniform piece size. These technically complex and labour intensive preprocessing methods are used for about 80% of scrapped vehicles and about 20% of scrapped domestic products, eg refrigerators. *Grinders* provide the same waste quality. They are just as technically complex, differing only in the method of pulverising used prior to material separation.

Float/sink testing can be linked to shredders and grinders for improved *separation* of non-ferrous and non-metallic parts. *Dropping weights* can be used to *reduce* large grey iron castings with large wall thicknesses. *Chemical preprocessing* can be used to separate harmful materials and alloys before they are used again in metal making.

Figure 7.130 illustrates the material flow in a modern shredding plant [7.329].

Because *plastics* now make up a large proportion of scrap, recycling these materials is becoming increasingly important [7.24. 7.215]. Preprocessing thermo-plastics can be done by shredding, washing, drying and granulating – provided this waste has been presorted. This is difficult for household waste. The prepro-cessing of mixed plastic waste can be done by mechanical separation, such as sort-ing, sizing and sieving, after it has been broken down into smaller parts. Other methods of separating include the use of electrostatics and floatation for density testing. Such preprocessing methods are still under development so the sorting of plastics prior to collection would provide an economically viable alternative. For thermosetting plastics and elastomers chemical preprocessing can be used.

The best waste and scrap quality, that is the highest material reutilisation rate, is achieved by *disassembling* the product prior to preprocessing. Such disassembly into appropriate material groups can either be undertaken by specialist companies or by the product manufactures themselves on dedicated disassembly lines.

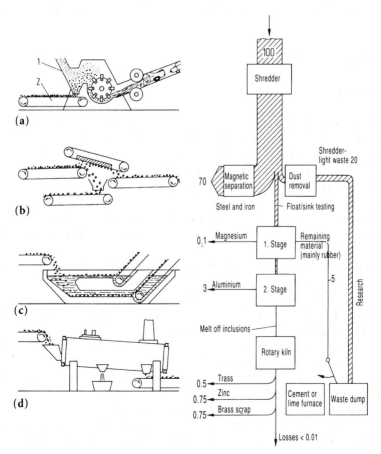

Figure 7.130. Operating principles and material flow in a shredding plant. (a) Shredder, *1* Dust removal, *2* Sorting conveyors; (b) Magnetic separation; (c) Float/sink testing; (d) Rotary kiln.

The prerequisites for economic disassembly should be established by designers through the selected *embodiment features* and *assembly methods* (see 7.5.8).

Economic preprocessing of scrap products and materials involves an appropriate combination of disassembly and preprocessing methods [7.198, 7.329].

Reconditioning

To be able to reuse or reprocess products *after* they have been used for the first time (in contrast to reusing products *during* their life–product recycling) a reconditioning process is required comprising the following steps [7.286, 7.287, 7.329, 7.344]:

- complete disassembly;
- cleaning;
- testing;
- reuse of worthwhile parts, repair of worm areas, rework of parts to be adapted, replacement of unusable parts by new ones;
- reassembly; and
- testing.

Two methods are used to recondition products, whether this is undertaken in special companies or by the product manufacturer [7.18]. The first method retains the identity of the original product, that is while changing and reworking parts their configuration is retained and the tolerances are matched to each other. An engine, for example, reconditioned using this method will retain its original engine number. The second method breaks up the original product in such a way that all parts are treated as new ones along with their individual tolerances. The result is that at the reassembly stage the reconditioned original parts and the new parts are combined as if they were all new. This method has a promising future because the same production and assembly facilities can be used for both the reconditioned and the original product.

Because reconditioning technology tends to employ conventional production and assembly processes, these processes will not be discussed further here (see 7.5.7 and 7.5.8).

3 Design for Recycling

To support preprocessing and reconditioning, designers can introduce specific measures during product development [7.19, 7.20, 7.154, 7.155, 7.198, 7.199, 7.329, 7.346, 7.347]. These measures, however, must not conflict with the other goals and requirements of the task (see Fig. 5.7). In particular the cost effectiveness of production and operation have to be guaranteed.

Recycling Considerations During the Design Process

Recycling possibilities should be considered during all stages of the design process. In the *requirements list,* demands and wishes relevant to recycling should be included (see Table 7.7) and recycling should be considered during the selection and arrangement of functions in the *function structure;* during the selection of the *working principle* and the *working structure* in *conceptual design;* and when establishing the *construction structure.*

In *embodiment design* the following preprocessing and reconditioning criteria should be considered:

- ease of disassembly and reassembly using suitable joining and fastening techniques;
- rework possibilities;
- choice of materials that do not conflict with reprocessing and minimise corrosion;
- ease of testing and sorting; and
- access for cleaning.

In *detail design* the planned recycling strategy and technology should be defined by labelling the recycling properties based on the product structure, for example by using a classification scheme [7.231, 7.329].

Embodiment Guidelines for Preprocessing

The following guidelines relate to the overall product or the individual assemblies. They can be applied singly or in combination with the aim of improving preprocessing or direct reprocessing.

Material compatibility: It is very difficult to design products made from a single material that can be reprocessed easily. For indivisible units, therefore, the aim should be to use materials that are compatible with regard to reprocessing. This

Table 7.7. Selection of criteria relevant to recycling for the requirements list

Function	Life span, product for long time operation, product for short time operation
Safety	Environmental regulations, properties of secondary materials, test requirements
Operation	Manuals with recycling and disposal regulations
Ergonomics	Design criteria for secondary materials
Production	Logistics of production waste, production properties of secondary materials
Assembly	Disassembly strategies, easy to disassemble connection technologies, accessibility of interfaces
Packaging, transport	Minimisation, reuse and or reprocessing of packaging and transport aids
Maintenance	Integration of maintenance logistics into the recycling strategy, easy identification and exchange of parts that wear
Recycling, disposal	Recycling strategy, product recycling, reconditioning, material recycling, preprocessing, product and part labelling
Cost	Acceptable manufacturing, operation, recycling and disposal costs

results in an output from the process that is more economical and has higher quality.

To fulfil this aim, the production requirements for reprocessing should be known. Here it is useful to define groups of compatible materials. Until such generally applicable groups are identified by materials scientists and the materials processing industry, designers should check the material compatibility in each case with experts. This is particularly important for large batch production with high recycling potential. Figure 7.131 shows a sample compatibility table for plastics.

Material separation: When material compatibility cannot be realised for inseparable parts or assemblies, additional interfaces should be introduced to break products down in such a way that during preprocessing, for example through disassembly, the incompatible materials can be separated.

Interfaces suitable for preprocessing: Interfaces that support high quality and economic preprocessing should be easily accessed and disassembled, and located near the outer edges of the product. Figure 7.132 shows types of connections that can be easily disassembled. For economical disassembly simple tools, automatic processes and untrained personnel are preferred, in particular for disassembly at scrap yards.

High value materials: Valuable and rare materials should be favourably positioned and labelled to facilitate separation.

Dangerous materials: Materials, liquids and gases that can be dangerous for humans and the environment during preprocessing or direct reprocessing should always be easy to separate or remove.

Embodiment Guidelines for Reconditioning

The following guidelines should be applied:

- ensure easy and damage free disassembly (see 7.5.8);

	Additive												
Important synthetic design materials	PE	PVC	PS	PC	PP	PA	POM	SAN	ABS	PBTP	PETP	PMMA	
PE	●	○	○	○	●	○	○	○	○	○	○	○	● Compatible
PVC	○	●	○	○	○	○	○	●	◐	○	○	●	
PS	○	○	●	○	○	○	○	○	○	○	○	○	
PC	○	◕	○	●	○	○	○	●	●	●	●	●	◐ Limited compatibility
PP	◕	○	○	○	●	○	○	○	○	○	○	○	
PA	○	○	◕	○	○	●	○	○	○	◕	◕	○	◕ Compatible in small quantities
POM	○	●	○	●	○	○	●	○	○	◕	○	○	
SAN	○	●	○	●	○	○	○	●	●	○	○	●	
ABS	○	◐	○	●	○	○	◕	○	●	◕	◕	●	○ Not compatible
PBTP	○	○	○	●	○	○	◕	○	○	●	○	○	
PETP	○	○	◕	●	○	○	◕	○	○	◕	●	○	
PMMA	○	●	◕	●	○	○	○	◕	●	●	○	●	

(Basic material)

Figure 7.131. Compatibility of plastics [7.159, 7.329].

Characteristics of connection / Connecting principle		Material closure: Plastic/metal gluing	Material closure: Welding	Force closure: Magnetic joint	Force closure: Velcro joint	Force closure: Threaded joint steel	Force closure: Threaded joint plastic	Force closure: Snap joint	Form closure: Snap joint	Form closure: Clamp joint	Form closure: Turn-lock fastner	Form closure: Turn-press fastner	Form closure: Press fastner	Form closure: Band with connector
Load capacity	Static	◑	●	◑	○	●	◑	◑	●	●	●	◑	◑	●
	Oscillating	◑	●	◑	○	●	◑	○	◑	◑	◑	◑	○	◑
Cost of assembly	Joining	◑	◑	●	●	◑	◑	●	●	●	●	●	●	●
	Inspection	○	○	◑	●	◑	◑	●	●	◑	●	●	●	◑
Cost of disassembly	Non-destructive disassembly	○	○	●	●	◑	◑	●	○	●	●	●	●	◑
	Destructive disassembly	◑	◑			◑	●		●	◑	◑	◑	◑	●
Recycling	Product recycling	○	○	◑	◑	◑	●	●	○	●	●	●	●	●
	Material recycling	◑	●	◑	◑	◑	●	●	●	●	●	◑	◑	●

● = Preferred ◑ = Suitable ○ = Less suitable

Figure 7.132. Interfaces suitable for reprocessing [7.262, 7.329].

- ensure that all reusable parts can be cleaned easily and without damage;
- facilitate testing and sorting through appropriate embodiment;
- ease the reworking of parts or the deposition of material by providing additional material and facilities for locating, clamping and measuring; and
- ease reassembly by using existing tools from one-off and small batch production.

To reduce the number of new parts that are needed the following measures are useful:

- limit wear to special purpose, easily adjustable or extendible parts (see 7.4.2);
- make it easy to identify the state of wear of a part and to decide whether it can be reused;
- ease material deposition on areas of wear by selecting appropriate base materials;
- minimise corrosion through embodiment and protective measures to increase the reusability of parts (see 7.5.4); and
- select connections that can be easily undone, based on the functions the product has to fulfil throughout its entire life. Prevent corrosion bonding, and also slackening through repetitive disconnecting [7.262, 7.314].

Labelling of Recycling Properties

The recycling properties and the required recycling procedures for assemblies and modules should be labelled in line with the proposed recycling strategy and the embodiment developed to fulfil that strategy. Figure 7.133 provides an example.

4 Examples of Design for Recycling

Recycling of Plain Pedestal Bearings

Plain pedestal bearings (see Fig. 8.25) are so common in machines that it is economic to consider recycling. The first possibility is to recycle by reworking the worn out parts, that is, to provide new or renewed cast bearing shells, lubrication rings and seals. The second is to exchange the bearing completely. Up until now about 99% of used pedestal bearings were recycled as a whole, though this has a low reprocessing efficiency. The reprocessing efficiency is determined by the purity of the material after the product has been preprocessed. This efficiency depends on the material combination in the product and the preprocessing technology used. Commonly available plain pedestal bearings, for example, consist of about 74% cast iron, 22.3% unalloyed steel, 3.5% non-ferrous metals and 0.2% non-metals. The weight percentages of the alloys in a bearing, similar to the one in Fig. 8.25, are compared to the alloy percentages allowed for the used material group "unalloyed steel" in Fig. 7.134 [7.198]. This figure shows that the percentages of lead

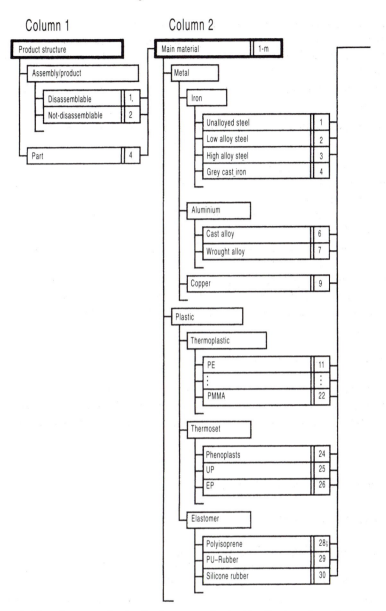

Figure 7.133a.   Structure of a classification system for the identification of recycling properties.

(Pb), which can produce a poisonous gas, and both copper (Cu) and antimony (Sb), which cannot be removed, are too high. Thus the recycling of the bearing as a whole has a negative effect on the reprocessing effort and resulting steel quality. Removal of the copper containing "lubrication ring" and "cast bearing shells" is not economic prior to preprocessing, for example, by shredding. A redesign of the bearing that takes into account recycling consists of choosing materials for these parts that are compatible with the other alloys in the main material group. The

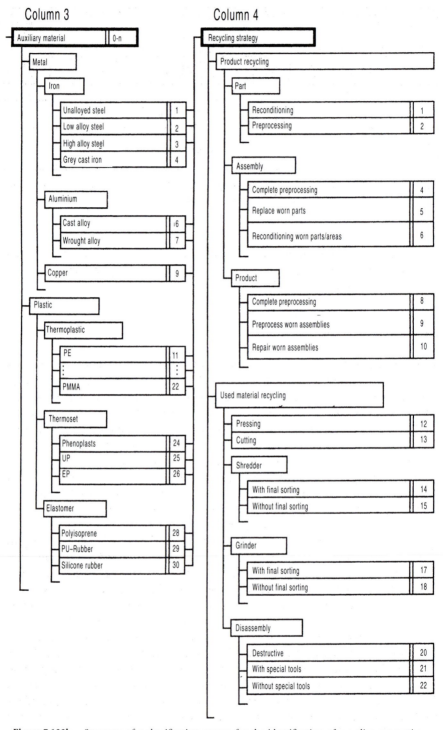

Figure 7.133b. Structure of a classification system for the identification of recycling properties.

% of admissible weight / Used metal group	Evaporating elements		Slag making elements										Addition to slag				Non separable elements								Evaporating elements (Toxic)			
	Li	C	Zn	Ca	Mg	Al	B	V	Ti	Si	Nb	Zr	Mn	Cr	P	S	As	Sb	Co	W	Mo	Ni	Sn	Cu	Cl	Cd	Be	Pb
Unalloyed steel	Any	2.0 (3.0) 4.0	Any	2.0 (3.0)	2.0 (3.0)	0.2 (0.5)	2.0 (3.0)	2.0 (3.0)	1.5 (2.5)	2.0 (3.0)	1.0 (1.5)		3.0 (4.0)	0.6	0.1 (0.3)	0.1 (0.2)	≪ 0.1	≪ 0.1	0.1	≪ 0.05	≪ 0.05	≪ 0.1	≪ 0.03	≪ 0.3 •		≪ 0.1	≪ 0.1	≪ 0.1
% of weight	-	2.12	0.01	-	-	0.08	-	-	-	1.16	-	-	0.17	-	0.16	0.16	0.01	0.4	-	-	-	0.01	0.15	0.8	-	0.02	-	2.07
Plain pedestal bearing ERNLB 18-180	% of weight more than 1%																											

Figure 7.134. Comparison of the weight percentages of the alloys in a plain pedestal bearing against the alloy percentages allowed for the used metal group "unalloyed steel" (Renk-Wülfel).

lubrication rings, for example, could be made out of an aluminium alloy with a low copper content (for example $AlMg_3$) and the bearing shell from grey cast iron, with or without a plastic coating.

Recycling of White Goods

White goods such as washing machines, dishwashers, refrigerators etc are very valuable for recycling because they are produced in large numbers and contain valuable materials. Figure 7.135 shows the weight percentages of the main materials in a dishwasher. There are numerous non-ferrous metals and non-metals, and a particularly high percentage of high alloy steels. Preprocessing the product as a whole by, for example, shredding is not economic because the high alloy steels cannot be reprocessed separately. In addition the non-ferrous metals complicate the reprocessing process, or at least increase the reprocessing effort. A product structure more suitable for reprocessing would comprise main assemblies that are easy to separate or disassemble so that they can be preprocessed separately by, for example, shredding, cutting or compacting.

In most white goods the housing components are connected using a variety of screw fasteners and rivets (see Fig. 7.136). Figures 7.137 and 7.138 show two examples of possible embodiment variants [7.198] that are easier to disassemble and are thus economic to recycle. The basic unit of the dishwasher in Fig. 7.137 has clamp fasteners and snap connectors instead of screws and rivets. This required a modification of the construction structure. The variant in Fig. 7.138 has the electrical components concentrated in a module that is easy to disassemble.

Another interesting variant for the dishwasher is shown is shown in Fig. 7.139. In this embodiment the base *1* contains all accessories including circulating pump *2*, water distribution pump *3*, washing detergent pump *4*, and the electronics *5*. This base assembly has been designed so that there are no connecting elements necessary for the components. They are simply kept in place by the lower part of casing *6*. The casing and the base can be opened and closed by means of hinge *7*.

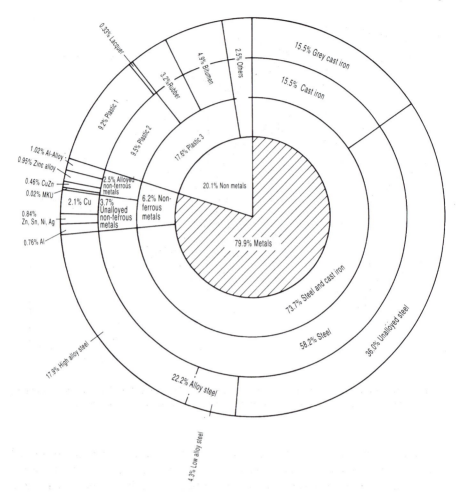

Figure 7.135. Material weight percentages of an AEG dishwhaser of 1979/80, after [7.198].

The maximum angle after tilting the casing is large enough to assemble all components and to remove them for recycling.

When developing products that are easier to recycle, particular care must be taken to ensure that they are not more expensive than traditional solutions as far as production and assembly are concerned.

5 Evaluating Recycling Potential

When developing new products it is necessary to evaluate solution variants against their potential for recycling. The established evaluation procedure discussed in 4.2.2 can be used. Such an analysis and evaluation based on recycling criteria can also be useful to improve the recycling potential of existing products. The flow chart in Fig. 7.140 shows this procedure for the analysis of electrical devices. Using

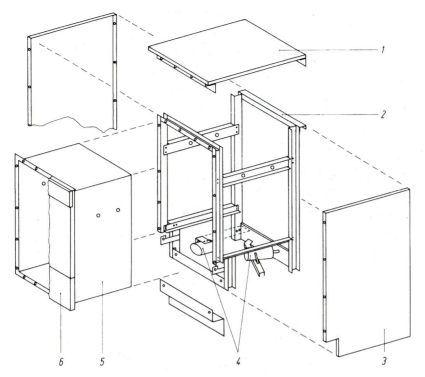

Figure 7.136. Construction structure of a dishwasher, after AEG-Telefunken. *1* Work surface, *2* Frame, *3* Side panels, *4* Circulating and washing detergent pumps, *5* container, *6* Front door.

evaluation values that are determined in the individual analysis steps, an overall value for recycling potential can be estimated. This overall value indicates the feasibility and cost effectiveness of the following options: predisassembly, disassembly, disintegration, reuse and reprocessing.

7.5.12 Design for Minimum Risk

Despite provisions against faults and disturbing factors (see 9), designers will still be left with gaps in their store of information and with evaluation uncertainties — for technical and economic reasons, it is not always possible to cover everything with theoretical or experimental analyses. Sometimes all designers can hope to do is to set limits. Thus despite the most careful approach, some doubt may remain whether the chosen solution invariably fulfils the function laid down in the requirements list or whether the economic assumptions are still justified in a rapidly changing market situation. In short, a certain risk remains.

One might be tempted always to design in such a way that the permitted limits are not exceeded, and to obviate any impairment of the function or early damage by running the equipment below full capacity. Experienced designers know that

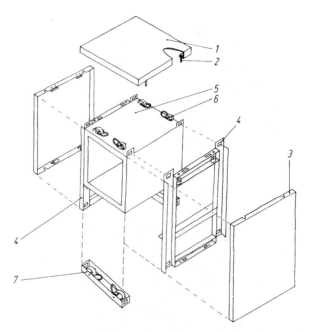

Figure 7.137. Embodiment variant for basic dishwasher unit, after [7.198]. *1* Work surface, *2* Snap connector, *3* Side panels, *4* Frame, *5* Container, *6* Clamp fastener, *7* Connecting section.

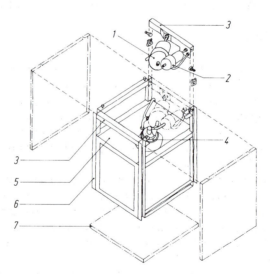

Figure 7.138. Embodiment variant for dishwasher, after [7.198], seen from below. *1* Circulating pump, *2* Washing detergent pump, *3* Support, *4* Collecting tank, *5* Container, *6* Frame, *7* Work surface.

with this approach they very quickly come up against another risk: the chosen solution becomes too large, too heavy or too expensive and can no longer compete in the market. The lower technical risk is offset by the greater economic risk.

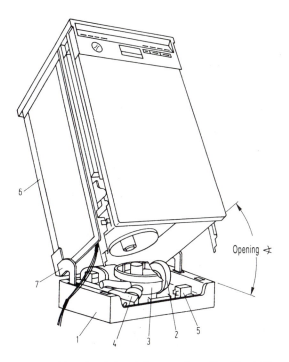

Figure 7.139. Dishwasher designed for ease of recycling (Bosch-Siemens).

1 Coping with Risks

Faced with this situation, designers must ask themselves what countermeasures they can take – provided, of course, that the solution was carefully chosen in the first place and the appropriate guidelines were scrupulously followed.

The essential approach, which we shall be examining in greater detail, is that designers must, on the basis of the analysis of faults, disturbing factors and weak spots, provide a substitute solution against the possibility that the original solution might not cover all uncertainties.

In the systematic search for solutions, several solution variants were elaborated and analysed. To that end, the advantages and disadvantages of individual solutions were discussed and compared. This comparison may have led to a new and improved solution. As a result, designers are familiar with the range of possible solutions; they have been able to rank them and also to take stock of the economic constraints.

In principle, the cheapest solution will be selected, provided only that it has sufficient technical merit for, though it may be more risky, it will afford the greatest economic leeway. The chances of marketing the resulting product, and hence of judging the validity of the solution, are greater than those of marketing a costlier product, which might jeopardise the entire development or, because of its "riskless" design, cannot provide information about performance limits.

While they are well advised to adopt this strategy, designers should assiduously

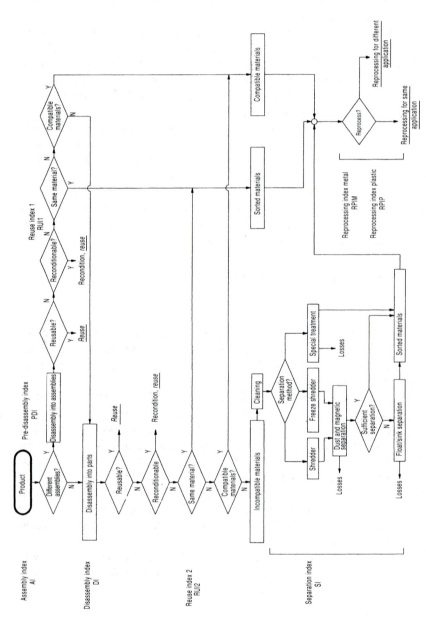

Figure 7.140. Product analysis: flow chart for evaluating recycling potential.

eschew reckless developments that might lead to damage, breakdowns and a great deal of unnecessary irritation. If risks cannot be eliminated by theoretical analyses or experiment in good time or with justifiable outlay, designers may be forced to opt for a cheaper and riskier solution, but they should always keep a more costly, less risky alternative in reserve.

To that end, the less cost-effective solution proposals elaborated in the conceptual and embodiment phases should be developed into a second or third solution reserved for critical design areas, and ready for immediate use in case of need. Provision for such development should be built into the chosen solution. If the latter should not meet all expectations, it can then be modified, if necessary step by step, without great outlay of money and time.

This systematic approach not only helps to reduce economic risks for a tolerable outlay, but also to introduce innovations one at a time, and to provide a detailed analysis of their performance, so that further developments can be made with minimum risk and at minimum cost. This approach must, of course, be coupled with a systematic follow-up of the practical experiences gained through it.

Through *design for minimum risk*, designers thus try to balance the technical against the economic hazards and so provide the manufacturer with valuable experience and the user with a reliable product.

2 Examples of Design for Minimum Risk

Example 1

A study of possible improvements in the performance of a stuffing box showed that, to increase the sealing pressure and the surface speed, the resulting frictional heat on the shaft must be removed rapidly in order to keep the temperature in the sealing areas below the limit.

To that end, it was suggested that the packing rings be mounted on the shaft so as to rotate with it and rub against the housing rather than the shaft. The heat generated by friction could then be extracted through the thin wall (see Fig. 7.141a). Theoretical and experimental studies showed that a marked improvement could be obtained if forced convection was substituted for natural convection cooling (see Figs. 7.141b and 7.142).

This raised the difficult question of whether natural convection cooling would nevertheless meet the required operational conditions and, if not, whether the more elaborate and more costly alternative with its additional cooling circuit would be accepted by the customers.

The "minimum risk" decision – that is, to construct the housing in such a way that either cooling system could easily be used – helped the designers gain experience for only a small increase in cost.

Example 2

In the development of a series of high-pressure steam valves operating at temperatures of more than 500°C, the question arose whether the customary method of

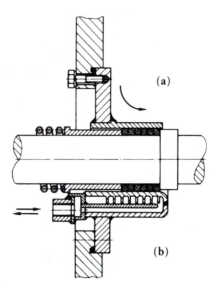

Figure 7.141.   Cooled stuffing box in which the packing revolves with the shaft. The appropriate design of shaft and press ring ensures the internal connection of the packing rings; a very short heat path facilitates good heat extraction:
(a) heat extraction by natural convection currents in the surrounding medium, dependent on the prevailing air flow;
(b) heat extraction by forced convection due to separate cooling air flow ensuring higher flow velocities and increased heat extraction.

nitriding the valve spindles and bushes should be retained despite the fact that the nitrided surface expands with temperature (thereby reducing the radial clearance), or whether very much more expensive stellite hard facing would have to be substituted. When the problem first arose, there was a lack of adequate information about the long-term behaviour of such layers at high temperatures. The "minimum risk" solution adopted was to select the wall thicknesses and the dimensions of the valve spindle and bushes so that, if necessary and without changing the other components, stellite-treated parts could be substituted for the others whenever necessary. As it turned out, the operating temperature range was considerably lower than had been anticipated, so that nitriding provided a satisfactory solution and also helped to identify the operational limits. Once these limits were known, the more expensive solution could be reserved for more demanding conditions.

Example 3

Reliable design calculations for large machine parts, particularly in one-off production, depend on the analytical methods and the postulated constraints.

It is not always possible to predict all characteristics with the necessary degree of accuracy. This applies, for instance, to the determination of the critical whirling speeds of shafts. Often, it is impossible to predict the precise flexibility of the

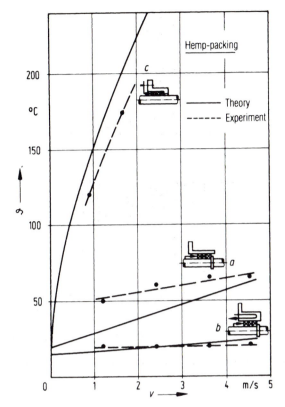

Figure 7.142. Theoretical and experimental temperature determinations at the seal plotted against the peripheral speed on the shaft:
(a) Layout as in Figure 7.141a;
(b) Layout as in Figure 7.141b;
(c) Conventional stuffing box with packing attached to the housing.

bearings and foundations. However, the difference between higher critical whirling speeds in high-speed installations is small within the range of flexibilities normally encountered. In the situation depicted in Fig. 7.143, "minimum risk" design can once again be applied to advantage because the spacing of the bearings, which has a major influence on the critical speed, can be adjusted (see Fig. 7.144). Interposed spring laminations (see Fig. 7.145), moreover, allow alteration of the effective flexibilities. Both measures, taken together or separately, will produce the required effect so that the second or third critical whirling speed can be eliminated from the operating speed range of the machine.

Example 4

Among the many suggestions put forward for a device to wind a strip into a double layered ring, two seemed particularly promising (see Figs. 7.146a and b).
The solution shown in Fig. 7.146a is the simpler and cheaper but also the riskier

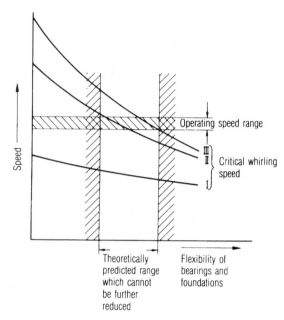

Figure 7.143. Critical whirling speeds (qualitative) for a shaft plotted against the flexibility of bearings and foundations.

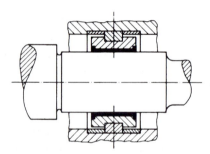

Figure 7.144. Support which, by selecting different spacers, allows the distances between the bearings to be varied.

of the two, because it is not certain whether the inner rotating mandrel *1* alone is invariably able, despite the increased friction produced by the knurling and the pressure of the springs *2*, to move the strip *3* forward.

The solution shown in Fig. 7.146b is less risky, because the pressure rollers attached to the ends of the springs and the feed roller *5*, which moreover can be power-driven, make the advance of the strip more certain. This solution, however, is the more costly of the two, and also more susceptible to wear because of the greater number of moving parts.

The "minimum risk" solution is that shown in Fig. 7.146a, but with a feed-in roller as in Fig. 7.146b, and arranged in such a way that, if need be, it can be driven without alteration of the other parts (see Figure 7.146c).

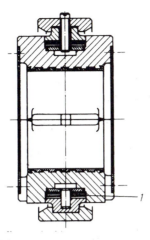

Figure 7.145. Plain bearings with laminated springs *l*, allowing adjustments of flexibility (BBC). (Laminated springs also have good clamping properties thus narrowing the critical range).

This additional element proved essential when the machine was tested, and was readily available.

Example 5

In complex ventilation systems it is often very difficult to precalculate the airflow and pressure losses precisely. An embodiment with "minimum risk" for ventilators might have blades that can be adjusted before they are welded to the disc. When enough experience has been gained, it is possible to substitute a non-adjustable and cheaper cast construction.

All these examples are intended to show that designers should meet risks not simply by considering the first step but also the second or third, which can often be done at relatively small cost. Experience has shown that emergency measures to correct unforeseen faults are many times more costly and time consuming.

7.6 Evaluating Embodiment Designs

In 4.2.2 we discussed the subject of design evaluation. The basic procedures outlined there apply equally well to the conceptual and to the subsequent phases. As embodiment progresses, the evaluation will, of course, rest on more and more concrete objectives and properties.

In the embodiment phase, the technical properties must be evaluated in terms of the *technical rating* R_t and the economic properties separately with the help of the calculated production costs in terms of the *economic rating* R_e. The two ratings can then be compared on a diagram (see Fig. 4.28).

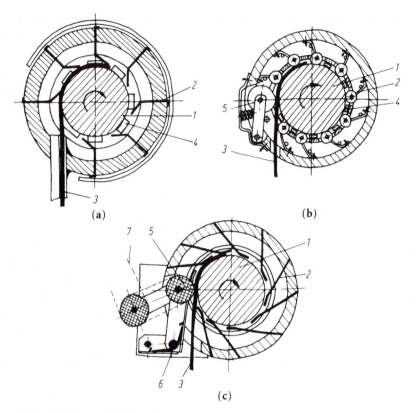

Figure 7.146. (a) Proposed winding device
1 rotating mandrel; *2* pressure springs; *3* strip to be wound; *4* parts of the ejection mechanism
(b) Proposed winding device
1 rotating mandrel; *2* springs with pressure rollers; *3* strip to be wound, *4* parts of ejection mechanism;
5 feed-in roller loaded by spring and possibly driven
(c) Chosen solution
1 rotating mandrel; *2* pressure springs; *3* strip to be wound; *5* feed-in roller tensioned by spring *6* and
driven by belt *7*.

The *prerequisites* of this approach are:

- That all the embodiment designs have the same degree of concreteness, that is, the same information content (for instance, rough designs must only be compared with rough designs). In many cases it suffices, while keeping the overall perspective in mind, to evaluate only those aspects that show marked differences from one another. Once that has been done, their relationship to the whole, of course, must be examined, for example the relationship between part costs and total costs.
- That the production costs (materials, labour and overheads) can be determined (see 10). If a particular solution introduces subsidiary production costs and demands special investment, then, depending on the point of view (producer's

or user's), these factors must be allowed for, if necessary by amortisation. In addition, optimisation can help to achieve a minimisation of production and operating costs.

If the determination of production costs is omitted, then the economic rating can only be evaluated qualitatively, as it was in the conceptual phase. In the embodiment phase, however, costs should, in principle, be determined more concretely (see 10).

As we mentioned in 4.2.2, the first step is to establish the *evaluation criteria*. They are derived from:

- the requirements list:
  - desirable improvement on minimum demands (how far exceeded),
  - wishes (fulfilled, not fulfilled, how well fulfilled).
- the technical properties (to what extent present and fulfilled).

The exhaustiveness of the evaluation criteria is tested against the headings of the checklist (see Fig. 7.147), specially adapted to the level of embodiment attained.

At least one significant evaluation criterion must be considered for each heading, though sometimes more will be needed. A heading may only be ignored if the corresponding properties are absent from, or identical in, all the variants. This approach avoids subjective over-valuations of individual properties. It must be followed by the procedural steps outlined in 4.2.2.

In the embodiment phase, evaluation also constitutes an essential search for weak spots.

Heading	Examples
Function	Fulfilment in accordance with the selected working principle: efficiency, risk, susceptibility to disturbances
Layout design	Space requirements, weight, arrangement, fits, scope for modifications
Form design	Durability, deformation, sealing, operating life, wear, shock resistance, stability, resonance
Safety	Direct safety methods, industrial safety, protection of the environment
Ergonomics	Human–machine relationship, handling, aesthetic considerations
Production	Risk-free methods, setting-up time, heat treatment, surface treatment, tolerances
Quality control	Testing possibilities
Assembly	Unambiguous, easy, comfortable, adjustable, resettable
Transport	Internal and external transportation, means of despatch, packing
Operation	Handling, operational behaviour, corrosion properties, consumption of energy
Recycling	Disassembly, reuse potential, reprocessing potential
Maintenance	Servicing, checking, repair and exchange
Costs	Evaluated separately (economic rating)
Schedules	Production schedule and completion date

Figure 7.147. Checklist for evaluating embodiment designs.

7.7   Example of Embodiment Design

The *conceptual design phase* involves a process that focuses mainly on functions and working structures and results in principle solutions (concepts). Based on the task, designers search for functions and physical effects along with combinations of geometry and materials (working principles) for their realisation. The methods used and the number of iterations in this process depend on the novelty and on the know-how of the designers. The basic solution path, however, is determined largely by the systematic approach along with the logical and physical interrelationships involved. It was therefore possible to propose a largely domain-independent approach in VDI Guidelines 2221 [7.321] and 2222 [7.322], and in chapters 4 and 6 of this book.

In the *embodiment design phase*, however, the emphasis is on determining construction structures of the individual assemblies and parts. The required steps are only described briefly in VDI 2221. In Figs. 3.3 and 7.1, an approach is proposed that is more detailed and has been tested in practice. The variation in approach and individual methods needed to deal with different tasks and problems are greater in embodiment design than in conceptual design. Embodiment design, characterised by a further elaboration of the selected principle solution, requires a more flexible approach, extensive knowledge of the relevant domain and greater experience.

Apart from the basic rules (see 7.3), the methods for embodiment design described in this chapter are mainly the principles (see 7.4) and the guidelines (see 7.5) that should be applied in all appropriate situations. For their application designers must have the ability and experience to recognise the problem and determine a suitable strategy. They are therefore useful as strategic guidelines for identifying and structuring embodiment tasks and problems, and for developing suitable embodiment solutions and finding applicable methods. Moreover they focus and abstract the engineering knowledge required for determining geometry and selecting materials in a directed way.

Explaining embodiment design using examples for different tasks would require too much space. It would also be dangerous because such examples might suggest that the specific approach described is the only correct one. The example used in the rest of this chapter is based on the principle solution discussed in chapter 6. Its only purpose is to show how the embodiment steps of Fig. 7.1 link together.

The embodiment task is the concretisation of the solution principle for the impulse-loading test rig for shaft-hub connections. After the clarification of the task; the elaboration of the requirements list; the identification of the essential problems through abstraction; the establishment of function structures (see Fig. 6.20); the search for working principles (see Fig. 6.14); the combination of working principles into working structures (see Fig. 6.21); the selection of suitable working structures (see Fig. 6.22); their concretisation into principle solution variants (see Figs. 6.27 and 6.28); and the evaluation of these solution variants (see Figs. 6.33, 6.34 and 6.35); designers should follow the steps shown in Fig. 7.1.

1st and 2nd Working Step: Identifying the Requirements that Determine Embodiment Features and Clarifying Spatial Constraints

The following items from the requirements list determined embodiment features:

- Determining layout:
  Test specimen held in position
  Loading applied to a stationary shaft, and only in one direction
  Hubside load take-off variable
  Torque input variable
  No special foundation.
- Determining dimensions:
  Diameter of shaft to be tested $\leq$ 100 mm
  Adjustable torque $T \leq$ 15 000 Nm in five steps (maintained for at least 3 s)
  Adjustable torque increase $dT/dt = 1.25 \cdot 10^3$ Nm/s in 5 steps
  Power consumption $\leq$ 5 kW.
- Determining material:
  Shaft and hub: 45C.
- Other requirements:
  Manufacture of test rig in own workshops
  Bought-out and standard parts wherever possible
  Easy to disassemble.

The requirements list did not contain specific spatial constraints.

3rd Step: Structuring into Main Function Carriers

The basis for this step was function structure variant No 4 (see Fig. 6.20) and the principle solution variant V_2 (see Fig. 6.27). Table 7.8 lists the main function carriers used in the selected solution variant to fulfil the various sub-functions, along with their main characteristics. The most important function carriers for determining the embodiment were:

- test specimen;
- lever between the cylindrical cam and the shaft of the test specimen; and
- cylindrical cam.

The other function carriers were:

- electric motor;
- flywheel;
- clutch;
- gearbox; and
- frame.

Table 7.8. Main function carriers.

Function	Function carrier	Characteristics
Transform energy; increase energy component	Electric motor	Power P_M Speed n_M Run-up time t_M
Store energy	Flywheel	Moment of Inertia J_F Speed n_F
Release energy	Clutch	Torque capacity T_{CL} Maximum speed n_{CL} Response time t_{CL}
Increase energy component	Gearbox	Power P_G Maximum output torque T_G at output speed n_G Gear ratio R_G
Control maginitude and time	Cylindrical cam	Power P_{CAM} Torque T_{CAM} Speed n_{CAM} Diameter D_{CAM} Cam angle α_{CAM} Rise h_{CAM}
Transform energy into torque	Lever	Length l_L Stiffness s_L
Load test connection	Test connection	Torque T Rate of torque increase dT/dt
Take up forces and torque	Frame	

4th Step: Rough Embodiment of the Main Function Carriers

Figure 7.148 shows a rough layout drawing for the three most important function carriers.

The embodiment of the test specimen in line with DIN 6885 and of the transmission lever, modelled and analysed as a cantilever, were relatively straightforward. The development and embodiment of the cylindrical cam, however, required a more detailed kinematic and dynamic analysis based on specific items in the requirements list.

A more precise analysis showed that the initial estimates undertaken in the conceptual phase of the cylindrical cam's performance were insufficient to proceed directly to embodiment. The following analysis therefore had to be carried out.

Figure 7.149 shows that:

Torque on the shaft: $T = s_L \cdot h_{CAM} \cdot l_L$

Torque increase: $dT/dt = \pi \cdot D_{CAM} \cdot n_{CAM} \cdot \tan \alpha_{CAM} \cdot s_L \cdot l_L$

Hold time: $t_L = \dfrac{U_{CAM}}{2\pi \cdot D_{CAM} \cdot n_{CAM}} = \dfrac{1}{2 \cdot n_{CAM}}$

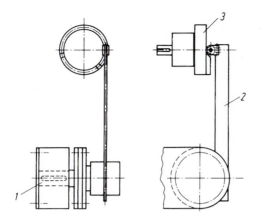

Figure 7.148. Main function carriers that determine the layout. *1* Test connection, *2* Transmission lever, *3* Cylindrical cam.

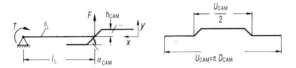

Figure 7.149. Geometric constraints for cylindrical cam and lever. s_L is the stiffness of the lever.

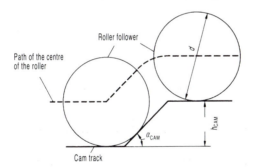

Figure 7.150. Cam path and lever movement.

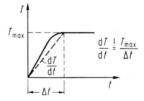

Figure 7.151. Torque increase.

The equation for the torque increase is only valid if the lever movement is parallel to the cam track. In order to minimise friction, a roller follower is required (see Fig. 7.150), so the actual torque increase is lower than calculated and also varies. We therefore used the average increase in our calculations (see Fig 7.151).

If, in line with the requirements list, the average torque increase dT/dt is used, then the calculation of dT/dt should not involve the full circumferential speed v_X, but instead the effective circumferential speed v_X^*, thus:

$$v_X^* = K \cdot v_X$$

The correction K depends on:

- cam angle α_{CAM};
- diameter of the roller follower d ; and
- rise of the cylindrical cam α_{CAM}.

The correction K can be derived from Fig. 7.152:

$$x = \frac{h_{CAM}}{\tan \alpha_{CAM}}$$

and

$$x = d/2 \cdot \left(\sin\alpha_{CAM} - \frac{1 - \cos \alpha_{CAM}}{\tan \alpha_{CAM}} \right)$$

giving

$$K = \frac{v_X^*}{v_X} = \frac{x}{x + \Delta x}$$

The formula is only valid when $d/2 \cdot (1 - \cos \alpha_{CAM}) \leq h_{CAM}$, for example;

$$K = \frac{\dfrac{h_{CAM}}{\tan \alpha_{CAM}}}{\dfrac{h_{CAM}}{\tan \alpha_{CAM}} + d/2 \cdot \left(\sin \alpha_{CAM} - \dfrac{1 - \cos \alpha_{CAM}}{\tan \alpha_{CAM}} \right)}$$

To obtain a value for K, the following estimates are made:

- cam angle $\alpha_{CAM} = 10 \ldots 45°$;
- the diameter of the roller follower $d = 60$ mm; and
- the rise of the cylindrical cam $h_{CAM} = 7.5$ mm and 30 mm respectively.

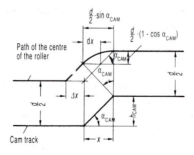

Figure 7.152.   Derivation of correction K.

Table 7.9. Reference values for corrections K

h_{CAM} mm	α_{CAM}	45	40	30	20	10
7.5	K	0.41	0.45	0.62	0.79	0.94
30	K	0.71	0.76	0.87	0.94	0.98

Table 7.9 contains the values of K based on the above formula.

After converting the cylindrical cam speed n_{CAM} and using the calculated correction value, the formula for the torque increase becomes:

$$n_{CAM} = \frac{\dfrac{dT}{dt}}{K \cdot \pi \cdot D_{CAM} \cdot \tan\alpha_{CAM} \cdot s_L \cdot l_L}$$

The speed controller range C

$$C = \frac{n_{CAMmax}}{n_{CAMmin}}$$

is determined as follows.

If the diameter of the cylindrical cam D_{CAM}, the stiffness s_L and the length l_L of the lever are considered constant for this solution concept, the above formula can be used to calculate the extremes of the speed n_{CAM} in relation to the other parameters dT/dt, K and α_{CAM} (see Table 7.10).

B is a constant that includes units and the other constants (π, D_{CAM}, s_L, l_L).

The speed control range C therefore becomes:

$$C = \frac{305 \cdot B}{116 \cdot B} = 2.6$$

This meant that:

- The function "control magnitude and time" could not be fulfilled by the cylindrical cam alone.
- The function structure had to change if we wished to maintain the principles underpinning the concept.
- The cylindrical cam had to have an adjustable drive with a speed control range of approximately $C = 2.6$.

Table 7.10. Determination of $n_{CAM\ min}$ and $n_{CAM\ max}$

	dT/dt	α_{CAM}	K	n_{CAM}
Minimum	20	10	0.98	$116 \cdot B$
Maximum	125	45	0.41	$305 \cdot B$

Figure 7.153 shows the adapted function structure variants (see Fig. 6.20). The sub-function "adjust speed" has been added. This could, for example, have been realised by a continuously adjustable drive motor. Several variants are possible (*4/1* to *4/3*).

The quantitative developments of the cylindrical cam on the basis of the formulas resulted in the following values for the main characteristics: spring stiffness of the lever $s_L = 700$ N/mm; lever length $l_L = 850$ mm; cylinder diameter $D_{CAM} = 300$ mm; cam angle $\alpha_{CAM} = 45 \ldots 10°$; constant $B = 0.107$ min^{-1} (see Table 7.10); speed range for the required rate of torque increase ($dT/dt_{max} = 125 \cdot 10^3$ Nm/s, $dT/dt_{min} = 20 \cdot 10^3$ Nm/s) $n_{CAM} = 12.4 \ldots 32.6$ min^{-1} for a control range $C = 2.6$.

The requirements for the adjustable torque increase dT/dt could thus be realised with the selected values.

This was not the case for the required hold time for the maximum torque. This value was $t_L = 0.5 \cdot n_{CAM} = 2.4 \ldots 0.92$ s, which was lower than the required value of 3 s. After a discussion with the client, the requirement was reduced to $t_L \geqslant 1$ s, which could be realised by using slightly more than half the circumference of the cylindrical cam.

Before a scale layout for the main function carriers that determine the embodiment could be drawn the following issues had to be resolved:

- What should be the spatial layout of the test specimen and the cylindrical cam?
- To what extent should auxiliary function carriers be considered?

It was decided that the test specimen should be positioned horizontally and as a consequence the cylindrical cam should rotate about a vertical axis for the following reasons:

- Easy exchange of test specimen and cylindrical cam
  – Design for ease of assembly.
- Easy access to the test specimen for measurements
  – Design for ease of operation.
- Smooth transmission of the clamping forces of the test specimen into the foundation
  – Short and direct force transmission paths.
- Easy resetting of the test rig for different types of specimen (in particular larger specimens)
  – Design for minimum risk.

The need for auxiliary function carriers was then estimated and the space requirements determined on the basis of experience. The result was, for example:

- That a separate bearing was needed for the cylindrical cam because of the axial force F_A and the tangential force F_T:

$$F_A = F_T = \frac{T_{max}}{l_L} = 17.6 \text{ kN}$$

- The outer diameter of the bolted joint between test specimen and lever should be about 400 mm to provide a torsionally stiff connection.

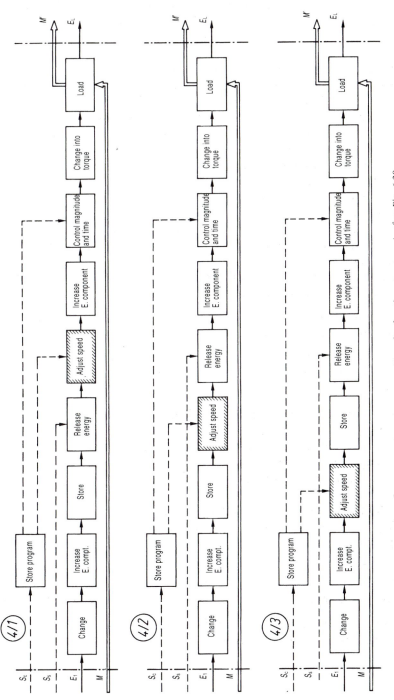

Figure 7.153.   Function structure variants for function structure 4, after Fig. 6.20.

The analysis showed that the auxiliary function carriers have only a marginal influence on the dimensions of the embodiment.

Figure 7.154a shows a rough layout based on function structure variant 4/1 where the speed control is achieved by means of an adjustable mechanism that is located behind the clutch in terms of the energy flow. Figure 7.154b shows a rough layout based on function structure variant 4/2 where the adjustable mechanism is located before the clutch. Variant 4/3 (see Fig. 7.154c) employs an adjustable geared motor.

5th Step: Selecting Suitable Layouts

Variant 4/3 was selected for further detailing because it takes up less space due to the adjustable geared motor (function integration).

6th Step: Rough Layout of the Other Main Function Carriers

The rough layout of the other main function carriers was based on the following requirements identified in step 4:

• Motor drive speed for cylindrical cam

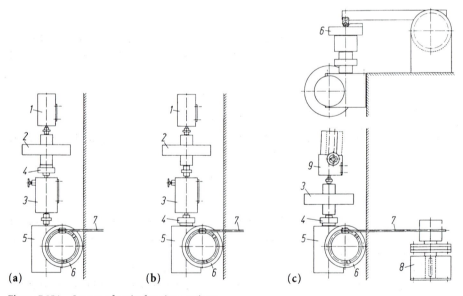

Figure 7.154. Layout of main function carriers:
(a) for function structure variant 4/1;
(b) for function structure variant 4/2;
(c) for function structure variant 4/3;
1 Motor, 2 Flywheel, 3 Adjustable gear, 4 Clutch, 5 Worm gear (angular), 6 Cylindrical cam, 7 Transmission lever, 8 Test connection, 9 Adjustable geared motor.

$$n_{CAM} = 12.4 \ldots 32.6 \text{ min}^{-1}$$

- Speed control range

$$C = 2.6$$

- Driving torque of cylindrical cam

$$T_{CAM} = F_T \cdot D_{CAM} / 2 \text{ and } F_T = F_A = T / l_L \text{ gives } T_{CAM} = 2650 \text{ Nm.}$$

- Driving power of cylindrical cam

$$P_{CAM} = T_{CAM} \cdot \omega_{CAM} \text{ thus } P_{CAM} = 9 \text{ kW}$$

For reasons of safety, the maximum flywheel speed n_F (and therefore also of the motor n_M) was chosen to be:

$$n_F = 1000 \text{ min}^{-1}$$

This required a transmission ratio of:

$$i = 80.7 \ldots 30.7$$

For the other main function carriers the characteristics were estimated as follows:

- Transferred torque of the coupling based on the driving torque of the cylindrical cam $T_{CAM} = 2650$ Nm and the actual transmission ratio i between the cylindrical cam and clutch

$$t_{CL} = T_{CAM} / i$$

- Moment of Inertia of the flywheel from the actual torque T_F taken up by the flywheel, the impact time Δt, the flywheel speed n_F and the allowable speed drop $\Delta n = 5\%$

$$J_F = \frac{T_F \cdot \Delta t}{2 \cdot \pi \cdot n_{CAM} \cdot \Delta n}$$

- The power of the electric motor P_M after calculating the required acceleration torque T_A

$$T_A = \frac{J_F \cdot 2 \cdot \pi \cdot n_M}{t_M} < T_{Amax}$$

from the Moment of Inertia J_F of the flywheel, the motor speed n_M, the run up time $t_M = 10$ s and the maximum acceleration torque of the motor T_{Amax} (from manufacturer's data).

Table 7.11 lists the calculated values for the main characteristics.

Apart from the flywheel, the main function carriers could all be selected from catalogues and bought directly from suppliers.

During the layout phase, the following characteristics were chosen for the flywheel:

$$\text{Speed } n_F = 1010 \text{ min}^{-1}$$
$$\text{Moment of Inertia } J_F = 1.9 \text{ kgm}^2$$

Table 7.11.   Calculated values for the characteristics of the main function carriers of variant 4/3

Function	Function carrier	Calculated values
Change energy Increase E-component Adjust speed	Electric motor with mechanical adjustment- variant 4/3	Power $P_M = 1.1$ kW Speed $n_M = 380...1000$ min^{-1} Speed control range $C = 2.6$
Store energy	Flywheel	Moment of inertia $J_F = 1.4$ kg m^2 Speed $n_F = 380...1000$ min^{-1}
Release energy	El. clutch	Transferred torque $T_{CL} = 86$ Nm
Increase E-component	Gear	Power $P_G = 9$ kW Nominal torque $T_G = 2650$ Nm at speed $n_G = 32.6$ min^{-1} Transmission ratio $i_G = 30.7$

Because losses such as friction had not been taken into account, the final value of J_F was substantially larger.

To save weight, the flywheel is made from a cylinder:

- Outer diameter $D_o = 480$ mm
- Inner diameter $D_I = 410$ mm
- Width $W = 100$ mm
- Mass $m = 38$ kg.

The final rough layout drawing was then produced on the basis of the main function carriers shown in Fig. 7.154c and by adding the frame.

Because the combined height of the lever bearing and the test specimen was much smaller than the combined height of the cylindrical cam and the entire drive system, the spatial constraints for the test rig shown in Fig. 7.155 were selected after a discussion with the client.

Steel channel sections were used for the frame for the following reasons:

- large second Moment of Area for a small cross-sectional area;
- no round corners;
- three flat reference surfaces available; and
- cheap.

Figure 7.156 shows the completed rough layout drawing.

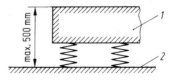

Figure 7.155.   Final spatial constraints. *1* Base plate for fixing the test machine, *2* Foundation.

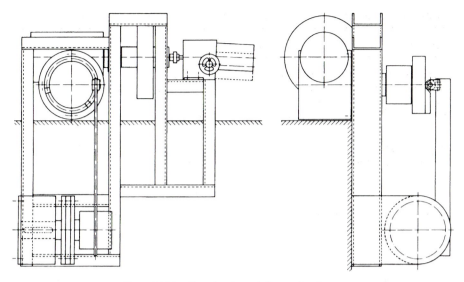

Figure 7.156. Completed rough layout drawing.

7th Step: Searching for Solutions for Auxiliary Functions

Producing a detailed layout drawing involved the following steps:

- searching for and selecting auxiliary function carriers;
- detailing the embodiment of the main function carriers based on the auxiliary function carriers; and
- detailing the embodiment of the auxiliary function carriers.

These steps were much more interrelated than those for the rough layout drawing. They influenced each other because they dealt with more concrete aspects which often required a repetition of previous steps on a higher information level.

The auxiliary function carriers were divided into three groups:

- carriers to connect the main function carriers together;
- carriers to support those main function carriers that move relative to the frame; and
- carriers to permanently connect main function carriers to the frame.

The following solutions were found.

Auxiliary function carriers to connect the main function carriers together were:

- Bolted joint between lever and test specimens; form fit membrane to avoid additional bending moments and to ensure easy assembly.
- Torsionally stiff connection between the worm gear pair and the cylindrical cam. This connection can be of two types, see Fig. 7.157:
  - Worm gear pair with hollow shaft—cylindrical cam
  - Worm gear pair—torsionally stiff connection – cylindrical cam.

The following arguments favour the torsionally stiff connection:

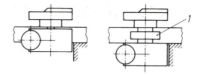

Figure 7.157. Connections between the worn gear pair and the cylindrical cam; *1* coupling.

- Separate assembly of worm gear pair and cylindrical cam possible
  – Design for ease of assembly,
- No interruption of the frame caused by a high shaft position
  – Simplification of embodiment,
- Easy centring of worm gear pair and cylindrical cam
  – Design for ease of production.
- Torsionally flexible connection between the flywheel and the electric motor.

Auxiliary function carriers to support those main function carriers that move relative to the frame were:

- Flywheel support. The requirements were: simple production (ie no accurate balancing needed); direct safety techniques to withstand the dynamic forces (safe life principle); and suspend from the frame. The use of bought-out parts (bearing housing with roller bearings) was not possible because these bearing housings are usually cast and are more suitable for standing rather that suspended applications. Because the flywheel was to be produced in-house, the magnitudes of the dynamic forces were relatively uncertain and so its support needed to be specially designed.
- Support for the cylindrical cam and lever; commercially available rolling element bearings.

Auxiliary function carriers to permanently connect main function carriers to the frame were:

- Simple half-finished products (welded sheet steel) to which the main function carriers were bolted.
- Special solution for connecting the test specimen to the lever (ie the frame). The requirements were: easy to assemble but separable connection; movable in the axial direction; free of play; and no tight tolerances. A Ringfeder connection was chosen.

8th Step: Detailing the Main Function Carriers Taking into Account the Auxiliary Function Carriers

The main function carriers had to be adapted as necessary to match the solutions selected for the auxiliary function carriers.

This resulted in the following:

- electric motor: bought-out part;

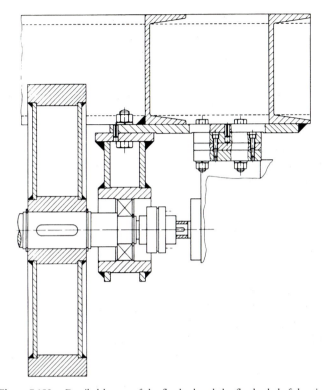

Figure 7.158.   Detailed layout of the flywheel and the flywheel shaft bearing.

- flywheel: see Fig. 7.158;
- clutch: bought-out part;
- gearbox: bought-out part;
- cylindrical cam: see Fig. 7.159;

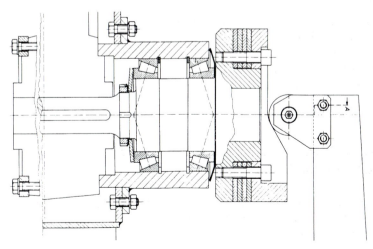

Figure 7.159.   Detailed layout of the bearing arrangement for the clyindrical cam.

- lever: see preliminary layout drawing in Fig. 7.160;
- test specimen: see preliminary layout drawing in Fig. 7.160; and
- frame: modified to suit the geometry of the selected motor.

9th Step: Detailing the Auxiliary Function Carriers and Completing the Preliminary Layout

Flywheel support bearing (as an example):

- Layout (see the guidelines for embodiment design in Fig. 7.3):
  The bearing forces were estimated as follows:

$$F_B = F_{dyn} + F_{stat}$$

With the weight being:

$$F_{stat} = m \cdot g = 400 \text{ N}$$

and the dynamic force being:

$$F_{dyn} = m \cdot e \cdot 4 \cdot \pi^2 \cdot n_F^2$$

and

mass $m = 40$ kg
speed $n_F = 1750$ min^{-1} (= max motor speed)
eccentricity of flywheel $e = 0.6$ mm, based on:
dimensional and shape accuracy of flywheel = 0.3 mm
play in flywheel shaft and bearings = 0.2 mm
unbalanced mass distribution = 0.1 mm,

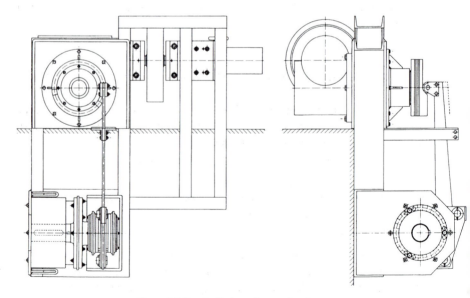

Figure 7.160. Preliminary layout drawing.

the bearing force is:

$$F_B = 1130 \text{ N}$$

This implies that even when additional gyroscopic forces occur, the bearing (dynamic capacity 65 000 N) and all the other parts that are in the force transmission path have adequate dimensions.

- Resonance:

  The embodiment of bearing and frame was very rigid so that resonance excited by the flywheel (maximum 30 Hz) was unlikely.

- Production:

  The embodiment allowed easy production because the flywheel support did not require tight tolerances for the frame.

- Assembly:

  The support for the flywheel could be assembled easily because of:
  - a simple bottom up approach:
  - easy accessibility to the connecting screws; and
  - simple adjustment of the clutch using a spacer after accurate location of the flywheel bearing support using dowel pins (possible without the flywheel).

- Maintenance:

  Maintenance free bearings were used.

Figure 7.160 shows the preliminary layout drawing of the test rig resulting from the embodiment steps discussed above.

10th Step: Evaluating Using Technical and Economic Criteria

Because only one embodiment was developed, no selection was involved but simply an assessment of the final embodiment based on criteria derived from the requirements list. The objective was to identify and improve weak spots.

The procedure involved the following steps in accordance with 4.2.2:

- identifying evaluation criteria;
- assessing whether the parameters meet the evaluation criteria;
- determining the overall rating; and
- searching for weak spots.

For the identified weak spots, improvements were proposed.

In this case we used 11 of the 13 criteria that were used for evaluating the concepts, see Fig. 7.161. The use of weightings was considered unnecessary.

The expected and calculated parameters of the test rig were evaluated against an ideal solution using a value range of 0 . . . 4 in line with VDI 2225. A more detailed evaluation did not seem beneficial. The result is shown in Fig. 7.161.

For the calculation of the overall rating, only the technical rating was used because there were no data for a formal assessment of the economic rating:

$$R = 29/44 = 0.66$$

This rating is rather low so a search for weak spots seemed useful.

First those parameters that had the lowest values were identified. A proposal was then made to improve those parameters that only received one or two marks:

- Few possible operator errors.
  Weak spot: motor speed:
  – The speed could be set at a value higher than necessary for the maximum rate of torque increase.
  – The run-up of the motor should only take place slowly because of the heat generated.
  Remedy: the allowed range for run-up and operation can be marked on the speed indicator of the motor. The machine can be shut down automatically if the speed becomes too high.
- Easy to change the load profile.
  Weak spot: exchange of the cylindrical cam was not possible because of the clamping pressure of the lever on the cam.
  Remedy: provide a means to lift the lever.
- High level of safety.
  Weak spot: rotating cylindrical cam was not protected.
  Remedy: provide protective cover.
- Quick exchange of test specimens (test connections).
  Weak spot: slow because of the number of screws in the Ringfeder connection.
  Remedy: no economic alternative possible.

The improvements have been added to the evaluation chart (see Fig. 7.161).

The remaining working steps to *define the overall layout* proposed in Fig. 7.1 are not discussed here. They were not very complex in the case of this test rig because it was a one-off product for a research institute and did not need a high degree of optimisation.

The *detail design* of the test rig following the working steps in 7.8 is also not discussed. It only involved conventional drawing and detail design steps.

Figure 7.162 shows the final impulse-loading test rig. It fulfilled the main expectations and confirmed the effectiveness of a systematic approach [7.125].

7.8 Detail Design

Detail design is that part of the design process which completes the embodiment of technical products with final instructions about the layout, forms, dimensions and surface properties of all individual components, the definitive selection of materials and a final scrutiny of the production methods, operating procedures and costs.

No.	Evaluation criteria	Wt	Parameters	Unit.	Variant 4/3 Magn	Value	Weighted value	Variant 4/3 mod. Magn	Value	Weighted value
1	Good reproducibility		Disturbing factors	–	low	4				
2				–						
3				–						
4	Tolerance of overloading		Overload reserve	%.	10	3				
5	High level of safety		Danger of injury	–	average	2		see text	4	
6	Few possible operator errors		Possibilities of operator errors	–	high	1		see text	3	
7	Small number of components		No. of components	–	low	3				
8	Low complexity of components		Complexity of components	–	low	3				
9	Many standard and bought-out parts		Proportion of standard and bought-out components	–	high	4				
10	Simple assembly		Simplicity of assembly	–	high	3				
11	Easy change of load profile		Change of load profile	–	bad	1		see text	2	
12	Quick exchange of test connections		Estimated time needed to exchange test connections	–	average	2		see text	2	
13	Good accessibility of measuring systems		Accessibility of measuring systems	–	good	3				
		$\Sigma Wi = 1.0$			$OV_1 = 29$ $R_1 = 0.66$			$OV_2 = 34$ $R_2 = 0.77$		

Figure 7.161. Extract from evaluation chart for embodiment based on Fig. 7.160.

Figure 7.162. Final impulse-loading test rig, after [7.200].

Another, and perhaps the most important, aspect of the detail design phase is the elaboration of production documents and especially of detailed component drawings (including workshop drawings), of assembly drawings and of appropriate parts lists. These activities are increasingly supported and, in some cases automated, by CAD software. This allows the direct use of product data for production planning and the control of CNC machine tools.

Depending on the type of product and production schedule (one-off, small batch, mass production), the design department must also provide the production department with assembly instructions, transport documentation and quality control measures (see 9), and the user with operating, maintenance and repair manuals. The documents drawn up at this stage are the basis for executing orders and for production scheduling, that is, for operations planning and control. In practice, the respective contributions of the design and production departments in this area may not be distinct.

The detail design phase involves the following steps (see Fig. 7.163).

Finalising the definitive layout, comprising the detailed drawing of components, and the detailed optimisation of shapes, materials, surfaces, tolerances and fits. To that end, designers should refer to the guidelines given in 7.5. Optimisation aims at maximum utilisation of the most suitable materials (uniform strength), at cost-effectiveness and at ease of production, due heed being paid to standards (including the use of standard parts and company repeat parts).

The *integration* of individual components into assemblies and through these into the overall product (fully documented with the help of drawings, parts lists and

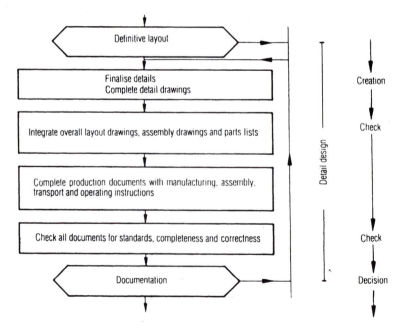

Figure 7.163. Steps of detail design.

numbering systems) is strongly influenced by production scheduling, delivery dates, and assembly and transport considerations.

The *completion* of production documents with production, assembly, transport and operating instructions is another crucial aspect of the detail design phase.

Equally important is the *checking* of all documents and especially of detail drawings and parts lists for:

- observance of general and in-house standards;
- accuracy of dimensions and tolerances;
- other essential production data; and
- ease of acquisition, for instance, the availability of standard parts.

Whether such checks are made by the design department itself or by a separate standards department will depend largely on the organisational structure of the company concerned, and plays a subordinate role in the actual execution of the task. In the same way as the steps of the conceptual and embodiment phases often overlap, so do the steps of the embodiment and detail design phases. Long lead-time parts, such as those involving forging and casting, should be dealt with first and their detail design and production instructions are often completed even before the definitive layout has been finalised. This overlapping of two design phases is particularly common in one-off production and in heavy engineering (see Fig. 7.106) and can be supported by CAD systems.

Detail design is very domain and product dependent and designers should refer to the many technical handbooks, suppliers catalogues and standards that deal with the detail design and selection of machine elements.

Corners must never be cut during the detail design phase, which has a critical effect on the technical function, on the production processes and on the elimination of production errors. Detail design has a major influence on production costs and product quality – and hence the success of a product in the market.

8 Developing Size Ranges and Modular Products

8.1 Size Ranges

Size ranges provide a rationalisation of design and production procedures [8.35]. For the *manufacturer* they have the following *advantages*:

- The design work can be done once and for all and can be used for a host of applications.
- The production of selected sizes can be repeated in batches and hence becomes more cost-effective.
- Higher quality is possible.

This implies the following *advantages* for the *user*:

- competitive and high quality products;
- short delivery times; and
- easy acquisition of replacement parts and fittings.

Disadvantages for both manufacturer and user are:

- limited choice of sizes, not always with optimum operational properties.

By size range we refer to technical artefacts (machines, assemblies or components) for a wide sphere of applications that:

- fulfil the same function;
- are based on the same solution principle;
- are made in varying sizes; and
- involve similar production processes.

If, in addition to the range of sizes, other associated functions have to be implemented, then *modular products* (see 8.2.2) will have to be developed side by side with size ranges. The development of size ranges may be original or based on an existing product but must, in either case, be carefully graded. We refer to the initial size as the *basic design* and to the derived sizes as *sequential designs* [8.35].

In the development of a size range it is essential to make use of similarity laws and helpful to make use of decimal-geometric preferred number series. These generic tools are discussed in the next two sections.

8.1.1   Similarity Laws

Geometric similarity ensures simplicity and clarity of design. Designers know, however, that technical artefacts stepped up in geometric proportions (the so-called pantograph constructions) are not satisfactory except in very rare cases. In particular, purely geometrical magnification is only permissible when similarity laws permit, which should always be checked. These laws are used very successfully in model testing [8.12, 8.20, 8.32, 8.34, 8.39, 8.43]. It is obvious to transfer this procedure to the development of size ranges. In general, however, the development of size ranges has a different objective from model technology, namely to achieve:

- the same level of material utilisation;
- with similar materials if possible; and
- with the same technology.

It follows that, if the function is to be fulfilled equally well throughout the range, the relative stresses must remain the same.

We speak of *similarity* if the relationship of at least one physical quantity in the basic and sequential designs is constant. It is possible to define basic similarities with the help of the fundamental quantities length, time, force, quantity of electricity (charge), temperature and luminous intensity (see Table 8.1).[1]

Thus we have *geometric similarity* if the ratio of all the lengths of any sequential design to all the lengths of the basic design is constant. Here, the non-dimensional parameter to be held constant is $\varphi_L = L_1/L_0$ where L_1 is any length of the first member of the size range (sequential design); and L_0 the corresponding length of the basic design. For the k-th sequential design $\varphi_{L_k} = \varphi_L^k$. In the same way, we can describe similarities in time, force, electricity, temperature and luminous intensity.

Table 8.1.   Basic similarities

Similarity	Basic quantity	Invariants
Geometric	Length	$\varphi_L = L_1/L_0$
Temporal	Time	$\varphi_t = t_1/t_0$
Force	Force	$\varphi_F = F_1/F_0$
Electrical	Charge	$\varphi_Q = Q_1/Q_0$
Thermal	Temperature	$\varphi_\vartheta = \vartheta_1/\vartheta_0$
Photometric	Luminous intensity	$\varphi_J = J_1/J_0$

---

[1] Fundamental physical quantities are as listed in the German text. The basic physical quantities selected for the SI system differ slightly and, along with their basic units shown in brackets, are: length (metre); time (second); mass (kilogram); electric current (ampere); thermodynamic temperature (kelvin); and luminous intensity (candela). The differences do not affect the principles described.

If two or more of the basic quantities are in constant proportion, then we have special similarities. Now, model technology has defined dimensionless parameters for important and recurring similarities. Thus, in the case of simultaneous invariance of length and time, we have *kinematic similarity*, and in the case of simultaneous invariance of length and force we speak of *static similarity*.

A very important similarity, namely *dynamic similarity*, appears when a constant force relationship is combined with geometric and temporal similarities. Depending on the forces involved, we arrive at different dimensionless parameters. *Thermal similarity* deserves special mention because, in the case of geometrically similar size ranges and the same utilisation of materials, it cannot be squared with dynamic similarity [8.37].

Table 8.2 lists important similarity relationships in the development of size ranges for mechanical systems. They are by no means exhaustive and must be supplemented from case to case, for instance in bearing developments by Sommerfeld's number and in hydraulic machines by the cavitation number and pressure index.

Similarity at Constant Stress

In heavy engineering systems, inertia forces (forces due to mass, acceleration etc) and elastic forces resulting from the stress-strain relationship play a predominant role.

Table 8.2.   Special similarity relationships

Similarity	Invariants	Group name	Definition	Description
Kinematic	φ_L, φ_t			
Static	φ_L, φ_F	Hooke	$Ho = \dfrac{F}{E \cdot L^2}$	Relative elastic force
Dynamic	φ_L, φ_t, φ_F	Newton	$Ne = \dfrac{F}{\rho \cdot v^2 \cdot L^2}$	Relative inertia
		Cauchy*	$Ca = \dfrac{Ho}{Ne} = \dfrac{\rho \cdot v^2}{E}$	Inertia force/elastic force
		Froude	$Fr = \dfrac{v^2}{g \cdot L}$	Inertia force/gravitational force
		NN**	$\dfrac{E}{\rho \cdot g \cdot L}$	Elastic force/gravitational force
		Reynolds	$Re = \dfrac{L \cdot v \cdot \rho}{\eta}$	Inertia force/frictional force in liquids and gases
Thermal	φ_L, φ_ϑ	Biot	$Bi = \dfrac{h \cdot L}{\lambda}$	Supplied or removed/conducted quantity of heat
	φ_L, φ_t, φ_ϑ	Fourier	$Fo = \dfrac{\lambda \cdot t}{c \cdot \rho \cdot L^2}$	Conducted/stored quantity of heat

* In some texts, we find $Ca = v \cdot \sqrt{\rho/E}$. This is appropriate if Ca is intended as a velocity ratio relationship.
** Not named.

If the stresses are to remain constant throughout a size range, then $\sigma = \varepsilon \cdot E =$ constant.

In that case the stress parameter becomes:

$$\varphi_\sigma = \frac{\sigma_1}{\sigma_0} = \frac{\varepsilon_1}{\varepsilon_0} \frac{E_1}{E_0} = 1$$

With the same material, that is at $\varphi_E = E_1/E_0 = 1$, we need:

$$\varphi_\varepsilon = \varepsilon_1/\varepsilon_0 = 1 \text{ , or } \varphi_\varepsilon = \frac{\Delta L_1}{\Delta L_0} \frac{L_0}{L_1} = 1 \text{ , or } \varphi_{\Delta L} = \varphi_L$$

With this so-called Cauchy condition, all changes in length must increase in the same ratio as the appropriate lengths (geometric similarity). The elastic force parameter then becomes:

$$\varphi_{FE} = \frac{\sigma_1 A_1}{\sigma_0 A_0} = \varphi_L^2 \text{ , with } \varphi_0 = \varphi_\varepsilon \cdot \varphi_E = 1 \text{ and } \varphi_A = \varphi_L^2$$

The inertia force parameter is:

$$\varphi_{FI} = \frac{m_1 a_1}{m_0 a_0} = \frac{\rho_1 V_1 a_1}{\rho_0 V_0 a_0}$$

With

$$\varphi_\rho = \rho_1/\rho_0 = 1 \text{ , } \varphi_v = V_1/V_0 = L_1^3/L_0^3 = \varphi_L^3$$

and

$$\varphi_a = \frac{L_1 t_0^2}{t_1^2 L_0} = \frac{\varphi_L}{\varphi_t^2}$$

we have

$$\varphi_{FI} = \varphi_L^4/\varphi_t^2$$

A dynamic similarity, that is a constant ratio between inertia and elastic forces with geometric similarity, can only be attained if $\varphi_t = \varphi_L$:

$$\varphi_{FE} = \varphi_L^2 = \varphi_{FI} = \varphi_L^4/\varphi_L^2 = \varphi_L^2$$

Hence the velocity ratio becomes:

$$\varphi_v = \varphi_L/\varphi_t = \varphi_L/\varphi_L = 1$$

With the same material, the same result can also be derived from the Cauchy number (see Table 8.2), for when ρ and E remain constant then the dynamic similarity will only remain constant if the velocity v also remains constant.

For all important quantities such as power, torque etc, and with $\varphi_L = \varphi_t =$ constant and $\varphi_\rho = \varphi_E = \varphi_\sigma = \varphi_v = 1$, it is now possible to establish the similarity relationships shown in Table 8.3.

It should be remembered that the utilisation of the materials and the safety level

Table 8.3. Similarity relationships for geometrical similarity and equal stresses: dependence of important quantities on length

With $Ca = \rho v^2/E =$ constant and the same material, that is $\rho = E =$ const., $v =$ const. In the case of geometrical similarity the following relationships occur

Speeds, n, ω Bending and torsional critical speeds n_{cr}, ω_{cr}	φ_L^{-1}
Strains ε, stresses σ, surface pressures p due to inertia and elastic forces, speeds v	φ_L^0
Spring stiffnesses s, elastic deformations ΔL Strains ε, stresses σ, surface pressures p due to gravity	φ_L^1
Forces F Powers P	φ_L^2
Weights W, torques T, torsion stiffnesses s_t Moments, M, M_t	φ_L^3
Second moments of area I, J	φ_L^4
Mass moments of inertia I', J'	φ_L^5

Note: The utilisation of the materials and safety level are only constant if the influence of the dimensions on the material properties can be ignored.

only remain constant if the influence of the dimensions on the material properties can be ignored throughout the size range.

Size ranges developed in accordance with these laws are geometrically similar and provide for the identical utilisation of the materials. Such developments are possible whenever gravity and temperature have no decisive influence on the design. If they have, the use of semi-similar series is advisable (see 8.1.5).

8.1.2 Decimal-Geometric Preferred Number Series

Once we are familiar with the most important similarity relationships, we still have to determine the best method of choosing the individual steps of a size range. Kienzle [8.26, 8.27] and Berg [8.5 to 8.9] have argued that a decimal-geometric series is the most useful.

A *decimal-geometric series* is based on multiplication by a constant factor φ and is developed within one decade. The constant factor φ determines the step sizes of the series and can be expressed as:

$$\varphi = \sqrt[n]{a_n/a_0} = \sqrt[n]{10}$$

where n is the number of steps within a decade. For 10 steps, the series would then have a factor:

$$\varphi = \sqrt[10]{10} = 1.25$$

and is called R 10. The number of terms in the series is $z = n + 1$.

Table 8.4 sets out the main values of four preferred number series [8.13].

The need for geometric scaling is often found in daily life and in technical practice. The resulting series conform with the Weber-Fechner law which states that the physiological sensation produced by a stimulus is proportional to the logarithm of the stimulus, eg sound and illumination.

Reuthe [8.40] has shown how, in the development of friction drives, designers instinctively choose the main dimensions by means of geometrical scaling. Our own work on turbine shaft oil scraper rings has confirmed these findings. In Fig. 8.1, shaft diameters are plotted using a logarithmic scale against the number of newly designed oil scraper rings (or rings on order) over a period of 10 years. The results show that there were 47 diameters with peaks at more or less regular intervals, which clearly demonstrates a geometrical scaling. However, the number of nominal sizes was disturbingly large – some differed by only a few millimetres and gave rise to very small production batches. Luckily, as Fig. 8.1 also shows, if preferred sizes are selected with the help of the R 20 series, the number of variants can be reduced to less than half, giving a considerably more balanced and higher requirement per nominal size. Had the designers chosen such preferred numbers deliberately, a much more suitable size range would have emerged by itself.

The use of preferred number series thus provides the following *advantages* [8.13]:

• Appropriate scaling leads to the selection of nominal sizes in accordance with

Table 8.4.   Main values of preferred numbers

Basic series				Basic series			
R 5	R 10	R 20	R 40	R 5	R 10	R 20	R 40
1.00	1.00	1.00	1.00		3.15	3.15	3.15
			1.06				3.35
		1.12	1.12			3.55	3.55
			1.18				3.75
	1.25	1.25	1.25	4.00	4.00	4.00	4.00
			1.32				4.25
		1.40	1.40			4.50	4.50
			1.50				4.75
1.60	1.60	1.60	1.60		5.00	5.00	5.00
			1.70				5.30
		1.80	1.80			5.60	5.60
			1.90				6.00
	2.00	2.00	2.00	6.30	6.30	6.30	6.30
			2.12				6.70
		2.24	2.24			7.10	7.10
			2.36				7.50
2.50	2.50	2.50	2.50		8.00	8.00	8.00
			2.65				8.50
		2.80	2.80			9.00	9.00
			3.00				9.50

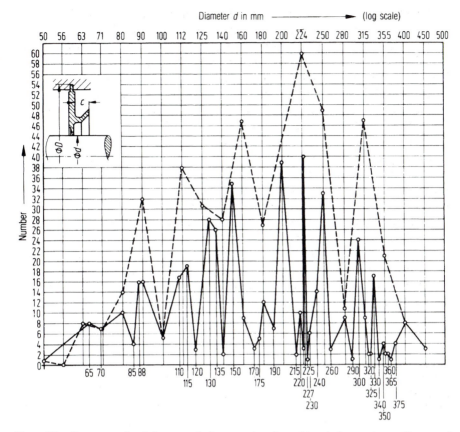

Figure 8.1.   Frequency of seal diameters d of scraper rings for turbine shafts; continuous line: actual situation; broken line: suggested size range.

demand. The finer series have common numerical values with the coarser. With proper gradation it is possible to approximate an arithmetical series. This facilitates jumping from row to row and hence provides the different steps needed for matching the distribution of the market requirement. The preferred number series contain both decimal powers and also doubles and halves.

- There is a reduction of the dimensional variants by the choice of dimensions based on preferred numbers with a consequent saving in production instructions, equipment and measuring tools.

- Since the products and quotients of terms of the series are in turn terms of a geometrical series, analyses and calculations reduce mainly to multiplication and division. As π is contained in the preferred number series with a good approximation, geometric gradation of component diameters will generate circumferences, circular areas, cylinder contents and spherical surfaces that are, in their turn, terms of the preferred number series.

- If the dimensions of a component or of a machine are terms of a geometrical series, then linear magnifications or diminutions will give rise to preferred

numbers in the same series provided, of course, that the magnification or diminution factor is also selected from the series.
- Automatic growth of the size range will be compatible with existing or future ranges.

8.1.3 Selection of Step Sizes

In general, when trying to rationalise a product size range, designers will select their increments once and for all. To that end they make an *appropriate selection of step sizes*, for instance in respect of power and torque. That selection can be based on several considerations. First of these is the market situation, which as a rule requires small increments so that the varied demands of customers can be met most effectively. The second consideration is efficient design and production. For technical and economic reasons, the selected step sizes must be fine enough to meet the technical demands (for instance, power), and yet coarse enough to allow large-batch production based on a simplified range. The selection of optimum step sizes thus involves an integrated approach to the "market – design – production – sales" system, and requires information about:

- market expectations (sales) in respect of individual sizes;
- market behaviour in respect of simplified ranges and the resulting gaps;
- production costs and times of the various step sizes (see 10.3.4 and [8.36]) and the effect on the overall production costs; and
- properties of each product in the size range.

Since the optimum selection of step sizes must be based on all the factors we have mentioned, it is not always possible to opt for a constant step factor; more often technical and economic considerations will demand the breakup of a particular range of sizes into several sets.

If we define a *characteristic number N* of a range such that:

$$N = \frac{\text{Greatest term of the range}}{\text{Smallest term of the range}} = \varphi^n$$

where n is the number of the steps in any particular range and $z = n + 1$ is the number of terms, then the factor:

$$\varphi = \sqrt[n]{N}$$

The range can be split up by means of a *constant* or a *variable factor*, that is, by steps within and between coarser or finer preferred number series (R 5 – R 40). The resulting step characteristics are shown in Fig. 8.2.

Type A has a constant factor (for instance $\varphi = 1.25$ corresponding to R 10) over the entire range.

In Type B, the lower part of the range is divided up coarsely (for instance $\varphi = 1.6$ corresponding to R 5) and the upper part more finely (for instance $\varphi = 1.25$ corresponding to R 10). Such degressive geometrical product ranges should be

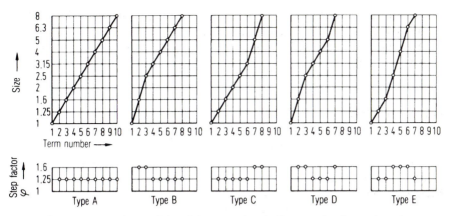

Figure 8.2. Step characteristics of size ranges [8.16]; (factors assigned to each step).

used whenever a coarser grading for the smaller product sizes is economically justifiable. If a degressive grading is used, it should always be based on the combination of several preferred number series with decreasing step factors and not on a constantly decreasing series since, on the one hand, what adaptations may be needed can be made accurate enough and, on the other hand, adherence to the preferred number series is advantageous for the reasons we have already set out.

Type C has a greater increment in the upper range and is used if demand is concentrated on smaller sizes. It is also known as a progressive geometrical range.

Type D has a smaller, and Type E a larger, step factor in the middle part of the range.

For simplicity, we can generally take it that the size gradation must be the finer the greater the demand and the more precisely certain technical stipulations have to be met. A different gradation can be chosen whenever the market demands and without great design effort. Needless to say, the effects on production must be taken into account as well.

In grading, a distinction must be made between *independent* and *dependent quantities*. As a rule, the task itself determines which sizes must be treated as dependent and which as independent. For example, geometric grading of the power output may be advantageous for market reasons and grading of sizes by preferred number series for production reasons. If the two are associated by a power law (see Fig. 8.3, curve *a*) then both can be graded by a preferred number series, either with exponent $p = 1$ (linear growth) or with $p \neq 1$ (non-linear growth). In Fig. 8.3, the dependent and independent quantities have been plotted logarithmically. If the preferred numbers have the same factor, then the spacing is constant (see Fig. 8.4).

However, technical systems may not involve power relationships between dependent and independent quantities. In that case, not all the sizes can be geometrically graded. Here designers must decide, depending on the task, whether they grade the independent or the dependent quantities in accordance with a preferred number series.

For economic reasons, it is often advisable to split the range into parts and to

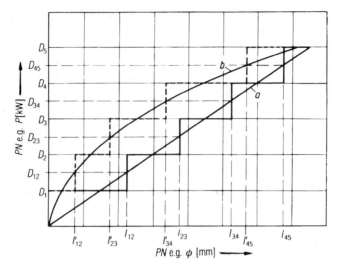

Figure 8.3. Grading of independent (I) and dependent (D) quantities. For a power function there is a linear relationship on the PN (preferred number) diagram (curve a); for others there is a non-linear relationship.

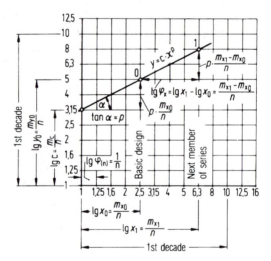

Figure 8.4. Technical relationships in the PN diagram; n step number in the finest underlying PN series; every intersection is a preferred number of this series; every integral exponent leads back to another preferred number.

replace several sizes with just one for each part (semi-similar size range), but care must be taken to ensure that the stepped line needed for such gradation is roughly equivalent to the continuous curve. Figure 8.3 shows the resulting stepped line for size relationships based on a power function (curve a) and for non-linear relationships which are not governed by that function (curve b).

Independent sizes I_{12}, I_{23} etc have been assigned to the geometrically graded part of the ranges $D_1 D_2$, $D_2 D_3$ etc. This correlation is obtained by replacing the $D_1 D_2$, $D_2 D_3$ etc with their geometric mean values $D_{12} = \sqrt{D_1 \cdot D_2}$ and then drawing the stepped line accordingly. This is preferable to fixing the line intuitively, which is done far too often. It will be seen that the dependent relationships based on curve *a* once again result in a geometric grading of the steps, while the non-linear relationships based on curve *b* do not (in other words, the I'-values are not geometrically graded). Here designers must again decide for what sizes a gradation based on preferred numbers is still appropriate.

Further deviations from strictly geometric gradings may, as we have already said, be imposed by production considerations. Practice has shown that it may be more economical to provide arithmetic or even irregular increments for some component dimensions, so that, in a product size range, semi-finished materials, which are not usually geometrically graded, can be exploited more fully, or the production process can be simplified (see 8.1.5). Even though grading based on preferred number series is generally advisable, designers should not use it rigidly, but decide each case individually after cost analysis (see 10.3.4).

Deviation from geometric grading will also occur if certain dimensions only have to be stepped, while others have to be adapted to specific customer demands. This is called a *sliding arrangement* [8.16], and may prove most effective when the special dimensioning involved does not lead to a significant increase in production costs. Thus for the ball valve series described in [8.16] the dimensions of the housing, drive shafts and bearings are firmly stepped, whereas the plugs and sealing rings have been given "sliding" dimensions for hydrodynamic reasons. A similar approach is used in the design of turbines and thermal equipment [8.28].

8.1.4 Geometrically Similar Size Ranges

If the basic design, the choice of materials and the necessary calculations are to hand, and if the nominal dimensions lie roughly in the middle of the intended size range, then nearly all technical relationships can be expressed in the general form:

$$y = cx^p \text{ or } \log y = \log c + p \log x$$

Every preferred number (PN) can be expressed by $PN = 10^{m/n}$ or:

$$\log (PN) = m/n$$

where m is the step number in the PN range and n the number of steps within one decade. Hence the technical relationship can also be expressed by:

$$\frac{m_y}{n} = \frac{m_c}{n} + p \frac{m_x}{n}$$

The basic design is assigned the index 0, the first successive member of the size range (sequential design) the index 1, the k-th member the index k.

All relationships can now be represented by straight lines in a double logarithmic graph, the slope of each line corresponding to the exponent p of the technical relationship (dependence) (see Fig. 8.4). For simplicity, we enter the preferred numbers instead of the logarithms and so obtain a very practicable visual tool for the development of size ranges, as Berg [8.7, 8.9] has pointed out. Every intersection represents a preferred number, and is always produced by lines with integral exponents. If the abscissa gives the nominal size x, then the factor $\varphi_x = x_1/x_0$. In geometrically similar size ranges it is equal to the length factor φ_L. Once the basic design has been fixed, all other magnitudes – dimensions, torques, power, speeds etc – can be derived from the known exponents of their physical or technical relationships (see Table 8.3) and can be drawn as straight lines with the appropriate slope (thus weight, $\varphi_W = \varphi_L^3$, will have a slope of 3:1).

As a result, the main dimensions of the product can be expressed in diagram form without the need for further drawings (see Figs 8.5 and 8.6).

Such data sheets enable designers, starting from the basic design, to provide the sales department, the purchasing department, the planning department and the production department with crucial information on every size in the range.

It should, however, be remembered that the measurements cannot be transferred directly from the data sheets to the drawings (which need only be made once an order has been received) unless the following factors have also been taken into consideration:

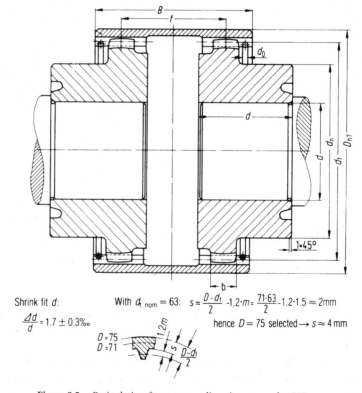

Shrink fit d:

$\dfrac{\Delta d}{d} = 1.7 \pm 0.3\%$

With $d_{t\,nom} = 63$: $s \approx \dfrac{D - d_t}{2} - 1.2 \cdot m = \dfrac{71 - 63}{2} - 1.2 \cdot 1.5 \approx 2\,\text{mm}$

hence $D = 75$ selected $\rightarrow s \approx 4\,\text{mm}$

$D = 75$
$D = 71$

Figure 8.5. Basic design for gear coupling size range, $d_t = 200$ mm.

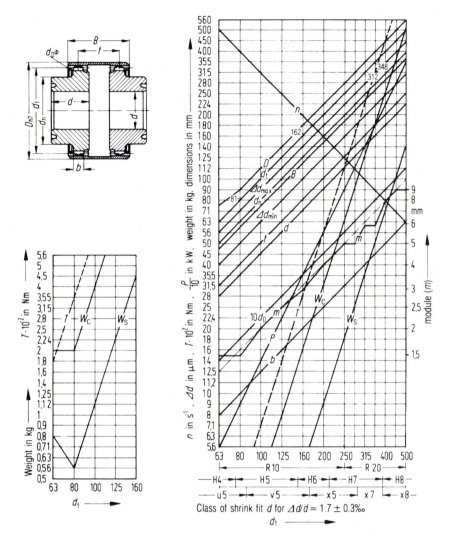

Figure 8.6. Data sheet for the gear coupling size range in the nominal diameter range d_t corresponding to the basic design shown in Fig. 8.5. Dimensions geometrically similar. Exceptions: outer sleeve diameter D of the smallest member for reasons of stiffness; standard modules m are not stepped in accordance with preferred numbers; special adaptation of pitch circle diameter because of the demand for an integral, even number of teeth. The class of shrink fit is shown under the abcissa.

1. *Fits and tolerances* are not in geometric step with the nominal sizes, the size of a tolerance unit i for a dimension D being given by $i = 0.45 \times \sqrt[3]{D} + 0.001D$, that is, the factor for the tolerance unit being determined by the relationship $\varphi_i = \varphi_L^{1/3}$.

   Particularly in the case of shrink and interference fits but also of function-determined bearing clearances etc, the tolerances must, because the elastic deformations tend towards φ_L, be adapted accordingly. In other words, smaller dimensions make more, and larger dimensions less, severe demands (see Fig. 8.6).

2. *Technological limitations* often demand deviations. Thus a cast wall cannot be reduced below a minimum thickness, and certain thicknesses cannot be completely hardened by quenching. In all such cases, the limiting dimensions must be ascertained, as was done, for instance, with the smallest sleeve for the gear coupling shown in Fig. 8.6, which had to be strengthened by an increase in the wall thickness ($D = 71$ mm to $D = 75$ mm). The same principle applies to measurement and machining provisions.

3. *Overriding standards* are not always based on preferred numbers, so the relevant components must be adapted accordingly (see Fig. 8.6, fixing the module).

4. *Overriding similarity laws* or *other requirements* may impose a more pronounced deviation from geometric similarity, in which case semi-similar series should be used (see 8.1.5).

Once the necessary deviations from geometric similarity have been determined, if necessary by checking drawings of the critical areas, they are entered in the data sheet. Production documents need not be prepared until actually needed. To illustrate the size range, say in catalogues or advertisements, displays of the type previously reserved for technical drawings have come into increasing use [8.7, 8.27]. Figure 8.7 shows an example based on a gearbox size range.

Figure 8.8 shows the basic design of a geometrical range of torque-limiters, providing for equal utilisation of materials. If the lining wears, the drop in torque must be kept as small as possible. This is done by means of a large number of peripheral coil springs with relatively flat characteristic curves. All sizes of the torque-limiter fulfil the similarity conditions mentioned in Table 8.3: relationships

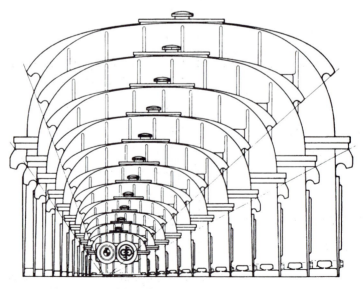

Figure 8.7.   Display of a gearbox size range [8.15] (Flender).

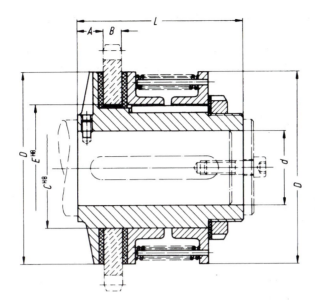

Figure 8.8. Basic design of a torque-limiter (Ringspann KG).

between forces are kept constant over the entire range and the utilisation of the materials is constant.

Figures 8.9a and b are the relevant data sheets. The identifiable deviation of dimension B is determined by the overriding standard width of the chain wheels (bought-out parts); the deviation of A by the use of standard screws and taps and also by technological factors (wall thickness). Figures 8.10a and b show the smallest and largest member of the size range respectively.

8.1.5 Semi-Similar Size Ranges

Geometrically similar size ranges based on a decimal-geometric series cannot always be realised. Significant deviations from geometrical similarity may be imposed by the following factors:

- overriding similarity laws;
- overriding task requirements; and
- overriding production requirements.

In all such cases, *semi-similar* size ranges must be developed.

1 Overriding Similarity Laws

Influence of Gravity

If inertia forces, elastic forces and weight act together, and if the latter cannot be neglected, then the relationships derived from the Cauchy condition no longer

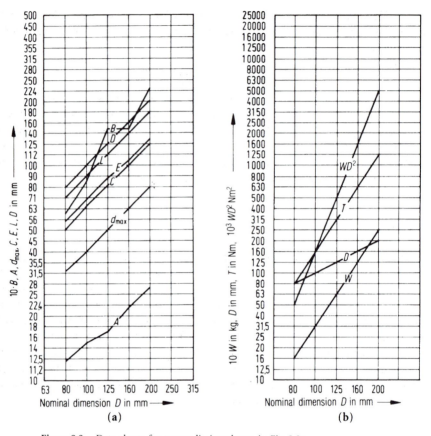

Figure 8.9. Data sheets for torque-limiter shown in Fig. 8.8:
(a) Dimensions adapted to overriding standards or the sizes of bought-out parts;
(b) Main parameters: torque T, weight W and moment of inertia WD^2.

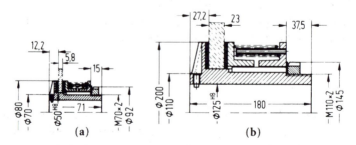

Figure 8.10. Layouts from the size range shown in Fig. 8.9 (Ringspann KG):
(a) smallest;
(b) largest.

apply. This, as we have explained, is because, while the inertia and elastic forces at constant speed depend on the length factor ($\varphi_{FI} = \varphi_{FE} = \varphi_L^2$), the weight increases as:

$$\varphi_{F_w} = \rho_1 \cdot g \cdot V_1 / (\rho_0 \cdot g \cdot V_0) = \varphi_\rho \varphi_L^3 \text{ and for } \varphi_\rho = 1 \text{ as } \varphi_{F_w} = \varphi_L^3$$

Table 8.2 shows that, if all other material properties and the speed remain constant, length is the only variable dimension. If it does vary, the relevant dimensionless parameter cannot remain constant – that is the relationship of the forces must change. Hence with similar cross-sections the stresses change as well and geometric similarity cannot be maintained. This is the case, for instance, with the construction of electrical machines and conveyor systems.

Influence of Thermal Processes

A similar series of problems arises with thermal processes. Constant temperature relationships φ_ϑ only apply when there is thermal similarity, regardless of whether the heat-flow is steady or fluctuating. The first case is represented by the so-called Biot number, $Bi = hL/\lambda$ [8.22], where h is the heat transfer coefficient and λ the coefficient of thermal conductivity of the heated wall. Here too it is obvious that, with approximately equal heat transfer coefficients (the velocity remaining the same) and with the same materials, only the length can vary, and indeed must vary in a size range. As a result the dimensionless parameter governing thermal similarity cannot itself remain unchanged [8.37]. The same is true of fluctuating heating or cooling processes represented by the Fourier number:

$$Fo = \lambda t / (c\rho L^2)$$

where λ is the coefficient of thermal conductivity, c the specific heat and ρ the density of the material. If the material remains the same, the time t and the length L are variable. For the Cauchy number to remain constant, the time must vary as a function of the length. Once again we are left only with the length, which must be variable in a size range. Hence the Fourier number can only remain constant if:

$$\varphi_t = \varphi_L^2$$

that is, if the time varies as the square of the length.

All other things being equal, therefore, thermal stresses due to temperature variations increase as the square of the wall thickness.

Other Similarity Relationships

If the function of a device is determined by physical processes that do not involve inertia or elastic forces, then the physical relationships must be taken into consideration in all designs based on similarity laws [8.20, 8.34, 8.39, 8.43].

In a plain bearing, for instance, the operating conditions are set by the Sommerfeld number:

$$S_0 = \bar{p}\psi^2/(\eta\omega)$$

where $\bar{p}$ is the mean pressure, ψ the non-dimensional clearance, η the dynamic viscosity and ω the rotational speed.

In a machine that otherwise obeys the Cauchy number, we have

$$\varphi_{S_0} = \frac{\bar{p}_1\psi_1^2\eta_0\omega_0}{\bar{p}_0\psi_0^2\eta_1\omega_1} = \varphi_{\bar{p}}\varphi_\psi^2 \frac{1}{\varphi_\eta} \frac{1}{\varphi_\omega}$$

With elastic forces we have $\varphi_{\bar{p}} = 1$; with weight we have $\varphi_{\bar{p}} = \varphi_L$; for the rest, we have:

$$\varphi_\psi = 1 \, , \; \varphi_\omega = 1/\varphi_L \, , \; \varphi_\eta = 1 \text{ at } \vartheta = \text{const.}$$

With elastic forces, therefore, we have $\varphi_{S_0} = \varphi_L$; with weight $\varphi_{S_0} = \varphi_L^2$. The Sommerfeld number increases with the overall size, the bearing becomes increasingly eccentric and, at a given size, may take up the clearance necessary for lubrication.

In a pipe with laminar flow, the loss of pressure is expressed by:

$$\Delta p = f\frac{l}{d}\frac{\rho}{2}v^2 = 32\eta\frac{l}{d^2}v$$

where $f = 64/R_e$ in the laminar region, $R_e = dv\rho/\eta$, $l = $ length of pipe, $d = $ diameter of pipe, $v = $ velocity in the pipe, $\rho = $ density of the fluid, and $\eta = $ dynamic viscosity of the fluid.

With $\eta = $ constant, the pressure loss function becomes:

$$\varphi_{\Delta p} = \varphi_v/\varphi_L$$

Thus, if the pressure loss is to remain constant, the velocity in the pipe must increase in proportion to the size. As a result, the Reynolds number may increase to such an extent that the transition region for turbulent flow is reached, in which case the above equations no longer hold.

Electric AC motors that have a discrete speed depending on the pole number cannot be used to adjust the speed of a finely stepped range of machines (for instance pumps) to maintain a constant Cauchy number. The consequences would be varying stresses and different outputs and the remedy is a suitably adapted semi-similar series.

2 Overriding Task Requirements

The choice of a semi-similar size range may be imposed, not only by similarity laws but also by overriding task requirements. This situation often arises in an ergonomic context. All components with which human beings come into contact in the course of their work – especially the controls, handles, standing and sitting places, and safety features – must fit physiological needs and physical dimensions. In general, none of these components can be changed with the nominal size of the range.

An overriding requirement may also appear for purely technical reasons, inasmuch as inputs and outputs may vary widely in size, as happens with paper and print products.

Figure 8.11 is a schematic representation of a lathe. Here, the size of the human-operated controls cannot be increased with the size of the range; indeed some cannot be altered at all. Thus the operating height must always be adapted to human dimensions, and there are some operations that require an exceptionally long turning length or an exceptionally large turning diameter. In all such cases the machine as a whole must be designed on semi-similar principles, while individual assemblies such as spindle drives, tail-stocks etc can be developed as geometrically similar series.

3 Overriding Production Requirements

The development of a size range is aimed at high cost-effectiveness. Within the range, especially if it is finely stepped, individual components and assemblies may be more coarsely stepped to provide larger batch sizes for even greater cost effectiveness.

Figure 8.12 is the data sheet of a geometrically similar turbine range consisting of seven sizes. Stuffing boxes and locating bolts are stepped more coarsely than the rest, ensuring greater batch sizes and greater economy. Figure 8.13 shows the increase in batch sizes for an assumed sales projection.

All these examples make it clear that it is not always possible to adhere to geometrically similar size ranges; instead, designers must strive, with the help of similarity laws, to arrive at that size range which provides the highest overall utilisation of the strength of every component. Depending on the physical constraints, each size will have to be individually selected. This is best done with the help of exponential equations, as we shall now go on to show.

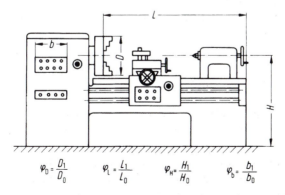

$$\varphi_D = \frac{D_1}{D_0} \qquad \varphi_l = \frac{l_1}{l_0} \qquad \varphi_H = \frac{H_1}{H_0} \qquad \varphi_b = \frac{b_1}{b_0}$$

Figure 8.11. Lathe with main dimensions and controls shown schematically; the diameter/length/height ratio may have to be varied to suit particular groups of products, that is $\varphi_D \neq \varphi_L \neq \varphi_H$, but if possible $\varphi_H = \varphi_b = 1$ for ergonomic reasons.

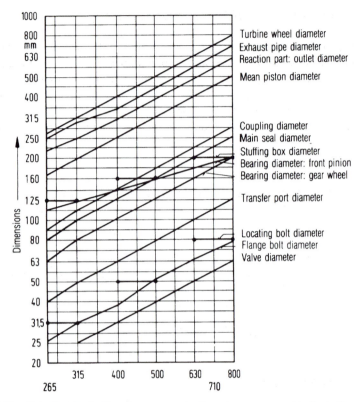

Figure 8.12. Data sheet for turbine size range: main dimensions are geometrically similar, deviations are determined by standards; stuffing boxes and locating bolts are in larger steps than the other components.

Sales forecast

Type	265	315	400	500	630	710	800
Number	6	9	9	6	3	2	1

3 locating bolts per turbine

Size	Φ25	Φ31.5	Φ40	Φ50	Φ63	Φ71	Φ80
Number	18	27	27	18	9	6	3

Combined to:

Size	Φ31.5	Φ50	Φ80
Number	45	45	18

Figure 8.13. Sales forecast in respect of turbine size range (Fig. 8.12) and the associated bolts. Because of the large step sizes, larger batch sizes are possible.

4 Adaptation with the Help of Exponential Equations

Exponential equations are a simple means of dealing with the requirements mentioned under the previous three sections and of developing semi-similar size ranges.

As we have pointed out, nearly all technical relationships can be expressed by power functions. When using preferred number diagrams only the exponent is important if one starts from a basic design.

A physical quantity of the k-th member of a size range can often be represented by:

$$y_k = c_k x_k^{p_x} z_k^{p_z}$$

The dependent variable y and the independent variables x and z can always be expressed by preferred numbers starting from the basic design (Index 0):

$$y_k = y_0 \varphi_L^{y_e k} ; \quad x_k = x_0 \varphi_L^{x_e k} ; \quad z_k = z_0 \varphi_L^{z_e k}$$

where φ_L is the chosen step factor of the dimension chosen as nominal in the size range, y_0, x_0, z_0 are the appropriate values of the basic design, k is the k-th step, and y_e, x_e and z_e are the associated exponents.

Since c_k is a constant, we have for all elements $c_k = c$:

$$y_k = y_0 \varphi_L^{y_e k} = c (x_0 \varphi_L^{x_e k})^{p_x} (z_0 \varphi_L^{z_e k})^{p_z}$$

$$y_k \qquad = c\, x_0^{p_x} z_0^{p_z} \cdot \varphi_L^{(x_e k p_x + z_e k p_z)}$$

With $y_0 = c x_0^{p_x} z_0^{p_z}$ we have:

$$y_0 \varphi_L^{y_e k} = y_0 \varphi_L^{(x_e k p_x + z_e k p_z)}$$

By equating the exponents, we obtain:

$$y_e = x_e p_x + z_e p_z$$

which is independent of k.

Here y_e, x_e and z_e are the exponents to be determined, and p_x and p_z the physical exponents of x and z.

The exponent y_e, must be determined independently of x_e and z_e.

Let us now consider a practical example: the provision of sprung elastic pipeline supports for a range of geometrically similar valves (see Fig. 8.14). The following requirements must be met:

- The stress in the spring due to the weight of the valve must be constant throughout the range.
- The stiffness of the spring must increase as the bending stiffness of the pipe.
- The mean spring diameter, $2R$, must preserve geometrical similarity with the increasing valve size (nominal dimension d).

What law must the spring wire diameter $2r$ and the number of active coils, n, obey?

First of all the appropriate relationships must be set down, so that the exponential equation can be determined (the subscript e shows that only the exponent of the corresponding quantity is involved):

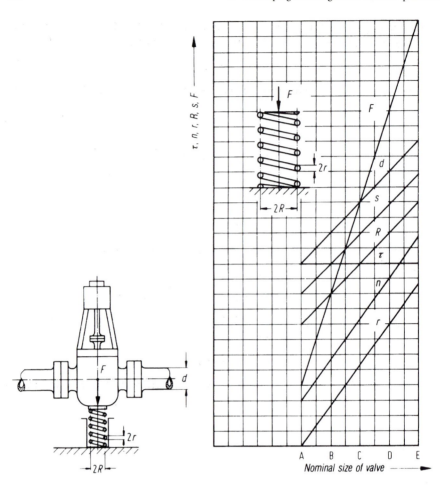

Figure 8.14.   Valve supported in pipe line by means of coil springs.

Figure 8.15.   Data sheet for semi-similar coil springs.

$$F = C\,d^3 \qquad (1)$$

$$\tau = \frac{F \cdot R}{r^3 \pi/2} \qquad (2)$$

$$s = \frac{G r^4}{4nR^3} \qquad (3)$$

$$F_e = 3d_e \qquad\qquad\qquad\qquad (1')$$

$$\tau_e = F_e + R_e - 3r_e = 0 \qquad (2')$$

$$s_e = 4r_e - n_e - 3R_e \qquad (3')$$

Let d be the independent variable.

Since the spring stress must remain constant, the factor $\varphi_\tau = 1$, and the exponent $\tau_e = 0$. The stiffness s of the spring must correspond to the bending stiffness of the pipes. According to Table 8.3 this is ensured by $\varphi_s = \varphi_L$. Since the basic dimension d of the valves increases geometrically, $\varphi_s = \varphi_d$, so that the exponent of s becomes:

$$s_e = d_e \qquad (4')$$

The loading is equal to the weight of the valve F; the weight dimension is related to the basic size d by $\varphi_F = \varphi_d{}^3$. The exponent of F referred to d is therefore:

$$F_e = 3d_e \qquad (5')$$

If the mean spring diameter is to increase in geometrical similarity, we must have $\varphi_R = \varphi_d$ or:

$$R_e = d_e \qquad (6')$$

Substituting equations (5') and (6') in equation (2') we obtain:

$$3d_e + d_e - 3r_e = 0$$

or

$$r_e = (4/3)d_e \qquad (7')$$

Substituting equations (4'), (6') and (7') in equation (3'), we obtain:

$$4r_e - n_e - 3d_e = d_e$$

$$n_e = 4r_e - 4d_e = 4(4/3)d_e - 4d_e = (4/3)d_e$$

Result: Spring wire diameter $2r$ and the number of active coils n must increase as $d^{4/3}$. In that case, the factor is:

$$\varphi_r = \varphi_n = \varphi_d^{4/3}$$

The spread of the individual sizes is shown qualitatively in the data sheet reproduced in Fig. 8.15.

5 Examples

Example 1

A range of high-pressure gear pumps is to consist of six sizes giving delivery volumes ranging from 1.6 to 250 cm^3 per revolution at a maximum operating pressure of 200 bar and a constant input speed of 1500 rev/min. In Fig. 8.16 the steps laid down for the six sizes are plotted against the delivery volumes. The following relationships are involved:

- The pitch circle diameters (each pump size has only one) are graded in accordance with R 10 with a factor of $\varphi_{d_0} = 1.25$, the sizes deviating very slightly from the preferred numbers by virtue of the constant, integral number of teeth and also because the standard values of the modules m differ very slightly from the R 10 series.
- The volume delivered per revolution resulting from the tooth geometry is

$$V = 2\pi \, d_0 mb, \text{ where } b = \text{gear tooth width.}$$

From one size to the next, and at geometrical similarity, the volume delivered therefore increases as:

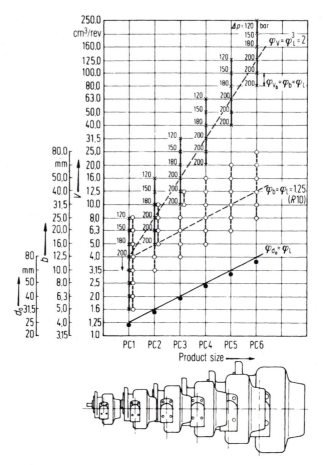

Figure 8.16. Data sheet for a size range of high-pressure gear pumps: V volume delivered per revolution; b gear-tooth width; d_0 pitch circle diameter of gears (Reichert, Hof).

$$\varphi_V = \varphi_{d_0} \varphi_m \varphi_b = \varphi_L^3 = 1.25^3 = 2$$

that is the volume delivered doubles from step to step (see Fig. 8.16). The pump power $P = \Delta p \cdot \dot{V}$ increases as:

$$\varphi_P = \varphi_{\Delta p}(\varphi_V/\varphi_t)$$

which, with:

$$\varphi_{\Delta p} = 1 \text{ and } \varphi_t = 1$$

becomes:

$$\varphi_P = \varphi_V = 2$$

Because of the constant rotational speed, the torque is stepped up accordingly.
• Every pump size has been provided with six tooth widths b, except the smallest size which has eight, so that smaller steps in the volume delivered can be

obtained. This means that for each pump size the geometrical volume delivered, $V = 2\pi\, d_0 mb$, will have a factor of $\varphi_{V_b} = \varphi_b = 1.25$, d_0 and m being constant and the chosen tooth width factor being $\varphi_b = 1.25$ (R 10). The power curve for any one pump size then becomes:

$$\varphi_{P_b} = \varphi_{V_b} = \varphi_b = 1.25$$

- To cope with the mechanical stresses (resulting from the increasing torques and the increasing bending moments due to increases in tooth width) with a shaft of constant diameter, the three pumps with the greatest tooth width in each size group must have their output pressure reduced. For overriding economic reasons (identical shaft diameter, identical bearings), the first two pumps of each size group do not have their strength fully exploited.
- The delivery volumes of the top three pumps in any size group correspond to the bottom three of the next group up. A delivery pressure of 200 bar can therefore be obtained over the entire delivery-volume range.

This particular size range was conceived as a semi-similar series with a small number of housing sizes and several tooth width sets, so that, at the same drive speed and pressure over the entire range ("overriding task requirements") and also at constant gear tooth size, constant gearwheel and shaft diameters per housing size ("overriding production requirements"), the maximum possible range of delivery volumes could be provided.

Example 2

In Fig. 8.17 the output P of a size range of electric motors with varying pole numbers (speeds) has been plotted against the various product sizes (shaft heights H). The shaft heights are in accordance with R 20 and have a step factor of $\varphi = 1.12$. The output of the electric motor is governed by $P \sim \omega JBbhtD$, that is at constant angular velocity ω or speed n, current density J and magnetic flux density B, the output is proportional to the conductor dimensions b, h, t and also to the distance $D/2$ of the conductor from the shaft axis.

The output factor is therefore given by:

$$\varphi_P = \varphi_L^4 = 1.12^4 = 1.6 \text{ (R 5)}$$

In the 4-pole motor (1500 rev/min) the output range is therefore 500-3150 kW. Because power output varies with speed, and also because the dimensions of the conductor, the diameter of the rotors and the heat removed by ventilation have to be varied, the slower 6-pole version must be reduced by three steps (355 to 2240 kW) and the 8-pole version by a further two steps (280 to 1800 kW).

To provide marketable and finer output increments and also to satisfy the overriding production requirements, four outputs are provided per shaft height or motor size, so that the output curve assumes the form of a stepped line. Smaller outputs are obtained by varying the size of the electrically active parts and fitting them into the same size of housing. In contrast to what happened in Example 1,

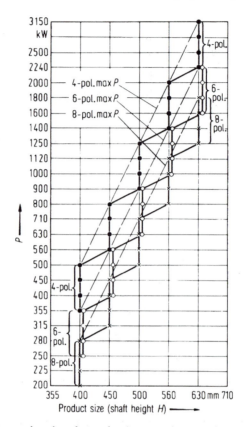

Figure 8.17.   Output data sheet for an electric motor size range (AEG Telefunken) [8.1].

the outputs for the different size groups (fixed pole number) do not overlap (although this has been done with other motor designs so as to maintain efficiency).

Figure 8.18 shows the welded housings of the motor range in greatly simplified form. The stepped sizes of several important dimensions are entered in a data sheet (see Fig. 8.19). It can be seen that the shaft height H, the housing height HC and the distance between the foundation bolts B and A are all stepped up by the factor $\varphi_L = \varphi_H = 1.12$, R 20. Just one housing length BC is provided for the four outputs per shaft size (see Fig. 8.18). This is possible because different sizes of the electrically active parts can be fitted easily into one housing size. Without this separation of the housings from the electrical components, the layout would not be economic and several housing lengths would have to be provided for each shaft height [8.31].

Because of "overriding similarity laws" on the electrical side (for instance in respect of the windings) the housing length step factor φ_{BC} cannot be kept constant over the entire range of the shaft heights. Figure 8.19 shows the increase in step factor for BC with increasing shaft height, the step size only approaching R 20 for the last two housings of the range.

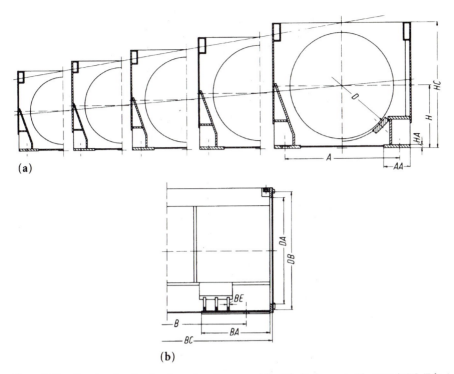

Figure 8.18. Housing for the electric motor size range (simplified) shown in Fig. 8.17 (AEG Telefunken: (a) cross sections; (b) elevation.

Let us now look at a few detailed measurements of this housing design. The baseplate dimensions AA and BA have been graded by a single step factor which lies between R 20 and R 40. This was done to save material while maintaining the minimum dimensions needed for assembly of fixing bolts. The baseplate thickness HA has been stepped in accordance with the usual semi-finished material dimensions, but by and large follows R 20. For the strengthening ribs an equal thickness BE is provided for four housing sizes. Only for the largest housing are thicker ribs required.

Because of overriding similarity laws, overriding task requirements and overriding production requirements, individual dimensions and nominal sizes may have to be stepped in accordance with laws that differ from those leading to geometric similarity. In every case, however, designers must, in the first instance, aim at size ranges based on the appropriate similarity laws and the preferred number series and only deviate from them after careful consideration of the costs involved.

8.1.6 Development of Size Ranges

Size-range development can be summed up as follows:

1. Prepare the basic design for the range. This can be completely new or derived from an existing product.

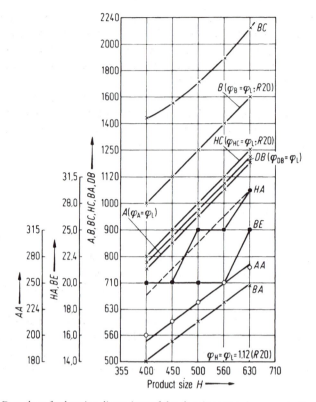

Figure 8.19. Data sheet for housing dimensions of the electric motor size range in Fig. 8.17 (symbols as in Fig. 8.18).

2. Determine the physical relationships (exponents) in accordance with similarity laws, using Table 8.3 for geometrically similar product ranges, or using exponential equations for semi-similar product ranges. Put down the results in the form of data sheets.
3. Determine the step sizes and add them to the data sheets.
4. Adapt the theoretically obtained ranges to satisfy overriding standards or technological requirements and record the deviations.
5. Check the product range against scale layouts of assemblies paying particular attention to critical areas for extreme dimensions.
6. Improve and perfect what documentation may be needed to determine the range and prepare production documents (when required).

The need for developing a semi-similar size range may not always appear from the requirements list or from a first survey of the physical relationships, but may only become clear during an actual development.

Section 8.1.5 describes how the individual components and dimensions can be determined during the development of semi-similar size ranges using exponential equations. The number of parameters and the number of equations to be solved may be quite large in complex applications. For that reason Kloberdanz [8.29] has developed a computer supported size range program. After formulating the physical relationships involved and adding the constraints, the program automatically determines the scaling rules and represents the results in the form of diagrams and data sheets. These can then be adapted interactively to other constraints such as company standards and stock lists.

Using parametric macros, the scaling rules are used to automatically generate rough geometrical layouts of sequential designs for a semi-similar size range (see Fig. 8.20). After the final size range has been determined, the details can then be defined using single part macros [8.29, 8.38].

8.2   Modular Products

In 8.1 we discussed the features and design potential of size ranges. Their aim is the rationalisation of product development by the implementation of the *same* function with the same solution principle and, if possible, with the same properties over a wide range of sizes.

Modular products provide rationalisation in a different situation. If a product is to fulfil *different* functions, then many variants will have to be provided, at great cost in design and production. Rationalisation is, however, possible if the particular *function variant* at any one time is based on a combination of fixed individual

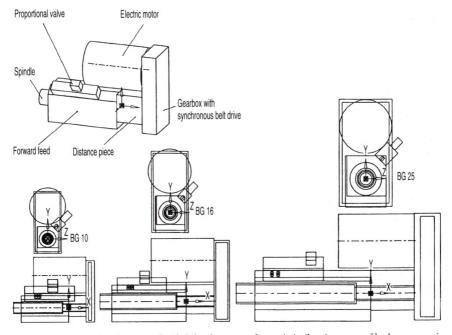

Figure 8.20.   Example of computed-aided development of a semi-similar size range of hydropneumatic forward feed units, after [8.29, 8.38].

parts and assemblies (function units), and this is precisely what a modular system sets out to achieve.

By *modular products* we refer to machines, assemblies and components that fulfil various overall functions through the combination of distinct building blocks or modules.

Because such modules may come in various sizes, modular products often involve size ranges. The modules should be produced by similar methods whenever possible. Since in a modular system the overall function results from a combination of discrete units, the development of modular products demands the elaboration of a corresponding function structure and this calls for greater design effort during the conceptual and embodiment phases than does the development of a pure size range.

The modular system can provide a favourable technical and economic solution whenever all or some different products are required in small batch numbers only, and whenever they can be based on a single unit or on only a few basic and additional units.

Besides fulfilling a variety of functions, modular systems can also serve to increase the production batch size of identical parts for use as building blocks in a variety of products. This additional objective, which greatly helps to rationalise the production procedure, is attained by the breakdown of the product into elementary components (see 7.5.7). Which of the two objectives is paramount depends largely on the product and on the task it has to perform. With a wide-ranging overall function, what matters most is the resolution (divisibility) of the product into function-oriented modules; with a small number of overall function variants, on the other hand, a production-oriented resolution is the paramount consideration.

Often, modular development is only initiated when what was originally conceived as an individual or size-range development is expected to yield a large number of variants. To that end, products that have already been marketed are often redesigned as a modular system. The disadvantage here is that the products are more or less predetermined; the advantage that their essential properties have already been tested so that an expensive new development can be dispensed with.

8.2.1 Modular Product Systematics

Modular product systematics are discussed in [8.10, 8.11, 8.30]. Basing ourselves on these findings, we shall first of all examine the principles and the most important concepts, and merely add a few amplifications [8.4].

Modular product systems are built up of separable or inseparable units, ie *modules*.

We must distinguish between *function modules* and *production modules*. Function modules help to implement technical functions independently or in combination with others. Production modules are designed independently of their function and are based on production considerations alone. Function modules in the narrower sense have been divided into equipment, accessory, connecting and other modules

[8.10, 8.11]. This division is neither clear-cut nor adequate for the development of modular systems.

For the classification of function modules it seems advantageous to define the various types of function that recur in modular systems and can be combined as sub-functions to fulfil different overall functions (overall function variants) (see Fig. 8.21).

Basic functions are fundamental to a system. They are not variable in principle. A basic function can fulfil an overall function simply or in combination with other functions. It is implemented in a basic module which may come in one or several sizes, stages and finishes. Basic modules are "essential".

Auxiliary functions are implemented by locating or joining *auxiliary modules* that are kept in step with the basic modules and are usually of the "essential" type.

Special functions are complementary and task-specific sub-functions that need not appear in all overall function variants. They are implemented by *special modules* of the "possible" type.

Adaptive functions are necessary for adaptation to other systems and to marginal conditions. They are implemented by *adaptive modules* whose dimensions are not fully fixed in advance and hence allow for unpredictable circumstances. Adaptive modules may be of the "essential" or the "possible" type.

Customer-specific functions not provided for in the modular system will recur time and again even in the most careful development. Such systems are implemented by *non-modules* which have to be designed individually for specific

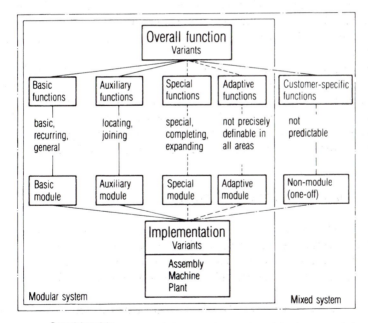

Figure 8.21. Function and module types in modular and mixed product systems.

tasks. If they are used, the result is a *mixed system*, that is a combination of modules and non-modules.

By the *importance of a module* we refer to its ranking within a modular system. Thus, function modules can be ranked as "essential" or as "possible" [8.14].

A production-oriented characteristic is the *complexity of a module*. Here we distinguish between *large modules* which, as assemblies, can be subdivided into components, and *small modules* that are components themselves.

A further aspect of module characterisation is their *type of combination*. Designers should always aim for technically advantageous combinations of similar modules. In practice, however, the combination of similar with different modules, and also with customer-specific non-modules, is often unavoidable. Non-modules, as mixed systems, can meet market requirements very economically.

For the characterisation of modular systems we can also consider their *resolution* – in other words, the extent to which a particular module can be broken down into individual parts for functional or production reasons. For the modular system as a whole, the resolution defines the number of individual units and their possible combinations.

Another characterisation is the level of concretisation of modular products. One-off products, such as turbines, pumps and compressors, often demand significant variations in performance and efficiency. They require their working geometries, for example blade passages and cylinder dimensions, to be adapted. However, many parts remain identical, for example the bearings, seals, and input and output sections. In this case a division into *geometric sections* (modules) is advantageous (see Fig. 1.10, working step 4 and Fig. 7.1, working step 3). The overall product is thus developed as a combined approach, that is as a size range made up from modular building blocks (see Fig. 8.22). The modules are generated in suitable step sizes. For the manufacturer, these modular products do not really exist until, based on specific requirements, the appropriate sets of drawings are combined as modules into a complete machine (see Fig. 8.23). Such a "fictitious" modular product is not only useful for the product development department, it can also provide the basis for fixed production modules that can be used to prepare production plans and software for CNC machines. Another use of the modules is to plan the optimum stock of casting patterns. These patterns can be divided into modules that are combined, where necessary, into more complex castings, such as housings. Depending on the need, the level of concretisation can be selected between having product modules available either in software (and paper) or in hardware.

For the application of closed modular systems, their *range and potential* can be expressed by *combinatorial plans* with a finite and predictable number of variants. Such plans make it possible to choose desired combinations directly. By contrast, open modular systems contain a great multiplicity of combinatorial possibilities, which cannot be fully planned or represented. A *specimen plan* provides examples of typical applications of the modular system.

The above-mentioned concepts of module development are summarised in Table 8.5.

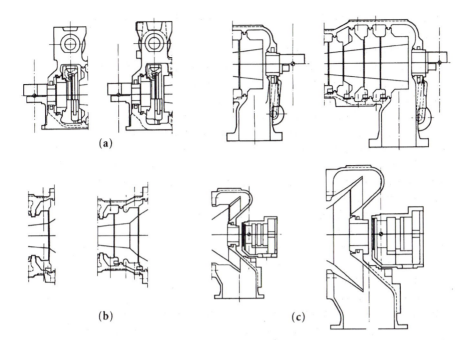

Figure 8.22. Building blocks for an industrial turbine size range generated using geometric sections (Siemens):
(a) Entry section;
(b) Middle section;
(c) Exit section.

8.2.2 Modular Product Development

In what follows the development of modular systems will be presented in accordance with the steps listed in Fig. 3.3.

Clarifying the Task

In their formulation of demands and wishes, for instance with the help of the checklist (see Fig. 5.7), designers must pay careful attention to the clarification of the various tasks to be performed by the product. A characteristic demand of the specification of a modular system is that it must fulfil several overall functions. This results in the *variants of the overall function* that a specific modular system has to fulfil.

Of particular importance for the economic analysis and application of modules are data about the market expectations of particular variants. Friedewald [8.18] speaks of the quantification of function variants for the technical and economic optimisation of modules. Whenever the implementation of rarely demanded variants increases the overall costs of the modular system, an attempt must be made to remove such variants. The more searching these analyses are before the actual

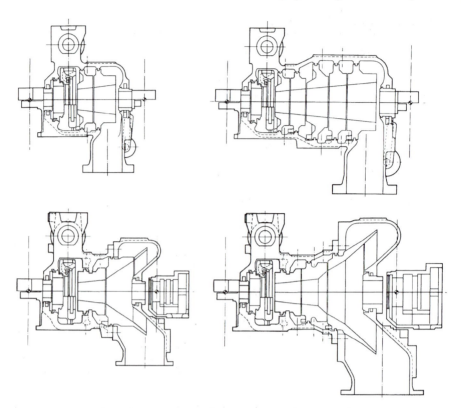

Figure 8.23.   Complete turbine designs for different pressure and flow requirements produced by combining the building blocks shown in Fig. 8.22 (Siemens).

development is begun, the greater are the chances of arriving at a cost-effective solution. However, the reduction of types by the removal of infrequently demanded and costly function variants cannot be finalised until the elaborated solution concept or even the embodiment design provides reliable information about the cost of the different variants and also about the influence of every individual variant on the cost of the modular system as a whole.

Establishing Function Structures

The establishment of function structures is of particular importance in the development of modular systems. With the function structure – that is the splitting up of the required overall function into sub-functions – the structure of the system is already laid down, at least in principle. From the outset, designers must try to subdivide the overall function variants into a minimum number of similar and recurring sub-functions (basic, auxiliary, special and adaptive functions, see Fig. 8.21). The function structures of the overall function variants must be logically and physically compatible, and the sub-functions determined by them must be

Table 8.5. Concepts of modular systematics

Classifying criteria	Distinguishing features
Types of module:	— Function modules
	• Basic modules
	• Auxiliary modules
	• Special modules
	• Adaptive modules
	• Non-modules
	— Production modules
Importance of modules	— Essential modules
	— Possible modules
Complexity of modules	— Large modules
	— Small modules
Combination of modules	— Similar modules only
	— Different modules only
	— Similar and different modules
	— Modules and non-modules
Resolution of modules	— Number of parts per module
	— Number of units and their possible combinations
Concretisation of modules	— Software/paper modules only
	— Mix of hardware and software modules
	— Hardware modules only
Application of modules	— Closed system with combinatorial plan
	— Open system with specimen plan

interchangeable. To that end, it is useful if, depending on the particular task, the overall function can be achieved by "essential" functions and by additional task-specific "possible" functions.

Figure 8.24 shows the function structure for the modular bearing system discussed in [8.3, 8.25]. The most frequently demanded overall functions, namely "non-locating bearing", "locating bearing" and "hydrostatic locating bearing", together with the appropriate basic, special, auxiliary and adaptive functions, are represented. By means of the sub-function "seal between rotating and stationary systems", we can show that it is often more cost-effective to combine several functions into one complex function; thus in the present case, the sealing function was combined with an adaptive function to satisfy various conditions. The production module, which performs this complex function, was accordingly specified as an unfinished one that could be completed during production as a simple line seal; as a line seal with an additional labyrinth; or as a seal with an additional coupling adapter (see Fig. 8.25). It should also be stressed that there are special functions (special modules) that occur in at least one overall function variant (here: "transfer axial force F_A from rotating to stationary system"), others that represent possible modules for all overall function variants (here: "set and measure oil pressure"), and yet others that only become necessary at a certain size (here: "feed high pressure oil").

In the setting up of function structures the following objectives should be borne in mind:

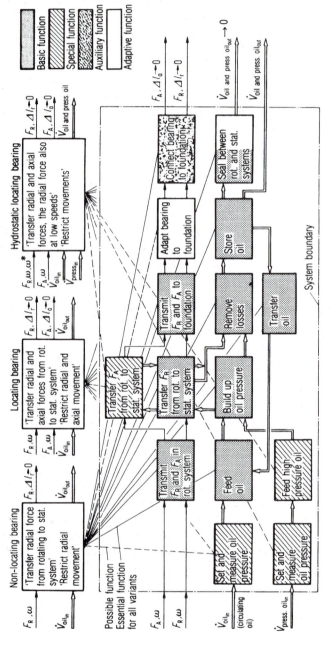

Figure 8.24. Function structure for a modular bearing system, after [8.25].

- Aim for the implementation of the required overall functions by the combination of the minimum number of easily implementable basic functions.
- Try to divide the overall functions into basic functions and if necessary into auxiliary, special and adaptive functions in accordance with Fig. 8.21, in such a way that variants in high demand are predominantly built-up with basic functions, and more rarely demanded variants with additional special and adaptive functions. For very rarely demanded function variants, mixed systems with additional functions (non-modules) are often more cost-effective.
- Try to combine several sub-functions in a single module if this increases cost-effectiveness. Such combinations are particularly recommended for the implementation of adaptive functions.

Searching for Solution Principles and Concept Variants

The next step is to find solution principles for the implementation of the various sub-functions. To that end, designers should, above all, look for such principles as provide variants without changes in working principle and basic design. As a rule, it is advantageous to stipulate similar types of energy and similar physical working principles for the individual function modules. Thus it is more cost effective and technically advantageous, in the combination of sub-solutions into overall solutions (solution variants), to implement various drive functions with a single type of energy rather than provide a single modular system with separate electrical, hydraulic and mechanical drives.

A satisfactory production solution is also ensured by the implementation of several functions by a single unfinished module that can be completed in various ways depending on the requirements.

However, so complex are the technical and economic factors involved that it is impossible to lay down hard and fast rules. Thus, in the case of the bearing system (see Fig. 8.25) it seems technically and economically advantageous to provide the bearing shell with lateral locating surfaces for taking up small axial forces. With larger axial forces, however, rolling bearings must be provided instead; it would be a mistake to try, for purely theoretical reasons, to transfer the radial and axial forces over the entire size range by means of plain bearings. The plain bearing system must be designed during the conceptual phase with two alternative lubrication systems (free ring or fixed ring) because their respective advantages and disadvantages can only be determined by later experiments [8.25]. The design of the ultimately chosen bearing system is shown in Fig. 8.25.

Selecting and Evaluating

If several concept variants have been found during the previous steps, each must now be evaluated with the help of technical and economic criteria so that the most favourable solution concept can be selected. Experience has shown that, since the

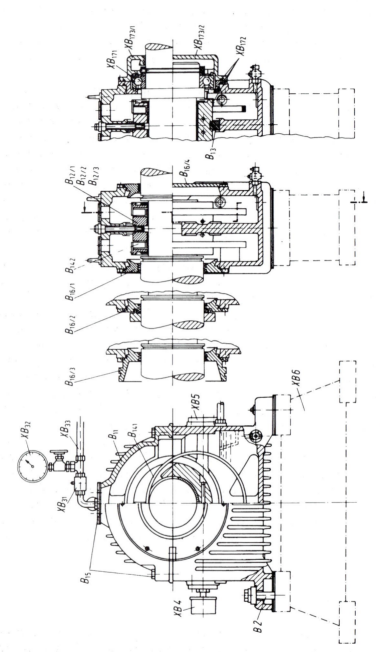

Figure 8.25. Layout of the modular bearing system shown in Fig. 8.24 (AEG Telefunken).

properties of any one variant are not yet sufficiently clear at this stage, such selections are very difficult to make.

Thus, in the case of the bearing system, preliminary evaluations have to be made even in the conceptual phase, for instance as to whether the axial forces should be taken up by plain or rolling bearings. However, the final choice of lubricating system can only be taken after the building of prototypes and experimentation with them.

Apart from the determination of the technical rating of individual concept variants, economic factors are of crucial importance in the design of modular systems. To come to grips with them, designers must estimate the production costs of the individual modules and their relative effect on the cost of the modular system as a whole. To that end, they will first of all determine the expected "function costs" of the sub-functions or of the modules fulfilling them. At the lower level of embodiment characteristic of the conceptual phase, they cannot usually hope to come up with more than very rough estimates. Since basic modules appear in all sorts of variants, they will select such solution principles as provide the most cost-effective basic modules. Special and adaptive modules take second place in the minimisation of costs.

For minimising the costs of a modular system, not only the modules themselves but also their interaction must be taken into account; in particular, the influence of special, auxiliary and adaptive modules on the *cost of the basic modules*. The influence of the cost of every overall function variant on the cost of the modular system as a whole must be fully determined. This may prove a complex task. Thus, in the bearing system we have been considering, the function variant "cool oil internally" would greatly influence the cost of the basic module "bearing housing", because the dimensions of the special module "water cooler" determine the dimensions of the housing and hence the overall costs. If there is only a small demand for this variant then it is certainly more cost-effective to fit the oil cooler to the outside of the housing and to put up with the extra cost of an oil pump.

In short, the layout of the basic modules must be adapted to the expected demand. To that end, the influence of the remaining modules is of great importance. If it is impossible to provide a marketable adaptation of the basic concept, the least cost-effective function variants should be eliminated from the modular system. It will often be more economical to replace unusual variants, which render the overall system more expensive, by making individual adaptations than to impose such adaptations on the whole modular system. An alternative is the use of mixed systems.

Preparing Dimensioned Layouts

Once a solution concept has been selected, the individual modules must be designed, in accordance both with their functions and also with the production requirements. In the design of modular systems, production and assembly considerations are of paramount economic importance. By paying heed to the embodiment design guidelines laid down in 7.5.7 and 7.5.8, designers must try to provide basic, auxiliary, special and adaptive modules with the maximum number of similar and recurring parts and the minimum number of unfinished parts and production processes.

When selecting step sizes, designers should aim at the optimum resolution of modules (modularity), and to that end they may well adopt the differential construction approach. The determination of the optimum number of modules is, however, a complex task, for it is influenced by the following factors:

- Requirements and quality must be maintained and the propagation of errors must be taken into account (see 7.4.5 – the principles for fault free design). Thus the greater the number of individual components, the greater the number of fits, and this may have untoward repercussions on the function, for instance on the vibration of a machine.
- Overall function variants must be created by simple assembly of modules (individual parts and assemblies).
- Modules may only be broken down to the extent that functions and quality permit and costs allow.
- In modular products marketed as overall systems, variants of which clients can assemble themselves by combinations of the modules [8.33], the most common modules must be designed for equal wear and tear and for easy replacement.
- In determining the most cost-efficient modularity, designers must pay special heed to the cost, not only of the design itself, but also of overall scheduling and of production processes including assembly, handling and distribution.

Figure 8.25 shows the scale layout of the bearing system we have been discussing. In Fig. 8.26 the structure of the overall function variants (see Fig. 8.24) is shown in the form of a family tree. In both these figures, only the most important assemblies and individual parts of the bearing system have been entered; the actual modularity is greater. If the function structure is compared with the final modular structure, it becomes clear that in the given modular system several functions are fulfilled by a single module or its variants. Table 8.6 shows the modules used and their assigned functions.

Preparing Production Documents

Production documents must be prepared in such a way that the execution of orders can be based on the simple, and if possible computer-aided, combination and further elaboration of modules for the required overall function variants.

Drawings require an appropriate part-numbering system and classification, two prerequisites of the optimum combination of modules (individual parts and assemblies).

The combination of individual modules into product variants must be recorded in the parts list. To build up a parts list, designers can refer to the so-called variant parts list [8.14] which is based on the structure of the product and in which a distinction is made between "essential" and "possible" modules.

Particularly suited to the numeration of drawings and parts lists is the method

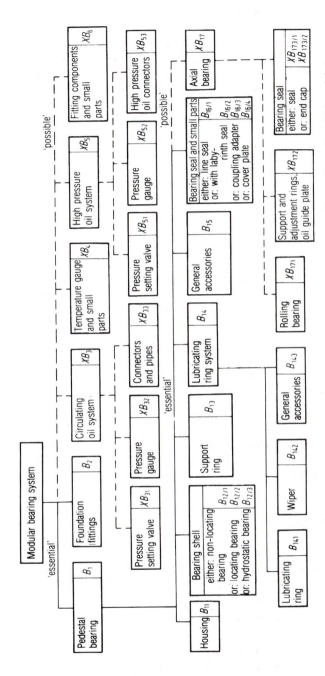

Figure 8.26. Family tree of the modular bearing system in accordance with Fig. 8.24 (prefix X indicates 'possible' modules).

Table 8.6.   Modules in the bearing system shown in Fig. 8.26

Module	No.	Type	Functions
Housing	B_{11}	Basic module	'Transmit F_R and F_A to foundation', 'Remove losses', 'Store oil'
Bearing shell	$B_{12/1}$	Basic module	'Transfer F_R from the rotating to the stationary system', 'Build up oil pressure'
	$B_{12/2}$	Variant of module $B_{12/1}$	additionally: 'Transfer F_A from the rotating to the stationary system'
	$B_{12/3}$	Variant of module $B_{12/1}$	additionally: 'Transfer hydro-static oil pressure to shaft'
Support ring between housing and bearing shell	B_{13}	Auxiliary module	'Connect bearing shell with housing'
Lubricating ring	B_{141}	Basic module	'Transfer oil'
Wiper	B_{142}	Basic module	'Feed oil'
General accessories	B_{143}	Basic module	'Control oil level' and 'Remove oil'
General accessories	B_{15}	Basic module/auxiliary module	'Accessory and connecting functions'
Bearing seal and small parts	$B_{16/1}$	Basic module	'Seal between rotating and stationary systems'
	$B_{16/2}$	Basic module/adaptive module	additionally: 'adapt to labyrinth seal'
	$B_{16/3}$	Basic module/adaptive module	additionally: 'Provide coupling adapter'
	$B_{16/4}$	Special module	'Seal housing in the absence of shaft'
Foundation fittings	B_2	Auxiliary module	'Connect bearing to foundations'
Pressure setting valve	XB_{31}	Special module	'Set pressure for circulating oil'
Pressure gauge	XB_{32}	Special module	'Measure oil pressure'
Connectors and pipes	XB_{33}	Auxiliary module	'Transfer circulating oil'
Temperature gauge and small parts	XB_4	Special module	'Measure temperature'
Pressure setting valve	XB_{51}	Special module	'Set pressure for high pressure oil'
Pressure gauge	XB_{52}	Special module	'Measure oil pressure'
High pressure oil connectors	XB_{53}	Auxiliary module	'Feed high pressure oil'
Fitting components and small parts	XB_6	Adaptive module	'Adapt bearing to foundation'
Rolling bearing	XB_{171}	Special module (for large axial forces)	'Transfer F_A from the rotating to the stationary system'
Support and adjustment rings, oil guide plate	XB_{172}	Auxiliary module	'Connect rolling bearing with housing', 'Supply oil to rolling bearing'
Bearing seal	$XB_{173/1}$	Special module	'Seal between rotating and stationary systems in case of rolling bearing variant'
	$XB_{173/2}$	Special module	'Seal housing in the absence of shaft'

of parallel encoding, which assigns identification numbers for the unequivocal and unmistakable description of components and assemblies, and classification numbers for the function-oriented recording and retrieval of these components and assemblies. The classification number is of particular importance in the modular product system, because it helps to detect functional and other similarities between components.

8.2.3   Advantages and Limitations of Modular Systems

For the *manufacturer*, modular systems provide *advantages* in nearly all areas of the company:

- Ready documentation is available for tenders, project planning and design; designing is done once and for all, though it may be more costly for that very reason.
- Additional design effort is needed for unforeseeable orders only.
- Combinations with non-modules are possible.
- Overall scheduling is simplified and delivery dates may be improved.
- The execution of orders by the design and production departments can be cut short through the production of modules in parallel; in addition parts can be supplied quickly.
- Computer-aided execution of orders is greatly facilitated.
- Calculations are simplified.
- Modules can be manufactured for stock with consequent savings.
- More appropriate subdivision of assemblies ensures favourable assembly conditions.
- Modular product technology can be applied at successive stages of product development, for example, in product planning, in the preparation of drawings and parts lists, in the purchase of raw materials and semi-finished materials, in the production of parts, in assembly work, and also in marketing.

For the *user* there are the following *advantages*:

- short delivery times;
- better exchange possibilities and easier maintenance;
- better spare parts service;
- possible changes of functions and extensions of the range; and
- almost total elimination of failures thanks to well-developed products.

For the *manufacturer* the *limit* of a modular product system is reached whenever the subdivision into modules leads to technical shortcomings and economic losses:

- Adaptations to special customer's wishes are not as easily made as they are with individual designs (loss of flexibility and market orientation).
- Once the system has been adopted, working drawings are made on receipt of orders only, with the result that the stock of drawings may be inadequate.
- Product changes can only be considered at long intervals because once-and-for-all development costs are high.

- The technical features and overall shape are more strongly influenced by the design of modules and the modularity than they would be by individual designs.
- Production costs are increased, for example because of the need for accurate locating surfaces; production quality must be higher because re-machining is impossible.
- Increased assembly effort and care are required.
- Since the user's as well as the producer's interests have to be taken into consideration, the determination of an optimal modular system may prove very difficult.
- Rare combinations needed to implement unusual requirements may prove much costlier than tailor-made designs.

For the *user* there are such *disadvantages* as:

- special wishes cannot be met easily;
- certain quality characteristics may be less satisfactory than they would be with special-purpose designs; and
- weights and structural volumes of modular products are usually greater than those of specially designed products, and so space requirements and foundation costs may increase.

Experience has shown that, while modular production helps to reduce general overheads (including administrative staff costs in particular), it may lead to an increase in shopfloor wages and especially in the cost of materials because, as mentioned earlier, it tends to involve greater weights and volumes. Only if a modular system is developed with the express intention of rendering every function variant more cost-effective than a specially designed product can there be a significant reduction in overall costs.

Figure 8.27. Modular high-output electric motor system with essential building blocks [8.41].

1 Rotor with windings, laminations and shaft
2 Stator with windings and laminations
3 Flanged plain bearing
4 Roller bearing
5 End caps
6 Housing for protective system IP23, with inner cooling
7 Housings for protective system IP54 with: air-water cooling; air-air cooling by external ventilation; and air-air cooling by self-ventilation
8 Blind cover
9 Cover
10 Air-water cooler
11 Air-air cooler
12 Connection box 315 A, 6 kV
13 Connection box 1980 A, 10 kV
14 External cooler with air inlet, protective grid and connecting parts.

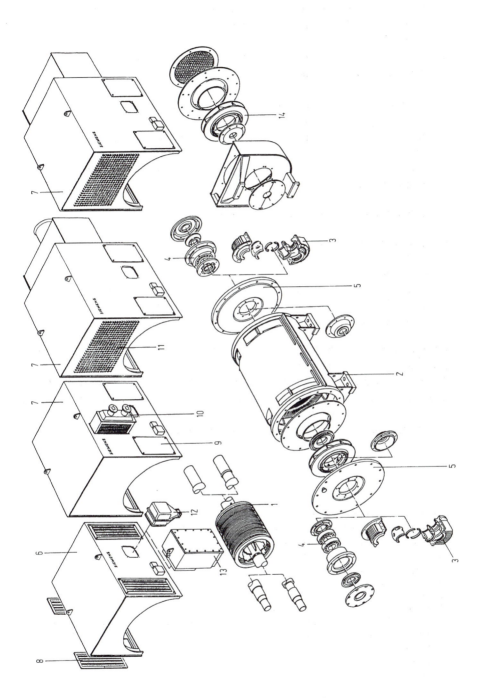

8.2.4   Examples

Electric Motor Systems

Modular systems are particularly cost-effective in the production of such universal drive systems as electric motors. Such systems fulfil a double purpose: they allow economic production of variants and can be easily adapted to specific customer requirements. Figure 8.27 shows a modular high-output electric motor system [8.41]. Of the identifiable modules, Numbers 5 and 9 are fixed basic modules; Numbers 1, 2 and 7 are basic modules with possible adaptations for specific requirements (for example Numbers 1 and 2 for adaptation to various voltages); and Numbers 3, 4, 6, 8, 10 to 14 are special modules for meeting certain safety provisions, types of cooling and types of connection.

This particular modular system is at one and the same time a size range, every module being available in different dimensions.

Gearboxes

Gearboxes are another familiar example of modular systems. They involve a multiplicity of market-determined function variants (for instance, the attachment of different input and output devices, various shaft positions and different gear ratios). For users it is a great advantage if they can build up gearbox configurations to suit their particular and often changing requirements or if, in the case of damage, they can undertake simple and speedy replacements. For the manufacturer the modular system provides a comprehensive gearbox system with only a few housings, gear wheels, shafts and bearings.

Figure 8.28 gives an example: the housing of a modular gearbox allowing for different input shaft positions and also for different gear arrangements (single and multiple stages, spur and bevel gears) [8.23]. The reader will notice the undivided housing *1*, which is closed on the output shaft side with an oval cover *2*, and on the other side, for easier gear assembly, with a large, circular cover *3* underneath a smaller oval cover *4*. While the circular cover *3* locates the bearings of the various input shafts, the oval cover *4* merely covers the bearing aperture when the input is through bevel gears; but if the input is through spur gears, it contains a shaft seal. The slow running output stage IV is always led through a bearing in the opposite side of the housing, with its shaft seal held in the oval cover *2*. The complete housing is therefore broken down into several function/production modules, the central housing section and the circular cover serving as basic modules, the two oval covers serving as modules adapted to particular arrangements and dimensions of the shafts, and the bearing of the bevel pinion and a rectangular blind cover serving as special modules. This arrangement has the advantage of providing, with a single central housing section, a multiplicity of possible variants by appropriate adaptation of the oval cover apertures to various spur gear combinations, by the inclusion of a special oval cover for the flange-mounting of a drive

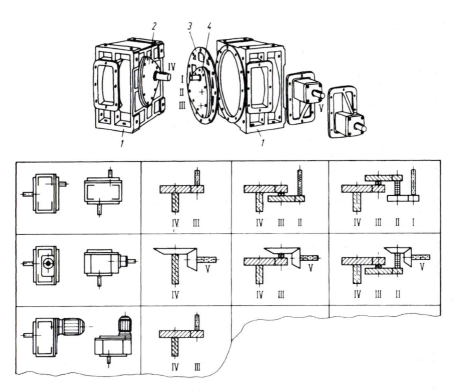

Figure 8.28.   'Hansen-Patent' gearbox [8.23].

motor, and by the provision for a bevel gear input stage. The disadvantage of this highly modular design is that it demands very accurate location of the various covers for proper alignment of the shafts and seals. Moreover, the housing is not fully utilised unless all the gear stages are built in. A further development, namely a split housing [8.24], obviates these problems (see Fig. 8.29). In that case the small individual covers that remain only carry the seals or only cover the bearing apertures.

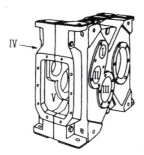

Figure 8.29.   Housing for 'Hansen-Patent' gearbox [8.24]; the housing is symmetrical.

Another modular gearbox design is shown in Fig. 8.30 [8.44]. It was developed for the express purpose of providing a broad gear system (in respect of shaft positions and gear ratios) with the fewest possible modules. In contrast to the example shown in Fig. 8.28, several housing variants were provided to match more closely different gearbox variants and hence to reduce weight and volume.

Another modular gearbox, in which inputs and outputs in particular can be varied, is described in [8.15].

Further Examples

Further examples taken from hydraulics, pneumatics and machine tool construction can be found in the literature [8.2, 8.21, 8.30, 8.42].

Modular Conveyor System

While all the systems discussed above are examples of "closed" modular systems, Fig. 8.31 shows the modules and a specimen plan of an "open" modular system. The fixed modules are shown under (a) and a sample combination under (b).

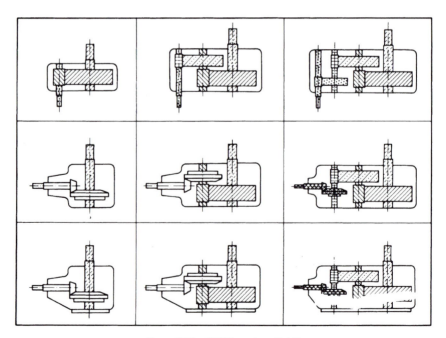

Figure 8.30.   WGW gearbox [8.44].

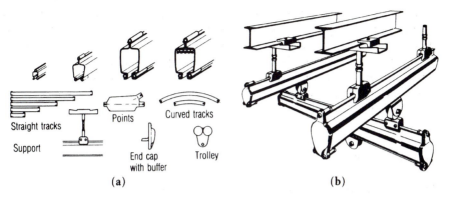

Figure 8.31.   Open modular system for conveyors (Demag, Duisburg), (a) fixed modules (b) sample combination.

9 Design for Quality

Achieving product quality appropriate for the market (see 2.1.8 and 7.2) starts with the design process [9.1, 9.9]. Quality cannot be achieved simply though testing and developing a product – it has to be built-in from the beginning of the design process and maintained throughout the production process. Just as design commits a large proportion of a product's costs (see 10), up to 80% of all faults can be traced back to insufficient planning and design work [9.12]. Furthermore, up to 60% of all breakdowns that occur within the warranty period are caused by incorrect or incomplete product development.

Ensuring quality and improving quality are team activities. These activities have to address all aspects of product development, starting with product planning and marketing. Quality is influenced decisively during design and development, and has to be realised during production. The basis for quality procedures is the international standard DIN ISO 9000-9004, which also defines the terminology [9.4 - 9.8].

The terms *Total Quality Management* TQM and *Total Quality Control* TQC [9.12] express a quality philosophy based on the idea that measures taken to avoid faults throughout the total product development process are the best way to achieve quality. Another term frequently used is *Quality Engineering*.

The main objective of a quality policy has to be fulfilment of customer requirements. For this purpose *Quality Function Deployment* QFD has been developed [9.12]. This method helps to translate the often vague customer requirements into requirements that are clearly formulated and quantified, and related to the different company departments. QFD thus helps to refine and complete the requirements list (see 5.2), making it an important part of a systematic product planning process (see 5.1).

The systematic approach along with the selection and evaluation methods described in this book support quality assurance in product design and development. Estimating product reliability, for which special analysis methods have been developed, is also important [9.2, 9.11]. The methods described in sections 9.2 and 9.3 aim at identifying possible disturbing factors and faults in a systematic way, and remedies and preventive measures are suggested.

The results of such analyses identify critical quality issues of a product under development, or those of its assemblies and components. The issues must be addressed in the quality plan for production and assembly. This plan is used to derive the quality control procedures to be used during production and testing. The complete set of tools for quality assurance is called the *House of Quality* HoQ.

9.1 Faults and Disturbing Factors

The design process involves a series of creative and corrective steps. Selection and evaluation methods (see 4.2) as well as tests and calculations help to identify and remove weak spots. Even so designers can make mistakes, or their knowledge may not be sufficient to identify or exclude links that are faulty or prone to disturbances. When designers are aware of the information they lack and the uncertainty in their decisions, they can avoid severe technical and economic consequences by designing to minimise risk (see 7.5.12).

Often malfunctions are not caused by design faults but by *disturbing factors*. According to Rodenacker [9.13] disturbing factors can be caused by variations in the input variables, that is by quality differences in the material, energy and signal flows entering the system (see Fig. 2.16). When these influence the output of a system adversely, it may be necessary to compensate for them by modifying the solution, eg through a control system.

Disturbing factors can be identified in the *function structure* when the allocation and connection of the sub-functions lack clarity. They show up in the *working principle* when the selected physical effects do not produce the expected results. Because of variations in *material properties* along with shape, position and surface deviations introduced during *production* and *assembly*, the selected *embodiment* can result in unexpected effects. Finally, *external disturbing factors* such as temperature, humidity, dust, vibration etc can cause damaging effects. It may therefore be necessary to suppress the effects of disturbing factors to avoid the danger of fault propagation.

Preventative measures can reduce the malfunctions caused by disturbing factors but cannot exclude them altogether. Examples of measures include the principles of fault free design (see 7.4.5) and the other embodiment principles (see 7.4) and guidelines (see 7.5).

Important prerequisites to prevent faults and disturbing factors, or at least limit their effects, are the identification and estimation of possible faults and disturbing factors as early as possible in the product development process. The following sections describe some established methods.

9.2 Fault-Tree Analysis

The influence of faults and disturbing factors can be determined systematically by *fault-tree analysis* [9.3].

From the conceptual phase, designers know what overall function and individual sub-functions have to be fulfilled. The established function structures can thus be used to identify all the functions to be checked. These functions are then negated one by one – that is assumed to be unfulfilled. By reference to the checklist (see

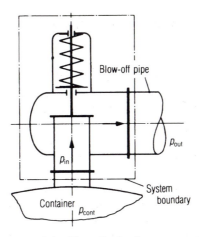

Figure 9.1.   Safety blow-off valve for a gas container.

7.2), designers can seek out the possible faults or disturbances that would cause particular functional failures. The OR or AND relationships of these faults and their effects can then be examined.

The conclusions help designers improve their designs and, if necessary, re-examine the solution concept or modify production, assembly, transport, operation and maintenance procedures. Let us take a concrete example.

The design of a safety blow-off valve for a gas container (see Fig. 9.1) must be checked for possible design faults during the *conceptual phase*. From the requirements list and the function structure, it is possible to specify the operating conditions depicted in Fig. 9.2. The blow-off valve is intended to open when the

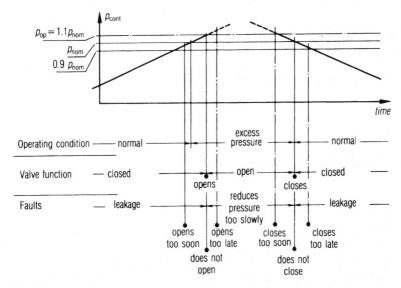

Figure 9.2.   Operating conditions, valve main functions and faults of the safety valve.

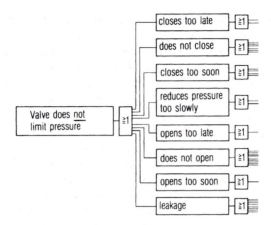

Figure 9.3.   Construction of fault-tree based on faults identified from Fig. 9.2.

operating pressure p_{op} exceeds 1.1 times the nominal working pressure p_{nom}, and to close when the container is again at nominal pressure. The main functions are therefore "open valve" and "close valve". The overall function can also be described as "limit pressure". Let us now assume a possible failure of the overall function, namely "valve does *not* limit pressure" (see Fig. 9.3). The valve functions shown in Fig. 9.2 and their timing are negated. Each has an OR relationship with the overall function. Each fault thus identified is next investigated in terms of its possible causes. The fault we have chosen to investigate in more detail is "does not open" (see Fig. 9.4).

An identified cause may be associated with further causes with which it has an OR or an AND relationship, and these may have to be scrutinised accordingly.

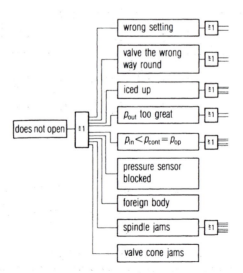

Figure 9.4.   Detail from completed fault-tree (Fig. 9.3) for the fault "does not open".

Figure 9.5 shows a selection of further causes and some of the remedies identified at this stage. Often these cannot be clarified in more detail until the embodiment phase. Grouping the remedial measures according to the departments involved (see Fig. 9.9) simplifies their execution. On the basis of the information gained from a fault-tree analysis, designers are able to improve and complete the requirements list (see Fig. 9.6) before they proceed to the embodiment phase. As a result, the design will be greatly improved and potential faults avoided.

The second example concerns the *embodiment phase*. A packing ring shaft seal in a generator connected to a turbine is used to prevent the leakage of pressurised cooling air (see Fig. 9.7). This large diameter seal interfaces with a sleeve that acts as a thermal barrier. The seal has to withstand a pressure difference of 1.5 bar. Possible malfunctions of this assembly have to be analysed.

The overall function is "prevent leakage of cooling air". At the beginning of an investigation, it is useful to clarify the sub-functions that have to be fulfilled by the various parts. When no function structure has been established, one can use a table such as the one shown in Fig. 9.8. For the "prevent" function the following sub-functions are essential:

- generate compression force;
- seal yet allow sliding; and
- remove frictional heat.

Next these sub-functions are negated and, at the same time, possible causes of malfunctions are sought (see Fig. 9.9).

The results of the fault-tree analysis point, first of all, to a malfunctioning of the thermal barrier *2* caused by unstable heat patterns (see 7.4.4-1). The frictional

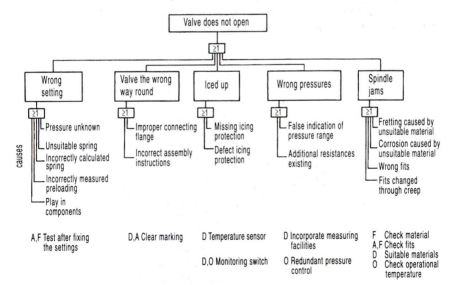

Figure 9.5.   Causes of and remedies of malfunctions, after Fig. 9.4:
D = design; P = production; A = assembly; O = operation (use and maintenance); F = formal procedure required.

Changes	D W	No.	Requirements ×)	
			Requirements List	*1st issue 1/9/73*
			for *Safety blow-off valve*	*Page* 1
				Responsible
1.9.73		22	Valve head with plane sealing surface (valve without taper)	
"		23	No rigid joint between valve head and spindle	
"		24	Easy maintenance or exchange of sealing surfaces	
"		25	Valve lift limited	
"		26	Damping of valve movement	
"	W	27	Installation in a closed, ice-proof area	
"		28	No sliding seals, avoid friction	
"		29	Ensure fool-proof mounting (e.g. different flange sizes for inlet and outlet)	
		×)	Requirements were revised after construction of fault-tree	
			Replaces issue of	

Figure 9.6.   Revision of requirements list after fault-tree analysis.

heat generated at the sliding interface can only flow away through the barrier into the shaft. This causes the barrier sleeve to heat up and expand. This increases the friction and at a certain temperature the barrier sleeve lifts off the shaft. This results in additional air leakage and damage to the shaft surface caused by the barrier sleeve slipping on the shaft. This layout is bad and the design principle needs improving: either the barrier should be removed and the seal connected to the shaft so it rotates with the shaft (removal of heat through the housing 5); or a sliding ring seal with radial sealing surfaces should be used.

Further design measures are necessary when using the packing ring shaft seal:

- The connection of housing 5 to frame 6 is insufficient and the housing can start rotating with the shaft due to the pre-loading of the packing ring seals against the shaft. The compression force from the pressure difference is too low for the O-ring seal 7 to transfer the moment through a frictional connection. *Remedy:* reposition seal 7 towards the outer diameter of housing 5; even better would be an additional form-fit connection to transfer the moment.

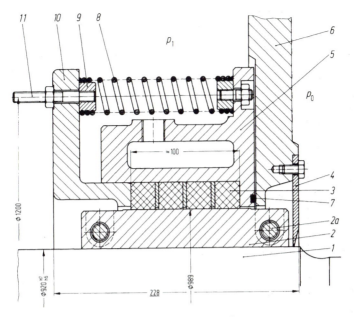

Figure 9.7. Packing ring shaft seal of a generator.

No.	Component	Function
1	Shaft	Transmit torque, carry sleeve, dissipate frictional heat
2, 2a	Sleeve (barrier)	Provide rotation and seal surface, protect shaft, dissipate frictional heat
3	Packing rings	Seal medium yet allow sliding, carry compression force and provide sealing pressure
4	Scraper ring	Protect against splashed oil
5	Gland housing	Carry packing rings, carry and transmit compression force
6	Frame	Carry components 4 and 5
7	O-ring	Seal p_1 from p_0
8	Tension spring	Generate compression force
9	Spring support	Transmit spring force
10	Transfer ring	Transmit compression force, carry tension spring
11	Bolt	Preload springs and adjust loading

Figure 9.8. Analysis of the components in Fig. 9.7 to identify their functions.

- With the current layout, the loading of springs *8* cannot be adjusted. *Remedy:* include sufficient space.
- For reasons of safety and simplicity, it is advantageous to use a compression rather than a tension spring.

Basically, designers should not only include design measures to improve the embodiment, but also measures to improve production, assembly and operation (use and maintenance) procedures, where these seem necessary. In certain cases it might be necessary to enforce specific test procedures (see Fig. 9.9).

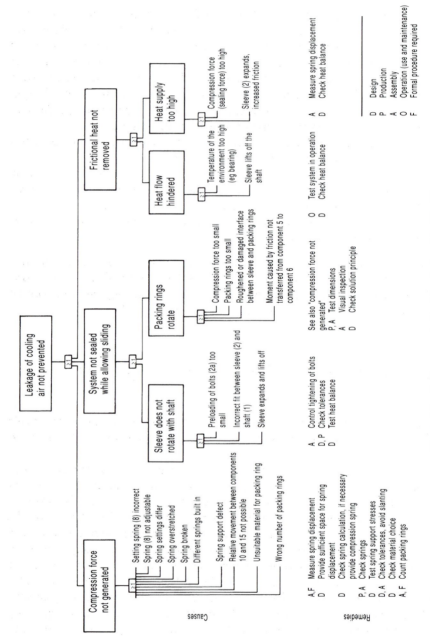

Figure 9.9. Fault-tree analysis of the shaft seal in Fig. 9.7.

In summary, the following procedure should be followed to identify and rectify faults and disturbing factors:

- Identify and negate functions.
- Search for causes of possible malfunctions from: a function structure that lacks clarity; a less than ideal working principle; a less than ideal embodiment; less than ideal materials; and less than ideal inputs caused by variations in the material, energy and signal flows. In line with the guidelines given for the embodiment design phase, further influences that might cause undesired system behaviours should be sought in the following areas: loading, shape changes, stability, resonance, wear, corrosion, sealing, safety, ergonomics, production, quality control, assembly, transport, operation and maintenance.
- Determine the prerequisites for malfunctions to occur, eg through AND relationships or even through OR relationships.
- Introduce suitable design measures by choosing another solution or making improvements to the existing solution. Quality control measures during production, assembly, transport, operation and maintenance can also be introduced. Preference, however, should always be given to the removal of a cause through an improved solution.

It has to be noted that because of the effort required to complete a full fault-tree analysis, this method is usually limited to important areas and critical processes. It is desirable that designers incorporate this way of thinking into their normal work patterns so that they can apply it almost unconsciously.

9.3   Failure Mode and Effect Analysis (FMEA)

FMEA is a formalised analytical method for the systematic identification of possible failures and the estimation of the related risks (effects) [9.9, 9.12, 9.14]. The main goal is to limit or avoid risk.

Figure 9.10 shows an FMEA chart with an example in which possible failures are listed together with their consequences, causes, risk numbers RN, proposed test measures, and suggested and applied remedial measures. The assessment of risk using risk numbers is important. They provide quantitative estimates of: the probability of occurrence; the significance of failure; and the difficulty of detection. FMEA is qualitative in nature and is a method of evaluating quality. Staff from design, development, production planning, quality control, purchasing, sales and customer service should be included in an FMEA team, for the same reasons as for their inclusion in a value analysis team (see 10.4). Apart from evaluating possible malfunctions caused by failures and disturbing factors, FMEA encourages early cooperation between the various departments involved in product development. Fault-tree analysis is intended to assist designers alone, whereas FMEA also functions as a means of handing over to production and supporting the overall quality assurance process.

Failure location/characteristic	Failure type	Failure consequence	Failure cause	Current situation: Proposed test steps	O	S	D	RN	Suggested remedial measures	Improved situation: Applied steps	O	S	D	RN
Shaft	Shaft fracture	Complete breakdown	Type of loading not identified correctly		3	10	10	300	Determine loading using suitable calculations	Proof of strength of the shaft	1	10	10	100
Bearing	Play in bearing assembly	Imprecise function fulfilment	Slacking of shaft nut during operation (impulse loading)		3	8	10	240	Additional locking of the shaft nut		1	8	10	80
	Sealing leakage	Early wear of bearings	Sealing not as required		2	5	10	100	Use of radial shaft seals recommended by DIN		1	5	10	50
Shaft-hub-connection (flange-bolt connection)	Insufficient frictional fit	Shear stress in bolts	Layout error (friction values neglected)		2	6	10	120	Application of a sufficiently high safety factor		1	6	10	60
	Precision of fittings	Joining not possible or centring insufficient	Design fault		2	5	1	10	Check tolerance calculation		1	5	1	5
	Failure of bolts	Complete breakdown	Type of loading not identified correctly		3	10	10	300	Suitable calculation for loading situation	Appropriate bolt dimensions	1	10	10	100
Cylindrical cam	Surface pressure too high	Pitting in the running surface	Lever pressure on surface too high		7	8	10	560	Suitable combination of materials and adapted geometry		2	8	10	160

TU-Berlin

Failure Mode and Effect Analysis Design (product)-FMEA ☒   **Process-FMEA** ☐

Name/ Department/ Supplier/ Telephone
Institute for Machine Design-Engineering Design

Component name
Cylindrical cam

By (Name/ Department/ Telephone)
Mr Wende

O: Occurrence   S: Significance   D: Detection   RN: Risk number

Probability of occurrence
(failure can exist)

very low	=	1
medium low	=	2-3
medium	=	4-6
medium high	=	7-8
high	=	9-10

Effect on customers

effects hardly noticeable	=	1
failures not important (little trouble to the customer)	=	2-3
reasonably serious failure	=	4-6
serious failure (annoying for the customer)	=	7-8
failure with large negative effects	=	9-10

Probability of detection
(before delivery to customers)

high	=	1
medium high	=	2-5
medium	=	6-8
medium low	=	9
low	=	10

high	=	1000
medium	=	125
no risk	=	1

After a period of use, the information in the FMEA records and analyses of the FMEA charts provide valuable insights into successful quality measures that can be used in subsequent products.

For the production process, an additional process FMEA is carried out using the same charts. This evaluation of the production steps, however, is often contained indirectly in the design (product) FMEA, because production issues should already have been taken into account during the design process.

Figure 9.10. FMEA chart with an example of the shaft, bearing and cylindrical cam of the design discussed in 7.7. (Fig. 7.159).

10 Design for Minimum Cost

10.1 Cost Factors

It is important to identify cost factors as early and as accurately as possible in the design process. This is true for all types of design, including the development of size ranges and modular products. It is well known that the majority of costs have been committed when the solution principle has been selected and its embodiment completed. During the production and assembly stages there are relatively few opportunities to reduce costs. It is important, therefore, to start cost optimisation as early as possible since any design changes that have to be made during production are usually very costly. This might prolong the design process, but overall it is more economical than a retrospective drive to reduce costs [10.18].

The *overall cost* of producing a product can be divided into direct costs and indirect costs (overheads). *Direct costs* are those costs that can be allocated directly to a specific cost carrier, for example material and labour costs for producing a specific component. *Indirect costs* are those costs that cannot be allocated directly, for example the costs of running the stores and illuminating the workshop.

Some costs depend on the number of items ordered, degree of facility utilisation or batch size. Material costs, production labour costs and consumable materials costs, for example, increase with higher turnover. In a cost calculation these are *variable costs*. *Fixed costs* are those that are incurred in a certain period and do not change, for example, management salaries, rent of space and interest on borrowings.

The *manufacturing cost* (see Fig. 10.1) is the total of the costs for material and production including additional costs such as for production tooling and fixtures, and for models and tests as far as they relate to a specific product. Manufacturing cost therefore consists of fixed and variable costs. For decision making during the design process, however, only variable costs are of interest [10.38]. This is because they are influenced directly by designers, for example, by the choice of material types, production times, batch sizes, production processes and assembly methods. Of interest, therefore, are the variable manufacturing costs which comprise direct costs and indirect costs (overheads).

Variable and fixed indirect costs are taken into account differently in different companies. Usually they are combined with the direct costs by using multiplication factors, such as a factor of 1.05 to 1.3 for indirect material costs, a factor 1.5 to

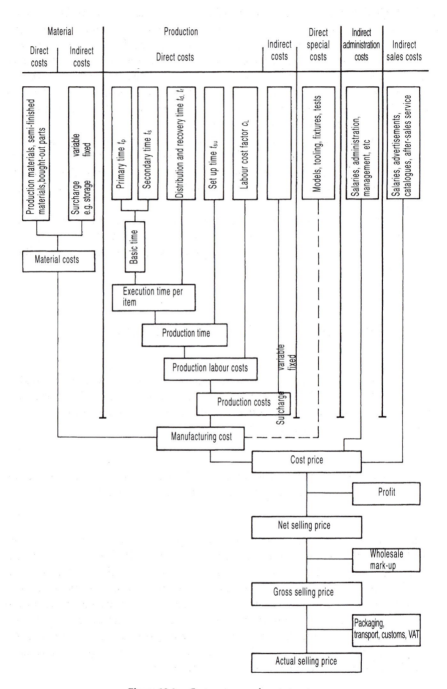

Figure 10.1. Cost sources and cost structure.

10 (or higher) for indirect production labour costs; and also by additions based on machine utilisation. The choice depends on the production processes and types of machine tool. When the factors are high or when machine utilisation is considered, it is useful to check whether it is possible, at least in theory, to reduce costs by using another production process. This avoidance strategy, however, increases the specific multiplication factors if the organisational structure does not change. This suggests a modification to factory planning to take into account the modified product and production structure.

Concluding from what has been said, variable overheads are usually taken into account through multiplication factors on the direct costs and in this way they influence the manufacturing costs. In general designers can limit themselves to calculating variable direct costs when comparing the costs of solution or product variants.

At an early stage it is important to make rough cost estimates rather than detailed calculations, which should only be undertaken when really necessary. New methods to identify costs early in the design process will be discussed in the following sections.

10.2   Fundamentals of Cost Calculations

In 10.1 the *variable part* of the manufacturing cost ($VMfC$) was recommended as a basis for decision making. It comprises direct material costs ($DMtC$) and production labour costs (PLC), including assembly costs. All costs for production and assembly operations have to be added:

$$VMfC = DMtC + \Sigma\ PLC \tag{1}$$

The *direct material costs* are determined by the weight W or volume V and the specific cost c, that is cost per unit weight or volume, as follows:

$$DMtC = c_w \cdot W = c_V \cdot V \tag{2}$$

The *direct production costs* can be calculated from the times needed for the individual production processes and assembly operations multiplied by a labour cost factor c_L (see Fig. 10.1). The production time is based on the primary time t_p, secondary time t_s and set-up time t_{su}, as well as distribution and recovery time. The last two times are generally taken into account as a constant factor on the basic time t_b, which is the sum of primary and secondary times and results in a time per unit. For the calculation of costs, therefore, primary, secondary and set-up times are important. Using the labour cost factor, the following simplified equation for a particular production operation can be used:

$$PLC \sim c_L\ (t_p + t_s + t_{su}) \tag{3}$$

For a more precise calculation, taking into account variable production overheads, see [10.21, 10.22, 10.40].

We have indicated that the *manufacturing cost* is the sum of direct and indirect costs. Direct costs are usually linked to the type of product, for example the primary time for turning [10.30] can be calculated from:

$$t_p = \frac{D \cdot \pi \cdot L \cdot i}{v_c \cdot f \cdot 1000} \quad \text{minutes} \tag{4}$$

with:

D = diameter in mm
L = length in mm
i = number of cuts
v_c = cutting speed in m/min
f = feed in mm/revolution.

The individual cost terms, for example turning cost (equation (4)), can thus be represented by exponential functions consisting of variables x with exponents p and constants K. The generic form of a cost equation, eg manufacturing cost, would be as follows:

$$VMfC = \sum_{i=1}^{n} K_i \cdot \prod_{j=1}^{m} x_{ij}^{P_{ij}} \tag{5}$$

with m equal to the number of variables x_j in cost term i and with n equal to the number of cost terms. In the case of three influencing variables, this results in:

$$VMfC = \sum_{i=1}^{n} K_i \, (x_{i1}^{P_{i1}}) \cdot (x_{i2}^{P_{i2}}) \cdot (x_{i3}^{P_{i3}}) \tag{6}$$

The variable direct costs of the manufacturing cost for a turned component, for example, comprise the direct material costs and the turning costs. For t_p equation (4) is used:

$$VMfC = \frac{\pi}{4} D^2 \cdot L \cdot c_V + \left(\frac{D \cdot \pi \cdot L \cdot i}{v_c \cdot f \cdot 1000} + t_s + t_{su} \right) c_L$$

in which the following approximations are made:

$D \approx D_{raw}$ (Diameter of raw material)
$L \approx L_{raw}$ (Length of raw material).

The variable terms of the manufacturing cost considered here can thus be represented as power series of different orders. When adding up several production operations an equivalent number of cost terms in the power series is created using equations (1) and (5) as appropriate.

For *cost estimations* based on quick or all-inclusive calculations, it is too much effort to determine the directs costs strictly according to their individual dependencies. A better way is to define *relative costs* that are more generic and have long-term validity (see 10.3.1). It is also possible to estimate costs based on the *share of the material costs* (see 10.3.2). This method is only valid when comparing items (components, assemblies) of similar size. Recently cost profiles in several companies have been analysed statistically. Using *regression analysis*, researchers have attempted to correlate variable costs and influencing factors (see 10.3.3).

To determine the regression function, a power series is selected whose exponents and coefficients are determined in such a way that the resulting equation deviates as little as possible from the findings. The selected exponents and coefficients generally do not represent the real dependencies, but only mathematical relations. In [10.27] it is shown that very different regression functions can provide good approximations for the same set of circumstances.

In cases where one influence dominates and this has been identified and introduced into the regression function by selecting the relevant variables, it is likely that this influence represents physical reality. When it is possible to relate back all cost factors to only one characteristic variable x, for example diameter or weight, the cost function can be reduced to a simple equation of the following form:

$$VMfC = a + bx^p \text{ (see 10.3.3).}$$

Extrapolation using *similarity relationships*, on the other hand, is based on the physical relationships involved in the particular technology and uses power series with the appropriate exponents taken from the equations for material costs, primary times, secondary times and times per item. The coefficients of the cost terms are derived from company specific data using a reference item (basic design or operation element) (see 10.3.4). The reason for developing this procedure was to make designers more aware of existing dependencies so that they can make more goal directed decisions. The application of similarity relationships requires the availability of a sufficiently similar item, eg part, assembly or production operation.

In the case of *geometric similarity* (see 8.1.4), very simple cost-growth laws with polynomials of maximum order 3 can be set up based on a geometric reference length. For *semi-similar variants* (see 8.1.5) (most geometric magnitudes remain constant, but some deviate from geometric similarity) the cost-growth laws consist of many terms comprising all geometric and material variables that are involved. They have the form of power series with functions that can have fractional exponents. They are often called differential cost-growth laws. They achieve relatively high precision and only require a reasonable effort to apply.

Much effort is currently going into the identification of costs at an early stage through making available such methods in combination with computer support, including CAD and Knowledge-Based Systems [10.14].

In the following sections, the different methods are explained in more detail. Which method should be used depends on the available time, required precision and the available data.

10.3   Methods for Estimating Costs

10.3.1 Comparing with Relative Costs

In this method prices or costs are related to a reference value. For this reason, the results are valid for much longer than when absolute costs are used. In [10.6]

	Name / No	Density ρ g/cm³	Young's modulus E N/mm²	Yield strength σ_y N/mm²	Tensile strength σ_T N/mm²	Strain to failure ε_F %	E/E_{St37}	σ_T/σ_{St37}	c^*_w	c^*_v	$\dfrac{c^*_w}{\sigma_T/\sigma_{T\,St37}}$ 7	$\dfrac{c^*_w}{E/E_{St37}}$	Relative costs for machining
General construction steels DIN17100	USt37-2 1.0112	7.85	$2.15 \cdot 10^5$	215...235	360...440	25	1	1	1	1	1 - 0,82	1	1
	St50-2 1.0532	7.85	$2.15 \cdot 10^5$	275...295	490...590	20	1	1.36...1.64	1.1	1.1	0.81 - 0.67	1.1	1
Cold drawn DIN1652	St37-2K+G 1.0161	7.85	$2.15 \cdot 10^5$	195...215	330...440	25	1	0.92...1.22	1.6	1.6	1.75 - 1.31	1.6	1
Machining steels DIN1651	10S20K+N 1.0721	7.85	$2.10 \cdot 10^5$	195...225	340...350	25	0.98	0.94...0.97	1.9	1.9	2.01 - 1.95	1.94	0.73
	9SMn28K+N 1.0715	7.85	$2.10 \cdot 10^5$	205...235	350...370	23	0.98	0.97...1.03	1.8	1.8	1.85 - 1.75	1.89	
	4S20K+N 1.0727	7.85	$2.10 \cdot 10^5$	305...335	570...700	14	0.98	1.58...1.94	2	2	1.26 - 1.03	2.05	
Tempering steels DIN17200	Ck35V 1.1181	7.85	$2.15 \cdot 10^5$	295...420	490...770	22...17	1	1.36...2.14	1.6	1.6	1.18 - 0.75	1.6	0.91
	Ck45V 1.1191	7.85	$2.15 \cdot 10^5$	380	630...780	17	1	1.75...2.17	1.78	1.78	1.02 - 0.82	1.78	1.05
	34Cr4V 1.7033	7.85	$2.15 \cdot 10^5$	470	700...850	15	1	1.94...2.36	2.13	2.13	1.1 - 0.9	2.13	1.43
	2CrMo4V .7225	7.85	$2.15 \cdot 10^5$	650	900...1100	12	1	2.30...3.05	2.24	2.24	0.9 - 0.73	2.24	1.73
	50CrV4V 1.8159	7.85	$2.15 \cdot 10^5$	700	900...1100	12	1	2.50...3.05	2.25	2.25	0.9 - 0.74	2.25	2.09
	C35K+N 1.0501	7.85	$2.15 \cdot 10^5$	275	490...590	22	1	1.36...1.64	1.7	1.7	1.25 - 1.04	1.7	
	Ck35K+V	7.85	$2.15 \cdot 10^5$	325...410	540...790	20...16	1	1.50...2.19	1.85	1.85	1.23 - 0.84	1.85	

Figure 10.2.   Characteristic values and relative material costs c^* for some materials (Reference USt 37.2 with $\sigma_T = 360$ N/mm²).

Case hardening steels DIN17210	C15 1.0401	7.85	2.15 · 10⁵	355...440	590...890	14...12	1	1.64...2.47	1.1	1.1	0.67 - 0.45	1.1	0.86
	Ck15 1.1141	7.85	2.15 · 10⁵	355...440	590...890	14...12	1	1.64...2.47	1.4	1.4	0.85 - 0.57	1.4	
Nitriding steels DIN17211	16MnCr5G 1.7131	7.85	2.15 · 10⁵	440...635	640...1190	11...9	1	1.78...3.30	1.7	1.7	0.96 - 0.51	1.7	1.14
	34CrAlNi7V 1.8550	7.85	2.15 · 10⁵	590	780...990	13	1	2.17...2.72	2.6	2.6	1.2 - 0.95	2.6	2.0
	41CrAlMo7V 1.8509	7.85	2.15 · 10⁵	635...735	830...1130	14...12	1	2.30...3.14	2.6	2.6	1.13 - 0.83	2.6	
	31CrMoV9 1.8519	7.85	2.15 · 10⁵	700	900...1050	13	1	2.50...2.92					2.0
Stainless steels DIN17440	X20Cr13 1.4021	7.70	2.10 · 10⁵	440...540	640...940	18...8	0.99	1.78...2.61	3.14	3.2	1.8 - 1.2	3.21	1.25
	X12CrNi188 1.4300	7.80	2.03 · 10⁵	220	500...700	50	0.94	1.39...1.94	8.45	8.4	6.06 - 4.35	8.95	
Casting materials without cores and recesses	GG - 25 0.6025	7.35	1.30 · 10⁵		250		0.60	0.69	2.0	1.2	2.88	3.3	1.24
	GS - 45 1.0443	7.85	2.15 · 10⁵	225	445...590	22	1	1.24...1.64	1.8	1.8	1.46 - 1.1	1.8	1.45
Non-ferrous metals	AlMg3F23 3.3535.26	2.66	0.70 · 10⁵	140	230	9	0.33	0.64	10.0	3.4	15.65	30.7	0.36
Light metals	AlMg5F26 3.3555.26	2.64	0.72 · 10⁵	150	250	8	0.33	0.69	11.6	3.9	16.70	34.6	
	AlMgSi1F32 3.2315.72	2.70	0.7 · 10⁵	250	310	10	0.33	0.86	8.72	3.0	10.13	26.8	0.51
Non-metals	Woven laminate Hgw 2088	1.25	7 · 10³		50		0.033	0.14	62.8	10	452.2	1928	(0.4)
	Glass fibre reinforced polyester HM 2472	1.60	10 · 10³		100		0.046	0.28					(0.71)
	Nylon 66 PA 66	1.14	2 · 10³		65		0.009	0.18	22.72	3.3	125.8	2442	(0.27)

Figure 10.2. Continued.

principles are described for creating relative cost catalogues. Catalogues for materials, semi-finished materials, and bought-out parts are common. The relative material costs c^* are usually based on a standard size of channel-section steel (USt 37-2) and can be calculated from the following equation, which uses the specific material costs c_W^* or c_V^* derived from weight and volume respectively:

$$c_{W,V}^* = \frac{c_{W,V}^*}{c_{W,V}^* \text{ (reference value)}}$$

It has to be noted that the resulting value is magnitude dependent. VDI Guideline 2225 Part 2 [10.35] therefore gives values for small, medium and large dimensions of all common materials. Material utilisation depends on the goals to be achieved. When strength requirements dominate a different material has to be selected than when stiffness requirements dominate.

Figure 10.2 lists the relative material costs c^* for some materials with medium dimensions including the relation with tensile strength σ_T (strength requirement) and with Young's modulus E (stiffness requirement). The cost relation for machining based on [10.28] is also listed. This shows, for example, that in the case of tempering steels and case hardening steels strength increases generally faster than the material cost. This indicates the economic advantages of using these materials. For stressed shapes that have to be stiff, grey cast iron and plastics are substantially more expensive than steel. However the relations listed in Fig. 10.2 change substantially in favour of cast or plastic parts when the shapes are complex and when there are additional corrosion resistance or surface finish requirements. In the case of highly alloyed materials, for example, obtaining a good surface finish can require very expensive machining.

Of particular interest are casting costs. In principle the overall cost is based on total weight, but the weight per item, number to be produced and item complexity play a role. Our own investigation [10.27] for steel castings resulted in the relationships shown in Fig. 10.3. This figure shows that for steel castings specific costs

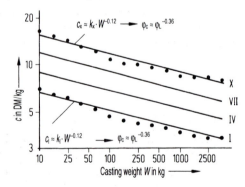

Figure 10.3. Costs for steel castings depending on weight per casting and level of complexity, after [10.27].

Level I: Solid castings without cores and recesses
Level IV: Solid castings with simple cores and recesses
Level VI: Hollow castings (cored) with simple webs and recesses
Level X: Hollow castings (cored) with complex cores.

reduce with increasing item weight, that is $\varphi_c = \varphi_w^{-0.12}$, so that the material costs increase by $\varphi_M = \varphi_w^{0.9}$, and not by $\varphi_M = \varphi_w^1$ (see cost-growth laws in 10.3.4).

For *semi-finished materials*, Fig. 10.4 shows that shape, that is whether round, square, flat plate or profiled, has little influence on specific price provided they are produced by rolling. Drawn materials are considerably more expensive (factor ≈ 1.6). Closed sections cost about twice as much for the same weight. Figure 10.4 also shows the material utilisation advantages of particular sections when subjected to bending moments. For carrying bending moments, the required sectional area, that is weight per unit length, of some sections is considerably smaller and therefore cheaper.

The relative costs of *bought-out parts* vary strongly with size (see also cost-growth laws in 10.3.4). Rieg [10.26] developed a procedure for determining and representing these costs. Figure 10.5 gives an example of such a relative cost diagram for rolling element bearings. A particular deep groove ball bearing from series 60 with $d = 50$ mm is used as a reference ($\varphi_d = 1$). The price of this bearing was 24.8 DM ($\varphi_P = 1$). The current price for a deep groove ball bearing 6007 with $d = 35$ mm is 18.33 DM. To find the price for bearing 6036 with $d = 180$ mm the procedure is as follows:

$d = 35$ mm $\qquad$ $\varphi_d = 35/50 = 0.7$ $\quad$ from Fig. 10.5: $\qquad$ $\varphi_{P_{6007}} = 0.61$

$d = 180$ mm $\qquad$ $\varphi_d = 180/50 = 3.6$ $\quad$ from Fig. 10.5: $\qquad$ $\varphi_{P_{6036}} = 28$

$P_{6036} = P_{6007} \, (\varphi_{P_{6036}} / \varphi_{P_{6007}}) = 18.33 \, (28/0.61) = 841$ DM.

Diagrams for screws, circlips, connectors and valves are given in [10.26, 10.27]. Figure 10.6 shows a cost comparison for different threaded connectors based on [10.3]. As described in 10.3.4 and 10.3.5, the cost relations can vary with size and this can be seen in Fig. 10.6.

Relative cost data has to be applied with caution, taking into account all the relevant circumstances [10.1]. When comparing and selecting items, not only must the cost relations be assessed but also the required functions, the application conditions and the space requirements. It is not sufficient simply to compare the costs of items without considering their effect on the rest of the design.

Section	Round 100 DIN 1013	Square 85 DIN 1014	L 160×17 DIN 1028	Pipe 159×5,6 DIN 2448	U 160 DIN 1026	I 160 DIN 1025
H/A in mm	12.5	14.1	20.8	37	48.3	54
I/A in mm²	625	601	2374	2944	3854	4323
c^*_w rolled	1	1.02	1.06	1.6 - 2.1	1	1
c^*_w drawn	1.6	1.6		2.8		

Figure 10.4. Specific material costs $\dot{c}_w$ for semi-finished materials. The reference first moment of area is $H \approx 10^5$ mm³ (the first moment of area of round 100 and of square 85). H/A: ratio of first moment of area/section area. I/A: ratio of second moment of area/section area.

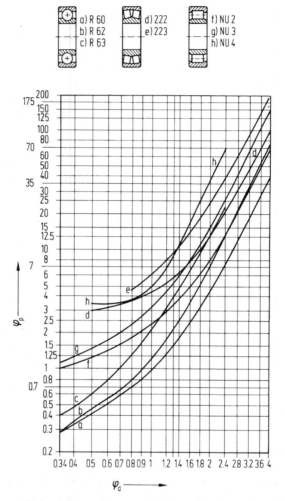

Figure 10.5. Relative costs for rolling element bearings based on [10.26]. Reference: a deep groove ball bearing from series 60 with $d = 50$ mm ($\varphi_d = 1$) and $P = 24.80$ DM ($\varphi_p = 1$).

10.3.2 Estimating Using Share of Material Costs

If in a particular application area the ratio m of material costs MtC to the manufacturing cost MfC is known and almost identical, it is possible to estimate the manufacturing cost after determining the material costs, $MfC = MtC / m$. This procedure is described in VDI Guideline 2225 [10.33]. The procedure cannot be used, however, when the cost structure changes, in particular with large size changes (see cost estimation using similarity relations in 10.3.4 and cost structures in 10.3.5).

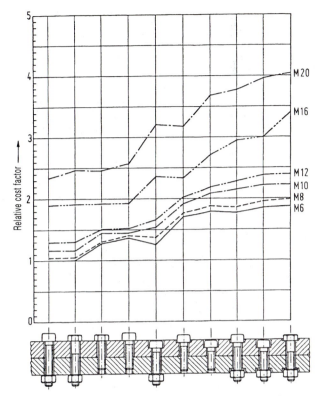

Figure 10.6. Relative cost factors for threaded connections using cap screws and hexagon bolts, M6 to M20, class 8.8, after [10.3, 10.4].

10.3.3 Estimating Using Regression Analysis

Based on a statistical analysis of data, the relation between costs or prices and the characteristic parameters (output, weight, diameter, shaft height etc) are determined. The results can be presented graphically for each of these parameters. Regression analysis is used to find a relationship that determines the regression equation using regression coefficients and exponents. Using this equation the costs can be calculated within certain limits. The effort needed to set up the equation can be considerable and usually involves computer support. The regression equation should be built up in such a way that parameters that may change, such as hourly labour rates, are represented as individual terms, or in the form of relative costs, so that they can be updated easily.

An example is the regression equation of Pacyna [10.23] for the cost of hand moulded grey iron castings:

$$C = 7.1479 \; B^{-0.0782} \; V^{0.8179} \; D^{-0.1124} \; T^{0.1655} \; P^{0.1786} \; N^{0.0387} \; \sigma_T^{0.2301} \; F^{1.0000}$$

in DM per unit, with:

$C = DMtC$, direct material costs of the cast item

B = batch size
V = material volume in litres
D = dimension ratio (see Fig. 10.7)
T = wall thickness ratio (see Fig. 10.7)
P = packing ratio (see Fig. 10.7)
N = number or cores (without cores = 0.5)
σ_T = tensile strength in N/mm²
F = factor of difficulty (normal = 1, main range 0.9 to 1.4).

This equation might need updating. Further guidelines for this procedure and examples of regression calculations can be found in [10.11, 10.13] and in VDI Guideline 2235 [10.38].

Regression analyses can also be used to set up more easily maintainable *cost functions* by introducing simplifications and similarity considerations (see 10.3.4). The following example from Klasmeier [10.19] shows the calculation of the costs for a pressure vessel for a high-voltage switch (the influencing parameters on the variable costs are shown in Fig. 10.8).

The regression equation for welded pressure vessels is:

$$VMfC = a + b \cdot d_v^{1.42} \cdot NP^{0.94} \cdot l^{0.21} \cdot t^{0.17}$$

The factors a and b cannot be given because they are commercial secrets.

We will now derive a special but simple cost function. Based on electro-technical laws:

Voltage $V \sim$ distance between electrodes $e \sim$ vessel diameter d_v, thus

$$V \sim e = k_1 \cdot d_v$$

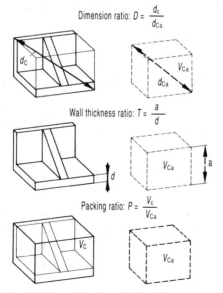

Figure 10.7.   Shape characteristics for castings [10.23]. The reference shape is a cube having the volume of the casting V_{Ca}.

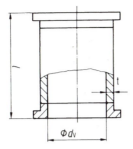

Figure 10.8. Geometric parameters of a pressure vessel for a high-voltage switch. Inner diameter of the vessel d_v. Length of the vessel l. Wall thickness of the vessel t. Nominal pressure NP.

where k_1 takes into account the conductor dimensions and the safety distance with constant gas pressure and constant temperature.

For thin walled vessels, such as the one discussed here, the standard thin-walled formula can be applied. Because the calculated wall thickness based on the required strength remains below the prescribed minimum wall thickness, we can take $t = t_{min}$ = constant.

We can also take l = constant, because the allowable voltage is independent of the vessel length.

In this way we can base the cost function simply on the variable parameter voltage:

$$VMfC = a_1 + b_1 \cdot V^{1.42}$$

10.3.4 Extrapolating Using Similarity Relations

1 Basic Design as Reference

When geometrically similar or semi-similar components are available in a size range or as variants of known components, it is useful to determine the cost-growth laws using similarity relations [10.27]. The step size of the manufacturing cost φ_{MfC} is equal to the ratio of the cost of the *sequential design* MfC_s (cost to be calculated) to the cost of the *basic design* MfC_0 (known cost) and is calculated using a similarity analysis (see 8.1.1):

$$\varphi_{VMfC} = \frac{VMfC_s}{VMfC_0} = \frac{DMtC_s}{DMtC_0} = \frac{\Sigma PLC_s}{\Sigma PLC_0}$$

The basic design (Index 0) is selected such that it can represent the largest possible size range. When the size of this design places it roughly in the centre of the range, the extrapolation errors are minimised. For the extrapolation to be valid, the sequential design should have sufficient similarity to the basic design in terms of production facilities, production processes etc.

The ratio of direct material costs to the manufacturing cost, and the ratio of the individual production costs or times (for example drilling, turning, grinding etc) to the manufacturing cost are calculated for the basic design and result in the following:

$$a_m = \frac{DMtC_0}{VMfC_0}; \quad a_{P_k} = \frac{PLC_{k_0}}{VMfC_0} \text{ for the } k\text{-th production operation.}$$

The ratios defined in this way are part of the variable manufacturing costs and represent the cost structure of the basic design (see 10.3.5).

When the cost-growth laws of the individual terms are known, the overall cost-growth law is:

$$\varphi_{VMfC} = a_m \cdot \varphi_{DMtC} + \sum_k a_{P_k} \cdot \varphi_{PLC_k}$$

When the length is the dependent characteristic parameter, this can be written generically as follows:

$$\varphi_{VMfC} = \sum_i a_i \cdot \varphi_L^{x_i}, \quad \varphi_L = \frac{L_s}{L_0} \text{ (see 8.1.1) with } \sum_i a_i = 1 \text{ and } a_i \geqslant 0$$

This procedure is not company specific. The results can be made company specific through the introduction of coefficient a_i derived from the basic design. This also ensures the use of up to date knowledge.

Determining exponents x_i that depend on the appropriate dimensions (characteristic parameter length) is easy for *geometrically similar components*. According to [10.27] one can use integer exponents for making *quick estimates*. This results in the following polynomial:

$$\varphi_{VMfC} = a_3 \cdot \varphi_L^3 + a_2 \cdot \varphi_L^2 + a_1 \cdot \varphi_L^1 + \frac{a_0}{\varphi_z} \text{ with } \varphi_z = \frac{z_s}{z_0}$$

with: z = batch size.

For material costs one can usually apply $\varphi_{DMtC} = \varphi_L^3$. For production operations Fig. 10.9 can be used [10.25, 10.27].

The terms a_i are calculated from the basic design and assigned to the individual integer exponents. The cost-growth law for the example shown in Figs 10.10 and 10.11 with $\varphi_z = 1$ becomes:

$$\varphi_{VMfC} = 0.49 \cdot \varphi_L^3 + 0.26 \cdot \varphi_L^2 + 0.20 \cdot \varphi_L + 0.05$$

A geometrically similar variant that is twice as large with $\varphi_L = 2$ would give a cost increase with step size $\varphi_{VMfC} = 5.41$.

This procedure can also be used for a more *precise extrapolation* and for *semi-similar variants* as shown in the following example of a drive shaft (see Fig. 10.12). The product is a friction welded shaft journal, with main dimensions d and l, with two drop forged disc-shaped elements welded together to form a cylinder. The component is finally turned to size.

The characteristic parameters cylinder diameter D and length B can be selected independently. The shaft journal diameter d and the journal length l have to be chosen in proportion to the cylinder diameter D.

| Machine type | Process | Exponent | | Accuracy |
		calculated	rounded	
Universal lathe	External and internal turning Threading Parting Groove turning Chamfer turning	2 ≈ 1 ≈ 1.5 ≈ 1	2 1 1 1	+ + + + +
Vertical boring and turning mill	External and internal milling	2	2	+ +
Radial drill	Drilling Threading Counter sinking	≈ 1	1	0
Drilling and milling machine	Turning Drilling Milling	≈ 1	1	0
Groove milling machine	Key groove milling	≈ 1.2	1	+
Universal cylindrical grinding machine	Surface grinding	≈ 1.8	2	+ +
Disc saw	Sawing section	≈ 2	2	0
Guillotine shears	Cutting sheet metal	1.5...1.8	2	+
Plate bending machine	Bending sheet metal	≈ 1.25	1	+
Press	Straightening section	1.6...1.7	2	+
Chamfering machine	Chamfering sheets	1	1	+ +
Flame cutting machine	Cutting sheets	1.25	1	+ +
MIG arc and manual electric arc welding	I-welds V, X, fillet, corner welds	2 2.5	2 2	+ + + +
Annealing		3	3	+ +
Sand blasting (depending on using weight or surface in the calculation)		2 o. 3	2 o. 3	+ +
Assembling		1	1	+ +
Tacking before welding		1	1	+ +
Trimming or cleaning by hand		1	1	+ +
Enamelling or coating		2	2	+ +

Figure 10.9. Exponent for the time per item for various production operations for geometrically similar items, after [10.25, 10.27].
Legend: ++ Accurate; + less than ++; 0 large deviations possible.

The relations between the times and each of the individual geometric parameters was based on an analysis of the primary and secondary times according to [10.24, 10.25, 10.27]. For turning, for example, the primary time is determined by the area of the surface to be turned, represented by the diameter and length of the component. The secondary time is constant for this size range. The welding costs, however, increase in relation to the seam thickness t, with $\varphi_t^{1.5}$, and linearly with the welding length l, that is with φ_l^1. The preparation cost for welding depends not only on the number of components but also on the square root of the weight which gives $\varphi_W^{0.5}$ or $\varphi_D \cdot \varphi_B^{0.5}$.

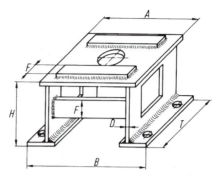

Figure 10.10. Basic design (welded) for geometrically similar series [10.27].

Operation		Costs increase with φ_L^3	Costs increase with φ_L^2	Costs increase with φ_L	Constant costs
material		800			
flame cutting				60	15
chamfering	joining			35	
tacking				105	
welding			500		
annealing		80			
sand blasting		40			
marking out				40	
horizontal boring	machin-ing			100	70
radial drilling				30	15
1890 DM = C_0		Σ_3 (= 920)	Σ_2 (= 500)	Σ_1 (370)	Σ_0(100)
		Σ_3/C_0	Σ_2/C_0	Σ_1/C_0	Σ_0/C_0
		$a_3 = 0{,}49$	$a_2 = 0{,}26$	$a_1 = 0{,}20$	$a_0 = 0{,}05$

Figure 10.11. Calculation scheme for determining cost contributions a_i to the basic design.

Figure 10.13 lists the cost contributions of the individual operations for a basic design with $D = 315$ mm and $B = 1000$ mm.

When the terms that have the same relations and parameters are brought together, the general form of the differential growth law for our example becomes:

$$\varphi_{MfC} = 0.164 \cdot \varphi_{cWC} \cdot \varphi_D^2 \cdot \varphi_B + 0.222 \cdot \varphi_{cWJ} \cdot \varphi_d^2 \cdot \varphi_l + \varphi_{cL} \, (0.081 \cdot \varphi_D^{2.5} +$$
$$0.075 \cdot \varphi_D \cdot \varphi_B + 0.113 \cdot \varphi_d \cdot \varphi_l + 0.038 \cdot \varphi_D^2 + 0.081 \cdot \varphi_D \cdot \varphi_B^{0.5} +$$
$$0.012 \cdot \varphi_D + 0.144) + 0.07$$

The diagram in Fig. 10.14 for the total cost range of the available variants is based on the use of $\varphi_D = \varphi_B = \varphi_d = \varphi_l$ for geometric similarity of the cylinder dimensions, and the use of $\varphi_D = \varphi_d = \varphi_l = $ constant with φ_B being variable for semi-similar variants. The terms φ_{cL} and φ_{cWJ} are constant for all sizes, whereas φ_{cWC} increases with 1.25 when $D = 355$ mm because of a price supplement due to a smaller batch size.

The cost curve for the variable manufacturing costs for a component or assembly size range is, even when drawn using a double logarithmic scale, curved and not linear (see Fig. 10.14). The reasons are that the direct costs always comprise constant terms, such as set-up costs for a specific batch size, and that some costs

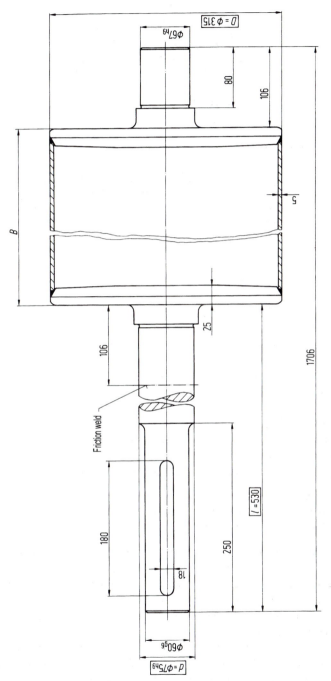

Figure 10.12. Drive shaft (basic design or reference design).

Material, production operation	Cost contribution	Cost growth law
Material		
Cylinder	0.164	$\varphi_{cWC} \cdot \varphi_D^2 \cdot \varphi_B$
Disc and journal	0.222	$\varphi_{cWJ} \cdot \varphi_d^2 \cdot \varphi_l$
Constant part	0.070	
Production operation		
Prepaing for welding	0.049	$\varphi_{cL} \cdot \varphi_D \cdot \varphi_B^{0.5}$
Welding	0.081	$\varphi_{cL} \cdot \varphi_D^{2.5}$
Trimming, cleaning	0.011	$\varphi_{cL} \cdot \varphi_D$
Turning, cylinder axial	0.054	$\varphi_{cL} \cdot \varphi_D \cdot \varphi_B$
Turning, journal axial	0.097	$\varphi_{cL} \cdot \varphi_d \cdot \varphi_l$
Turning, journal radial	0.038	$\varphi_{cL} \cdot \varphi_d^2$
Turning (constant)	0.114	φ_{cL}
Milling	0.016	$\varphi_{cL} \cdot \varphi_d \cdot \varphi_l$
Milling (constant)	0.021	φ_{cL}
Surface treating	0.021	$\varphi_{cL} \cdot \varphi_D \cdot \varphi_B$
Preparing for surface treatment	0.032	$\varphi_{cL} \cdot \varphi_D \cdot \varphi_B^{0.5}$
Cutting	0.001	$\varphi_{cL} \cdot \varphi_D$
Cutting (constant)	0.009	φ_{cL}
	1.000	

Figure 10.13. Cost contribution for the basic design of the drive shaft (see Fig. 10.12); $D = 315$ mm, $B = 1000$ mm, $\varphi_{cW} =$ step size of the specific material costs, $\varphi_{cL} =$ step size of the labour cost.

increase with high powers, such as material costs which increase with the third power of the characteristic parameter length.

The comparison between a conventional calculation and an extrapolation using cost-growth laws shows that the latter gives a sufficiently precise estimate of the costs. The estimation of the manufacturing cost is quite accurate because the large number of individual terms balance out the errors. The error is, in general, smaller than ± 10%. The individual operations, however, can have larger errors [10.17, 10.27].

Further examples can be found in [10.19, 10.24, 10.25, 10.27].

2 Operation Element as Reference

According to Beelich [10.24] a so-called *operation element* representing a specific production procedure can be used instead of a basic design. The main idea is to define a normalised, relatively simple element that has to be subjected to all the essential partial operations of the specific production operation, for example turning, grinding, welding etc. The cost of a real component is extrapolated from this simple element. The normalisation involves setting all dimension-determining geometric parameters equal to 1 so that specific production times result. For the operation element, the required production times are determined from the specific technology involved.

The next step is to determine the cost-growth law of this operation element as described before, only now using the step size $\varphi_i = X_{iP} / X_{iO}$ ($X_{iP} =$ parameter of

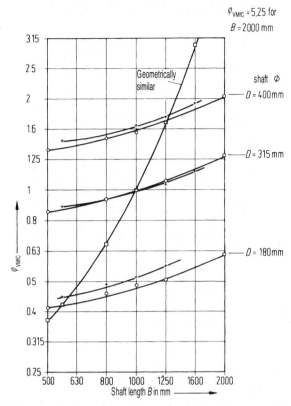

Figure 10.14.   Relative manufacturing costs for geometrically similar and semi-similar drive shafts as shown in Fig. 10.12. Basic design $D = 315$ mm, $B = 1000$ mm. Curves indicated "+" are calculated in the conventional way.

the actual component or part, and X_{iO} = parameter of the operation element).

The use of operation elements is particularly advantageous when one main production operation is involved. On the other hand, operation elements for various different production operations allow extrapolation into complex components or assemblies.

For the production operation "manual electric arc welding" Beelich [10.24] describes the generation of an operation element.

Example of the *creation of an operation element*:

The analysis of the production operation "manual electric arc welding" resulted in the following times for the various partial operations.

Times to combine, align and clamp parts into a welding assembly can be determined based on Ruckes [10.29] as follows:

$$t_{wr} = C_r \cdot \alpha \cdot \sqrt{W} \cdot \sqrt{x}$$

with:

α = factor of difficulty (see Fig. 10.15)
W = overall weight
x = number of parts.

The primary time for seam welding can be calculated from the time necessary to fill a specific seam volume with a specific volume of electrode as follows [10.24]:

$$t_{wS} = C_p \cdot t^{1.5} \cdot l$$

with:

t = seam thickness (= plate thickness for a V-weld)
l = seam length.

The secondary times for changing electrodes and initiating the welding sequence (t_{wci}), and for removing slag and cleaning the seam (t_{wrc}) relate to the number of electrodes n_e and the number of weld runs n. Both parameters can be linked and compared with the volumes and cross-sections of seam and electrode [10.24]. Analyses also revealed the influence of the factor of difficulty α. It was considered useful to include this factor as a square root:

$$t_{wci} + t_{wrc} = C_s \cdot \sqrt{\alpha} \cdot t^{1.5} \cdot l$$

The material costs of the welding material can be calculated from the specific weld seam weight W_s^* and the specific cost factor c_W :

$$MC_w = W_s^* \cdot t^2 \cdot l \cdot c_W$$

This results in the following formula for the total production cost of welding:

$$PC_w = c_L [C_r \cdot \alpha \cdot \sqrt{W} \cdot \sqrt{x} + (C_p + C_s \cdot \sqrt{\alpha})\, t^{1.5} \cdot l] + W_s^* \cdot t^2 \cdot l \cdot c_W$$

For the operation element "manual electric arc welding" (see Fig. 10.16), the welding costs can be calculated with the following normalised data and company-specific production times:

$\alpha = 1$ $c_L = 1$ MU/min (labour cost)
$W = 1$ kg $c_W = 10$ MU/kg (specific material cost)
$x = 1$ $C_r = 1$ min/kg$^{0.5}$
$t = 1$ mm $C_p = 0.8$ min/mm$^{1.5} \cdot$ m   } Specific production times
$l = 1$ m $C_s = 1.2$ min/mm$^{1.5} \cdot$ m

Type and shape of the seam		V-weld 60°	Fillet weld 90°
Type of construction	Shape of the part / Length of the seam		
2D	Tank shape / Long seam	1	2
3D	Plate, sheet metal / Short seam	1.5	2.5
3D	Sections such as U, L / Pipe	2	3
	Sections such as T, I	2.5	4

Figure 10.15. Factor of difficulty α for normal tolerances and basically right angles. (In case of higher precision and oblique angles, the factors have to be increased by 1 to 2 points.)

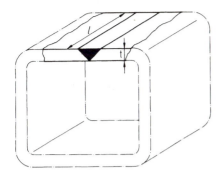

Figure 10.16. Operation element "welding".

$W_S{}^* = 0.0095 \text{ kg/mm}^2 \cdot \text{m}$ (specific seam weight with a raised seam
of $k_{rS} = 1.21$)

$PC_{wO} = 3.095$ MU (monetary units) (1 MU is approximately £0.5 in 1995)

With this information, the cost terms for operation element 0 can be calculated as follows:

$$a_r = \frac{PC_{wr0}}{PC_{w0}} = \frac{1}{3.095} = 0.32$$

$$a_{Sp} = \frac{PC_{wS0}}{PC_{w0}} = \frac{0.8}{3.095} = 0.26$$

$$a_{Ss} = \frac{PC_{wci0} + PC_{wrc0}}{PC_{w0}} = \frac{1.2}{3.095} = 0.39$$

$$a_M = \frac{MC_{w0}}{PC_{w0}} = \frac{0.095}{3.095} = 0.03$$

The resulting cost-growth law for the operation element "manual electric arc welding" thus becomes:

$$\varphi_{PCw} = \varphi_{cL}\ \underbrace{(0.32 \cdot \varphi_\alpha \cdot \varphi_w{}^{0.5} \cdot \varphi_x{}^{0.5}}\ +\ \underbrace{(0.26}\ +\ \underbrace{0.39 \cdot \varphi_\alpha{}^{0.5}}\ ) \cdot \varphi_t{}^{0.5} \cdot \varphi_l)\ +\ \underbrace{0.03 \cdot \varphi_t{}^2 \cdot \varphi_i \cdot \varphi_{cw}}$$

preparing:	welding:	changing electrodes,	welding
combining	seam	initiating welding	material
aligning	welding	sequence, removing	
clamping		slag	

With this formula the cost of components that have to be welded can be extrapolated using the relevant parameter values to determine step sizes φ.

Example of *using the operation element:*

The cost of the welded frame shown in Fig. 10.17 has to be estimated. The welding costs can be calculated with the data for the assembly and the step sizes related to the operation elements as shown in Fig. 10.18. When the values are substituted into the above equation separately for V-welds and fillet welds, the step size is:

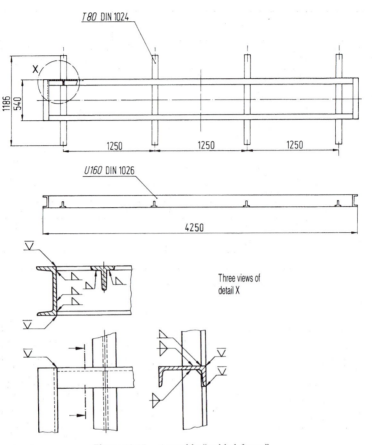

Figure 10.17.   Assembly "welded frame".

$$\varphi_{PCw} = 163.67 \quad \text{(see Fig. 10.19)}.$$

The manufacturing cost for the production operation "manual electric arc welding" thus becomes:

$$MfC = PC_{w0} \cdot \varphi_{PCw} = 3.095 \cdot 163.67 = 506 \text{ DM}$$

when one MU = 1 DM.

As the equation shows the thickness of the weld has a large influence. If the V-weld, for example, could be reduced from 10 to 8 mm, this would result in a considerable cost saving, since φ_t would be 8 rather than 10. Because of the exponents 2 and 1.5 respectively, the values of φ_S and φ_M are lower (see Fig. 10.19) and the manufacturing cost is reduced significantly:

$$MfC = 3.095 \cdot 143.22 = 443 \text{ DM}.$$

		Table for welded frame					
Production operation Material	Label			Dimension	Data from the welded assembly	Data of the operational element	Step size φ
Preparing for welding	Weight of the assembly		W	kg	226	1	226
	Number of parts		x		16	1	16
	Factor of difficulty		α		3	1	3
	Labour cost factor		c_L	DM/min	1	1	1
Seam welding	Fillet weld	Seam thickness	a	mm	4	1	4
		Seam length	l_f	m	4.52	1	4.52
		Factor of difficulty	α		3	1	3
	V-weld	Sheet thickness	t	mm	10	1	10
		Seam length	l_v	m	2.44	1	2.44
		Factor of difficulty	α		2	1	2
Welding material	Specific material costs		c_w	DM/kg	10	10	1
Date: 1/85	Author: BI						

Figure 10.18.   Table for calculating step sizes.

Production operation Material		Growth law	Calculation	Stepsize φ	
				t =10mm	t =8mm
Preparation		$\varphi_{pw} = 0.32 \cdot \varphi_{cL} \cdot \varphi_\alpha \cdot \varphi_W^{0.5} \cdot \varphi_x^{0.5}$	$0.32 \cdot 1 \cdot 3 \cdot 226^{0.5} \cdot 16^{0.5}$	57.73	57.73
Welding	Fillet weld	$\varphi_s = (0.26 + 0.39 \cdot \varphi_\alpha^{0.5}) \cdot \varphi_t^{1.5} \cdot \varphi_l \cdot \varphi_{cL}$	$(0.26+0.39 \cdot 3^{0.5}) \cdot 4^{1.5} \cdot 4.52 \cdot 1$	33.83	33.83
	V-weld		$(0.26+0.39 \cdot 2^{0.5}) \cdot 10^{1.5} \cdot 2.44 \cdot 1$	62.62	44.81
Welding material	Fillet weld	$\varphi_M = 0.03 \cdot \varphi_t^2 \cdot \varphi_l \cdot \varphi_{cw}$	$0.03 \cdot 4^2 \cdot 4.52 \cdot 1$	2.17	2.17
	V-weld		$0.03 \cdot 10^2 \cdot 2.44 \cdot 1$	7.32	4.68
			$\varphi_{PCw} =$	163.67	143.22

Figure 10.19.   Calculating the step size for the welding costs of the operation element "welding" for the welded assembly in Fig. 10.17.

3   Regression Analysis as Reference

In cases where a physically recognisable regularity of certain cost factors is not available and can only be determined statistically, the results of a regression analysis can be used to support similarity considerations. The following example is taken from the work of Klasmeier [10.19].

The production costs of a pressure vessel for a heat exchanger shown in Fig. 10.20 are to be estimated. Both flat end plates are machined and then welded into

Figure 10.20.   Pressure vessel for a heat exchanger (with flat end plates and submerged arc welding). d = 500 mm (constant), l = 2000 mm (constant).

a cylinder. A regression analysis resulted in the following time per item for the machining of the end plates:

$$t_{i_E} = z_1 \left[100.02 + 1.23 \cdot 10^{-4} \left(\frac{s_E}{4} \right)^{0.88} \cdot d^{1.87} \right] \text{ in minutes,}$$

with:

s_E = end plate thickness in mm
d = inner diameter of the cylinder (selected for strength requirements)
z_1 = allowance factor (= 1.32).

This calculation includes primary and secondary times for a specialised production workshop. The set-up time is 40 minutes.

The time for submerged arc welding of the relatively thick circular end plates is also calculated using regression analysis and results in:

$$t_{i_w} = z_1 \left(144.35 + 0.011 \cdot s_C^{1.07} \cdot d^{1.04} \cdot n_{cw}^{0.13} \right)$$

with:

s_C = cylinder wall thickness
n_{cw} = number of circular welds
Set-up time = 30 minutes.

Using these equations, the cost for a basic design can be calculated. Here we use s_{C0} = 14.3 mm, d_0 = 500 mm and s_{E0} = 42.2 mm (see Fig. 10.20). At the same time only s_E and s_C are taken as variables (ie d = constant) for the further development of the cost-growth law. It can be seen that with increasing pressure p, the end plate thickness is governed by the following equation for strength requirements:

$$\varphi_{s_E} = \varphi_p^{1/2}$$

For the cylinder wall thickness, the pressure vessel formula can be used:

$$\varphi_{s_C} = \varphi_p$$

The production costs can now be reduced to these cost effects related to the pressure p. Taking into account the indirect production costs (overheads) by using the multiplication factor g_L = 1.9, this results in:

$$PC_{E0} = 150.1 \langle p^0 \rangle + 127.2 \langle p^{0.44} \rangle \text{ DM}$$
$$PC_{w0} = 108.9 \langle p^0 \rangle + 81.5 \langle p^{1.07} \rangle \text{ DM}$$

The $\langle \rangle$ indicates the dependency of these terms on pressure.

Using the scheme in Fig. 10.11, the cost-growth law can be summarised as follows:

	$\varphi_p^{1.07}$	$\varphi_p^{0.44}$	φ_p^0
PC_{E0}		127.3	150.1
PC_{w0}	81.5		108.9
	81.5	127.3	259.0
	0.174	0.272	0.554 (Terms)

$$\varphi_{PC} = 0.174 \varphi_p^{1.07} + 0.272 \varphi_p^{0.44} + 0.554.$$

A regression analysis in which the mechanical machining and welding were included gave, for example, the following cost function:

$$\varphi_{PC} = 0.45\varphi_p^{0.701} + 0.55$$

This shows a considerable similarity to the cost-growth law for an adapted "middle" exponent for p.

10.3.5 Cost Structures

In the previous discussion it became clear that the *cost structure* changes with the overall dimensions and with semi-similar variants. Dominating are the cost terms that increase with φ_L^3 and φ_L^2 such as material costs and surface finish costs. Figure 10.21 shows the change in manufacturing cost structure in relation to overall dimensions and batch size based on Ehrlenspiel [10.11]. With increasing batch size, the one-off costs and the terms that are independent of the dimensions, which are mainly the set-up costs, are reduced. Figure 10.22 shows the cost structure in relation to the overall dimensions for the example shown in Fig. 10.10. This figure shows that when the overall dimensions vary from $\varphi_L = 0.4$ to $\varphi_L = 2.5$, ie with a factor 6.25, the cost structure changes from an emphasis on production costs to an emphasis on material costs. Cost structures for cast items can be found in [10.15].

Without knowledge of the cost structure, that is, without knowledge of the contributions of direct material costs and production labour costs to the variable manufacturing costs, designers cannot identify the measures that would lead to cost reduction. Therefore it is important to provide the appropriate data. For original designs, estimates based on rough calculations or similarity relations are useful. For adaptive designs, useful data are final calculations from previous designs. Figure 10.23 shows an example of the cost distribution for a synchronous generator [10.20]. This shows, for example, that it is not advantageous to redesign rotor

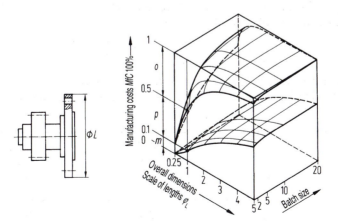

Figure 10.21. Manufacturing cost structure for gear boxes depending on overall dimensions, or length step size φ_L, and batch size, after [10.11, 10.38]. m = material costs contribution, p = production costs contribution, o = costs that occur only once (set-up costs).

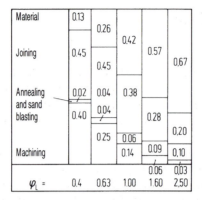

Figure 10.22.   Cost structure for the example in Fig. 10.10, showing a large change in the contributions of the various factors when the overall dimensions change.

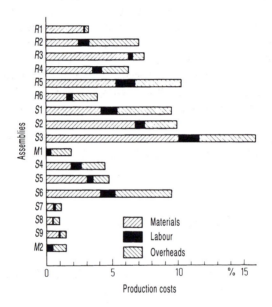

Figure 10.23.   Cost structure of a synchronous generator, from [10.20] (Siemens). Examples *R1*: rotor shaft; *R2*: rotor body; *S3*: stator housing; *S5*: bearing; *S6*: spider; *M2*: mountings etc.

shaft *R1* to reduce production labour costs and indirect production costs. A weight reduction or a more suitable choice of material, however, could lead to substantial cost reductions because of the high contribution of material costs. The situation is different for stator housing *S3* because the high contribution of indirect production costs indicates advantages in changing the production process by modifying the design.

10.4 Value Analysis

The objective of value analysis is to reduce costs (see 1.2.3-3). Value analysis applies various methods to capture and evaluate cost data during the further development and improvement of products [10.10, 10.42, 10.43]. It requires teamwork and function-oriented cost decisions. This results in two main features. The first is interdisciplinary collaboration between relevant company departments, that is, communication between experts in sales, purchase, design, production and cost estimating (value analysis team). This has been found to be essential for discussing and evaluating given solutions. The combination of a wide range of knowledge and experience is more likely to lead to a solution suitable for production with acceptable cost, and to do so quickly, than when individual designers undertake this procedure.

The second feature is the division of the overall function into sub-functions of reduced complexity and the allocation of function carriers, that is, of assemblies and components. From the costs calculated for the individual components, it is possible to estimate what the cost would be to realise the required overall function and necessary sub-functions. Such "function costs" can then be an important basis for the evaluation of design variants because they take into account equally the sales viewpoint (necessity of all functions), design viewpoint (selection of concepts and layouts), and production viewpoint (embodiment of the individual components).

The content and aims of the guidelines for the development of solutions given in this book are in line with those of value analysis.

10.5 Rules for Minimising Costs

In addition to the statements made in 7.5.7 and 7.5.8, the following general rules to minimise costs can be stated [10.13, 10.38]:

- Aim for low complexity, that is, a low number of parts and few production processes.
- Aim for small overall dimensions to reduce material costs, because these costs increase disproportionately with size, most frequently diameter.
- Aim for large numbers (large batch size) to spread the once-only costs, because, for example, set-up costs can be spread, high performance production processes can be used, and benefits of repetition can be exploited.
- Aim for minimising precision requirements, that is specify, where possible, large tolerances and rough surface finishes.

In applying these rules one has to take into account the task and the size of the artefact.

With respect to costs, it has been shown that the economic viewpoint does not have to contradict the environmental viewpoint – in fact they can be mutually supportive [10.2]. This is particularly true when energy and material reducing measures are taken into account during the search for solutions and the embodiment. This results both in a reduction of costs and a reduction of resources and environmental load. This is illustrated with the following checklist.

Save energy by:

- avoiding energy transformations (see 6.3: Establishing Function Structures);
- reducing flow losses;
- reducing friction losses (see 7.4.1: Principle of Balanced Forces);
- using waste energy (see 7.4.3: Principle of Self-Help);
- using machine sizes suitable for the process;
- dividing the system into sub-systems to achieve a higher overall efficiency; and
- using machine components with reduced losses.

Save material by:

- selecting suitable materials (see 10.3.1: Relative Costs);
- adopting tension/compression force transfer (see 7.4.1: Principle of Short and Direct Force Transmission Paths);
- selecting the best sections for the loading (see 10.3.1: Relative Costs);
- distributing and channelling flows effectively (see 7.4.2: Principle of Division of Tasks);
- increasing speed;
- adopting integral construction and function integration (see 7.3.2: Simplicity and 7.5.7: Design for Production);
- avoiding overdesign while maintaining safety (see 7.3.3: Safety, Principle of Damage Limitation); and
- producing components using material saving processes such as casting, forging, deep drawing etc.

11 Summary

11.1 The Systematic Approach

After examining the historical background, the fundamentals and generally applicable problem solving and evaluation methods, this book describes a step-by-step approach to design, starting with product planning and clarification of the task and proceeding to the conceptual and embodiment design phases. As a special contribution to cost-effective design, the development of size ranges and modular products is discussed.

Conceptual design and *embodiment design* are the two crucial phases in the creation of technical products or systems. Their respective steps are shown in Figs. 11.1 and 11.2, where the various methods are correlated with their main or supporting applications (see also [11.7] for an overview of methods). The figures also chart the *progress* of engineering design work and the *importance* and the *timing* of the various steps. The proposed correlations are not, of course, absolutely fixed, because tasks and problems differ from product to product, and because some of the steps may have to be omitted.

The various methods should only be applied when they are required and useful. Work should never be done for the sake of systematics or for pedantic reasons alone. Depending on their inclinations, experiences and skills, designers will tend to prefer certain methods. This is particularly true when several methods are appropriate for a particular step, as this helps to switch between different viewpoints and thinking levels. Switching can be achieved by large jumps forward (executing concrete steps early in the process) and subsequently returning to the original step (analysing the results and creating new ideas). When searching for solutions, switching between thinking levels plays a particularly important role.

The ability to abstract, to work systematically and to think logically and creatively complement the professional knowledge of designers. In the various design steps, these abilities are demanded to varying degrees. *Abstraction* is needed particularly for identifying essential problems, for setting up function structures, for determining the characteristics of classification schemes and for applying the principles and rules of embodiment design. *Systematic and logical thinking* help in elaborating function structures, in setting up classification schemes, in analysing systems and processes, in combining elements, in identifying faults, and in evaluating solutions. *Creative ability* helps in varying function structures, in searching for

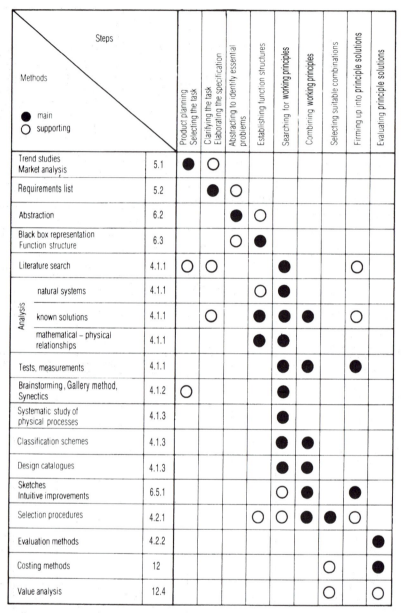

Methods (● main, ○ supporting)		Product planning / Selecting the task	Clarifying the task / Elaborating the specification	Abstracting to identify essential problems	Establishing function structures	Searching for working principles	Combining working principles	Selecting suitable combinations	Firming up into principle solutions	Evaluating principle solutions
Trend studies / Market analysis	5.1	●	○							
Requirements list	5.2		●	○						
Abstraction	6.2			●	○					
Black box representation / Function structure	6.3			○	●					
Literature search	4.1.1	○	○			●			○	
Analysis — natural systems	4.1.1				○	●				
Analysis — known solutions	4.1.1		○			●	●	●	○	
Analysis — mathematical – physical relationships	4.1.1					●	●			
Tests, measurements	4.1.1					●	●		●	
Brainstorming, Gallery method, Synectics	4.1.2	○				●				
Systematic study of physical processes	4.1.3					●				
Classification schemes	4.1.3					●	●			
Design catalogues	4.1.3					●	●			
Sketches / Intuitive improvements	6.5.1					○	●		●	
Selection procedures	4.2.1					○	○	●	●	○
Evaluation methods	4.2.2									●
Costing methods	12								○	●
Value analysis	12.4								○	○

Figure 11.1. Correlation of methods with the various steps of the conceptual design phase (numbers refer to chapters and sections).

solutions with the help of intuitive methods, in combining elements with the help of classification schemes or design catalogues, and in applying the basic rules, principles and guidelines. *Professional knowledge* is needed particularly for drawing up requirements lists, for searching for weak links, for selecting and evaluating, and for checking using the various checklists and fault tracing methods.

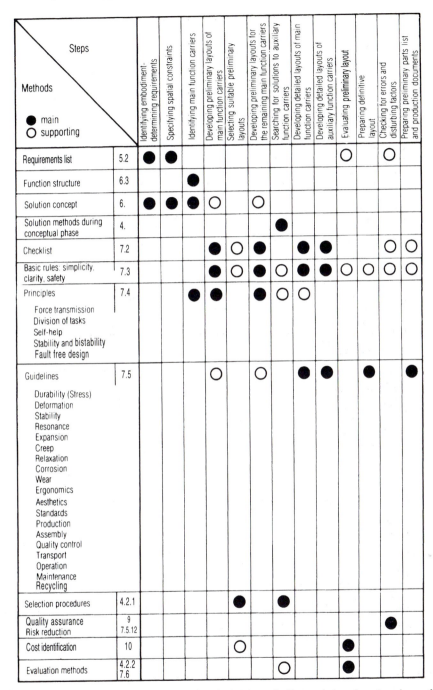

Figure 11.2. Correlation of methods with the steps of the embodiment design phase (numbers refer to chapters and sections).

Figure 11.3 lists the *guidelines* and their main characteristics that support the creative and corrective activities in the various design phases. The lists are in accordance with the general suggestions given in 2.1.8, and ensure that the technical function is implemented economically and safely. To find solutions, the relations between function structure, working structure and construction structure should be considered, as well as the general and task specific constraints. The characteristics are adapted to the level of concretisation.

Before a requirements list can be drawn up, the requirements must be known in detail so that the functions and important constraints can be identified. For that reason, the main characteristic "function" makes way for the associated

Clarifying the task	Conceptual design		Embodiment design	
Elaborating the requirements list	Selecting	Evaluating	Embodying	Evaluating
Determining requirements	Identifying the best combination of principles	Discovering optimum concept	Checking embodiment Determining layout, forms and materials	Identifying optimum embodiment
(Figure 5.7)	(Figure 4.20)	(Figure 6.29)	(Figure 7.3)	(Figure 7.147)
Geometry	Compatible with the overall task	Function	Function	Function, working principle
Kinematics		Working principle	Working principle	Layout design
Forces	Fulfils demands of the specification	Embodiment	Layout Durability	Form design
Energy			Deformation Stability	
Material	Realisable in principle		Resonance Expansion	
Signals			Corrosion Wear	
Safety	Incorporates direct safety measures	Safety	Safety	Safety
Ergonomics		Ergonomics	Ergonomics	Ergonomics
Production		Production	Production	Production
Quality control		Quality control	Quality control	Quality control
Assembly	Preferred by designer's company	Assembly	Assembly	Assembly
Transport		Transport	Transport	Transport
Operation		Operation	Operation	Operation
Maintenance	Within permissible costs	Maintenance	Maintenance	Maintenance
Recycling		Recycling	Recycling	Recycling
Costs		Costs	Costs	Costs
Schedules			Schedules	Schedules

Figure 11.3. Summary of checklists with main characteristics and related working steps.

characteristics "geometry", "kinematics", "forces", "energy", "material" and "signals", all of which facilitate the identification and description of the overall function. Similarly in the embodiment phase, the characteristic "embodiment" is replaced with the appropriate "layout" characteristics. Similar characteristics apply to evaluation; they have a welcome redundancy which ensures that they cover all contingencies. Methods of cost identification should be applied as early as possible, but during the embodiment phase at the very latest.

Some of the methods we have examined are applicable at different levels of embodiment and can therefore be used *repeatedly*. This is particularly the case with documentation (for example, requirements lists, function structures, selection and evaluation charts). Moreover, it has been found that systematically elaborated documents for certain product groups have a wider application in that they can be used for other products, thus reducing the overall effort of the systematic approach.

11.2 Experiences of Applying the Systematic Approach in Practice

The overall approach and the specific methods described in this book have been applied many times in the last few years to solve problems in industry. They have been applied by engineering students working on projects in industry, by faculty assisting with industrial projects, and by practising designers in industry. The experiences gained from these activities have been analysed and published [11.1, 11.2, 11.6]. With respect to the individual methods, the following conclusions can be drawn:

- Task clarification and drawing up the *requirements list* prove to be essential and important methods.
- Abstracting and creating *function structures* often causes difficulties because of the abstract representation. Designers are more used to thinking in objects and visual images [11.4]. Nevertheless it is necessary and helpful that at least the main functions are identified and listed.
- *Intuitive search methods* are mainly applied when no solution seems achievable using conventional methods. For embodiment issues the gallery method is more effective than brainstorming. Both, however, can only encourage ideas. A careful analysis and further development of the results are necessary.
- *Discursive solution methods*, such as classification schemes and morphological matrices, initially cause some difficulties because the appropriate but abstract classifying criteria and their characteristics are not, or not fully, recognised. This suggests insufficient training in the systematic approach. However when such systematics are recognised and applied, they help to provide a more fundamental overview leading to better solutions and more patent possibilities. They also help to compare the solutions of competitors.

- *Selection and evaluation methods* are frequently used, but often, from a systematic point of view, they are combined in ways that are not recommended, resulting in individual approaches. Despite how they are often applied, selection and evaluation methods nevertheless help designers to make more objective decisions. In most cases these methods are essential.

The objection is often raised that applying a systematic approach during the *conceptual design phase* takes too much time. It is true that the time needed for this phase increases for original designs. However, the time normally needed in this phase for concretising ideas into principle solutions, for example through rough calculations, developing solutions, and analysis of various layouts, is about the same as when a systematic approach is not used, that is, around 60 to 70%. Experience shows that any apparent increase in the time required is repaid several times over in the subsequent embodiment and detail design phases because irritations, sidetracks and renewed searches for solutions are avoided. Design work becomes more goal directed and efficient.

The *embodiment design phase* can also benefit directly from a systematic approach. Applying the basic rules, principles and guidelines of embodiment design usually reduces work effort, avoids errors and disturbances, and improves material utilisation and product quality. Checking solutions using the methods for identifying faults and unnecessary costs also improves product quality, and only takes too much time when not limited to the essential. Evaluations do not take much time in comparison to the benefits, in particular when searching for weak spots.

In summary:

- Industrial companies express a clear interest in systematic design especially when they are involved in original design or plan to introduce CAD.
- The systematic approach is being widely accepted in industry, although this may involve only the application of individual methods as the needs arise.
- In particular, a systematic approach is being adopted for developing new designs where it is necessary to generate unconventional ideas, that is, to fulfil new functions with new solutions.
- The approach has hardly been introduced at all for adaptive or variant design [11.2, 11.3]. This is understandable because working with functions and function structures is not the most important task in these types of design. Adaptive and variant design benefit more from computer support.

Industrial companies applying a systematic approach state:

- The number of patents, in particular defensive patents, increases.
- The overall duration of development projects is shorter, despite longer conceptual design phases.
- The probability of finding good solutions is greater.
- It is easier to manage the increasing complexity of problems and products.
- Creativity increases while maintaining realistic deadlines [11.5].
- A transfer effect is noticeable, that is, staff work more systematically in other areas.

The following side effects are observed:

- Information flow is improved.
- Teamwork and motivation benefit.
- Communication with clients increases.

A particular success of systematic design is that young engineers taught the approach and methods can contribute to a company surprisingly quickly, without first having to gain extensive experience.

The following aspects have been criticised by industry:

- Procedures for estimating costs are insufficiently developed.
- The approach can only be successfully applied when designers and managers have both been trained to use it; and both groups consistently require the other to apply it.
- Intuition and creativity cannot be replaced by a systematic approach, it can only support them.

The overall conclusion is clear: the benefits of applying a systematic approach to design far outweigh any disadvantages.

References

Chapter 1

1.1. Altshuler, G.S.: Erfinden – Wege zur Lösung technischer Probleme. Technik 1984.

1.2. Andreasen, M.M.; Kähler, S.; Lund, T.: Design for Assembly. Berlin: Springer 1983. Deutsche Ausgabe: Montagegerechtes Konstruieren. Berlin: Springer 1985.

1.3. Andreasen, M.M.; Hein, L.: Integrated Product Development. Bedford, Berlin: IFS (Publications) Ltd, Springer 1987.

1.4. Andreasen, M.M.: Methodical Design Framed by New Procedures. Proceedings of ICED 91, Schriftenreihe WDK 20, Zürich: HEURISTA 1991.

1.5. Archer, L.B.: The Implications for the Study of Design Methods of Recent Developments in Neighbouring Disciplines. Proceedings of ICED 85, Schriftenreihe WDK 12, Zürich: HEURISTA 1985.

1.6. Bach, C.: Die Maschinenelemente. Stuttgart: Arnold Bergsträsser Verlasbuchhandlung, 1. Aufl. 1880, 12. Aufl. 1920.

1.7. Bauer, C.-O.: Anforderungen aus der Produkthaftung an den Konstrukteur. Beispiel: Verbindungstechnik. Konstruktion 42 (1990) 261–265.

1.8. Beitz, W.: Systemtechnik im Ingenieurbereich. VDI-Berichte Nr. 174. Düsseldorf: VDI-Verlag 1971 (mit weiteren Literaturhinweisen).

1.9. Beitz, W.: Systemtechnik in der Konstruktion. DIN-Mitteilungen 49 (1970) 295–302.

1.10. Birkhofer, H.: Konstruieren im Sondermaschinenbau – Erfahrungen mit Methodik und Rechnereinsatz. VDI-Berichte Nr. 812, Düsseldorf: VDI-Verlag 1990.

1.11. Birkhofer, H.: Von der Produktidee zum Produkt – Eine kritische Betrachtung zur Auswahl und Bewertung in der Konstruktion. Festschrift zum 65. Geburtstag von G. Pahl. Herausgeber: F.G. Kollmann, TU Darmstadt 1990.

1.12. Bischoff, W.; Hansen, F.: Rationelles Konstruieren. Konstruktionsbücher Bd. 5. Berlin: VEB-Verlag Technik 1953.

1.13. Bjärnemo, R.: Evaluation and Decision Techniques in the Engineering Design Process – In practice. Proceedings of ICED 91, Schriftenreihe WDK 20. Zürich: HEURISTA 1991.

1.14. Blass, E.: Verfahren mit Systemtechnik entwickelt. VDI-Nachrichten Nr. 29 (1981).

1.15. Bock, A.: Konstruktionssystematik – die Methode der ordnenden Gesichtspunkte. Feingerätetechnik 4 (1955) 4.

1.16. Brader, C.; Höhl, G.: Rechnereinsatz in der Konzeptphase des Konstruktionsprozesses. VDI-Berichte 219, Düsseldorf: VDI-Verlag 1974.

1.17. Brankamp, K.; Wiendahl, H.P.: Rechnergestütztes Konstruieren – Voraussetzung und Möglichkeiten. Konstruktion 23 (1971) 168–178.

1.18. Breiing, A.: Analyse des methodischen Vorgehens im Konstruktionsprozeß. Bericht IKB-B-006/90, Zürich: ETH 1990.

1.19.   Büchel, A.: Systems Engineering: Industrielle Organisation 38 (1969) 373–385.

1.20.   Chestnut, H.: Systems Engineering Tools. New York: Wiley & Sons Inc. 1965, 8ff.

1.21.   Cross, N.: Engineering Design Methods. Chichester: J Wiley & Sons Ltd. 1989.

1.22.   Daenzer, W.F.: Systems Engineering. Köln: P. Haunstein-Verlag 1978–79.

1.23.   De Boer, S.J.: Decision Methods and Techniques in Methodical Engineering Design. De Lier: Academisch Boeken Centrum 1989.

1.24.   Dietrych, J.; Rugenstein, J.: Einführung in die Konstruktionswissenschaft. Gliwice: Politechnika Slaska IM. W. Pstrowskiego 1982.

1.25.   Dixon, J.R.: Design Engineering: Inventiveness, Analysis, and Decision Making. New York: McGraw-Hill 1966.

1.26.   Dixon, J.R.: On Research Methodology Towards – A Scientific Theory of Engineering Design. In Design Theory 88 (ed. by S.L. Newsome, W.R. Spillers, S. Finger). New York: Springer 1988.

1.27.   Ehrlenspiel, K.: Kostengünstig konstruieren. Berlin: Springer 1985.

1.28.   Ehrlenspiel, K.; Figel, K.: Applications of Expert Systems in Machine Design. Konstruktion 39 (1987) 280–284.

1.29.   Elmaragh, W.H.; Seering, W.P.; Ullman, D.G.: Design Theory and Methodology-DTM 89. AS-ME DE – Vol. 17, New York 1989.

1.30.   Erkens, A.: Beiträge zur Konstruktionserziehung. Z. VDI 72 (1928) 17–21.

1.31.   Eversheim, W.: Eine analytische Betrachtung von Konstruktionsaufgaben. Industrieanzeiger 91 (1969) H. 87.

1.32.   Federn, K.: Wandel in der konstruktiven Gestaltung. Chem.-Ing.-Techn. 42 (1970) 729–737.

1.33.   Findeisen, D.: Dynamisches System Schwingprüfmaschine. Fortschrittberichte der VDI-Z, Reihe 11, Nr. 18. Düsseldorf: VDI-Verlag 1974.

1.34.   Finger, S.; Dixon, J.R.: A Review of Research in Mechanical Engineering Design. Research in Engineering Design (1989) Vol. 1, Nr. 1/2, 51–67 und 121–137.

1.35.   Flemming, M.: Die Bedeutung von Bauweisen für die Konstruktion. Proceedings of ICED 91, Schriftenreihe WDK 20. Zürich: HEURISTA 1991.

1.36.   Flursheim, C.: Engineering Design Interfaces: A Management Philosophy. London: The Design Council 1977.

1.37.   Flursheim, C.: Industrial Design and Engineering. London: The Design Council 1985.

1.38.   Franke, H.J.: Konstruktionsmethodik und Konstruktionspraxis – eine kritische Betrachtung. In: Proceedings of ICED '85 Hamburg. Zürich: HEURISTA 1985.

1.39.   Franke, H.M.: Der Lebenszyklus technischer Produkte. VDI-Berichte Nr. 512. Düsseldorf: VDI-Verlag 1984.

1.40.   Franke, R.: Vom Aufbau der Getriebe. Düsseldorf: VDI-Verlag 1948/1951.

1.41.   Franz, L.; Hofmann, M.: CAD/CAM-Systeme im Maschinenbau. Leipzig: Fachbuch 1989.

1.42.   French, M.J.: Conceptual Design for Engineers. London, Berlin: The Design Council, Springer 1985.

1.43.   French, M.J.: Invention and Evolution: Design in Nature and Engineering. Cambridge: C.U.P. 1988.

1.44.   Frick, R.: Integration der industriellen Formgestaltung in den Erzeugnis-Entwicklungsprozeß Habilitationsschrift TH Karl-Marx-Stadt 1979.

1.45.   Frick, R.: Arbeit des Industrial Designers im Entwicklungsteam. Konstruktion 42 (1990) 149–156.

1.46.   Gasparski, W.: On Design Differently. Proceedings of ICED 87, Schriftenreihe WDK 13. New York: ASME 1987.

1.47.   Gierse, F.J.: Wertanalyse und Konstruktionsmethodik in der Produktentwicklung. VDI-Berichte Nr. 430. Düsseldorf: VDI-Verlag 1981.

1.48.   Gierse, F.J.: Funktionen und Funktionen-Strukturen, zentrale Werkzeuge der Wertanalyse. VDI-Berichte Nr. 849, Düsseldorf: VDI-Verlag 1990.

1.49.   Glegg, G.L.: The Design of Design. Cambridge: C.U.P. 1969.

1.50.   Glegg, G.L.: The Development of Design. Cambridge: C.U.P. 1981.

1.52.   Gregory, S.A.: Creativity in Engineering. London: Butterworth 1970.

1.53.   Groeger, B.: Ein System zur rechnerunterstützten und wissensbasierten Bearbeitung des Konstruktionsprozesses. Konstruktion 42 (1990) 91–96.

1.54. Groeger, B.: Die Einbeziehung der Wissensverarbeitung in den rechnerunterstützten Konstruktionsprozeß. Diss. TU Berlin 1991.

1.55. Hales, C.; Wallace, K.M.: Systematic Design in Practice. Proceedings of ICED 91, Schriftenreihe WDK 20. Zürich: HEURISTA 1991.

1.56. Hales, C.: Analysis of the Engineering Design Process in an Industrial Context. Eastleigh/Hampshire: Gants Hill Publications 1987.

1.57. Hansen, F.: Konstruktionssystematik. Berlin: VEB-Verlag Technik 1956.

1.58. Hansen, F.: Konstruktionssystematik, 2. Aufl. Berlin: VEB-Verlag Technik 1965.

1.59. Hansen, F.: Konstruktionswissenschaft – Grundlagen und Methoden. München: Hanser 1974.

1.60. Heinrich, W.: Eine systematische Betrachtung der konstruktiven Entwicklung technischer Erzeugnisse. Habilitationsschrift TU Dresden 1976.

1.61. Heinrich, W.: Kreatives Problemlösen in der Konstruktion. Konstruktion 44 (1992) 57–63.

1.62. Hennig, J.: Ein Beitrag zur Methodik der Verarbeitungsmaschinenlehre, Habilitationsschrift TU Dresden 1976.

1.63. Höhne, G.: Struktursynthese und Variationstechnik beim Konstruieren. Habilitationsschrift, TH Ilmenau 1983.

1.64. Hongo, K.; Nakajima, N.: Relevant Features of the Decade 1981–91 of the Theories of Design in Japan. Proceedings of ICED 91, Schriftenreihe WDK 20. Zürich: HEURISTA 1991.

1.65. Hubka, V.: Theorie technischer Systeme. Berlin: Springer 1984.

1.66. Hubka, V.; Eder, W.E.: Theory of Technical Systems – A Total Concept Theory for Engineering Design. Berlin: Springer 1988.

1.67. Hubka, V.; Schregenberger, J.W.: Eine Ordnung konstruktionswissenschaftlicher Aussagen. VDI-Z 131 (1989) 33–36.

1.68. Jakobsen, K.: Functional Requirements in the Design Process. In: "Modern Design Principles". Trondheim: Tapir 1988.

1.69. Jung, A.: Funktionale Gestaltbildung – Gestaltende Konstruktionslehre für Vorrichtungen, Geräte, Instrumente und Maschinen. Berlin: Springer 1989.

1.70. Kannapan, S.M.; Marshek, K.M.: Design Synthetic Reasoning: A Methodology for Mechanical Design. Research in Engineering Design (1991), Vol. 2, Nr. 4, 221–238.

1.71. Kesselring, F.: Die starke Konstruktion. Z VDI 86 (1942) 321–330, 749–752.

1.72. Kesselring, F.: Technische Kompositionslehre. Berlin: Springer 1954.

1.73. Kesselring, F.: Bewertung von Konstruktionen. Düsseldorf: VDI-Verlag 1951.

1.74. Klose, J.: Zur Entwicklung einer speicherunterstützten Konstruktion von Maschinen unter Wiederverwendung von Baugruppen. Habilitationsschrift TU Dresden 1979.

1.75. Klose, J.: Konstruktionsinformatik im Maschinenbau. Berlin: Technik 1990.

1.76. Koller, R.: Konstruktionslehre für den Maschinenbau. Grundlagen, Arbeitsschritte, Prinziplösungen. Berlin: Springer 1985. 3. Aufl. 1994.

1.77. Koller, R.: Entwicklung und Systematik der Bauweisen technischer Systeme – ein Beitrag zur Konstruktionsmethodik. Konstruktion 38 (1986) 1–7.

1.78. Koller, R.: CAD – Automatisiertes Zeichnen, Darstellen und Konstruieren. Berlin: Springer 1989.

1.79. Kostelic, A.: Design for Quality. Proceedings of ICED 90, Schriftenreihe WDK 19. Zürich: HEURISTA 1990.

1.80. Kuhlenkamp, A.: Konstruktionslehre der Feinwerktechnik. München: C. Hanser 1971.

1.81. Laudien, K.: Maschinenelemente. Leipzig: Dr. Max Junecke Verlagsbuchhandlung 1931.

1.82. Lehmann, C.M.: Wissensbasierte Unterstützung von Konstruktionsprozessen. Reihe Produktionstechnik, Bd. 76. München: Hanser 1989.

1.83. Leyer, A.: Maschinenkonstruktionslehre. Hefte 1–6 technica-Reihe. Basel: Birkhäuser 1963–1971.

1.84   Martyrer, E.: Der Ingenieur und das Konstruieren. Konstruktion 12 (1960) 1–4.

1.85   Matousek, R.: Konstruktionslehre des allgemeinen Maschinenbaus. Berlin: Springer 1957 Reprint.

1.86   Müller, J.: Grundlagen der systematischen Heuristik. Schriften zur soz. Wirtschaftsführung. Berlin: Dietz 1970.

1.87   Müller, J.; Koch, P. (Hrsg.): Programmbibliothek zur systematischen Heuristik für Naturwissenschaften und Ingenieure. Techn.-Wiss. Abhandlungen des Zentralinstituts für Schweißtechnik, Nr. 97–99. Halle 1973.

1.88   Müller, J.: Probleme Schöpferischer Ingenieurarbeit. Manuskriptdruck TH Karl-Marx-Stadt 1984.

1.89   Müller, J.: Arbeitsmethoden der Technikwissenschaften – Systematik, Heuristik, Kreativität. Berlin: Springer 1990.

1.90   Nadler, G.: The Planning and Design Approach. New York: Wiley 1981.

1.91   Niemann, G.: Maschinenelemente, Bd. 1. Berlin: Springer 1. Aufl. 1950, 2 Aufl. 1965, 3 Aufl. 1975 (unter Mitwirkung von M. Hirt).

1.92   N.N.: Leonardo da Vinci. Das Lebensbild eines Genies. Wiesbaden: Vollmer 1955, 493–505.

1.93   Odrin, W.M.: Morphologische Synthese von Systemen: Aufgabenstellung, Klassifikation, Morphologische Suchmethoden. Kiew: Institut f. Kybernetik, Preprints 3 und 5, 1986.

1.94   O'Grady, P.; Young, R.E.: Constraint Nets for Life Cycle Engineering: Concurrent Engineering. Proceedings of National Science Foundation Grantees Conference, 1992.

1.95   Opitz, H. und andere: Die Konstruktion – ein Schwerpunkt der Rationalisierung. Industrie-Anzeiger 93 (1971) 1491–1503.

1.96   Ostrofsky, B.: Design, Planning and Development Methodology. New Jersey: Prentice-Hall, Inc. 1977.

1.97   Pahl, G.: Entwurfsingenieur und Konstruktionslehre unterstützen die moderne Konstruktionsarbeit. Konstruktion 19 (1967) 337–344.

1.98   Pahl, G.; Beitz, W.: Für die Konstruktionspraxis. Aufsatzreihe in der Konstruktion 24 (1972), 25 (1973) und 26 (1974).

1.99   Pahl, G.; Beitz, W.: A gréptervezés elmélete és gyakorlata (transl. by M. Kozma, J. Straub, edit. by T. Bercsey, L. Varga). Budapest: Müszaki Könyvkiadó 1981.

1.100   Pahl, G.; Beitz, W.: NAUKA konstruowania (transl. by A. Walczak). Warszawa: Wydawnictwa Naukowo-Techniczne 1984.

1.101   Pahl, G.; Beitz, W. (transl. and edited by K. Wallace). Engineering Design – A Systematic Approach. London/Berlin: The Design Council/Springer 1988.

1.102   Pahl, G.; Beitz, W.: Koneensuunittluoppi (transl. by U. Konttinen). Helsinki: Metalliteollisuuden Kustannus Oy 1990.

1.103   Patsak, G.: Systemtechnik. Berlin: Springer 1982.

1.104   Penny, R.K.: Principles of Engineering Design. Postgraduate 46 (1970) 344–349.

1.105   Pighini, U.: Methodological Design of Machine Elements. Proceedings of ICED 90, Schriftenreihe WDK 19; Zürich: HEURISTA 1990.

1.106   Polovnikin, A.I.: Untersuchung und Entwicklung von Konstruktionsmethoden. MBT 29 (1979) 7, 297–301.

1.107   Polovnikin, A.I. (Hrsg.): Automatisierung des suchenden Konstruierens. Moskau: Radio u. Kommunikation 1981.

1.108   Proceedings of ICED 1981–1991 (ed. by V. Hubka and others), Schriftenreihe WDK 7, 10, 12, 13, 16, 18, 19, 20. Zürich: HEURISTA 1981–1991.

1.109   Pugh, S.: Total Design; Integrated Methods for Successful Product Engineering. Reading: Addison-Wesley 1990.

1.110   Redtenbacher, F.: Prinzipien der Mechanik und des Maschinenbaus. Mannheim: Bassermann 1852, 257–290.

1.111   Reuleaux, F.; Moll, C.: Konstruktionslehre für den Maschinenbau. Braunschweig: Vieweg 1854.

1.112   Richter, A.: Nichtlineare Optimierung signalverarbeitender Geräte. VDI-Berichte 219. Düsseldorf: VDI-Verlag 1974 (mit weiteren Literaturhinweisen).

1.113   Richter, A.; Kranz, G.: Ein Beitrag zur nichtlinearen Optimierung und dynamischen Programmierung in der rechnerunterstützten Konstruktion. Konstruktion 26 (1974) 361–367.

1.114   Riedler, A.: Das Maschinenzeichnen. Berlin: Springer 1913.

1.115   Rinderle, J.R.: Design Theory and Methodology. – DTM 90 ASME DE – Vol. 27, New York 1990.

1.116   Rodenacker, W.G.: Methodisches Konstruieren. Konstruktionsbücher, Bd. 27. Berlin: Springer 1970, 2. Aufl. 1976, 3 Aufl. 1984, 4 Aufl. 1991.

1.117   Rodenacker, W.G.: Neue Gedanken zur Konstruktionsmethodik. Konstruktion 43 (1991) 330–334.

1.118   Rodenacker, W.G.; Claussen, U.: Regeln des Methodischen Konstruierens. Mainz: Krausskopf 1973/74.

1.119   Rötscher, F.: Die Maschinenelemente. Berlin: Springer 1927.

1.120 Roozenburg, N.; Eekels, J.: EVAD Evaluation and Decision in Design. Schriftenreihe WDK 17. Zürich: HEURISTA 1990.

1.121 Roozenburg, N.F.M.; Eekels, J.: Produktontwerpen, Structurr en methoden. Utrecht: Uitgeverij Lemma B.V. 1991.

1.122 Roth, K.: Konstruieren mit Konstruktionskatalogen. Berlin: Springer 1982. 2. Aufl. 1994.

1.123 Roth, K.: Modellbildung für das methodische Konstruieren ohne und mit Rechnerunterstützung. VDI-Z (1986) 21–25.

1.124 Rugenstein, J.: Arbeitsblätter Konstruktionstechnik. TH Magdeburg 1978/79.

1.125 Saling, K.-H.: Prinzip- und Variantenkonstruktion in der Auftragsabwicklung – Voraussetzungen und Grundlagen. VDI-Berichte Nr. 152. Düsseldorf: VDI-Verlag 1970.

1.126 Schlottmann, D.: Konstruktionslehre. Berlin: Technik 1987.

1.127 Schregenberger, J.W.: Methodenbewußtes Problemlösen – Ein Beitrag zur Ausbildung von Konstrukteuren. Bern: Haupt 1981.

1.128 Seeger, H.: Design technischer Produkte, Programme und Systeme. Anforderungen Lösungen und Bewertungen. Berlin: Springer 1992.

1.129 Stauffer, L.A. (Edited) Design Theory and Methodology – DTM 91. ASME DE – Vol. 31, Suffolk (UK): Mechanical Engineering Publications Ltd. 1991.

1.130 Suh, N.P.: The Principles of Design. Oxford/UK: Oxford University Press 1988.

1.131 Taguchi, G.: Introduction of Quality Engineering. New York: UNIPUB 1986.

1.132 Tropschuh, P.: Rechnerunterstützung für das Projektieren mit Hilfe eines wissensbasierten Systems: München: Hanser 1989.

1.133 Tschochner, H.: Konstruieren und Gestalten. Essen: Girardet 1954.

1.134 Ullman, D.G.; Stauffer, L.A.; Dietterich, T.G.: A Model of the Mechanical Design Process Based on Empirical Data. AIEDA, Academic Press (1988), H. 1, 33–52.

1.135 Ullman, D.G.: The Mechanical Design Process. New York: McGraw-Hill 1992.

1.136 Ullman, D.G.: A Taxonomy for Mechanical Design. Res. Eng. Des. 3 (1992) 179–189.

1.137 Ulrich, K.T.; Seering, W.: Synthesis of Schematic Descriptions in Mechanical Design. Research in Engineering Design (1989), Vol. 1, Nr. 1, 3–18.

1.138 Van den Kroonenberg, H.H.: Design Methodology as a Condition for Computer Aided Design. VDI-Berichte Nr. 565, Düsseldorf: VDI-Verlag 1989.

1.139 VDI: Simultaneous Engineering—heue Wege des Projektmanagements. VDI-Tagung Frankfurt, Tagungsband. Dusseldorf: VDI-Verlag 1989.

1.140 VDI: Anforderungen an Konstruktions- und Entwicklungsingenieure – Empfehlungen der VDI-Gesellschaft Entwicklung – Konstruktion – Vertrieb (VDI-EKV) zur Ausbildung. Jahrbuch 92. Düsseldorf: VDI-Verlag 1992.

1.141 VDI-Berichte 775: Expertensysteme in Entwicklung und Konstruktion – Bestandsaufnahme und Entwicklungen. Düsseldorf: VDI-Verlag 1989.

1.142 VDI Design Handbook 2221: Systematic Approach to the Design of Technical Systems and Products (transl. by K. Wallace). Düsseldorf: VDI-Verlag 1987.

1.143 VDI-Fachgruppe Konstruktion (ADKI): Engpaß Konstruktion. Konstruktion 19 (1967) 192–195.

1.144 VDI-Richtlinie 2221: Methodik zum Entwickeln und Konstruieren technischer Systeme und Produkte. Düsseldorf: VDI-Verlag 1993.

1.145 VDI-Richtlinie 2222 Blatt 1: Konzipieren technischer Produkte: Düsseldorf: VDI-Verlag 1977.

1.146 VDI-Richtlinie 2222 Blatt 2: Erstellung und Anwendung von Konstruktionskatalogen. Düsseldorf: VDI-Verlag 1982.

1.147 VDI-Richtlinie 2225: Technisch-wirtschaftliches Konstruieren. Düsseldorf: VDI-Verlag 1977, Blatt 3: 1990.

1.148 Wächtler, R.: Die Dynamik des Entwickelns (Konstruierens). Feinwerktechnik 73 (1969) 329–333.

1.149 Wächtler, R.: Beitrag zur Theorie des Entwickelns (Konstruierens). Feinwerktechnik 71 (1967) 353–358.

1.150 Wallace, K.; Hales, C.: Detailed Analysis of an Engineering Design Project. Proceedings ICED '87, Schriftenreihe WDK 13. New York: ASME 1987.

1.151 Wögerbauer, H.: Die Technik des Konstruierens. 2. Aufl. München: Oldenbourg 1943.

1.152 Yoshikawa, H.: Automation in Thinking in Design. Computer Applications in Production and Engineering. Amsterdam: North-Holland 1983.

1.153 Zangemeister, C.: Zur Charakteristik der Systemtechnik. TU Berlin: Aufbauseminar Systemtechnik 1969.

Chapter 2

2.1.   Beitz, W.: Kreativität des Konstrukteurs. Konstruktion 37 (1985) 381–386.
2.2.   Brankamp, K.: Produktivitätssteigerung in der mittelständigen Industrie NRW. VDI-Taschenbuch. Düsseldorf: VDI-Verlag 1975.
2.3.   Brunthaler, St.: Konstruktion von Anwendersoftware in Analogie zum methodischen Konstruieren. Konstruktion 37 (1985), H 7, 289–290, 325–326, 360–362, 405–406, 448–450, 489–494.
2.4.   DIN 40 900 T 12: Binäre Elemente, IEC 617–12 modifiziert. Berlin: Beuth.
2.5.   DIN 43 300: Informationsverarbeitung – Begriffe. Berlin: Beuth.
2.6.   DIN 44 301: Informationstheorie – Begriffe. Berlin: Beuth.
2.7.   DIN 69 910: Wertanalyse, Begriffe, Methode. Berlin: Beuth.
2.8.   Dörner, D.: Problemlösen als Informationsverarbeitung. Stuttgart: W. Kohlhammer. 2. Aufl. 1979.
2.9.   Dörner, D.; Kreuzig, H.W., Reither, F.; Stäudel, T.: Lohhausen. Vom Umgang mit Unbestimmtheit und Komplexität. Bern: Verlag Hans Hubert 1983.
2.10. Ehrenspiel, K.; Dylla, N.: Untersuchung des individuellen Vorgehens beim Konstruieren. Konstruktion 43 (1991) 43–51.
2.11. Frick, H.; Müller, J.: Graphisches Darstellungsvermögen von Konstrukteuren. Konstruktion 42 (1990) 321–324.
2.12. Fricke, G.; Pahl, G.: Zusammenhang zwischen personenbedingtem Vorgehen und Lösungsgüte. Proceedings of ICED '91, Zürich.
2.13. Gerber, H.: Ein Konstruktionsverfahren für Geräte mit logischer Funktionsweise. Konstruktion 25 (1973) 13–17.
2.14. Hansen, F.: Konstruktionssystematik. Berlin: VEB Verlag Technik 1966.
2.15. Holliger, H.: Handbuch der Morphologie – Elementare Prinzipien und Methoden zur Lösung kreativer Probleme. Zürich: MIZ Verlag 1972.
2.16. Holliger, H.: Morphologie – Idee und Grundlage einer interdisziplinären Methodenlehre. Kommunikation 1. Vol. V1. Quickborn: Schnelle 1970.
2.17. Hubka, V.: Theorie Technischer Systeme. Berlin: Springer 1984.
2.18. Hubka, V.; Eder, W.E.: Theory of Technical Systems. Berlin: Springer 1988.
2.19. Janis, I.L.; Mann, L.: Decisions making. Free Press of Glencoe. New York: 1977.
2.20. Klaus, G.: Wörterbuch der Kybernetik. Handbücher 6142 und 6143. Frankfurt: Fischer 1971.
2.21. Koller, R.: Konstruktionslehre für den Maschinenbau. Berlin: Springer 1976, 2. Aufl. 1985. 3. Aufl. 1994.
2.22. Koller, R.: Kann der Konstruktionsprozeß in Algorithmen gefaßt und dem Rechner übertragen werden? VDI-Berichte Nr. 219. Düsseldorf: VDI-Verlag 1974.
2.23. Kroy, W.: Abbau von Kreativitätshemmungen in Organisationen. In: Schriftenreihe Forschung, Entwicklung, Innovation Bd. 1: Personal-Management in der industriellen Forschung und Entwicklung. Köln: C. Heymanns 1984.
2.24. Krumhauer, P.: Rechnerunterstützung für die Konzeptphase der Konstruktion. Diss. TU Berlin 1974, D 83.
2.25. Mewes, D.: Der Informationsbedarf im konstruktiven Maschinenbau. VDI-Taschenbuch T49. Düsseldorf: VDI-Verlag 1973.
2.26. Miller, G.A.; Galanter, E.; Pribram, K.: Plans and the Structure of Behavior. New York: Holt, Rinehardt & Winston, 1960.
2.27. Müller, J.: Grundlagen der systematischen Heuristik. Schriften zu soz. Wirtschaftsführung. Berlin: Dietz 1970.
2.28. Müller, J.: Arbeitsmethoden der Technikwissenschaften. Berlin: Springer 1990.

2.29. Nadler, G.: Arbeitsgestaltung – zukunftsbewußt. München: Hanser 1969. Amerikanische Originalausgabe: Work Systems Design: The Ideals Concept. Homewood, Illinois: Richard D. Irwin Inc. 1967.

2.30. Nadler, G.: Work Design. Homewood, Illinois: Richard D. Irwin Inc. 1963.

2.31. N.N.: Lexikon der Neue Brockhaus. Wiesbaden: F.A. Brockhaus 1958.

2.32. Pütz, J.: Digitaltechnik. Düsseldorf: VDI-Verlag 1975.

2.33. Rodenacker, W.G.: Methodisches Konstruieren. Konstruktionsbücher Bd. 27. Berlin: Springer 1970, 2. Aufl. 1976, 3 Aufl. 1984, 4. Aufl. 1991.

2.34. Rodenacker, W.G.; Claussen, U.: Regeln des Methodischen Konstruierens, Bd. I u. II. Mainz: Krauskopf 1973 und 1975.

2.35. Roth, K.: Konstruieren mit Konstruktionskatalogen. Berlin: Springer 1982. 2. Aufl. 1994.

2.36. Roth, K.; Franke, H.-J.; Simonek, R.: Die allgemeine Funktionsstruktur, ein wesentliches Hilfsmittel zum methodischen Konstruieren. Konstruieren 24 (1972) 277–282.

2.37. Rutz, A.: Konstruieren als gedanklicher Prozeß. Diss. TU München 1985.

2.38. Schmidt, H.G.: Heuritische Methoden als Hilfen zur Entscheidungsfindung beim Konzipieren technischer Produkte. Schriftenreihe Konstruktionstechnik, H. 1. Herausgeber W. Beitz. Technische Universität Berlin, 1980.

2.39. VDI-Richtlinie 2221: Methodik zum Entwickeln und Konstruieren technischer Systeme und Produkte. Düsseldorf: VDI-Verlag 1993.

2.40. VDI-Richtlinie 2222: Blatt 1: Konstruktionsmethodik – Konzipieren technischer Produkte. Düsseldorf: VDI-Verlag 1977.

2.41. VDI-Richtlinie 2242. Blatt 1: Ergonomiegerechtes Konstruieren. Düsseldorf: VDI-Verlag 1986.

2.42. VDI-Richtlinie 2801. Blatt 1–3: Wertanalyse. Düsseldorf: VDI-Verlag 1970/71.

2.43. Voigt, C.D.: Systematik und Einsatz der Wertanalyse, 3. Aufl. München: Siemens-Verlag 1974.

2.44. Weizsäcker von, C.F.: Die Einheit der Nature – Studien. München: Hanser 1971.

Chapter 3

3.1. Beelich, K.H.; Schwede, H.H.: Denken Planen Handeln. 3. Aufl. Würzburg: Vogelbuchverlag 1983.

3.2. Dörner, D.: Problemlösen als Informationsverarbeitung. 2. Aufl. Stuttgart: Kohlhammer 1979.

3.3. Dörner, D.; Kreuzig, H.W.; Reither, F., Ständel, T.: Lohhausen. Vom Umgang mit Unbestimmtheit und Komplexität. Bern: Verlag Hans Huber 1983.

3.4. Krick, V.: An Introduction to Engineering and Engineering Design, Second Edition. New York, London, Sydney, Toronto: Wiley & Sons Inc. 1969.

3.5. Leyer, A.: Zur Frage der Aufsätze über Maschinenkonstruktion in der "technika". Technika 26 (1973) 2495–2498.

3.6. Miller, G.A.; Gralanter, E., Pribram, K.: Plans and the Structure of Behavior. New York: Holt, Rinehart u. Winston 1960.

3.7. Müller, J.: Arbeitsmethoden der Technikwissenschaften. Berlin: Springer 1990.

3.8. Pahl, G.: Die Arbeitsschritte beim Konstruieren. Konstruktion 24 (1972) 149–153.

3.9. Penny, R.K.: Principles of Engineering Design. Postgraduate J. 46 (1970) 344–349.

3.10. VDI-Richtlinie 2221: Methodik zum Entwickeln und Konstruieren technischer Systeme und Produkte. Düsseldorf: VDI-Verlag 1993.

3.11. VDI-Richtlinie 2222 Blatt 1: Konzipieren technischer Produkte. Düsseldorf: VDI-Verlag 1977.

3.12. VDI-Richtlinie 2223: Begriffe und Bezeichnungen im Konstruktionsbereich. Düsseldorf: VDI-Verlag 1969.

3.13. Aus der Arbeit der VDI-Fachgruppe Konstruktion (ADKI). Empfehlungen für Begriffe und Bezeichnungen im Konstruktionsbereich. Konstruktion 18 (1966) 390–391.

3.14. Wahl, M.P.: Grundlagen eines Management-Informationssystemes. Neuwied, Berlin: Luchterhand 1969.

Chapter 4

4.1.   Baatz, U.: Bildschirmunterstütztes Konstruieren. Diss. RWTH Aachen 1971.

4.2.   Bengisu, Ö.: Elektrohydraulische Analogie. Ölhydraulik und Pneumatik 14 (1970) 122–127.

4.3.   Dalkey, N.D.; Helmer, O.: An Experimental Application of the Delphi Method to the Use of Experts. Management Science Bd. 9, No. 3, April 1963.

4.4.   Diekhöner, G.: Erstellen und Anwenden von Konstruktionskatalogen im Rahmen des methodischen Konstruierens. Fortschrittsberichte der VDI-Zweitschriften Reihe 1, Nr. 75. Düsseldorf: VDI-Verlag 1981.

4.5.   Diekhöner, G.; Lohkamp, F.: Objektkataloge – Hilfsmittel beim methodischen Konstruieren. Konstruktion 28 (1976) 359–364.

4.6.   Dreibholz, D.: Ordnungsschemata bei der Suche von Lösungen. Konstruktion 27 (1975) 233–240.

4.7.   Ersoy, M.: Gießtechnische Fertigungsverfahren – Konstruktionskatalog für Fertigungsverfahren. wt-Z. in der Fertigung 66 (1976) 211–217.

4.8.   Ersoy, M.: Klemmverbindungen zum Spannen von Werkstücken. VDI-Berichte 493. Düsseldorf: VDI-Verlag 1983.

4.9.   Ewald, O.: Lösungssammlungen für das methodische Konstruieren. Düsseldorf: VDI-Verlag 1975.

4.10.   Feldmann, K.: Beitrag zur Konstruktionsoptimierung von automatischen Drehmaschinen. Diss. TU Berlin 1974.

4.11.   Föllinger, O.; Weber, W.: Methoden der Schaltalgebra. München: Oldenbourg 1967.

4.12.   Fuhrmann, U.; Hinterwaldner, R.: Konstruktionskatalog für Klebeverbindungen tragender Elemente. VDI-Berichte 493. Düsseldorf: VDI-Verlag 1983.

4.13.   Gießner, F.: Gesetzmäßigkeiten und Konstruktionskataloge elastischer Verbindungen. Diss. Braunschweig 1975.

4.14.   Gordon, W.J.J.: Synectics, the Development of Creative Capacity. New York: Harper 1961.

4.15.   Grandt, J.: Auswahlkriterien von Nietverbindungen im industriellen Einsatz. VDI-Berichte 493. Düsseldorf: VDI-Verlag 1983.

4.16.   Hellfritz, H.: Innovation via Galeriemethode. Königstein/Ts.: Eigenverlag 1978.

4.17.   Herrmann, J.: Beitrag zur optimalen Arbeitsraumgestaltung an numerisch gesteuerten Drehmaschinen. Diss. TU Berlin 1970.

4.18.   Hertel, H.: Biologie und Technik – Struktur, Form, Bewegung. Mainz: Krauskopf 1963.

4.19.   Hertel, H.: Leichtbau. Berlin: Springer 1969.

4.20.   Jung, R.; Schneider, J.: Elektrische Kleinmotoren. Marktübersicht mit Konstruktionskatalog. Feinwerktechnik und Meßtechnik 92 (1984) 153–165.

4.21.   Kesselring, F.: Bewertung von Konstruktionen, ein Mittel zur Steuerung von Konstruktionsarbeit. Düsseldorf: VDI-Verlag 1951.

4.22.   Koller, R.: Konstruktionslehre für den Maschinenbau. Berlin: Springer 1976, 2. Aufl. 1985. 3. Aufl. 1994.

4.23.   Kollmann, F.G.: Welle-Nabe-Verbindungen. Konstruktionsbücher Bd. 32. Berlin: Springer 1983.

4.24.   Kopowski, E.: Einsatz neuer Konstruktionskataloge zur Verbindungsauswahl. VDI-Berichte 493. Düsseldorf: VDI-Verlag 1983.

4.25.   Krumhauer, P.: Rechnerunterstützung für die Konzeptphase der Konstruktion. Diss. TU Berlin 1974.

4.26.   Lowka, D.: Methoden zur Entscheidungsfindung im Konstrutionsprozeß; Feinwerktechnik und Meßtechnik 83 (1975) 19–21.

4.27.   Neudörfer, A.: Gesetzmäßigkeiten und systematische Lösungssammlung der Anzeiger und Bedienteile. Düsseldorf: VDI-Verlag 1981.

4.28.   Konstruktionskatalog für Gefahrstellen. Werkstatt und Betrieb 116 (1983) 71–74.

4.29.   Neudörfer, A.: Konstruktionskatalog trennender Schutzeinrichtungen. Werkstatt und Betrieb 116 (1983) 203–206.

4.30.   NN: Kreativität. Dokumentation der 43 Kreativ-Methoden. Manager Magazin 11 (1972) 51–57.

4.31. Osborn, A.F.: Applied Imagination – Principles and Procedures of Creative Thinking. New York: Scribner 1957.
4.32. Pahl, G.: Rückblick zur Reihe "Für die Konstruktionspraxis". Konstruktion 26 (1974) 491–495.
4.33. Pahl, G.; Beelich, K.H.: Lagebericht. Erfahrungen mit dem methodischen Konstruieren. Werkstatt und Betrieb 114 (1981) 773–782.
4.34. Raab, W.: Schneider, J.: Gliederungssystematik für getriebetechnische Konstruktionskataloge. Antriebstechnik 21 (1982) 603.
4.35. Richter, A.; Aschoff, H.-J.: Problemstellungen bei der funtionsorientierten Konstruktionssynthese signalverarbeitender Geräte aus der Sicht der Systemdynamik. Feinwerktechnik 75 (1971) 374–379.
4.36. Richter, A.; Aschoff, H.-J.: Über die funtionsorientierte konstruktive Gestaltung von signalverarbeitenden Geräten nach statischen und dynamischen Gesichtspunkten. Feinwerktechnik 75 (1971) 443–446.
4.37. Rodenacker, W.G.: Methodisches Konstruieren. Konstruktionsbücher Bd. 27. Berlin: Springer 1970, 2. Aufl. 1976, 3. Aufl. 1984, 4. Aufl. 1991.
4.38. Rohrbach, B.: Kreativ nach Regeln – Methode 635, eine neue Technik zum Lösen von Problemen. Absatzwirtschaft 12 (1969) 73–75.
4.39. Roozenburg, N.; Eckels, J. (Editors): Evaluation and Decision in Design. Schriftenreihe WDK 17. Zürich: HEURISTA 1990.
4.40. Roth, K.: Konstruieren mit Konstruktionskatalogen. Berlin: Springer 1982. 2. Aufl. 1994.
4.41. Roth, K.: Einheitliche Systematik der Verbindungen. VDI-Berichte 493. Düsseldorf: VDI-Verlag 1983.
4.42. Roth, K.; Birkhofer, H.; Ersoy, M.: Methodisches Konstruieren neuer Sicherheitsgurtschlösser. VDI-Z. 117 (1975) 613–618.
4.43. Schlösser, W.M.J.; Olderaan, W.F.T.C.: Eine Analogontheorie der Antriebe mit rotierender Bewertung. Ölhydraulik und Pneumatik 5 (1961) 413–418.
4.44. Schneider, J.: Konstruktionskataloge als Hilfsmittel bei der Entwicklung von Antrieben. Diss. Darmstadt 1985.
4.45   Stabe, H.; Gerhard, E.: Anregungen zur Bewertung technischer Konstruktionen. Feinwerktechnik und Meßtechnik 82 (1974) 378–383 (einschließlich weiterer Literaturhinweise).
4.46. Stahl, U.: Überlegungen zum Einfluß der Gewichtung bei der Bewertung von Alternativen. Konstruktion 28 (1976) 273–274.
4.47. VDI-Richtlinie 2222 Blatt 2: Konstruktionsmethodik, Erstellung und Anwendung von Konstruktionskatalogen. Düsseldorf: VDI-Verlag 1982.
4.48. VDI-Richtlinie 2225: Technisch-wirtschaftliches Konstruieren. Düsseldorf: VDI-Verlag 1977.
4.49. VDI-Richtlinie 2227 Blatt 1 und 2: Lösung von Bewegungsaufgaben mit Getrieben. Düsseldorf: VDI-Verlag 1991.
4.50. VDI-Richtlinie 2740 (Entwurf): Greifer für Handhabungsgeräte und Industrieoboter. Düsseldorf: VDI-Verlag 1991.
4.51. Withing, Ch.: Creative Thinking. New York: Reinhold 1958.
4.52. Wölse, H.; Kastner, M.: Konstruktionskataloge für geschweißte Verbindungen an Stahlprofilen. VDI-Berichte 493. Düsseldorf: VDI-Verlag 1983.
4.53. Zangemeister, Ch.: Nutzwertanalyse in der Systemtechnik. München: Wittenmannsche Buchhandlung 1970.
4.54. Zwicky, F.; Entdecken, Erfinden, Forschen im Morphologischen Weltbild. München: Droemer-Knaur 1966-1971.

Chapter 5

5.1.   Barrenscheen, J.; Drebing, U.; Sieverding, H.: Rechnerunterstützte Erstellung von Anforderungslisten. VDI-Z. 131 (1989), Nr. 4, 84–89.

5.2.   Brankamp, K.: Produktplanung – Instrument der Zukunftssicherung im Unternehmen. Konstruktion 26 (1974) 319–321.

5.3.   Franke, H.-J.: Methodische Schritte beim Klären konstruktiver Aufgabenstellungen. Konstruktion 27 (1975) 395–402.

5.4.   Gälweiler, A.: Unternehmensplanung. Frankfurt: Herder & Herder 1974.

5.5.   Geschka, H.: Produktplanung in Großunternehmen. Proceedings ICED 91, Schriftenreihe WDK 20. Zürich: HEURISTA 1991.

5.6.   Geyer, E.: Marktgerechte Produktplanung und Produktentwicklung. Teil I: Produkt und Markt, Teil II: Produkt und Betrieb. RKW-Schriftenreihe Nr. 18 and 26. Heidelberg: Gehlsen 1972 (mit zahlreichen weiteren Literaturstellen).

5.7.   Groeger, B.: Ein System zur rechnerunterstützten und wissenbasierten Bearbeitung des Konstruktionsprozesses. Konstruktion 42 (1990) 91–96.

5.8.   Hansen, F.: Konstruktionssystematik. Berlin: VEB Verlag Technik 1966.

5.9.   Kehrmann, H.: Die Entwicklung von Produktstrategien. Diss. TH Aachen 1972.

5.10. Kehrmann, H.: Systematik und Finden und Bewerten neuer Produkte. wt-Z. ind. Fertigung 63 (1973) 607–612.

5.11. Kesselring, F.; Arn, E.: Methodisches Planen. Entwickeln und Gestalten technischer Produkte. Konstruktion 23 (1971) 212–218.

5.12. Kramer, F.: Erfolgreiche Unternehmensplanung. Berlin: Beuth 1974.

5.13. Kramer, F.: Anpassung der Produkt- und Marktstrategien an veränderte Umweltsituationen. VDI-Berichte Nr. 503: Produktplanung und Vertrieb. Düsseldorf: VDI-Verlag 1983.

5.14. Kramer, F.: Unternehmensbezogene Erfolgsstrategien. VDI-Berichte Nr. 538: Besser als der Wettbewerb – Marktzwänge und Lösungswege. Düsseldorf: VDI-Verlag 1984.

5.15. Kramer, F.: Innovative Produktpolitik, Strategie – Planung – Entwicklung – Einführung. Berlin: Springer 1986.

5.16. Kramer, F.: Produktplanung in der mittelständischen Industrie, Wettbewerbsvorteile durch Differenzierungs-Management. Proceedings ICED 91, Schriftenreihe WDK 20. Zürich: HEURISTA 1991.

5.17. Pahl, G.: Klären der Aufgabenstellung und Erarbeitung der Anforderungsliste. Konstruktion 24 (1974) 195–199.

5.18. Pahl, G.: Wege zur Lösungsfindung. Industrielle Organisation 39 (1970) 155–161.

5.19. Plinke, W.: Marketing und Produktplanung. ICED 91 – Proceedings, Schriftenreihe WDK 20. Zürich: HEURISTA 1991.

5.20. Roth, K.; Birkhofer, H.; Ersoy, M.: Methodisches Konstruieren neuer Sicherheitsschlösser. VDI-Z. 117 (1975) 613–618.

5.21. VDI-Berichte Nr. 229: Produktinnovation – Herausforderung und Aufgabe. Düsseldorf: VDI-Verlag 1976.

5.22. VDI-Richtlinie 2220: Produktplanung, Ablauf, Begriffe und Organisation. Düsseldorf: VDI-Verlag 1980.

5.23. VDI-Taschenbuch T 46: Systematische Produktplanung – ein Mittel zur Unternehmenssicherung. Düsseldorf: VDI-Verlag 1975.

Chapter 6

6.1.   Beitz, W.: Methodisches Konzipieren technischer Systeme, gezeigt am Beispiel einer Kartoffel-Vollerntemaschine. Konstruktion 25 (1973) 65–71.

6.2.   Brunthaler, St.: Konstruktion von Anwendersoftware in Analogie zum methodischen Konstruieren. Konstruktion 37 (1985), 289–290, 325–326, 360–362, 405–406, 448–450, 489–494.

6.3.   Grabowski, H.: Elektronische Datenverarbeitung. In: Dubbel, Teil Y, 17. Aufl. Berlin: Springer 1990.

6.4.   Grote, K.-H.: Untersuchungen zum Tragverhalten von Mehrschraubenverbindungen. (Schriftenreihe Konstruktionstechnik, H.6) Berlin: TU Berlin 1984.

6.5.   Hansen, F.: Konstruktionssystematik, 2. Aufl. Berlin: VEB-Verlag 1965.

6.6.   Keller, K.: Entwicklung eines Fergigungsautomaten zur Fertigung von Filigranträgern. Unveröffentlichte Diplomarbeit, TU Berlin 1971.

6.7.   Koller, R.: Konstruktionslehre für den Maschinenbau. Berlin: Springer 1976; 2. Aufl. 1985. 3. Aufl. 1994.

6.8.   Kramer, F.: Produktinnovations- und Produkteinführungssystem eines mittleren Industriebetriebes. Konstruktion 27 (1975) 1–7.

6.9.   Krick, E.V.: An Introduction to Engineering and Engineering Design; 2nd Edition. New York: Wiley & Sons, Inc. 1969.

6.10.   Lehmann, M.: Entwicklungsmethodik für die Anwendung der Mikroelektronik im Maschinenbau. Konstruktion 37 (1985) 339–342.

6.11.   Pahl, G.; Beitz, W.: Konstruktionslehre. Berlin: Springer 1977, 1. Aufl.; 1986, 2. Aufl.

6.12.   Pahl, G.; Konstruieren mit 3D-CAD-Systemen. Grundlagen, Arbeitstechnik, Anwendungen. Berlin: Springer 1990.

6.13.   Richter, W.: Gestalten nach dem Skizzierverfahren. Konstruktion 39 (1987), H.6, 227–237.

6.14.   Roth, K.: Konstruieren mit Konstruktionskatalogen. Berlin: Springer 1982. 2. Aufl. 1994.

6.15.   Steuer, K. Theorie des Konstruierens in der Ingenieurausbildung. Leipzig: VEB-Fachbuchverlag 1968.

6.16.   Stürmer, U.: PC BOLT, Programmbeschreibung. Berlin: TU Berlin 1986.

6.17.   VDI-Richtlinie 2222 Blatt 2: Konstruktionsmethodik, Erstellung und Anwendung von Konstruktionskatalogen. Düsseldorf: VDI-Verlag 1982.

6.18.   VDI-Richtlinie 2225: Technisch-wirtschaftliches Konstruieren. Düsseldorf: VDI-Verlag 1977.

6.19.   VDI-Richtlinie 2230, Blatt 1: Systematische Berechnung hochbeanspruchter Schraubenverbindungen. Düsseldorf: VDI-Verlag 1986.

Chapter 7

7.1.   AEG-Telefunken: Biegen, Werknormblatt 5 N 8410 (1971).

7.2.   Andreasen, M.M.; Kähler, S.; Lund, T.: Design for Assembly. Berlin: Springer 1983. Deutsche Ausgabe: Montagegerechtes Konstruieren. Berlin: Springer 1985.

7.3.   Andresen, U.: Die Rationalisierung der Montage beginnt im Konstruktionsbüro. Konstruktion 27 (1975) 478–484. Ungekürzte Fassung mit weiterem Schrifttum; Ein Beitrag zum methodischen Konstruieren bei der montagegerechten Gestaltung von Teilen der Großserienfertigung. Diss. TU Braunschweig 1975.

7.4.   Bänninger, E.: Normung – Zwansjacke oder unentbehrlicher Helfer des Konstrukteurs? Technica 21 (1970) 1947–1972 und 22 (1970) 2111–2117.

7.5.   Bautz, W.; Thum, A.: Die Gestaltfestigkeit. Stahl und Eisen 55 (1935) 1025–1029, Schweizer. Bauzeitung 106 (1935) 25–30.

7.6.   Beckstroem, J.: Eigenspannungen und Betriebsverhalten von Schweißkonstruktionen. Konstruktion 17 (1965) 10–15.

7.7.   Beelich, K.H.: Kriech- und relaxationsgerecht. Konstruktion 25 (1973) 415–421.

7.8.   Behnisch, H.: Thermisches Trennen in der Metallbearbeitung – wirtschaftlich und genau. ZwF 68 (1973) 337–340.

7.9.   Beitz, W. u.a.: Vorlesungsunterlagen zur Normung. Herausgeber: DIN Deutsches Institut für Normung. Berlin: Beuth 1981.

7.10.   Beitz, W.: Technische Regeln und Normen in Wissenschaft und Technik. DIN-Mitt. 64 (1985) 114–115.

7.11.   Beitz, W.: Was ist unter "normungsfähig" zu verstehen? Ein Standpunkt aus der Sicht der Konstruktionstechnik. DIN-Mitt. 61 (1982) 518–522.

7.12.   Beitz, W.: Normung und Innovation, ein Spannungsfeld? DIN-Mitt. 65 (1986) 86–89.

7.13.   Beitz, W.: Moderne Konstruktionstechnik im Elektromaschinenbau. Konstruktion 21 (1969) 461–468.

7.14. Beitz, W.: Die normgerechte Konstruktion. Konstruktion 25 (1973) 319–327.

7.15. Beitz, W.: Fertigungs- und montagegerecht. Konstruktion 25 (1973) 489–497.

7.16. Beitz, W.: Möglichkeiten zur material- und energiesparenden Konstruktion. Konstruktion 42 (1990) 12, 378–384.

7.17. Beitz, W.; Hesser, W.: Nutzung von Normen und Regelwerken beim Konstruieren. VDI-Berichte Nr. 311. Düsseldorf: VDI-Verlag 1978.

7.18. Beitz, W.; Hove, U.; Poushirazi, M.: Altteileverwendung im Automobilbau. FAT-Schriftenreihe Nr. 24. Frankfurt: Forschungsvereinigung Automobiltechnik 1982 (mit umfangreichem Schrifttum).

7.19. Beitz, W.; Meyer, H.: Untersuchungen zur recyclingfreundlichen Gestaltung von Haushaltsgroßgeräten. Konstruktion 33 (1981) 257–262, 305–315.

7.20. Beitz, W.; Pourshirazi, M.: Ressourcenbewußte Gestaltung von Produkten. Wissenschaftsmagazin TU Berlin, Heft 8: 1985.

7.21. Beitz, W.; Wende, A.: Konzept für ein recyclingorientiertes Produktmodell. VDI-Berichte 906. Düsseldorf: VDI-Verlag 1991.

7.22. Beitz, W.: Staudinger, H.: Guß im Elektromaschinenbau. Konstruktion 21 (1969) 125–130.

7.23. Biezeno, C.B.; Grammel, R.: Technische Dynamik, Bd. 1 und 2, 2 Aufl. Berlin: Springer 1953.

7.24. Birnkraut, H.W.: Wiederverwerten von Kunststoff-Abfällen. Kunststoffe 72 (1982) 415–419.

7.25. Bode, K.-H.: Konstruktions-Atlas "Werkstoff- und verfahrensgerecht konstruieren". Darmstadt: Hoppenstedt 1984.

7.26. Böcker, W.: Künstliche Beleuchtung: ergonomisch und energiesparend. Frankfurt/M.: Campus 1981.

7.27. Brandenberger, H.: Fergigungsgerechtes Konstruieren. Zürich: Schweizer Druck- und Verlagshaus.

7.28. Broichhausen, J.: Beeinflussung der Dauerhaltbarkeit von Konstruktionswerkstoffen und Werkstoffverbindungen durch konstruktive Kerben, Oberflächenkerben und metallurgische Kerben. VDI-Fortschritts-Berichte, Reihe 1, Nr. 20. Düsseldorf: VDI-Verlag 1970.

7.29. Bronner, A.: Wertanalyse als integrierte Rationalisierung. wt-Z. 58 (1968) H. 1.

7.30. Budde, E.; Reihlen, H.: Zur Bedeutung technischer Regeln in der Rechtsprechungspraxis der Richter. DIN-Mitt. 63 (1984) 248–250.

7.31. Bullinger, -J.; Solf, J.J.: Ergonomische Arbeitsmittelgestaltung. 1. Systematik; 2. Handgeführte Werkzeuge, Fallstudien; 3. Stellteile an Werkzeugmaschinen, Fallstudien. Bremerhaven: Wirtschaftsverl. NW. 1979.

7.32. Burandt, U.: Ergonomie für Design und Entwicklung. Köln: Verlag Dr. Otto Schmidt 1978.

7.33. Burgdorf, M.: Fließgerechte Gestaltung von Werkstücken. wt-Z. ind. Fertig. 63 (1973) 387–392.

7.34. Clausmeyer, H.: Kritischer Spannungszustand und Trennbruch unter mehrachsiger Beanspruchung. Konstruktion 21 (1969) 52–59.

7.35. Compes, P.: Sicherheitstechnisches Gestalten. Habilitationsschrift TH Aachen 1970.

7.36. Cornelius, E.A.; Marlinghaus, J.: Gestaltung von Hartlotkonstruktionen hoher Tragfähigkeit. Konstruktion 19 (1967) 321–327.

7.37. Fornu, O.: Ultraschallschweißen. Z. Technische Rundschau 37 (1973) 25–27.

7.38. Czerwenka, G.; Schnell, W.: Einführung in die Rechenmethoden des Leichtbaus I und II: Hochschultaschenbuch 124/124a. Mannheim: Bibliographies Institut 1967.

7.39. Czichos, H.; Habig, K.-H.: Tribologie Handbuch – Reibung und Verschleiß. Braunschweig: Vieweg 1992.

7.40. Dangl, K.; Baumann, K.; Ruttmann, W.: Erfahrungen mit austenitischen Armaturen und Formstücken. Sonderheft VGB-Werkstofftagung 1969, 98.

7.41. Dey, W.: Notwendigkeiten und Grenzen der Normung aus der Sicht des Maschinenbaus unter besonderer Berücksichtigung rechtsrelevanter technischer Regeln mit sicherheitstechnischen Festlegungen. DIN-Mitt. 61 (1982) 578–583.

7.42. Dilling, H.-J.; Rauschenbach, Th.: Rationalisierung und Automatisierung der Montage (mit umfangreichem Schrifttum). Düsseldorf: VDI-Verlag 1975.

7.43. DIN 820 Teil 2: Gestaltung von Normblättern. Berlin: Beuth.

7.44. DIN 820 Teil 3: Normungsarbeit – Begriffe. Berlin: Beuth.

7.45. DIN 820 Teil 21 bis 29: Gestaltung von Normblättern. Berlin: Beuth.

7.46. DIN ISO 1101: Form- und Lagetolerierung. Berlin: Beuth.
7.47. DIN ISO 2768: Allgemeintoleranzen. Teil 1 – Toleranzen für Längen- und Winkelmaße. Teil 2 – Toleranzen für Form und Lage. Berlin: Beuth.
7.48. DIN ISO 2692: Form- und Lagetolerierung; Maximum – Material – Prinzip. Berlin: Beuth.
7.49. DIN 4844 Teil 1: Sicherheitskennzeichnung. Begriffe, Grundsätze und Sicherheitszeichen. Berlin: Beuth.
7.50. DIN 4844 Teil 2: Sicherheitskennzeichnung. Sicherheitsfarben. Berlin: Beuth.
7.51. DIN 4844 Teil 3: Sicherheitskennzeichnung; Ergänzende Festlegungen zu Teil 1 und Teil 2. Berlin: Beuth.
7.52. DIN 5034: Innenraumbeleuchtung mit Tageslicht; Leitsätze.
        Beiblatt 1 – Berechnung und Messung.
        Beiblatt 2 – Vereinfachte Bestimmung lichttechnisch ausreichender Fensterabmessungen.
        Teil 1 – Tageslicht in Innenräumen; Leitsätze.
        Teil 4 (Entwurf) – Tageslich in Innenräumen; Vereinfachte Bestimmung von Mindestfenster-größen für Wohnräume. Berlin: Beuth 1981.
7.53. DIN 5035: Innenraumbeleuchtung mit künstlichem Licht.
        Teil 1 – Begriffe und allgemeine Anforderungen.
        Teil 2 – Richtwerte für Arbeitsstätten.
        Teil 3 – Spezielle Empfehlungen für die Beleuchtung in Krankenhäusern.
        Teil 4 – Spezielle Empfehlungen für die Beleuchtung von Unterrichtsstätten.
        Teil 5 – Notbeleuchtung.
        Teil 6 – Messung und Bewertung. Berlin: Beuth.
7.54. DIN 5040: Leuchten für Beleuchtungszwecke.
        Teil 1 – Lichttechnische Merkmale und Einteilung.
        Teil 2 – Innenleuchten, Begriffe, Einteilung.
        Teil 3 – Außenleuchten, Begriffe, Einteilung. Berlin: Beuth.
7.55. DIN 7521–7527: Schmiedestücke aus Stahl. Berlin: Beuth.
7.56. DIN ISO 8015: Tolerierungsgrundsatz. Berlin: Beuth.
7.57. DIN 8577: Fertigungsverfahren; Übersicht. Berlin: Beuth.
7.58. DIN 8580: Fertigungsverfahren; Einteilung. Berlin: Beuth.
7.59. DIN 8593: Fertigungsverfahren; Fügen – Einordnung, Unterteilung, Begriffe. Berlin: Beuth.
7.60. DIN 9005: Gesenkschmiedestücke aus Magnesium-Knetlegierungen. Berlin: Beuth.
7.61. DIN 31 000: Sicherheitsgerechtes Gestalten technischer Erzeugnisse. Allgemeine Leitsätze. Berlin: Beuth. Teilweise ersetzt durch DIN EN 292 Teil 1 u. 2: Sicherheit von Maschinen, Grundbegriffe, allgemeine Gestaltungsleitsätze 1991.
7.62. DIN 31 001 Teil 1, 2 u. 3: Schutzeinrichtungen. Berlin: Beuth.
7.63. DIN 31 001 Teil 2: Schutzeinrichtungen. Werkstoffe, Anforderungen, Anwendung. Berlin: Beuth.
7.64. DIN 31 001 Teil 5: Schutzeinrichtungen. Sicherheitstechnische Anforderungen an Verriege-lungen. Berlin: Beuth.
7.65. DIN 31 004 (Entwurf): Begriffe der Sicherheitstechnik. Grundbegriffe. Berlin: Beuth 1982. Ersetzt durch DIN VDE 31 000 Teil 2: Allgemeine Leitsätze für das sicherheitsgerechte Gestalten techn-ischer Erzeugnisse; Begriffe der Sicherheitstechnik; Grundbegriffe (1987).
7.66. DIN 31 051, Teil 1: Instandhaltung; Begriffe. Berlin: Beuth.
7.67. DIN 33 400: Gestalten von Arbeitssystemen nach arbeitswissenschaftlichen Erkenntnissen; Begriffe und Allgemeine Leitsätze. Beiblatt 1 – Beispiel für höhenverstellbare Arbeitsplattformen. Berlin: Beuth.
7.68. DIN 33 401: Stellteile; Begriffe, Eignung, Gestaltungshinweise. Beiblatt 1 – Erläuterungen zu Ersatzmöglichkeiten und Eignungshinweisen für Hand-Stellteile. Berlin: Beuth.
7.69   DIN 33 402: Körpermaße des Menschen;
        Teil 1 – Begriffe, Meßverfahren.
        Teil 2 – Werte.
        Beiblatt 1 – Anwendung von Körpermaßen in der Praxis;
        Teil 3 – Bewegungsfreiraum bei verschiedenen Körperhaltungen. Berlin: Beuth.
7.70. DIN 33 403: Klima am Arbeitsplatz und in der Arbeitsumgebung;
        Teil 1 – Grundlagen zur Klimaermittlung.

Teil 2 – Einfluß des Klimas auf den Menschen.
Teil 3 – Beurteilung des Klimas im Erträglichkeitsbereich. Berlin: Beuth.

7.71. DIN 33 404: Gefahrensignale für Arbeitsstätten;
Teil 1 – Akustische Gefahrensignale; Begriffe, Anforderungen, Prüfung, Gestaltungshinweise.
Beiblatt 1 – Akustische Gefahrensignale; Gestaltungsbeispiele.
Teil 2 – Optische Gefahrensignale; Begriffe, Sicherheitstechnische Anforderungen, Prüfung.
Teil 3 – Akustische Gefahrensignale; Einheitliches Notsignal, Sicherheitstechnische Anforderungen, Prüfung. Berlin: Beuth.

7.72. DIN 33 408: Körperumrißschablonen.
Teil 1 – Seitenansicht für Sitzplätze.
Beiblatt 1 – Anwendungsbeispiele. Berlin: Beuth.

7.73. DIN 33 411: Körperkräfte des Menschen.
Teil 1 – Begriffe, Zusammenhänge, Bestimmungsgrößen. Berlin: Beuth.

7.74. DIN 33 412 (Entwurf): Ergonomische Gestaltung von Büroarbeitsplätzen; Begriffe, Flächenermittlung, Sicherheitstechnische Anforderungen. Berlin: Beuth 1981.

7.75. DIN 33 413: Ergonomische Gesichtspunkte für Anzeigeeinrichtungen.
Teil 1 – Arten, Wahrnehmungsaufgaben, Eignung. Berlin: Beuth.

7.76. DIN 33 414: Ergonomische Gestaltung von Warten.
Teil 1 – Begriffe; Maße für Sitzarbeitsplätze. Berlin: Beuth.

7.77. DIN 40 041: Zuverlässigkeit elektrischer Bauelemente. Berlin: Beuth.

7.78. DIN 40 042: (Vornorm): Zuverlässigkeit elektrischer Geräte, Anlagen und Systeme. Berlin: Beuth 1970.

7.79. DIN IEC-73/VDE 0199. Kennfarben für Leuchtmelder und Druckknöpfe. Berlin: Beuth 1978.

7.80. DIN 43 602: Betätigungssinn und Anordnung von Bedienteilen. Berlin: Beuth.

7.81. DIN 50 900 Teil 1: Korrosion der Metalle. Allgemeine Begriffe. Berlin: Beuth.

7.82. DIN 50 900 Teil 2: Korrosion der Metalle. Elektrochemische Begriffe. Berlin: Beuth.

7.83. DIN 50 960: Korrosionsschutz, galvanische Überzüge. Berlin: Beuth.

7.84. DIN 66 233: Bildschirmarbeitsplätze; Begriffe. Berlin: Beuth.

7.85. DIN 66 234: Bildschirmarbeitsplätze.
Teil 1 – Geometrische Gestaltung der Schriftzeichen.
Teil 2 (Entwurf) – Wahrnehmbarkeit von Zeichen auf Bildschirmen.
Teil 3 – Gruppierungen und Formatierung von Daten.
Teil 5 – Codierung von Information; Berlin: Beuth.

7.86. DIN: Handbuch der Normung – Innerbetriebliche Normungsarbeit, Bd. 1: Überbetriebliche Normung, 5. Aufl. Berlin: Beuth 1981.

7.87. DIN: Handbuch der Normung – Innerbetriebliche Normungsarbeit, Bd. 2: Grundlagen, 4. Aufl. Berlin: Beuth 1980.

7.88. DIN: Handbuch der Normung – Innerbetriebliche Normungsarbeit, Bd. 3: Aspekte und Systeme, 5. Aufl. Berlin: Beuth 1980.

7.89. DIN: Handbuch der Normung – Innerbetriebliche Normungsarbeit, Bd. 4: Führungswissen, 2. Aufl. Berlin: Beuth 1976.

7.90. DIN-Katalog für technische Regeln, Bd. 1 bis 3. Berlin: Beuth.

7.91. DIN – Normungsheft 10: Grundlagen der Normungsarbeit des DIN. Berlin: Beuth 1982.

7.92. DIN – Taschenbuch 1: Grundnormen 20. Aufl. Berlin: Beuth 1983.

7.93. DIN – Taschenbuch 3: Normen für Studium und Praxis, 7. Aufl. Berlin: Beuth 1980.

7.94. DIN – Taschenbuch 22: Einheiten und Begriffe für physikalische Größen, 6, Aufl. Berlin: Beuth 1984.

7.95. Dittmayer, S.: Leitlinien für die Konstruktion arbeitsstrukturierter und montagegerechter Produkte. Industrie-Anzeiger 104 (1982) 58–59.

7.96. Dobeneck, v., D.: Die Elektronenstrahltechnik – ein vielseitiges Fertigungsverfahren. Feinwerktechnik und Micronic 77 (1973) 98–106.

7.97. Ehrlenspiel, K.: Mehrweggetriebe für Turbomaschinen. VDI-Z. 111 (1969) 218–221.

7.98. Ehrlenspiel, K.: Planetengetriebe – Lastausgleich und konstruktive Entwicklung. VDI-Berichte Nr. 105, 57–67. Düsseldorf: VDI-Verlag 1967.

7.99.   Endres, W.: Wärmespannungen beim Aufheizen dickwandiger Hohlzylinder. Brown-Boveri-Mitteilungen (1958) 21–28.

7.100. Erker, A.; Mayer, K.: Relaxations- und Sprödbruchverhalten von warmfesten Schraubenverbindungen. VGB Kraftwerkstechnik 53 (1973) 121–131.

7.101. Eversheim, W.; Pfeffekoven, K.H.: Aufbau einer anforderungsgerechten Montageorganisation. Industrie-Anzeiger 104 (1982) 75–80.

7.102. Eversheim, W.; Pfeffekoven, K.H.: Planung und Steuerung des Montageablaufs komplexer Produkte mit Hilfe der EDV. VDI-Z. 125 (1983), 217–222.

7.103. Eversheim, W.: Müller, W.: Beurteilung von Werkstücken hinsichtlich ihrer Eignung für die automatisierte Montage. VDI-Z. 125 (1983) 319–322.

7.104. Eversheim, W.; Müller, W.: Montagegerechte Konstruktion. Proc. of the 3rd Int. Conf. on Assembly Automation in Böblingen (1982) 191–204.

7.105. Eversheim, W.; Ungeheuer, U.; Pfeffekoven, K.H.: Montageorientierte Erzeugnisstrukturierung in der Einzel- und Kleinserienproduktion – ein Gegensatz zur funktionsorientierten Erzeugnisgliederung? VDI-Z. 125 (1983) 475–479.

7.106. Fachverband Pulvermetallurgie: Sinterteile – ihre Eigenschaften und Anwendung. Berlin: Beuth 1971.

7.107. Falk, K.: Theorie und Auslegung einfacher Backenbremsen. Konstruktion 19 (1967) 268–271.

7.108. Feldmann, H.D.: Konstruktionsrichtlinien für Kaltfließpreßteile aus Stahl. Konstruktion 11 (1959) 82–89.

7.109. Florin, C.; Imgrund, H.: Über die Grundlagen der Warmfestigkeit. Arch. Eisenhüttenwesen 41 (1970) 777–778.

7.110   Föller, D.: Maschinenakustische Berechnungsgrundlagen für den Konstrukteur. VDI-Berichte Nr. 239. Düsseldorf: VDI-Verlag 1975.

7.111. Friedewald, H.J.: Normung – integrierender Bestandteil einer Firmenkonzeption. DIN-Mitteilungen 49 (1970) 3–12.

7.112. Gairola, A.: Montagegerechtes Konstruieren – Ein Beitrag zur Konstruktionsmethodik. Diss. TU Darmstadt 1981.

7.113. Gassner, E.: Ermittlung von Betriebsfestigkeitskennwerten auf der Basis der reduzierten Bauteildauerfestigkeit. Materialprüfung 26 (1984) Nr. 11.

7.114. Geißlinger, W.: Montagegerechtes Konstruieren. Wt-Zeitschrift für industrielle Fertigung 71 (1981) 29–32.

7.115. Gesetz über technische Arbeitsmittel (Gerätesicherheitsgesetz), zuletzt geändert durch BBerg G vom 13. Aug. 1980. Gesetz zur Änderung des Gesetzes über technische Arbeitsmittel und der Gewerbeordnung (In: BGBl I, 1979). Allgemeine Verwaltungsvorschrift zum Gesetz über technische Arbeitsmittel vom 11. Juni 1979. Zu beziehen durch: Deutsches Informationszentrum für technische Regeln (DITR), Berlin.

7.116. Gnilke, W.: Lebensdauerberechnung der Maschinenelemente. München: C. Hanser 1980.

7.117. Gräfen, H.: Berücksichtigung der Korrosion bei der konstruktiven Gestaltung von Chemieapparaten. Werkstoffe und Korrosion 23 (1972) 247–254.

7.118. Gräfen, H.; Gerischer, K.; Horn, E.M.: Die Bedeutung der Werkstoffauswahl für die Gebrauchstauglichkeit von Chemieapparaten – Auswahlkriterien und Prüfverfahren. Z. f. Werkstofftechnik 4 (1973) 169–186.

7.119. Gräfen, H.; Spähn, H.: Probleme der chemischen Korrosion in der Hochdrucktechnik. Chemie-Ingenieur-Technik 39 (1967) 525–530.

7.120. Grandjean, E.: Physiologische Arbeitsgestaltung, Leitfaden der Ergonomie, 3. erw. Aufl. Thun, München: Ott 1979.

7.121. Grunert, M.: Stahl- und Spannbeton als Werkstoff im Maschinenbau. Maschinenbautechnik 22 (1973) 374–378.

7.122. Günther, T.: Schadensfälle an Apparaten und deren Berücksichtigung für neue Konstruktionen. Chemie-Ingenieur-Technik 42 (1970) 774–780.

7.123. Günther, W.: Schwingfestigkeit. Leipzig: VEB-Verlag Technik 1972.

7.124. Habig, K.-H.: Verschleiß und Härte von Werkstoffen. München: C. Hanser 1980.

7.125. Hähn, G.: Entwurf eines Stoßprüfstandes mit Hilfe konstruktionssystematischer Methoden. Studienarbeit TU Berlin.

7.126. Hänchen, R.: Gegossene Maschinenteile. München: Hanser 1964.

7.127. Hänchen, R.; Decker, K.H.: Neue Festigkeitsberechnung für den Maschinenbau. München: Hanser 1967.

7.128. Händel, S.: Kostengünstigere Gestaltung und Anwendung von Normen (manuell und rechnerunterstützt). DIN-Mitt. 62 (1983) 565–571.

7.129. Häusler, N.: Der Mechanismus der Biegemomentübertragung in Schrumpfverbindungen. Diss. TH Darmstadt 1974.

7.130. Haibach, E.: Betriebsfestigkeit – Verfahren und Daten zur Bauteilberrechnung. Düsseldorf: VDI-Verlag 1989.

7.131. Handbuch der Arbeitsgestaltung und Arbeitsorganisation. Düsseldorf: VDI-Verlag 1980.

7.132. Handbuch der Ergonomie. Mit ergonomischen Konstruktionsrichtlinien. Hrsg. vom Bundesamt für Wehrtechnik und Beschaffung. Wiss. Bearb. H. Schmidtke. München: Hanser 1975 ff.

7.133. Hartlieb, B.; Nitsche, H.; Urban, W.: Systematische Zusammenhänge in der Normung. DIN-Mitt. 61 (1982) 657–662.

7.134. Hartmann, A.: Die Druckgefährdung von Absperrschiebern bei Erwärmung des geschlossenen Schiebergehäuses. Mitt. VGB (1959) 303–307.

7.135. Hartmann, A.: Schaden am Gehäusedeckel eines 20-atü-Dampfschiebers. Mitt. VGB (1959) 315–316.

7.136. Heckl, M.: Minderung der Körperschallentstehung und Körperschallfortleitung bei Maschinen und Maschinenelementen. VDI-Berichte 239. Düsseldorf: VDI-Verlag 1975.

7.137. Heckl, M.: Konstruktive Möglichkeiten zur Minderung der Luftschallabstrahlung. VDI-Berichte 239. Düsseldorf: VDI-Verlag 1975.

7.138. Heesch, H.; Kienzle, O.: Flächenschluß Buchreihe Wissenschaftliche Normung. Berlin: Springer 1963.

7.139. Heinz, K.; Tertilt, G.: Montage- und Handhabungstechnik. VDI-Z. 126 (1984) 151–157.

7.140. Hertel, H.: Ermüdungsfestigkeit der Konstruktionen. Berlin: Springer 1969.

7.141. Hertel, H.: Leichtbau. Berlin: Springer 1960.

7.142. Herzke, I.: Technologie und Wirtschaftlichkeit des Plasma-Abtragens. ZwF 66 (1971) 284–291.

7.143. Hesse, S.; Zapf, H.: Automatisches Fügen. Berlin: VEB Verlag Technik 1972.

7.144. Hesser, W.: Untersuchungen zum Beziehungsfeld zwischen Konstruktion und Normung. Diss. TU Berlin. DIN-Normungskunde, Bd. 16. Berlin: Beuth 1981.

7.145. Hoenow, G.; Mann, H.: Roboter – montagegerechtes Konstruieren. Maschinenbautechnik 30 (1981) 202–204.

7.146. Hömig, H.: Metall und Wasser – Eine kleine Korrosionskunde. Essen: Vulkan 1971.

7.147. Hüskes, H.; Schmidt, W.: Unterschiede im Kriechverhalten bei Raumtemperatur von Stählen mit und ohne ausgeprägter Streckgrenze. DEW-Tech. Berichte 12 (1972) 29–34.

7.148. Illgner, K.-H.: Werkstoffauswahl im Hinblick auf wirtschaftliche Fertigungen. VDI-Z. 114 (1972) 837–841, 992–995.

7.149. Iredale, R.: Automatic Assembly. Metalworking 8 (1964), S. 55–60.

7.150. Jacobi, P.; Volmer, J.: Fügemechanismen für die automatisierte Montage mit Industrierobotern. Maschinenbautechnik 31 (1982) 451–453.

7.151. Jaeger, T. A.: Zur Sicherheitsproblematik technologischer Entwicklungen. QZ 19 (1974) 1–9.

7.152. Jenner, R.-D.; Kaufmann, H.; Schäfer, D.: Planungshilfen für die ergonomische Gestaltung – Zeichenschablonen für die menschliche Gestalt, Maßstab 1 : 10. Esslingen: IWA-Riehle 1978.

7.153. Johnson, Kenneth, Lester: Grundlagen der Netzplantechnik. VDI-Taschenbücher T 53. Düsseldorf: VDI-Verlag 1974.

7.154. Jorden, W.: Recyclinggerechtes Konstruieren – Utopie oder Notwendigkeit. Schweizer Maschinenmarkt (1984) 23–25, 32–33.

7.155. Jorden, W.: Recyclinggerechtes Konstruieren als vordringliche Aufgabe zum Einsparen von Rohstoffen. Maschinenmarkt 89 (1983) 1406–1409.

7.156. Jorden, W.: Der Tolerierungsgrundsatz – eine unbekannte Größe mit schwerwiegenden Folgen. Konstruktion 43 (1991) 170–176.

7.157. Jorden, W.: Toleranzen für Form, Lage und Maß München: Hanser 1991.

7.158. Jung, A.: Schmiedetechnische Überlegungen für die Konstruktion von Gesenkschmiedestücken aus Stahl. Konstruktion 11 (1959) 90–98.

7.159. Käufer, H.: Recycling von Kunststoffen, integriert in Konstruktion und Anwendungstechnik. Konstruktion 42 (1990) 415–420.

7.160. Keil, E.; Müller, E.O.; Bettziehe, P.: Zeitabhängigkeit der Festigkeits- und Verformbarkeitswerte von Stählen im Temperaturbereich unter 400°C. Eisenhüttenwesen 43 (1971) 757–762.

7.161. Kesselring, F.: Technische Kompositionslehre. Berlin: Springer 1954.

7.162. Kienzle, O.: Normen und Konstruieren. Konstruktion 19 (1967) 121–125.

7.163. Kienzle, O.: Normung und Wissenschaft, Schweiz. Techn. Z. (1943) 533–539.

7.164. Klein, M.: Einführung in die DIN-Normen, 10. Aufl. Stuttgart: Teubner 1989.

7.165. Klöcker, I.: Produktgestaltung, Aufgabe – Kriterien – Ausführung. Berlin: Springer 1981.

7.166. Kloos, K.H.: Werkstoffoberfläche und Verschleißverhalten in Fertigung und konstruktive Anwendung. VDI-Berichte Nr. 194. Düsseldorf: VDI-Verlag 1973.

7.167. Kloss, G.: Einige übergeordnete Konstruktionshinweise zur Erzielung echter Kostensenkung. VDI-Fortschrittsberichte, Reihe 1, Nr. 1. Düsseldorf: VDI-Verlag 1964.

7.168. Klotter, K.: Technische Schwingungslehre, Bd. 1. Teil A und B, 3. Aufl. Berlin: Springer 1980/81.

7.169. Knappe, W.: Thermische Eigenschaften von Kunststoffen. VDI-Z. 111 (1969) 746–752.

7.170. Köhler, G.; Rögnitz, H.: Maschinenteile, Bd. 1 u. Bd. 2, 6. Aufl. Stuttgart: Teubner 1981.

7.171. Koenig, W.: Wechselwirkung zwischen Konstruktion und rationeller Fertigung. VDI-Z. 95 (1953) 896–903.

7.172. Korrosionsschutzgerechte Konstruktion – Merkblätter zur Verhütung von Korrosion durch konstruktive und fertigungstechnische Maßnahmen. Herausgeber Dechema Deutsche Gesellschaft für chemisches Apparatewesen e.V. Frankfurt am Main 1981.

7.173. Kragelskii, I.V.: Friction and Wear. London: Butterworth 1965.

7.174. Krause, W. (Hrsg.): Gerätekonstruktion, 2. Aufl. Berlin: VEB Verlag Technik 1986.

7.175. Krieg, K.G.; Heller, W.; Hunecke, G.; Zemlin, H.: Leitfaden der DIN-Normen. Berlin: Beuth 1983.

7.176. Kroemer, K.H.: Was man von Schaltern, Kurbeln und Pedalen wissen muß. Berlin: Beuth 1967.

7.177. Kühnpast, R.: Das System der selbsthelfenden Lösungen in der maschinenbaulichen Konstruktion. Diss. TH Darmstadt 1968.

7.178. Lang, K.; Voigtländer, G.: Neue Reihe von Drehstrommaschinen großer Leistung in Bauform B 3. Siemens-Z. 45 (1971) 33–37.

7.179. Lambrecht, D.; Scherl, W.: Überblick über den Aufbau moderner wasserstoffgekühlter Generatoren. Berlin: Verlag AEG 1963, 181–191.

7.180. Leipholz, H.: Festigkeitslehre für den Konstrukteur. Konstruktionsbücher Bd. 25. Berlin: Springer 1969.

7.181. Leyer, A.: Grenzen und Wandlung im Produktionsprozeß, technica 12 (1963) 191–208.

7.182. Leyer, A.: Kraft- und Bewegungselemente des Maschinenbaus, technica 26 (1973) 2498–2510, 2507–2520, technica 5 (1974) 319–324, technica 6 (1974) 435–440.

7.183. Leyer, A.: Maschinenkonstruktionslehre, Hefte 1–7. technica-Reihe. Basel: Birkhäuser 1963–1978.

7.184. Lotter, B.: Arbeitsbuch der Montagetechnik. Mainz: Fachverlage Krausskopf-Ingenieur Digest 1982.

7.185. Lotter, B.: Montagefreundliche Gestaltung eines Produktes. Verbindungstechnik 14 (1982) 28–31.

7.186. Lüpertz, H.: Neue zeichnerische Darstellungsart zur Rationalisierung des Konstruktionsprozesses vornehmlich bei methodischen Vorgehensweisen. Diss. TH Darmstadt 1974.

7.187. Macherauch, E.; Müller, H.: Grundzüge der Versagensbetrachtung bei quasi-statischer Beanspruchung. Der Maschinenschaden 56 (1983) 86–97.

7.188. Maduschka, L.: Beanspruchung von Schraubenverbindungen und zweckmäßige Gestaltung der Gewindeträger. Forsch. Ing. Wes. 7 (1936) 299–305.

7.189. Magnus, K.: Schwingungen, 3. Aufl. Stuttgart: Teubner 1976.

7.190. Magyar, J.: Aus nichtveröffentlichtem Unterrichtsmaterial der TU Budapest, Lehrstuhl für Maschinenelemente.

7.191. Mahle-Kolbenkunde, 2. Aufl. Stuttgart: 1964.

7.192. Marré, T.; Reichert, M.: Anlagenüberwachung und Wartung. Sicherheit in der Chemie. Verl. Wiss. u. Polit. 1979.

7.193. Matousek, R.: Konstruktionslehre des allgemeinen Maschinenbaus. Berlin: Springer 1957, Reprint 1974.

7.194. Matting, A.; Ulmer, K.: Spannungsverteilung in Metallklebverbindungen. VDI-Z. 105 (1963) 1449–1457.

7.195. Mauz, W.; Kies, H.: Funkenerosives und elektrochemisches Senken. ZwF 68 (1973) 418–422.

7.196. Melan, E.; Parkus, H.: Wärmespannungen infolge stationärer Temperaturfelder. Wien: Springer 1953.

7.197. Menges, G.; Taprogge, R.: Denken in Verformungen erleichert das Dimensionieren von Kunststoffteilen. VDI-Z. 112 (1970) 341–346, 627–629.

7.198. Meyer, H.: Recyclingorientierte Produktgestaltung. VDI-Fortschrittsberichte Reihe 1, Nr. 98. Düsseldorf: VDI-Verlag 1983.

7.199. Meyer, H., Beitz, W.: Konstruktionshilfen zur recyclingorientierten Produktgestaltung. VDI-Z. 124 (1982) 255–267.

7.200. Militzer, O.M. Rechenmodell für die Auslegung von Wellen-Naben-Paßfederverbindungen. Diss. TU Berlin 1975.

7.201. Möhler, E.: Der Einfluß des Ingenieurs auf die Arbeitssicherheit, 4. Aufl. Berlin: Verlag Tribüne 1965.

7.202. Müller, K.: Schrauben aus thermoplastischen Kunststoffen. Werkstattblatt 514 und 515. München: Hanser 1970.

7.203. Müller, K.: Schrauben aus thermoplastischen Kunststoffen. Kunststoffe 56 (1966) 241–250, 422–429.

7.204. Müller, R.: Das Wolfsburger Modell der Schwingfestigkeit. VDI-Z. 122 (1980) 761–768, 841–847.

7.205. Müllner, E.: Entwicklungstendenzen im Bau von Turbogeneratoren. BBC Nachrichten (1960) 279–286.

7.206. Munz, D.; Schwalbe, K.; Mayr, P.: Dauerschwingverhalten metallischer Werkstoffe. Braunschweig: Vieweg 1971.

7.207. Murell, K.F.: Ergonomie, Grundlagen und Praxis der Gestaltung optimaler Arbeitsverhältnisse. Düsseldorf: Econ 1971.

7.208. Neuber, H.: Kerbspannungslehre, 3, Aufl. Berlin: Springer 1985.

7.209. Neuber, H.: Über die Berücksichtigung der Spannungskonzentration bei Festigkeitsberechnungen. Konstruktion 20 (1968) 245–251.

7.210. Neudörfer, A.: Anzeiger und Bedienteile – Gesetzmäßigkeiten und systematische Lösungssammlungen. Düsseldorf: VDI-Verlag 1981.

7.211. Neumann, J.; Timpe, K.P.: Psychologische Arbeitsgestaltung. Berlin: VEB Deutscher Verlag der Wissenschaften 1976.

7.212. Niemann, G.: Maschinenelemente, Bd. 1. Berlin: Springer 1963, 2. Aufl. 1975.

7.213. N.N.: Ergebnisse deutscher Zeitstandversuche langer Dauer. Düsseldorf: Stahleisen 1969.

7.214. N.N.: Nickelhaltige Werkstoffe mit besonderer Wärmeausdehnung. Nickel-Berichte D 16 (1958) 79–83.

7.215. N.N.: Forschungsprogramm Wiederverwertung von Kunststoffabfällen. VKE (Hrsg.) Frankfurt: 1979–1981.

7.216. Oehler, G.; Weber, A.: Steife Blech- und Kunststoffkonstruktionen. Konstruktionsbücher, Bd. 30. Berlin: Springer 1972.

7.217. Pahl, G.: Ausdehnungsgerecht. Konstruktion 25 (1973) 367–373.

7.218. Pahl, G.: Bewährung und Entwicklungsstand großer Getriebe in Kraftwerken. Mitteilungen der VGB 52, Kraftwerkstechnik (1972) 404–415.

7.219. Pahl, G.: Entwurfsingenieur und Konstruktionslehre unterstützen die moderne Konstruktionsarbeit. Konstruktion 19 (1967) 337–344.

7.220. Pahl, G.: Grundregeln für die Gestaltung von Maschinen und Apparaten. Konstruktion 25 (1973) 271–277.

7.221. Pahl, G.: Konstruktionstechnik im thermischen Maschinenbau. Konstruktion 15 (1963) 91–98.

7.222. Pahl, G.: Prinzip der Aufgabenteilung. Konstruktion 25 (1973) 191–196.

7.223. Pahl, G.: Prinzipien der Kraftleitung. Konstruktion 25 (1973) 151–156.

7.224. Pahl, G.: Das Prinzip der Selbsthilfe. Konstruktion 25 (1973) 231–237.

7.225. Pahl, G.: Sicherheitstechnik aus konstruktiver Sicht. Konstruktion 23 (1971) 201–208.

7.226. Pahl, G.: Vorgehen beim Entwerfen. ICED 1983. Schweizer Maschinenmarkt. 84. Jahrgang 1984, Heft 25, 35–37.

7.227. Pahl, G.: Konstruktionsmethodik als Hilfsmittel zum Erkennen von Korrosionsgefahren. 12. Konstr.-Symposium Dechema, Frankfurt 1981.

7.228. Paland, E.G.: Untersuchungen über die Sicherungseigenschaften von Schraubenverbindungen bei dynamischer Belastung. Diss. TH Hannover 1960.

7.229. Peters, O.H.; Meyna, A. Handbuch der Sicherheitstechnik. München: C. Hanser 1985.

7.230. Pflüger, A.: Stabilitätsprobleme der Elastostatik. Berlin: Springer 1964.

7.231. Pourshirazi, M.: Recycling und Werkstoffsubstitution bei technischen Produkten als Beitrag zur Ressourcenschonung. Schriftenreihe Konstruktionstechnik Heft 12 (Hrsg. W. Beitz). Berlin: TU Berlin 1987.

7.232. Rabinowicz, E.: Friction on Wear of Materials. New York: Wiley and Sons, Inc. 1965.

7.233. REFA: Methodenlehre des Arbeitsstudiums, Teil 1, Grundlagen, 7. Aufl. München: Hanser 1984.

7.234. Reinhardt, K.-G.: Verbindungskombinationen und Stand ihrer Anwendung. Schweißtechnik 19 (1969) Heft 4.

7.235. Rembold, U.; Blume, Ch.; Dillmann, R.; Mörkel, G.: Technische Anforderungen an zukünftige Industrieroboter – Analyse von Montagevorgängen und montagegerechtes Konstruieren. VDI-Z. 123 (1981) 763–772.

7.236. Reuter, H.: Die Flanschverbindung im Dampfturbinenbau. BBC-Nachrichten 40 (1958) 355–365.

7.237. Reuter, H.: Stabile und labile Vorgänge in Dampfturbinen. BBC-Nachrichten 40 (1958) 391–398.

7.238. Richter, E.; Schilling, W.; Weise, M.: Montage im Maschinenbau. Berlin: VEB-Verlag Technik 1974.

7.239. Rixmann, W.: Ein neuer Ford-Taunus 12 M. ATZ 64 (1962) 306–311.

7.240. Rodenacker, W.G.: Methodisches Konstruieren. Berlin: Springer 1970. 2. Auflage 1976, 3. Auflage 1984, 4. Auflage 1991.

7.241. Rögnitz, H.; Köhler, G.: Fertigungsgerechtes Gestalten im Maschinen- und Gerätebau. Stuttgart: Teubner 1959.

7.242. Rohmert, W.: Maximalkräfte von Männern im Bewegungsraum der Arme und Beine. Forschungsbericht Nr. 1616 des Landes NRW. Köln: Westdeutscher Verlag 1966.

7.243. Rohmert, W.; Hettinger, Th.: Körperkräfte im Bewegungsraum. RKW – Reihe Arbeitsphysiologie – Arbeitspsychologie. Berlin: Beuth 1963.

7.244. Rohmert, W.; Jenik, P.: Maximalkräfte von Frauen im Bewegungsraum der Arme und Beine. REFA – Schriftenreihe Arbeitswissenschaft und Praxis 22, Berlin: Beuth 1972.

7.245. Rohmert, W.; Rutenfranz, J. (Hrsg.): Praktische Arbeitsphysiologie. Stuttgart: Thieme Verlag 1983.

7.246. Rosemann, H.: Zuverlässigkeit und Verfügbarkeit technischer Anlagen und Geräte. Berlin: Springer 1981.

7.247. Roth, K.: Die Kennlinie von einfachen und zusammengesetzten Reibsystemen. Feinwerktechnik 64 (1960) 135–142.

7.248. Rubo, E.: Der chemische Angriff auf Werkstoffe aus der Sicht des Konstrukteurs. Der Maschinenschaden (1966) 65–74.

7.249. Rubo, E.: Die Wirkung der Erosion bei der Strömungskorrosion. Cz-Chemie-Technik 1 (1972) 177–179.

7.250. Rubo, E.: Höhere Sicherheit chemisch beanspruchter Bauteile durch konstruktive Korrosionsbewertung am Beispiel von Druckapparaten. Konstruktion 12 (1960) 490–498.

7.251. Rubo, E.: Kostengünstiger Gebrauch ungeschützter korrosionsanfälliger Metalle bei korrosiven Angriff. Konstruktion 37 (1985) 11–20.

7.252. Salm, M.; Endres, W.: Anfahren und Laständerung von Dampfturbinen. Brown-Boveri-Mitteilungen (1958) 339–347.

7.253. Sandager, Markovits, Bredtschneider: Piping Elements for Coal-Hydrogenations Service. Trans. ASME May 1950, 370ff.

7.254. Schacht, M.: Methodische Neugestaltung von Normen als Grundlage für eine Integration in den rechnerunterstützten Konstruktionsprozeß DIN-Normungskunde, Bd. 28. Berlin: Beuth 1991.

7.255. Schacht, M.: Rechnerunterstützte Bereitstellung und methodische Entwicklung von Normen. Konstruktion 42 (1990) 1, 3–14.

7.256. Schier, H.: Fototechnische Fertigungsverfahren. Feinwerktechnik + Micronic 76 (1972) 326–330.
7.257. Schilling, K.: Konstruktionsprinzipien der Feinwerktechnik. Proceedings ICED '91, Schriftenreihe WDK 20. Zürich: Heurista 1991.
7.258. Schilling, W.: Montagegerechtes Konstruieren, Fertigungstechnik u. Betrieb 18 (1968) 731–740.
7.259. Schmid, E.: Theoretische und experimentelle Untersuchung des Mechanismus der Drehmomentübertragung von Kegel-Preß-Verbindungen. VDI-Fortschrittsberichte Reihe 1, Nr. 16. Düsseldorf: VDI-Verlag 1969.
7.260. Schmidt, K.P.; Schröder, P.J.: Konstruktive Möglichkeiten zur Minderung der Geräuschentstehung. VDI-Berichte 239. Düsseldorf: VDI-Verlag 1975.
7.261. Schmidt, E.: Sicherheit und Zuverlässigkeit aus konstruktiver Sicht. Ein Beitrag zur Konstruktionslehre. Diss. TH Darmstadt 1981.
7.262. Schmidt-Kretschmer, M.; Beitz, W.: Demontagefreundliche Verbindungstechnik – ein Beitrag zum Produktrecycling. VDI-Berichte 906. Düsseldorf: VDI-Verlag 1991.
7.263. Schmidtke, H. (Hrsg.): Lehrbuch der Ergonomie. München: Hanser 1981.
7.264. Schott, G.: Ermüdungsfestigkeit – Lebensdauerberechnung für Kollektiv- und Zufallsbeanspruchungen. Leipzig: VEB, Deutscher Verlag f. Grundstoffindustrie 1983.
7.265. Schraft, R.D.: Montagegerechte Konstruktion – die Voraussetzung für eine erfolgreiche Automatisierung. Proc. of the 3rd Int. Conf. on Assembly Automation in Böblingen (1982) 165–176.
7.266. Schraft, R.D.; Bäßler, R.: Die montagegerechte Produktgestaltung muß durch systematische Vorgehensweisen umgesetzt werden. VDI-Z. 126 (1984) 843–852.
7.267. Schütz, W.; Zenner, H.: Schadensakkumulationshypothesen zur Lebensdauervorhersage bei schwingender Beanspruchung – Ein kritischer Überblick. Z. Werkstofftechnik (1973) 25–33, 97–102.
7.268. Schütz, W.: Lebensdauervorhersage schwingend beanspruchter Bauteile. Der Maschinenschaden 56 (1963) 221–230.
7.269. Schwaigerer, S.: Festigkeitsberechnung von Bauelementen des Dampfkessel-Behälter- und Rohrleitungsbaus, 4. Aufl. Berlin: Springer 1983.
7.270. Schweizer, W.; Kiesewetter, L.: Moderne Fertigungsverfahren der Feinwerktechnik. Berlin: Springer 1981.
7.271. Schwenk, W.: Stand der Kenntnisse über die Korrosion von Stahl. Stahl und Eisen 89 (1969) 535–547.
7.272. Seeger, O.W.: Sicherheitsgerechtes Gestalten technischer Erzeugnisse. Berlin: Beuth 1983.
7.273. Seeger, O.W.: Maschinenschutz, aber wie? Schriftenreihe Arbeitssicherheit, Heft 8. Köln: Aulis 1972.
7.274. Seeger, H.: Technisches Design. Grafenau: Expert Verlag 1980.
7.275. Seeger, H.: Industrie-Designs. Grafenau: Expert Verlag 1983.
7.276. Seeger, H.: Design technischer Produkte, Programme und Systeme. Anforderungen, Lösungen und Bemerkungen. Berlin: Springer 1992.
7.277. Shreir, L.L.: Corrosion. London: George Newnes Ltd. 1963 und 1965.
7.278. Sicherheitsregeln für berührungslos wirkende Schutzeinrichtungen an kraftbetriebenen Arbeitsmitteln. ZH 1/597. Köln: Heymanns 1979.
7.279. Sieck, U.: Kriterien der montagegerechten Gestaltung in den Phasen des Montageprozesses. Automatisierungspraxis 10 (1973) 284–286.
7.280. Simon, H.; Thoma, M.: Angewandte Oberflächentechnik für metallische Werkstoffe. München: C. Hanser 1985.
7.281. Spähn, H.; Fäßler, K.: Kontaktkorrosion. Grundlagen – Auswirkung – Verhütung. Werkstoffe und Korrosion 17 (1966) 321–331.
7.282. Spähn, H.; Fäßler, K.: Zur konstruktiven Gestaltung korrosionsbeanspruchter Apparate in der chemischen Industrie. Konstruktion 24 (1972) 249–258, 321–325.
7.283. Spähn, H.; Rubo, E.; Pahl, G.: Korrosionsgerechte Gestaltung. Konstruktion 25 (1973) 455–459.
7.284. Spur, G.; Stöferle, Th. (Hrsg.): Handbuch der Fertigungstechnik. Bd. 1: Urformen, Bd. 2: Umformen, Bd. 3: Spanen, Bd. 4: Abtragen, Beschichten, Wärmebehandeln, Bd. 5: Fügen, Handhaben, Montieren, Bd. 6: Fabrikbetrieb. München: C. Hanser 1979–1986.
7.285. Steinack, K.; Veenhoff, F.: Die Entwicklung der Hochtemperaturturbinen der AEG. AEG-Mitt. 50 (1960) 433–453.

7.286. Steinhilper, R.: Produktrecycling im Maschinenbau. Berlin: Springer 1988.

7.287. Steinhilper, R.: Der Horizont bestimmt den Erfolg beim Recycling. Konstruktion 42 (1990) 396–404.

7.288. Stöferle, Th.; Dilling, H.-J.; Rauschenbach, Th.: Rationelle Montage – Herausforderung an den Ingenieur. VDI-Z. 117 (1975) 715–719.

7.289. Stöferle, Th.; Dilling, H.-J.; Rauschenbach, Th.: Rationalisierung und Automatisierung in der Montage. Werkstatt und Betrieb 107 (1974) 327–335.

7.290. Susanto, A.: Methodik zur Entwicklung von Normen. DIN-Normungskunde, Bd. 23. Berlin: Beuth 1988.

7.291. Suter, F.; Weiss, G.: Das hydraulische Sicherheitssystem S74 für Großdampfturbinen. Brown Boveri-Mitt. 64 (1977) 330–338.

7.292. Swift, K.; Redford, H.: Design for Assembly. Engineering (1980) 799–802.

7.293. Tauscher, H.: Dauerfestigkeit von Stahl und Gußeisen. Leipzig: VEB-Verlag 1982.

7.294. ten Bosch, M.: Berechnung der Maschinenelemente. Reprint. Berlin: Springer 1972.

7.295. TGL 19340: Dauerfestigkeit der Maschinenteile. DDR-Standards. Berlin: 1984.

7.296. Thomé-Kozkiensky, K.-J. (Hrsg.): Materialrecycling durch Abfallaufbereitung. Tagungsband TU Berlin 1983.

7.297. Thumb, N.: Grundlagen und Praxis der Netzplantechnik. München: Moderne Industrie 1969.

7.298. Thum, A.: Die Entwicklung von der Lehre der Gestaltfestigkeit. VDI-Z. 88 (1944) 609–615.

7.299. Tietz, H.: Ein Höchsttemperatur-Kraftwerk mit einer Frischdampftemperatur von 610°C. VDI-Z. 96 (1953) 802–809.

7.300. Tjalve, E.: Systematische Formgebung für Industrieprodukte. Düsseldorf: VDI-Verlag 1978.

7.301. Tochtermann, W.; Bodenstein, F.: Konstruktionselemente des Maschinenbaues, Teil 1 und 2, 9. Aufl. Berlin: Springer 1979.

7.302. Trapp, H.-J.: Beitrag zum rechnerischen Betriebsfestigkeitsnachweis für Bauteile in Kranhubwerken. Konstruktion 27 (1975) 112–149.

7.303. Tschochner, H.: Konstruieren und Gestalten. Essen: Girardet 1954.

7.304. Tuffenstammer, K.: Lärmarm Konstruieren – ein Beitrag zur Humanisierung des Arbeitslebens. VDI-Berichte 239. Düsseldorf: VDI-Verlag 1975.

7.305. Uhlig, H.H.: Korrosion und Korrosionsschutz. Berlin: Akademie-Verlag 1970.

7.306. Veit, H.-J.; Scheermann, H.: Schweißgerechtes Konstruieren. Fachbuchreihe Schweißtechnik Nr. 32. Düsseldorf: Deutscher Verlag für Schweißtechnik 1972.

7.307. van der Mooren, A.L.: Instandhaltungsgerechtes Konstruieren und Projektieren. Konstruktionsbücher Bd. 37. Berlin: Springer 1991.

7.308. VDI/ADB-Ausschuß Schmieden: Schmiedstücke – Gestaltung, Anwendung. Hagen: Informationsstelle Schmiedstück-Verwendung im Industrieverband Deutscher Schmieden 1975.

7.309. VDI-Berichte Nr. 129: Kerbproblem. Düsseldorf: VDI-Verlag 1968.

7.310. VDI-Berichte Nr. 214: Werkstoffe und Bauteilfestigkeit. Vorträge der VDI-Tagung Düsseldorf. Düsseldorf: VDI-Verlag 1974.

7.311. VDI-Berichte Nr. 362: Ideen verwicklichen mit Guß – Konstruktionsmethodik, Gestaltung, Anwendung. Düsseldorf: VDI-Verlag 1980.

7.312. VDI-Berichte Nr. 389: Lärmarm Konstruieren. Düsseldorf: VDI-Verlag 1981.

7.313. VDI-Berichte Nr. 420: Schmiedeteile konstruieren für die Zukunft. Düsseldorf: VDI-Verlag 1981.

7.314. VDI-Berichte Nr. 493: Spektrum der Verbindungstechnik – Auswählen der besten Verbindungen mit neuen Konstruktionskatalogen. Düsseldorf: VDI-Verlag 1983.

7.315. VDI-Berichte Nr. 523: Konstruieren mit Blech. Düsseldorf: VDI-Verlag 1984.

7.316. VDI-Berichte Nr. 544: Das Schmiedeteil als Konstruktionselement – Entwicklungen – Anwendungen – Wirtschaftlichkeit. Düsseldorf: VDI-Verlag 1985.

7.317. VDI-Berichte Nr. 556: Automatisierung der Montage in der Feinwerktechnik. Düsseldorf: VDI-Verlag 1985.

7.318. VDI-Berichte Nr. 563: Konstruieren mit Verbund- und Hybridwerkstoffen. Düsseldorf: VDI-Verlag 1985.

7.319. VDI-Richtlinie 2006: Gestalten von Spritzgußteilen aus thermoplastischen Kunststoffen. Düsseldorf: VDI-Verlag 1979.

7.320. VDI-Richtlinie 2057: Beurteilung der Einwirkung mechanischer Schwingungen auf den

Menschen. Blatt 1 (Entwurf) – Grundlagen, Gliederung, Begriffe (1983). Blatt 2: Schwingungs-einwirkung auf den menschlichen Körper (1981). Blatt 3 (Entwurf): Schwingungsbeanspruchung des Menschen (1979). Düsseldorf: VDI-Verlag.

7.321. VDI-Richtlinie 2221: Methodik zum Entwickeln technischer Systeme und Produkte. Düsseldorf: VDI-Verlag 1986.

7.322. VDI-Richtlinie 2222: Konstruktionsmethodik; Konzipieren technischer Produkte; Düsseldorf: VDI-Verlag 1993.

7.323. VDI-Richtlinie 2224: Formgebung technischer Erzeugnisse. Empfehlungen für den Konstrukteur. Düsseldorf: VDI-Verlag 1972.

7.324. VDI-Richtlinie 2225 Blatt 1 und Blatt 2: Technisch-wirtschaftliches Konstruieren. Düsseldorf: VDI-Verlag 1977. VDI 2225 (Entwurf): Vereinfachte Kostenermittlung 1984.

7.325. VDI-Richtlinie 2226: Empfehlung für die Festigkeitsberechnung metallischer Bauteile. Düsseldorf: VDI-Verlag 1965.

7.326. VDI-Richtlinie 2227 (Entwurf): Festigkeit bei wiederholter Beanspruchung, Zeit- und Dauerfestigkeit metallischer Werkstoffe, insbesondere von Stählen (mit ausführlichem Schrifttum). Düsseldorf: VDI-Verlag 1974.

7.327. VDI-Richtlinie 2242, Blatt 1: Konstruieren ergonomiegerechter Erzeugnisse. Düsseldorf: VDI-Verlag 1986.

7.328. VDI-Richtlinie 2242, Blatt 2: Konstruieren ergonomiegerechter Erzeugnisse. Düsseldorf: VDI-Verlag 1986.

7.329. VDI-Richtlinie 2243: Konstruieren recyclinggerechter technischer Produkte. Düsseldorf: VDI-Verlag 1993.

7.330. VDI-Richtlinie 2244 (Entwurf): Konstruktion sicherheitsgerechter Produkte. Düsseldorf: VDI-Verlag 1985.

7.331. VDI-Richtlinie 2570: Lärmminderung in Betrieben; Allgemeine Grundlagen. Düsseldorf: VDI-Verlag 1980.

7.332. VDI-Richtlinie 2802: Wertanalyse. Düsseldorf: VDI-Verlag 1976.

7.333. VDI-Richtlinie 3237, Bl. 1 und Bl. 2: Fertigungsgerechte Werkstückgestaltung im Hinblick auf automatisches Zubringen, Fertigen und Montieren. Düsseldorf: VDI-Verlag 1967 und 1973.

7.334. VDI-Richtlinie 3239: Sinnbilder für Zubringefunktionen. Düsseldorf: VDI-Verlag 1966.

7.335. VDI-Richtlinie 3720, Bl. 1 bis Bl. 6: Lärmarm konstruieren. Düsseldorf: VDI-Verlag 1978 bis 1984.

7.336. VDI-Richtlinie 4004, B. 2: Überlebenskenngrößen. Düsseldorf: VDI-Verlag 1986.

7.337. VDI: Wertanalyse. VDI-Taschenbücher T 35. Düsseldorf: VDI-Verlag 1972.

7.338. Wagner, K.; Pfeil, B.; Keil, G.: Zur Einteilung von Verschleißvorgängen, Schmierungstechnik 6 (1975) 299–302, 325–330.

7.339. Wahl, W.: Abrasive Verschleißschäden und ihre Verminderung. VDI-Berichte Nr. 243, "Methodik der Schadenuntersuchung". Düsseldorf: VDI-Verlag 1975.

7.340. Walczak, A.: Selbstjustierende Funktionskette als kosten- und montagegünstiges Gestaltungsprinzip, gezeigt am Beispiel eines mit methodischen Hilfsmitteln entwickelten Leseräts. Konstruktion 38 (1986) 1, 27–30.

7.341. Walter, J.: Möglichkeiten und Grenzen der Montageautomatisierung. VDI-Z. 124 (1982) 853–859.

7.342. Wanke, K.: Wassergekühlte Turbogeneratoren. In "AEG-Dampfturbinen, Turbogeneratoren". Berlin: Verlag AEG (1963) 159–168.

7.343. Warnecke, H.J.; Löhr, H.-G.; Kiener, W.: Montagetechnik. Mainz: Krausskopf 1975.

7.344. Warnecke, H.J.; Steinhilper, R.: Instandsetzung, Aufarbeitung, Aufbereitung – Recyclingverfahren und Produktgestaltung. VDI-Z. 124 (1982) 751–758.

7.345. Weber, A.: Werkstoff- und fertigungsgerechtes Konstruieren mit thermoplastischen Kunststoffen. Konstruktion 16 (1964) 2–11.

7.346. Weber, R.: Recycling bei Kraftfahrzeugen. Konstruktion 42 (1990) 410–414.

7.347. Weege, R.-D.: Recyclinggerechtes Konstruieren. Düsseldorf: VDI-Verlag 1981.

7.348. Welch, B.: Thermal Instability in High-Speed-Gearing. Journal of Engineering for Power 1961. 91 ff.

7.349. Wellinger, K.; Dietmann, H.: Festigkeitsberechnung. Stuttgart: Kröner 1969.

7.350. Wiedemann, J.: Leichtbau. Bd. 1: Elemente; Bd. 2: Konstruktion. Berlin: Springer 1986/1989.

7.351. Wiegand, H.; Beelich, K.H.: Einfluß überlagerter Schwingungsbeanspruchung auf das Verhalten von Schraubenverbindungen bei hohen Temperaturen. Draht-Welt 54 (1968) 566–570.

7.352. Wiegand, H.; Beelich, K.H.: Relaxation bei statischer Beanspruchung von Schraubenverbindungen. Draht-Welt 54 (1968) 306–322.

7.353. Wiegand, H.; Flemming, G.: Hochtemperaturverhalten von Schraubenverbindungen. VDI-Z. 113 (1971) 1239–1244.

7.354. Wiegand, H.; Kloos, K.-H.; Thomala, W.: Schraubenverbindungen. Konstruktionsbücher (Hrsg. G. Pahl) Bd. 5, 4. Aufl. Berlin: Springer 1988.

7.355. Wiegand, H.; Illgner, K.H.; Beelich, K.H.: Einfluß der Federkonstanten und der Anzugsbedingungen auf die Vorspannung von Schraubenverbindungen. Konstruktion 20 (1968) 130–137.

7.356. Witte, K.-W.: Konstruktion senkt Montagekosten. VDI-Z. 126 (1984) 835–840.

7.357. Woodson, W.E.: Human Factors Design Handbook. New York: McGraw Hill Book Co. 1981.

7.358. ZGV-Lehrtafeln: Erfahrungen, Untersuchungen, Erkenntnisse für das Konstruieren von Bauteilen aus Großwerkstoffen. Düsseldorf: Gießerei-Verlag.

7.359. ZGV-Mitteilungen: Fertigungsgerechte Gestaltung von Gußkonstruktionen. Düsseldorf: Gießerei-Verlag.

7.360. ZGV: Konstruieren und Gießen. Düsseldorf: Gießerei-Verlag.

7.361. ZHI-Verzeichnis: Richtlinien, Sicherheitsregeln und Merkblätter der Träger der gesetzlichen Unfallverordnung. Köln: Heymanns (wird laufend erneuert).

7.362. Zienkiewicz, O.G.: Methode der finiten Elemente, 2. Aufl. München: Hanser 1984.

7.363. Zünkler, B.: Gesichtspunkte für das Gestalten von Gesenkschmiedeteilen. Konstruktion 14 (1962) 274–280.

Chapter 8

8.1.   AEG-Telefunken: Hochspannungs-Asynchron-Normmotoren, Baukastensystem, 160 kW–3150 kW. Druckschrift E 41.01.02/0370.

8.2.   Achenbach, H.-P.: Ein Baukastensystem für pneumatische Wegeventile. wt-Z. ind. Fertigung 65 (1975) 13–17.

8.3.   Beitz, W.; Keusch, W.: Die Durchführung von Gleitlager-Variantenkonstruktionen mit Hilfe elektronischer Datenverarbeitungsanlagen. VDI-Berichte Nr. 196. Düsseldorf: VDI-Verlag 1973.

8.4.   Beitz, W.; Pahl, G.: Baukastenkonstruktionen. Konstruktion 26 (1974) 153–160.

8.5.   Berg, S.: Angewandte Normzahl. Berlin: Beuth 1949.

8.6.   Berg, S.: Die besondere Eignung der Normzahlen für die Größtstugung. DIN-Mitteilungen 48 (1969) 222–226.

8.7.   Berg, S.: Konstruieren in Größenreihen mit Normzahlen. Konstruktion 17 (1965) 15–21.

8.8.   Berg, S.: Die NZ, das allgemeine Ordnungsmittel. Schriftenreihe der AG für Rat. des Landes NRW (1959) H. 4.

8.9.   Berg, S.: Theorie der NZ und ihre praktische Anwendung bei der Planung und Gestaltung sowie in der Fertigung. Schriftenreihe der AG für Rat. des Landes NRW (1958) H. 35.

8.10. Borowski, K.-H.: Das Baukastensystem der Technik. Schriftenreihe Wissenschaftliche Normung, H. 5. Berlin: Springer 1961.

8.11. Brankamp, K.; Herrmann, J.: Baukastensystematik – Grundlagen und Anwendung in Technik und Organisation. Ind.-Anz. 91 (1969) H. 31 und 50.

8.12. Dietz, P.: Baukastensystematik und methodisches Konstruieren im Werkzeugmaschinenbau. Werkstatt u. Betrieb 116 (1983) 185–189 und 485–488.

8.13. DIN 323, Blatt 2: Normzahlen und Normzahlreihen (mit weiterem Schrifttum). Berlin: Beuth 1974.

8.14. Eversheim, W.; Wiendahl, H.-P.: Rationelle Auftragsabwicklung im Konstruktionsbüro. Girardet Taschenbücher, Bd. 1. Essen: Girardet 1971.

8.15.   Flender: Firmenprospekt Nr. K 2173/D. Bocholt 1972.

8.16.   Franzmann, K.: Interner Entwicklungsbericht der Fa. Borsig Berlin 1975.

8.17.   Friedewald, H.-J.: Normzahlen – Grundlage eines wirtschaftlichen Erzeugnisprogramms. Handbuch der Normung, Bd. 3. Berlin: Beuth 1972.

8.18.   Friedewald, H.-J.: Normung integrieren – der Bestandteil einer Firmenkonzeption. DIN-Mitteilungen 49 (1970) H. 1.

8.19.   Gerhard, E.: Ähnlichkeitsgesetze beim Entwurf elektromechanischer Geräte. VDI-Z. 111 (1969) 1013–1019.

8.20.   Gerhard, E.: Baureihenentwicklung. Konstruktionsmethode Ähnlichkeit. Grafenau: Expert 1984.

8.21.   Gläser, F.-J.: Baukastensysteme in der Hydraulik. wt-Z. ind. Fertigung 65 (1975) 19–20.

8.22.   Gregorig, R.: Zur Thermodynamik der existenzfähigen Dampfblase an einem aktiven Verdampfungskeim. Verfahrenstechnik (1967) 389.

8.23.   Hansen Transmissions International: Firmenprospekt Nr. 6102–62/D. Antwerpen 1969.

8.24.   Hansen Transmissions International: Firmenprospekt Nr. 202D. Antwerpen 1976.

8.25.   Keusch, W.: Entwicklung einer Gleitlagerreihe im Baukastenprinzip. Diss. TU Berlin 1972.

8.26.   Kienzle, O.: Die NZ und ihre Anwendung VDI-Z. 83 (1939) 717.

8.27.   Kienzle, O.: Normungszahlen. Berlin: Springer 1950.

8.28.   Kiesow, H.; Mihm, H.; Rosenbusch, R.; Automatisierung von Entwurf, Konstruktion und Auftragsbearbeitung im Anlagenbau, dargestellt am Beispiel des Wärmetauscherbaus. IBM-Nachrichten 20 (1970) 147–153.

8.29   Kloberdanz, H.: Rechnerunterstützte Baureihenentwicklung. Fortschritt-Berichte VDI, Reihe 20, Nr. 40. Düsseldorf: VDI-Verlag 1991.

8.30.   Koller, R.: Entwicklung und Systematik der Bauweisen technischer Systeme – ein Beitrag zur Konstruktionsmethodik. Konstruktion 38 (1986) 1–7.

8.31.   Lang, K.; Voigtländer, G.: Neue Reihe von Drehstrommaschinen großer Leistung in Bauform B 3. Siemens-Z. 45 (1971) 33–37.

8.32.   Lehmann, Th.: Die Grundlagen der Ähnlichkeitsmechanik und Beispiele für ihre Anwendung beim Entwerfen von Werkzeugmaschinen der mechanischen Umformtechnik. Konstruktion 11 (1959) 465–473.

8.33.   Maier, K.: Konstruktionsbaukästen in der Industrie. wt-Z. ind. Fertigung 65 (1975) 21–24.

8.34.   Matz, W.: Die Anwendung des Ähnlichkeitsgesetzes in der Verfahrenstechnik. Berlin: Springer 1954.

8.35.   Pahl, G.; Beitz, W.: Baureihenentwicklung; Konstruktion 26 (1974) 71–79 und 113–118.

8.36.   Pahl, G.; Rieg, F.: Kostenwachstumgesetze für Baureihen. München: C. Hanser 1984.

8.37.   Pahl, G.; Zhang, Z.: Dynamische und thermische Ähnlichkeit in Baureihen von Schaltkupplungen. Konstruktion 36 (1984) 421–426.

8.38.   Pahl, G.: Konstruieren mit 3D-CAD-Systemen. Kap. 8.8: Baureihenentwicklung. Berlin: Springer 1990.

8.39.   Pawlowski, J.: Die Ähnlichkeitstheorie in der physikalisch-technischen Forschung. Berlin: Springer 1971.

8.40.   Reuthe, W.: Größenstufung und Ähnlichkeitsmechanik bei Maschinenelementen, Bearbeitungseinheiten und Werkzeugmaschinen. Konstruktion 10 (1958) 465–476.

8.41.   Siemens: Drehstrommotoren für Hochspannung. Katalog M 27 1990).

8.42.   Schwarz, W.: Universal-Werkzeugfräs- und -bohrmaschinen nach Grundprinzipien des Baukastensystems. wt-Z. ind. Fertigung 65 (1975) 9–12.

8.43.   Weber, M.: Das allgemeine Ähnlichkeitsprinzip der Physik und sein Zusammenhang mit der Dimensionslehre und der Modellwissenschaft. Jahrb. der Schiffsbautechn. Ges., H. 31 (1930) 274–354.

8.44.   Westdeutsche Getriebewerke: Firmenprospekt. Bochum 1975.

Chapter 9

9.1.   Beitz, W.: Qualitätsorientierte Produktgestaltung. Konstruktion 43 (1991) 177–184.

9.2.   Bertsche, B.; Lechner, G.: Zuverlässigkeit im Maschinenbau. Berlin: Springer 1990.

9.3.   DIN 25 424 Teil 1: Fehlerbaumanalyse; Methode und Bildzeichen. Berlin: Beuth.

9.4.   DIN ISO 9000: Qualitätsmanagement – Qualitätssicherungsnormen; Leitfaden zur Auswahl und Anwendung. Berlin: Beuth.

9.5.   DIN ISO 9001: Qualitätssicherungssysteme – Modell zur Darlegung der Qualitätssicherung in Design/Entwicklung, Produktion, Montage und Kundendienst. Berlin: Beuth.

9.6.   DIN ISO 9002: —; Modell zur Darlegung der Qualitätssicherung in Produktion und Montage. Berlin: Beuth.

9.7.   DIN ISO 9003: —; Modell zur Darlegung der Qualitätssicherung bei der Endprüfung. Berlin: Beuth.

9.8.   DIN ISO 9004: Qualitätsmanagement und Elemente eines Qualitätssicherungssystems; Leitfaden. Berlin: Beuth.

9.9.   Franke, W.D.: Fehlermöglichkeits- und -einflußanalyse in der industriellen Praxis. Landsberg: Moderne Industrie 1987.

9.10.   Kamiske, G.: Qualitätsorientierte Produktgestaltung. Konstruktion 43 (1991) 177–184.

9.11.   Maeguchi, Y.; Lechner, G.; v. Eiff, H.; Brodbeck, P.: Zuverlässigkeitsanalyse eines Trochoiden-Getriebes, Konstruktion 45 (1993) H. 1.

9.12.   Masing, W. (Hrsg.): Handbuch der Qualitätssicherung, 2. Aufl. München: C. Hanser 1988.

9.13.   Rodenacker, W.G.: Methodisches Konstruieren, 4. Aufl. Berlin: Springer 1991.

9.14.   VDA: Qualitätskontrolle in der Automobilindustrie, Bd. 4 – Sicherung der Qualität vor Serieneinsatz, 2. Aufl. Frankfurt: VDA 1986.

Chapter 10

10.1.   Bauer, C.-O.: Relativkosten-Kataloge – wertvolles Hilfsmittel oder teure Sackgasse? DIN-Mitt. 64, 1985, Nr. 5, 221–229.

10.2.   Beitz, W.: Möglichkeiten zur material- und energiesparenden Konstruktion. Konstruktion 42 (1990) 378–384.

10.3.   Busch, W.; Heller, W.: Relativkosten-Kataloge als Hilfsmittel zur Kostenfrüherkennung.

10.4.   DIN 32 990 Teil 1: Kosteninformationen; Begriffe zu Kosteninformationen in der Maschinenindustrie. Berlin: Beuth 1989.

10.5.   DIN 32 991 Teil 1: Kosteninformationen; Kosteninformations-Unterlagen; Gestaltungsgrundsätze. Berlin: Beuth 1987.

10.6.   DIN 32 991 Teil 1 Beiblatt 1: Kosteninformationen; Gestaltungsgrundsätze für Kosteninformationsunterlagen; Beispiele für Relativkosten-Blätter. Berlin: Beuth 1990.

10.7.   DIN 32 992 Teil 1: Kosteninformationen; Berechnungsgrundlagen; Kalkulationsarten und -verfahren. Berlin: Beuth 1989.

10.8.   DIN 32 992 Teil 2: Kosteninformationen; Berechnungsgrundlagen; Verfahren der Kurzkalkulation. Berlin: Beuth 1989.

10.9.   DIN 32 992 Teil 3: Kosteninformationen; Berechnungsgrundlagen; Ermittlung von Relativkosten-Zahlen. Berlin: Beuth 1987.

10.10. DIN 69 910: Wertanalyse. Berlin: Beuth 1987.

10.11. Ehrlenspiel, K.; Kiewert, A.; Lindemann, U.: Kostenfrüherkennung im Konstruktionsprozeß VDI-Berichte Nr. 347. Düsseldorf: VDI-Verlag 1979.

10.12. Ehrlenspiel, K.: Genauigkeit, Gültigkeitsgrenzen, Aktualisierung der Erkenntnisse und Hilfsmittel zum kostengünstigen Konstruieren. Konstruktion 32 (1982) 487–492.

10.13. Ehrlenspiel, K.: Kostengünstig Konstruieren. Konstruktionsbücher, Bd. 35. Berlin: Springer 1985.

10.14. Ehrlenspiel, K.: Kostengesteuertes Design – Konstruieren und Kalkulieren am Bildschirm. Konstruktion 40 (1988) 359–364.

10.15. Ehrlenspiel, K.; Pickel, H.: Konstruieren kostengünstiger Gußteile – Kostenstrukturen, Konstruktionsregeln und Rechneranwendung (CAD). Konstruktion 38 (1986) 227–236.

10.16. Fischer, D.: Kostenanalyse von Stirnzahnrädern. Erarbeitung und Vergleich von Hilfsmitteln zur Kostenfrüherkennung. TU München: Dissertation 1983.

10.17. Kiewert, A.: Kurzkalkulationen und die Beurteilung ihrer Genauigkeit. VDI-Z. 124 (1982) 443–446.

10.18. Kiewert, A.: Wirtschaftlichkeitsbetrachtungen zum kostengerechten Konstruieren. Konstruktion 40 (1988) 301–307.

10.19. Klasmeier, U.: Kurzkalkulationsverfahren zur Kostenermittlung beim methodischen Konstruieren. Schriftenreihe Konstruktionstechnik, H. 7. TU Berlin: Dissertation 1985.

10.20. Kloss, G.: Einige übergeordnete Konstruktionshinweise zur Erzielung echter Kostensenkung. VDI-Fortschrittsberichte, Reihe 1, Nr. 1. Düsseldorf: VDI-Verlag 1964.

10.21. Maurer, C.: Standardkosten- und Deckungsbeitragsrechnung in Zulieferbetrieben des Maschinenbaus. Darmstadt: S. Toeche-Mittler-Verlag 1980.

10.22. Mellerowicz, K.: Kosten und Kostenrechnung, Bd. 1. Berlin: Walter de Gruyter 1974.

10.23. Pacyna, H.; Hillebrand, A.; Rutz, A.: Kostenfrüherkennung für Gußteile. VDI-Berichte Nr. 457: Konstrukteure senken Herstellkosten – Methoden und Hilfsmittel. Düsseldorf: VDI-Verlag 1982.

10.24. Pahl, G.; Beelich, K.H.: Kostenwachstumsgesetze nach Ähnlichkeitsbeziehungen für Baureihen. VDI-Berichte Nr. 457. Düsseldorf: VDI-Verlag 1982.

10.25. Pahl, G.; Rieg, F.: Kostenwachstumsgesetze nach Ähnlichkeitsbeziehungen für Baureihen. VDI-Berichte Nr. 457. Düsseldorf: VDI-Verlag 1982.

10.26. Pahl, G.; Rieg, F.: Relativkostendiagramme für Zukaufteile. Approximationspolynome helfen bei der Kostenabschätzung von fremdgelieferten Teilen. Konstruktion 36 (1984) 1–6.

10.27. Pahl, G.; Rieg, F.: Kostenwachstumsgesetze für Baureihen. München: C. Hanser 1984.

10.28 Rauschenbach, T.: Kostenoptimierung konstruktiver Lösungen. Möglichkeiten für die Einzel- und Kleinserienproduktion. Düsseldorf: VDI-Verlag 1978.

10.29. Ruckes, J.: Betriebs- und Angebotskalkulation im Stahl- und Apparatebau. Berlin: Springer 1973.

10.30. Siegerist, M.; Langheinrich, G.: Die neuzeitliche Vorkalkulation der spangebenden Fertigung im Maschinenbau. Berlin: Technischer Verlag Herbert Cram 1974.

10.31. Schneider, P.: Verbilligte Konstruktion durch Kostenvergleiche in Werknormen. DIN-Mitt. 46 (1967) 141–145.

10.32. Steinwachs, H.O.: Konstrukteure senken Herstellkosten. Eschborn: RKW 1985.

10.33. VDI-Richtlinie 2225 Blatt 1: Konstruktionsmethodik; Technisch-wirtschaftliches Konstruieren; Anleitung und Beispiele. Düsseldorf: VDI-Verlag 1977.

10.34. VDI-Richtlinie 2225 Blatt 1 (Entwurf): Konstruktionsmethodik; Technisch-wirtschaftliches Konstruieren; Vereinfachte Kostenermittlung. Düsseldorf: VDI-Verlag 1984.

10.35. VDI-Richtlinie 2225 Blatt 2: Konstruktionsmethodik; Technisch-wirtschaftliches Konstruieren; Tabellenwerk. Berlin: Beuth 1977.

10.36. VDI-Richtlinie 2225 Blatt 3: Konstruktionsmethodik; Technisch-wirtschaftliches Konstruieren; Technisch-wirtschaftliche Bewertung. Düsseldorf: VDI-Verlag 1990.

10.37. VDI-Richtlinie 2234: Wirtschaftliche Grundlagen für den Konstrukteur. Düsseldorf: VDI-Verlag 1990.

10.38. VDI-Richtlinie 2235: Wirtschaftliche Entscheidungen beim Konstruieren; Methoden und Hilfen. Düsseldorf: VDI-Verlag 1987.

10.39. VDI-Richtlinie 3237 Blatt 1: Fertigungsgerechte Werkstückgestaltung im Hinblick auf automatisches Zubringen, Fertigen und Montieren. Düsseldorf: VDI-Verlag 1967.

10.40. VDI-Richtlinie 3258 Blatt 1: Kostenrechnung mit Maschinenstundensätzen: Begriffe, Bezeichnungen, Zusammenhänge. Düsseldorf: VDI-Verlag 1962.

10.41. VDI-Richtlinie 3258 Blatt 2: Kostenrechnung mit Maschinenstundensätzen: Erläuterungen und Beispiele. Düsseldorf: VDI-Verlag 1964.

10.42. VDI: Wertanalyse. VDI-Taschenbücher T 35. Düsseldorf: VDI-Verlag 1972.

10.43. Vogt, C.-D.; Systematik und Einsatz der Wertanalyse. Berlin: Siemens Verlag 1974.

10.44. Warnecke, H. JK.; Bullinger, H.J.: Kostenrechnung für Ingenieure. München: C. Hanser 1990.

Chapter 11

11.1. Beitz, W., Birkhofer, H.; Pahl, G.: Konstruktionsmethodik in der Praxis. Konstruktion 44 (1992) Heft 12.

11.2. Birkhofer, H.: Methodik in der Konstruktionspraxis – Erfolge, Grenzen und Perspektiven. Proceedings of ICED '91, HEURISTA 1991. Vol. 1, S. 224–233.

11.3. Franke, H.-J.: Konstruktionsmethodik und Konstruktionspraxis – Eine kritische Betrachtung ICED '85 Hamburg. Schriftenreihe WDK 12 910–924. Edition Heurista.

11.4. Jorden, W.; Havenstein, G.; Schwartzkopf, W.: Vergleich von Konstruktionswissenschaft und Praxis – Teilergebnisse eines Forschungsvorhabens. ICED '85 Hamburg. Schriftenreihe WDK 12 957–966. Edition Heurista.

11.5. Pahl, G.: Denkpsychologische Erkenntnisse und Folgerungen für die Konstruktionslehre. Proceedings of ICED '85 Hamburg. Schriftenreihe WDK 12 817–832. Edition Heurista.

11.6. Pahl, G.; Beelich, K.H.: Lagebericht. Erfahrungen mit dem methodischen Konstruieren. Werkstatt und Betrieb 114 (1981) 773–782.

11.7. VDI-Richtlinie 2221: Methodik zum Entwickeln und Konstruieren technischer Systeme und Produkte. Düsseldorf: VDI-Verlag 1993.

English Bibliography

Adams, J. L., 1974 Conceptual Blockbusting, Freeman, San Francisco.

Alexander, C., 1962 Notes on the Synnthesis of Form, Dissertation, Harvard University, Harvard.

Alger, J. R. M. and Hays, C. V., 1964 Creative Synthesis in Design, Prentice-Hall, Englewood Cliffs, N.J.

Andreasen, M. M., Kahler, S. and Lund, T. 1983 Design for Assembly, Springer, Berlin.

Andreasen, M. M. and Hein, L., 1987 Integrated Product Development, IFS Publications/Springer-Verlag, Bedford/Berlin.

Archer, L. B., 1971 Technological Innovation – A Methodology, Inforlink, Frimley, Surrey.

Archer, L. B., 1974 Design Awareness and Planned Creativity, Design Council, London.

Ashby, M. F., 1992 Materials Selection in Mechanical Design, Prentice Hall, Englewood Cliffs, N.J.

Asimow, M., 1962 Introduction to Design. Prentice-Hall, Englewood Cliffs, N.J.

Bailey, R. L., 1978 Disciplined Creativity for Engineers, Ann Arbor Science, Ann Arbor, Mich.

Blessing, L. T. M., 1994 A Process-based Approach to Computer-supported Engineering Design, Blessing, Cambridge.

Bradbury, J. A. A., 1989 Product Innovations – Idea to Exploitation, Wiley, Chichester.

British Standards Institution, 1989 BS7000: Guide to Managing Product Design, BSI, London.

Burr, A. H., 1981 Mechanical Analysis and Design, Elsevier, Amsterdam.

Buur, J., 1990 A Theoretical Approach to Mechatronics Design, Institute for Engineering Design, Technical University Denmark, Lyngby.

Carter, D. E., 1992 Concurrent Engineering: The Development Environment for the 1990s, Addison-Wesley, New York.

Ciampa, D., 1992 Total Quality, Addison-Wesley, New York.

Clausing, D., 1994 Total Quality Development: a step-by-step guide to concurrent engineering, ASME Press, New York.

Collen, A. and Gasparski, W. W., 1995 Design & Systems – Praxiology: The International Annual of Practical Philosophy & Methodology, Vol. 3, Transaction Publishers, New Brunswick.

Cross, N., 1984 Developments in Design Methodology, Wiley, Chichester.

Cross, N., 1989 Engineering Design Methods, Wiley, Chichester.

Cross, N., Dorst, K. and Roozenburg, N. (eds), 1992 Research in Design Thinking, Delft University Press, Delft.

de Bono, E., 1971 The Use of Lateral Thinking, Penguin, Harmondsworth.

de Bono, E., 1976 Teaching Thinking, Temple Smith, London.

Dieter, G.E.,1991 Engineering Design, 2nd Edition, McGraw-Hill, New York.

Deutschman, A. D., Michels, W. J. and Wilsom, C. E., 1975 Machine Design, Macmillan, London.

Dixon, J. R., 1966 Design Engineering: Inventiveness, Analysis and Decision Making, McGraw-Hill, New york.

Dym, C. L., 1994 Engineering Design: A Synthesis of Views, Cambridge University Press, New York.

Eder, W. E. and Gosling, W., 1965 Mechanical System Design, Pergamon, Oxford.

Faires, V. M., 1972 Design of Machine Elements, Macmillan, London.

Ferguson, E. S., 1992 Engineers and the Mind's Eye, MIT Press, Cambridge, MA.

Flursheim, C. H., 1977 Engineering Design Interfaces, Design Council, London.

Fox, J., 1993 Quality Through Design, McGraw-Hill, New York.

French, M. J., 1985 Conceptual Design for Engineers, Design Council/Springer-Verlag, London/Berlin.

French, M. J., 1992 Form, Structure and Mechanism, MacMillan, Basingstoke.

French, M. J., 1994 Invention and Evolution: Design in Nature and Engineering, C.U.P., Cambridge.

Furman, T. T., 1981 Approximate Methods in Engineering Design, Academic Press, New York.

Glegg, G. L., 1969 The Design of Design, C.U.P., Cambridge.

Glegg, G. L., 1972 The Selection of Design, C.U.P., Cambridge.

Glegg, G. L., 1973 The Science of Design, C.U.P., Cambridge.

Glegg, G. L., 1981 The Development of Design, C.U.P., Cambridge.

Gordon, W. J. J., Synectics, Harper & Row, New York.

Gregory, S. A. (ed.), 1966 The Design Method, Butterworth, London.

Gregory, S. A. 1970 Creativity in Engineering, Butterworth, London.

Hales, C., 1991 Analysis of the Engineering Design Process in an Industrial Context, Gants Hill Publications, Eastleigh.

Hales, C., 1993 Managing Engineering Design, Longman, Harlow.

Hall, A. D., 1962 A Methodology for Systems Engineering, Van Nostrand Company, Princeton, N.J.

Harrisberger, L., 1966 Engineermanship: a Philosophy of Design, Brooks/Cole, Belmont, Ca.

Hubka, V., 1982 Principles of Engineering Design, Butterworth, London.

Hubka, V. and Eder, W. E., 1988 Theory of Technical Systems – a Total Concept Theory for Engineering Design, Springer, Berlin.

Hubka, V., Andreasen, M. M. and Eder, W. E., 1988 Practical Studies in Systematic Design, Butterworths, London.

Johnson, R. C., 1980 Optimum Design of Mechanical Elements, Wiley, New York.

Johnson, R. C., 1978 Mechanical Design Synthesis, Krieger, Huntington, New York.

Jones, J. C. and Thornley, D. (eds), 1963 Conference on Design Methods, Pergamon, Oxford.

Jones, J. C. and Thornley, D. 1970 Design Methods: Seeds of Human Futures, Wiley, New York.

Krick, E. V., 1969 An Introduction to Engineering and Engineering Design, Wiley, New York.

Lanigan, M., 1992 Engineers in Business – The Principles of Management and Product Design, Addison-Wesley, New York.

Leech, D. J. and Turner, B. T., 1985 Engineering Design for Profit, Ellis Horwood, Chichester.

Lewis, W. and Samuel, A., 1989 Fundamentals of Engineering Design, Prentice Hall, New York.

Leyer, A., 1974 Machine Design, Blackie, London.

Marples, D. L. 1960 The Decisions of Engineering Design, Inst. Eng. Des., London.

Matchett, E., 1963 The Controlled Evolution of Engineering Design, Inst. Eng. Des., London.

Matousek, R., 1963 Engineering Design: A Systematic Approach, Blackie, London.

Mayall, W. H., 1979 Principles in Design, Design Council, London.

Middendorf, W. H., 1969 Engineering Design, Allyn & Bacon, Boston, Mass.

Middendorf, W. H., 1981 What Every Engineer Should Know About Inventing, Dekker, New York.

Morrison, D., 1969 Engineering Design, McGraw-Hill, New York.

Nadler, G., 1963 Work Design, Irwin, Homewood, Ill.

Nadler, G., 1967 Work Systems Design: The Ideals Concept, Irwin, Homewood, Ill.

National Research Council, 1991 Improving Engineering Design: Designing for Competitive Advantage, National Academic Press, Washington D.C.

Oakley, M. (ed.), 1990 Design Management: a Handbook of Issues and Methods, Blackwell, Oxford.

Osborn, A. F., 1953 Applied Imagination, Scribner's, New York.

Ostrofsky, B., 1977 Design, Planning, and Development Methodology, Prentice-Hall, Englewood Cliffs, N.J.

Parr, R. E., 1970 Principles of Mechanical Design, McGraw-Hill, New York.

Peace, G. S., 1992 Taguchi Methods – a Hands-on Approach, Addison-Wesley, New York.

Petroski, H., 1994 Design Paradigms – Case Histories of Error and Judgement, Cambridge University Press, Cambridge.

Pitts, G., 1973 Techniques in Engineering Design, Butterworth, London.

Polak, P., 1976 A Background to Engineering Design, Macmillan, London.

Pugh, S., 1991 Total Design – Integrated Methods for Successful Product Engineering, Addison-Wesley, Wokingham.

Redford, G. D., 1975 Mechanical Engineering Design, Macmillan, London.

Roozenburg, N. F. M. and Eekels, J., 1995 Product Design: Fundamentals and Methods, Wiley, Chichester.

Ruiz, C. and Koenigsberger, F., 1970 Design for Strength and Production, Macmillan, London.

Sherwin, K., 1982 Engineering Design for Performance, Horwood, Chichester.

Shigley, J. G., 1981 Mechanical Engineering Design, McGraw-Hill, Oakland.

Simon, H. A. 1969 The Science of the Artificial, MIT Press, Cambridge, Mass.

Simon, H. A., 1975 A Student's Introduction to Engineering Design, Pergamon, Oxford.

Spotts, M. F., 1978 Design of Machine Elements, Prentice-Hall, Englewood Cliffs, N.J.

Starkey, C. V., 1992 Engineering Design Decisions, Arnold, Cambridge, MA.

Svensson, N. L., 1976 Introduction to Engineering Design, Pitman, London.

Thring, M. W. and Laithwaite, E. R., 1977 How to Invent, Macmillan, London.

Tjalve, E., 1979 Short Course in Industrial Design, Newnes-Butterworth, London.

Tjalve, E., Andreasen, M. M. and Schmidt, F. F., 1979 Engineering Graphic Modelling, Newnes-Butterworth, London.

Ullman, D., 1992 The Mechanical Design Process, McGraw-Hill, New York.

Ulrich, T. K. and Eppinger, S. D., 1995 Product Design and Development, McGraw-Hill, New York.

VDI, 1987 VDI Design Handbook 2221: Systematic Approach to the Design of Technical Systems and Products (translation of 1986 German edition), Verein Deutscher Ingenieure Verlag, Düsseldorf.

Walker, D. J., Dagger, B. K. J. and Roy, R., 1991 Creative Techniques in Product and Engineering Design – a practical workbook, Woodhead, Cambridge.

Wallace, P. J., 1952 The Techniques of Design, Pitman, London.

Woodson, T. T., 1966 Introduction to Engineering Design, McGraw-Hill, New York.

Conference proceedings

The Design Theory and Methodology Conference (DTM, since 1987), The American Society of Mechanical Engineers, New York.

The International Conference on Engineering Design (ICED, since 1981), WDK series, Heurista, Zürich.

Index